New Frontiers in Condensed Matter Physics

New Frontiers in Condensed Matter Physics

Editor: Raymond Stevens

NY RESEARCH PRESS
New York

Published by NY Research Press
118-35 Queens Blvd., Suite 400,
Forest Hills, NY 11375, USA
www.nyresearchpress.com

New Frontiers in Condensed Matter Physics
Edited by Raymond Stevens

International Standard Book Number: 978-1-63238-765-3 (Hardback)

Cataloging-in-Publication Data

New frontiers in condensed matter physics / edited by Raymond Stevens.
p. cm.
Includes bibliographical references and index.
ISBN 978-1-63238-765-3
1. Condensed matter. 2. Solid state physics. 3. Matter. 4. Physics. I. Stevens, Raymond.
QC173.454 .N49 2020
530.41--dc23

Printed in China.

Contents

Preface

Condensed matter physics is the sub-field of physics that is concerned with the study of macroscopic and microscopic physical properties of matter. It is involved in measuring various material properties by using experimental probes and methods from theoretical physics. It also focuses on studying the behavior of condensed phases such as solids and liquids by using physical laws. This discipline is broadly divided into experimental condensed matter physics and theoretical condensed matter physics. Some of the laws which are used for conducting research in this field are laws of quantum mechanics, statistical mechanics and electromagnetism. This book unravels the recent studies in the field of condensed matter physics. The various studies that are constantly contributing towards advancing technologies and evolution of this field are examined in detail. It is a vital tool for all researching or studying condensed matter physics as it gives incredible insights into emerging trends and concepts of this field.

This book unites the global concepts and researches in an organized manner for a comprehensive understanding of the subject. It is a ripe text for all researchers, students, scientists or anyone else who is interested in acquiring a better knowledge of this dynamic field.

I extend my sincere thanks to the contributors for such eloquent research chapters. Finally, I thank my family for being a source of support and help.

Editor

Design and Fabrication of a Novel Wideband DNG Metamaterial with the Absorber Application in Microwave X-Band

Ahmed Mahmood,[1,2] Gölge Ögücü Yetkin,[2] and Cumali Sabah[3,4]

[1] *Department of Electrical Engineering, Salahaddin University, Erbil, Iraq*
[2] *Department of Electrical and Electronics Engineering, University of Gaziantep, 27310 Gaziantep, Turkey*
[3] *Department of Electrical and Electronics Engineering, Middle East Technical University-Northern Cyprus Campus (METU-NCC), Kalkanli, 99738 Guzelyurt, TRNC, Mersin 10, Turkey*
[4] *Kalkanli Technology Valley, Middle East Technical University-Northern Cyprus Campus (METU-NCC), Kalkanli, 99738 Guzelyurt, TRNC, Mersin 10, Turkey*

Correspondence should be addressed to Gölge Ögücü Yetkin; golge.ogucu@gmail.com

Academic Editor: Gary Wysin

A novel metamaterial, which exhibits a wideband double negative behavior in X-band, is proposed, designed, and investigated in this paper. The metamaterial is composed of modified S-shaped split-ring resonators (S-SRR). The periodic structure is designed and simulated using CST MWs. Next, the experiments are carried out, and it is shown that the simulation and the experimental results agree well and the designed structure has a wide bandwidth in X-band. An absorber application of this metamaterial is also provided, and the structure can be used as an absorber with absorption rate of over 80% for the polarization angles between 0° and 40°.

1. Introduction

Metamaterials are based on periodic or nonperiodic artificial engineered structures that exhibit unusual characteristic behaviors, like negative permeability ε and negative permittivity μ, simultaneously over a common resonance frequency band [1, 2]. From the first realization of the metamaterials by periodic wire and split-ring resonator (SRR) structures [3, 4], there are various modified shapes and geometries such as square, triangle, Λ-, Ω-, L-, S-, and P-shaped resonators [5–11] utilized in the designs in order to meet certain specifications in the applications. These geometries can be set in one-, two- or three-dimensional arrangements to control the electromagnetic wave propagation behavior through the medium.

The wideband metamaterials in X-band have been designed in [10, 11]. S-shaped SRRs were printed on opposite side of the host material and transmitted power was provided to show the DNG characteristics of the designed configuration in [10]. The double P-shaped resonators in [11] were printed on single side unlike that in [10]. The S-parameters were presented; however there were no possible metamaterial applications found.

In this paper, a novel wideband metamaterial comprising a modified S-shaped SRR is proposed for X-band frequencies. The entire periodic structure is placed on one side of the substrate, which simplifies the manufacturing process and can be considered as an advantage of the designed structure. The retrieved constitutive parameters and the S-parameters obtained via simulations and experiments demonstrate that the proposed structure exhibits a wideband left-handed characteristic in X-band frequencies. An absorption application is selected and also studied for demonstration of the performance of the metamaterial. The structure can be used as an absorber with absorption rate of over 80% for the polarization angles between 0° and 40°.

2. Design and Simulation Conditions

A unit cell of the proposed structure is composed of a modified S-shaped SRR printed on the front side of a square substrate of a side length of 16 mm and thickness of 1.6 mm as

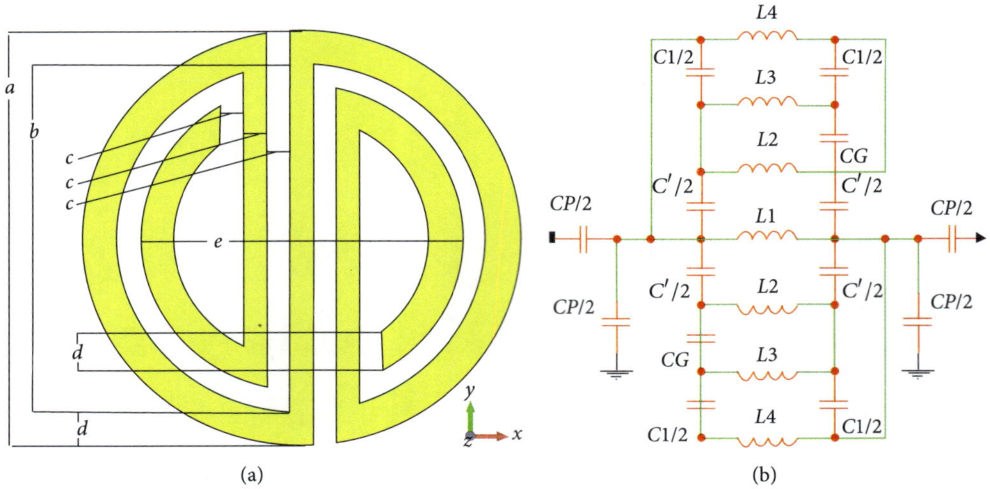

FIGURE 1: (a) The proposed unit cell with $a = 9.69$ mm, $b = 8.78$ mm, $c = 0.51$ mm, $d = 0.93$ mm, and $e = 8.25$ mm. (b) Equivalent circuit model of proposed unit cell.

shown in Figure 1(a). The two ends of the S-shaped SRR are elongated in such a way that they take the form of C-shaped SRRs. In the design, FR-4 Epoxy of permittivity $\varepsilon_r = 4.3$ and loss tangent $\delta = 0.025$ is chosen as the substrate, and the copper of electrical conductivity $\sigma = 5.8 \times 10^7$ S/m with coating thickness of 0.017 mm is chosen for printing the SRRs. The simulations are carried out by using the commercial program CST Microwave Studio 2016. Figure 1(b) shows the LC parameters that represent the electric topology of a modified S-SRR unit cell with the (L, L_1, L_2, L_3) inductors and (C', C_1, C_p, C_G) capacitors. Explicit expressions for L and C can be found in [12–14].

A unit cell of the proposed structure together with the ports applied in the simulations is shown in Figure 2. In the simulations, the unit cell is placed between two waveguide ports, which are perpendicular to the direction of the wave propagation, which is along negative z-direction. To imitate the infinite structure, perfect electric boundary (PEB) conditions are set at the boundary surfaces perpendicular to the E-field while perfect magnetic boundary (PMB) conditions are at the boundary surfaces perpendicular to the H-field (see Figure 2). The frequency band is chosen as the X-band region (8–12 GHz).

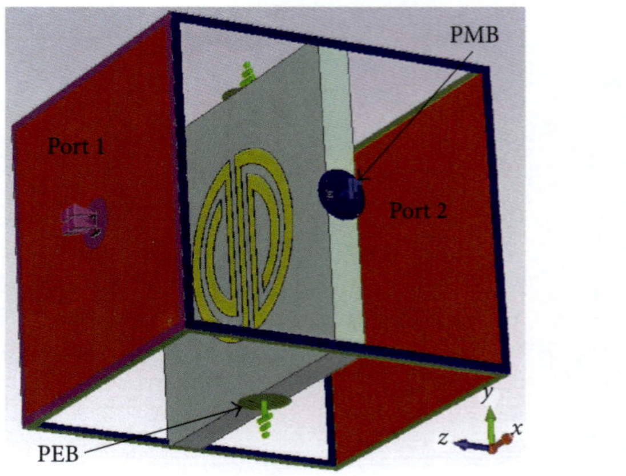

FIGURE 2: The boundary conditions of the proposed unit cell used in the simulations.

3. Electric Field, Magnetic Field, and Surface Current Distributions

Basically, the nonhomogenous metamaterial structures provide a resonant RLC circuit behavior in some certain frequency bands. For the particular geometry in this paper, the modified S-shaped SRRs behave like inductors and the gaps as capacitors. To illustrate how the structure works when placed in an electromagnetic field region, the surface current distribution is presented in Figure 3 and the magnetic and electric field densities in Figure 4 at the operation frequency of 10.75 GHz, at which the structure behaves like a left-handed material. It can be observed from Figure 3 that the current is highly concentrated around the central wire and the copper shape structure. The magnetic field shows similar distribution as the current as shown in Figure 4(a), which supports that the resonance obtained is of magnetic type and the modified S-shaped elements are strongly coupled to this magnetic resonance. Moreover, a close examination of the current distribution in Figure 3 leads to the observation of some virtual gaps in addition to the actual gaps. These virtual gaps occur at current minimum points and this provides an additional capacitance-like response by being the voltage maxima points at the same time. The electric field is highly localized around these actual and virtual gaps as seen in Figure 4(b). The reason of high concentration around the actual gaps is the capacitive effect at these regions [13, 14].

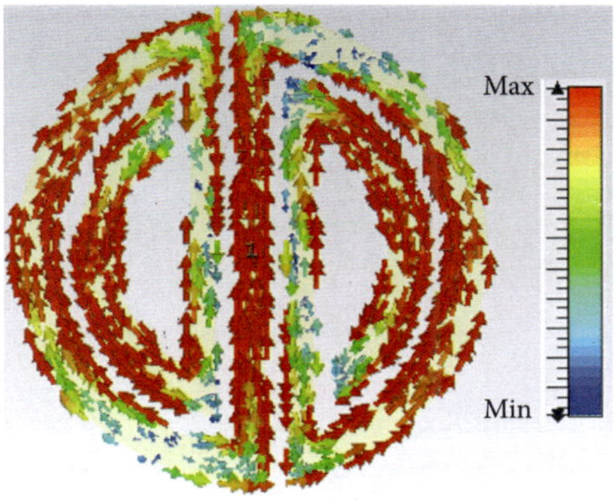

FIGURE 3: The surface current distribution at 10.75 GHz.

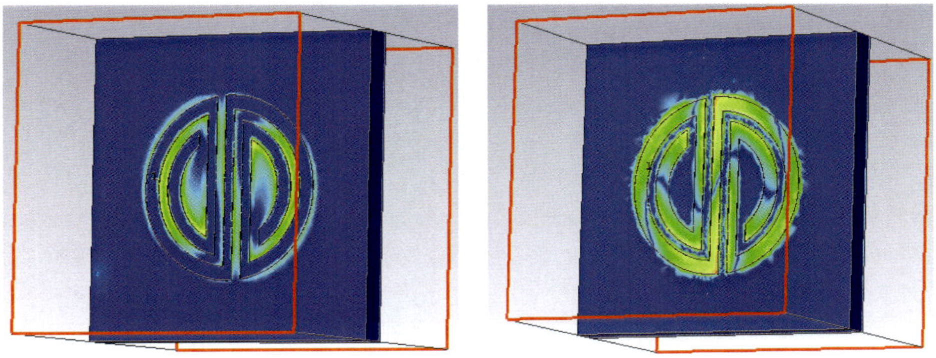

(a) (b)

FIGURE 4: (a) Magnetic field and (b) electric field distributions for the proposed unit cell at 10.75 GHz.

(a) (b)

FIGURE 5: (a) Vector network analyzer, two horn antennas with coaxial cables. (b) Manufactured sample.

4. Simulated and Experimental Results

Before presenting the comparison of the simulated and the experimental results, it would be better to give some information about the experimental set-up. Experiment layout is prepared by placing the designed sample between two horn antennas, which are connected to a vector network analyzer (VNA) through coaxial cables as in Figure 5(a). By this way, the surface of the sample, which is shown in Figure 5(b), is guaranteed to be perpendicular to the direction of wave propagation. The metamaterial prototype is manufactured at in-house facilities using a circuit plotter with standard milling tool sets.

The incident electromagnetic wave propagates along negative z-axis to excite the combined materials of the sample. The electric field vector and magnetic field vector are in the y-axis and negative x-axis, respectively (see Figure 2). For calibration purposes, a measurement is conducted without the sample. After that, the sample is placed between the two horn antennas and the S-parameters are measured via VNA.

The simulated and the measured S-parameter results, which are between 8 and 12 GHz, are presented in Figure 6.

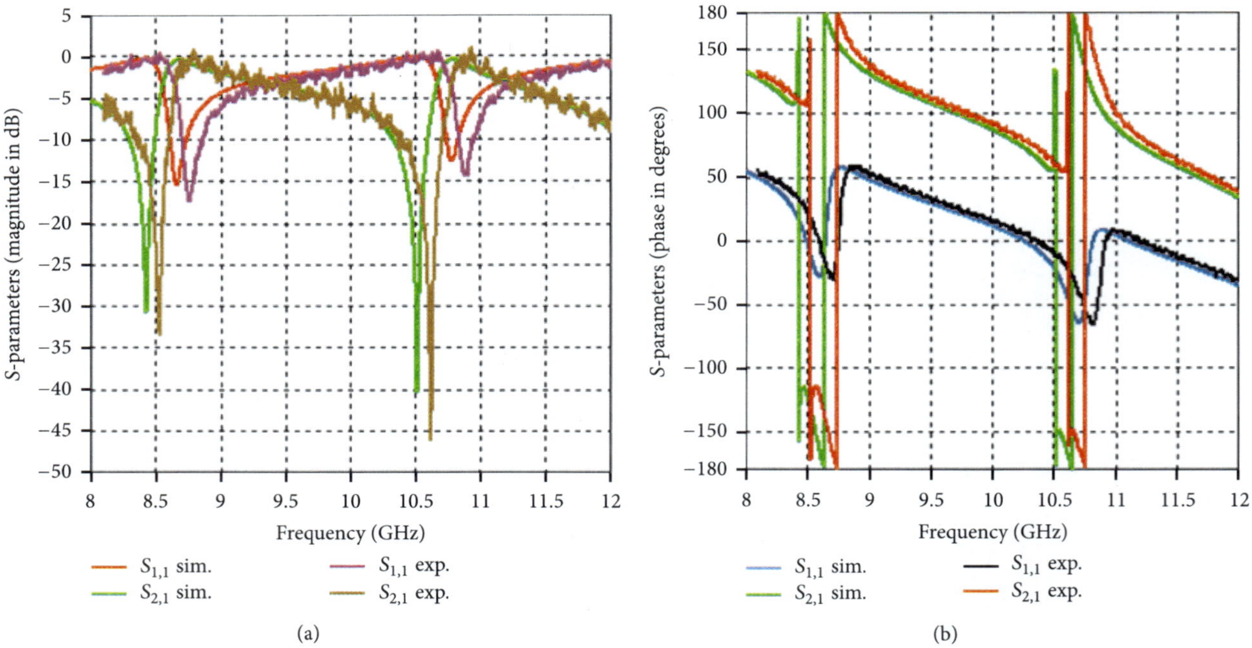

FIGURE 6: Simulated and measured S-parameters: (a) the magnitude (dB) and (b) the phase (degrees).

There are some small differences between the two types of obtained data, which may be due to the manufacture tolerances related to the dielectric dispersion and the etching process.

The effective permittivity and permeability values are obtained from the S-parameters given in Figure 6 by the use of the Nicolson-Ross-Weir (NRW) retrieval method [6, 15–17]. Using this method, the effective permittivity and permeability can be written as $\varepsilon = (k/k_0)((1-\Gamma)/(1+\Gamma))$ and $\mu = (k/k_0)((1+\Gamma)/(1-\Gamma))$, respectively, where k_0 is the propagation constant of free space. Here, $\Gamma = [(S_{11}^2 + S_{21}^2 + 1)/2S_{11}^2] \pm \sqrt{[(S_{11}^2 + S_{21}^2 + 1)/2S_{11}^2]^2 - 1}$, and the sign is chosen such that $|\Gamma| < 1$. Furthermore, $k = [\log(1/|T|)] + j(2m\pi - \text{phase}(T))/d$, where $T = [S_{11} + S_{21} - \Gamma]/[1 - \Gamma(S_{11} + S_{21})]$, $m = 0, \pm1, \pm2, \pm3, \ldots$, and d is the thickness of the slab. It clarifies that the integer m has multiple choices, and it is not straightforward to assign the exact solution for the permittivity and permeability. To find the exact value of the integer m, the group delay is calculated at each frequency, and then by taking the difference in phase of the S_{21} data at adjacent frequency points and comparing it with the group delay caused with the assumed value of m, the convergence on the correct value of m is achieved. Thus, NRW method can be applied with justification.

Figure 7 presents the simulated and the experimental results for the real parts of the effective permittivity and permeability together with the real part of the refractive index of the structure. As can be observed, the results compare well, and the structure has negative real parts in the frequency bands ~8–10.45 GHz and ~10.65–11.75 GHz for the permittivity and in the frequency band ~8.5–12 GHz for the permeability. Thus the proposed structure exhibits a wideband metamaterial behavior.

5. Absorber Application

To investigate the performance of the proposed S-shaped SRR, the absorption characteristics of the structure are studied both experimentally and through simulations. The absorption of a material is function of frequency and can be formulated by $A(\omega) = 1 - R(\omega) - T(\omega)$, where $R(\omega) = |S_{11}|^2$ and $T(\omega) = |S_{21}|^2$ represent the reflection and the transmission characteristics, respectively [18]. In order to maximize the absorption, both the reflection and the transmission can be minimized at the resonance frequency. In this study, a copper layer on the backside of the substrate material is added, so that the absorption expression is reduced to $A(\omega) \approx 1 - R(\omega)$, and it almost depends only on reflection now [18–20].

Figure 8(a) shows the VNA and the experiment alignment setting. After the calibration test without the testing sample, horn antenna is used to illuminate the sample in the X-band frequency regime, which is connected with VNA by a coaxial cable. As can be seen in Figure 8(b), the measurement results show good agreement with the simulation results. A maximum in the absorption ratio is about 99.99% at 8.94 GHz experimentally. Both the experimental and the simulation results suggest that the proposed structure can be used for perfect absorption applications. Note that the discrepancies between the experimental and simulation data are imputed to fabrication tolerances and dielectric dispersion of the substrate. The misalignment during the experiment may also be considered as another source of error.

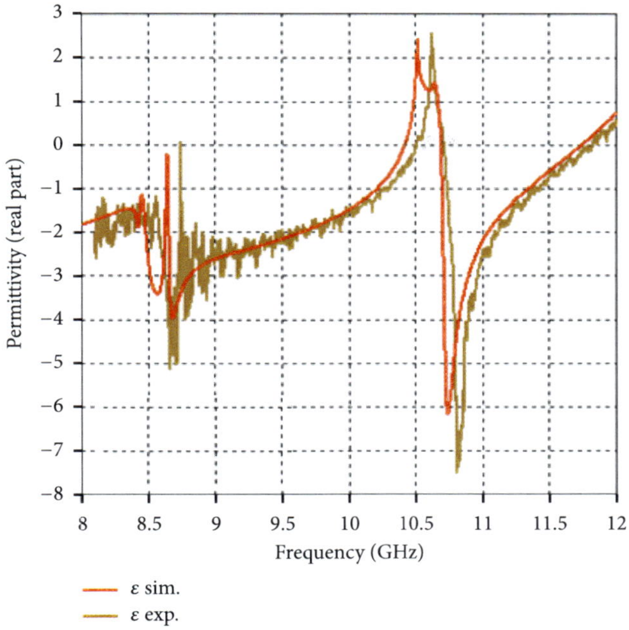

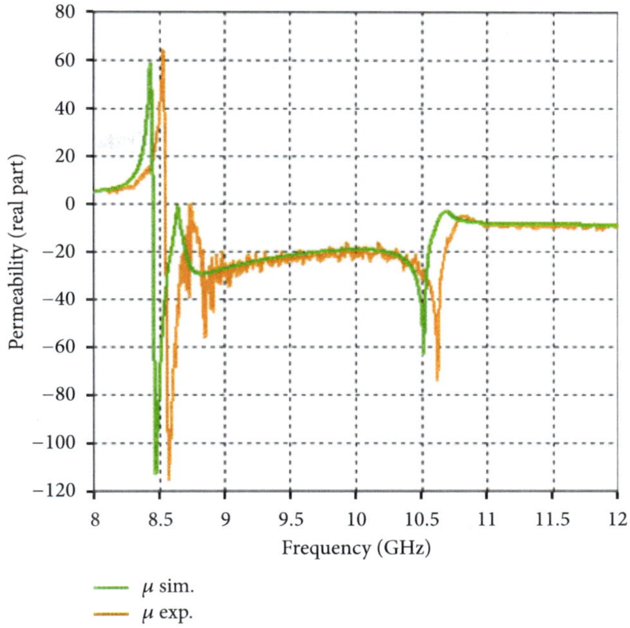

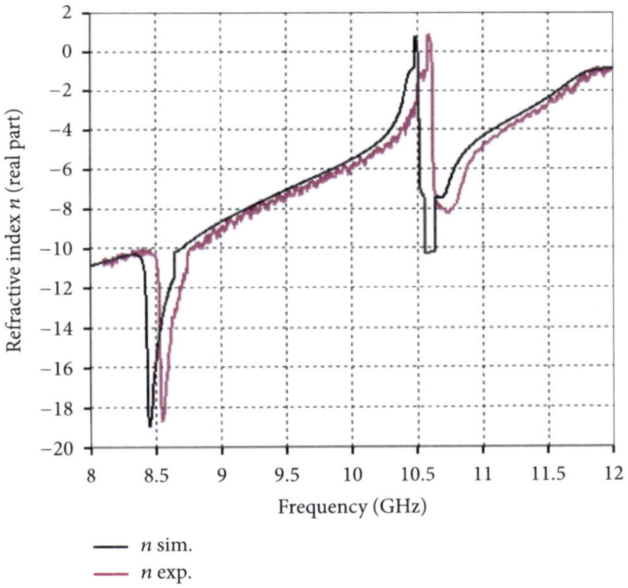

FIGURE 7: The simulated and experimental results for the real parts of (a) permittivity, (b) permeability, and (c) refractive index.

Furthermore, the S-parameter and the absorption characteristics are simulated for different polarization angles in 8.75–9.05 GHz frequency band. In Figure 9, S_{11} is presented for normal and oblique incidences. It is observed that reflection achieves its lowest value in normal incidence case and the resonance occurs around 8.912 GHz. As the polarization varies from normal to oblique incidence, the reflection also varies and a stable frequency response can be obtained for $\theta < 40^\circ$ when $\phi = 0^\circ$.

In Figure 10, the absorption characteristics are provided in the frequency band of 8.75–9.05 GHz for the polarization angle θ between 0° and 40°. It can be deduced that the proposed metamaterial can be used as a narrow band, 175 MHz perfect absorber with absorption over 90% in 8.825–9.0 GHz. At the resonance frequency around 8.912 GHz, S_{11} is 0.0039. Correspondingly, the absorption rate is 99.99% at the resonance and the device can be used as a perfect absorber at the mentioned frequency.

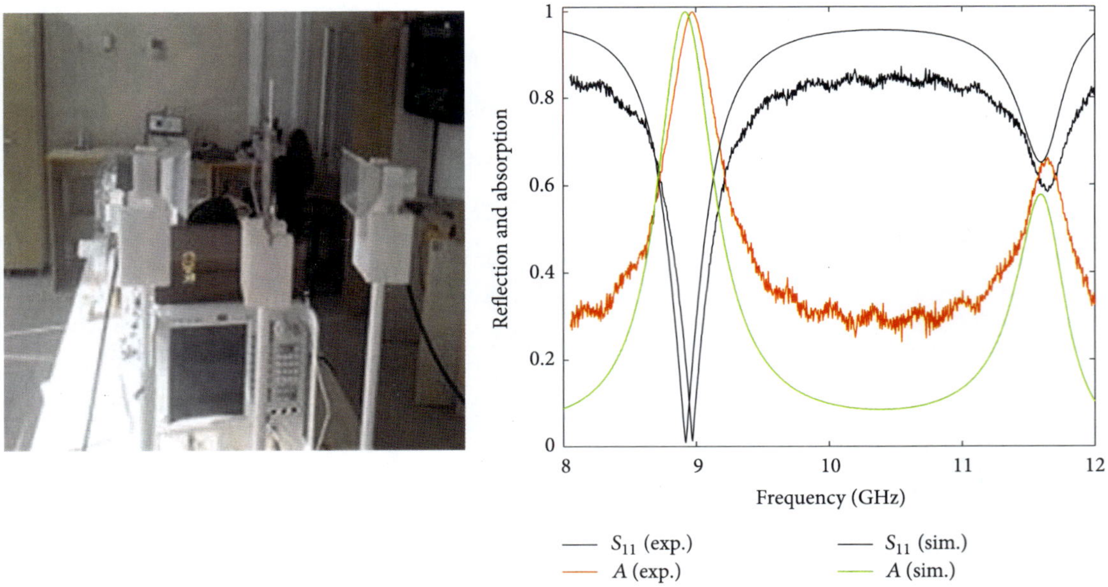

(a) (b)

FIGURE 8: (a) Absorption experiment setting. (b) Simulated and measured results of reflection and absorption ratio.

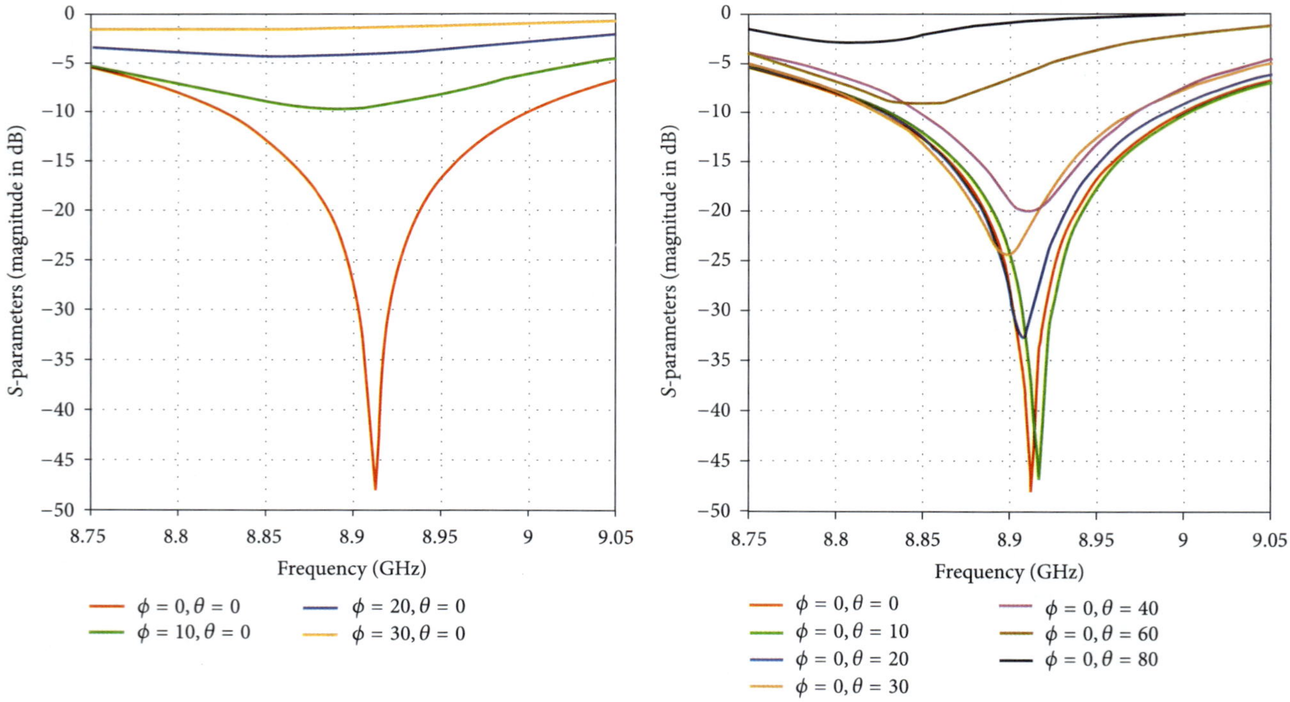

(a) (b)

FIGURE 9: Frequency response of S_{11} for different polarization angles: (a) $\theta = 0^\circ$ and ϕ varies and (b) θ varies and $\phi = 0^\circ$.

6. Conclusion

In this work, a modified S-shape ring resonator in periodic arrangement is introduced and investigated both numerically and experimentally. Based on the experiment results obtained, it has been demonstrated that the proposed structure exhibits wideband DNG characteristics in the frequency band of interest. In other words, the experimented novel metamaterial is well designed and successfully operates around the resonance frequency to provide simultaneously

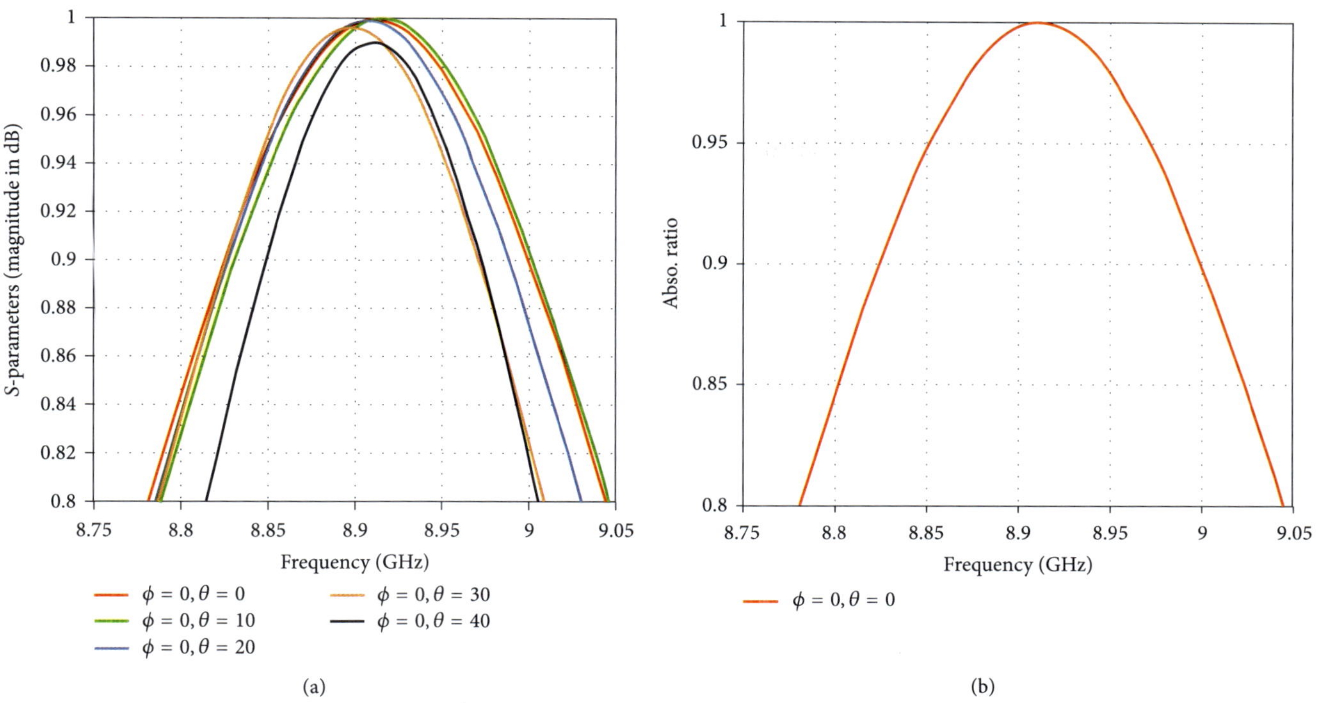

FIGURE 10: Absorption for different polarization angles: (a) θ varies and $\phi = 0^\circ$and (b) $\theta = 0^\circ$ and $\phi = 0^\circ$.

double negative permittivity and permeability in X-band. Additionally, simple fabrication of this novel metamaterial is another design advantage.

An absorber application has also been studied and measured. The measurement results show good agreement with the simulation results. Additionally, the simulated results for S-parameters and the absorption rate are obtained at different polarization angles. It has been shown that high absorption can be achieved for oblique incidences up to 40° and with absorption peak more than 90% between 8.825 and 9 GHz. A perfect absorber can be obtained at the resonance frequency around 8.912 GHz.

Conflicts of Interest

The authors declare that there are no conflicts of interest regarding the publication of this paper.

References

[1] V. G. Veselago, "The electrodynamics of substances with simultaneously negative values of ε and μ," *Soviet Physics Uspekhi*, vol. 10, no. 4, pp. 509–514, 1968.

[2] J. B. Pendry, A. J. Holden, W. J. Stewart, and I. Youngs, "Extremely low frequency plasmons in metallic mesostructures," *Physical Review Letters*, vol. 76, no. 25, pp. 4773–4776, 1996.

[3] D. R. Smith, W. J. Padilla, D. C. Vier, S. C. Nemat-Nasser, and S. Schultz, "Composite medium with simultaneously negative permeability and permittivity," *Physical Review Letters*, vol. 84, no. 18, pp. 4184–4187, 2000.

[4] R. A. Shelby, D. R. Smith, and S. Schultz, "Experimental verification of a negative index of refraction," *Science*, vol. 292, no. 5514, pp. 77–79, 2001.

[5] P. Markoš and C. M. Soukoulis, "Numerical studies of left-handed materials and arrays of split ring resonators," *Physical Review E*, vol. 65, no. 3, Article ID 036622, 2002.

[6] C. Sabah, "Electric and magnetic excitations in anisotropic broadside-coupled triangular-split-ring resonators," *Applied Physics A: Materials Science and Processing*, vol. 108, no. 2, pp. 457–463, 2012.

[7] C. Sabah and F. Urbani, "Experimental analysis of Λ-shaped magnetic resonator for mu-negative metamaterials," *Optics Communications*, vol. 294, pp. 409–413, 2013.

[8] J. H. Lv, X. W. Hu, M. H. Liu, B. R. Yan, and L. H. Kong, "Negative refraction of a double L-shaped metamaterial," *Journal of Optics A: Pure and Applied Optics*, vol. 11, no. 8, Article ID 085101, 2009.

[9] J. Huangfu, L. Ran, H. Chen et al., "Experimental confirmation of negative refractive index of a metamaterial composed of Ω-like metallic patterns," *Applied Physics Letters*, vol. 84, no. 9, pp. 1537–1539, 2004.

[10] H. Chen, L. Ran, J. Huangfu et al., "Left-handed materials composed of only S-shaped resonators," *Physical Review E*, vol. 70, no. 5, Article ID 57605, 2004.

[11] D.-L. Jin, J.-S. Hong, and H. Xiong, "A novel single-sided wideband metamaterial," *Applied Computational Electromagnetics Society Journal*, vol. 27, no. 12, pp. 971–976, 2012.

[12] J. Naqui, J. Coromina, A. Karami-Horestani, C. Fumeaux, and F. Martín, "Angular displacement and velocity sensors based on coplanar waveguides (CPWs) loaded with S-shaped split ring resonators (S-SRR)," *Sensors*, vol. 15, no. 5, pp. 9628–9650, 2015.

[13] H. S. Chen, L. X. Ran, J. T. Huangfu et al., "Magnetic properties of S-shaped split-ring resonators," *Progress in Electromagnetics Research*, vol. 51, pp. 231–247, 2005.

[14] A. K. Horestani, M. Duran-Sindreu, J. Naqui, C. Fumeaux, and F. Martin, "Coplanar waveguides loaded with s-shaped split-ring resonators: modeling and application to compact microwave filters," *IEEE Antennas and Wireless Propagation Letters*, vol. 13, pp. 1349–1352, 2014.

[15] A. M. Nicolson and G. F. Ross, "Measurement of the intrinsic properties of materials by time-domain techniques," *IEEE Transactions on Instrumentation and Measurement*, vol. 19, no. 4, pp. 377–382, 1970.

[16] W. B. Weir, "Automatic measurement of complex dielectric constant and permeability at microwave frequencies," *Proceedings of the IEEE*, vol. 62, no. 1, pp. 33–36, 1974.

[17] V. Milosevic, B. Jokanovic, and R. Bojanic, "Effective electromagnetic parameters of metamaterial transmission line loaded with asymmetric unit cells," *IEEE Transactions on Microwave Theory and Techniques*, vol. 61, no. 8, pp. 2761–2772, 2013.

[18] N. I. Landy, S. Sajuyigbe, J. J. Mock, D. R. Smith, and W. J. Padilla, "Perfect metamaterial absorber," *Physical Review Letters*, vol. 100, no. 20, Article ID 207402, 2008.

[19] F. Dinçer, M. Karaaslan, and C. Sabah, "Design and analysis of perfect metamaterial absorber in GHz and THz frequencies," *Journal of Electromagnetic Waves and Applications*, vol. 29, no. 18, pp. 2492–2500, 2015.

[20] F. Dincer, O. Akgol, M. Karaaslan, E. Unal, and C. Sabah, "Polarization angle independent perfect metamaterial absorbers for solar cell applications in the microwave, infrared, and visible regime," *Progress in Electromagnetics Research*, vol. 144, pp. 93–101, 2014.

2

An Improved Theoretical Approach to Study Electromagnetic Waves through Fiber Bragg Gratings

Pedro Pereyra

Departamento de Ciencias Básicas, UAM Azcapotzalco, 02200 Mexico City, Mexico

Correspondence should be addressed to Pedro Pereyra; pereyrapedro@gmail.com

Academic Editor: Joseph S. Poon

We show that using the theory of finite periodic systems we obtain an improved approach to calculate transmission coefficients and transmission times of electromagnetic waves propagating through fiber Bragg gratings. We discuss similarities, advantages, and differences between this approach and the well known less accurate one coupled mode approximation and the pseudo-Floquet Mathieu functions approach.

1. Introduction

A complete theoretical description of the electromagnetic waves motion through fiber Bragg gratings (FBG) remains to be achieved. Despite the numerous numerical calculations and simulations (see, e.g., [1, 2]) the full analytical solution of the wave equation in the presence of a periodic modulation of the refractive index is still an open problem. In the last years, a large number of fiber Bragg gratings devices were developed for different types of applications, among others, for optical communications devices, tunable wavelength filters and temperature, and pressure sensors [3–8]. In the theoretical approaches proposed to study these systems, one has also to deal, besides of the spatial modulation of the refractive index, with model limitations and the finiteness of the actual Bragg systems. In the small modulation amplitude limit, the wave equation becomes a Mathieu equation, and one has to face divergency problems in the numerical evaluation of these functions. For standard wavelengths of FBG, written with ultraviolet lasers, 1 cm fiber contains thousands of refractive-index modulations. An extensive literature has been published on coupled modes approximation [9–13], and a simplified one-dimensional model of two counterrunning and synchronous modes was introduced by Kogelnik and Shank [14, 15]. This single mode approximation, called also the one coupled mode approximation (OCMA), has been a useful and simplified approach [16]. However this model works well only for frequencies in the neighborhood of the Bragg frequency ω_B. For this reason an alternative method that will make it possible to study the transmission properties of electromagnetic waves in the whole range of frequencies is very much called for.

In this paper, we will present a theoretical approach, based on the theory of finite periodic systems [17–23] (TFPS) that will allow a precise calculation of transmission coefficients and transmission phase times through finite fiber Bragg gratings. We will show that, at least, the predictions of the well-established models are fully reproduced.

In the next sections we will outline the one coupled mode approximation (OCMA), the Mathieu functions approximation, and the theory of finite periodic systems (TFPS). We will derive the relevant results using the transfer matrix method and we will apply these results to study the evolution of longitudinal electromagnetic waves across a fiber Bragg grating. We will evaluate transmission coefficients and transmission times of specific Bragg grating fibers and compare the specific results of the well-established one coupled mode approximation with those of the theory of finite periodic systems (TFPS) and the Mathieu functions approximation. To overcome divergencies using Mathieu function we will also show results obtained in a combined approach of Mathieu functions and the TFPS.

2. Definition of the Fiber Bragg Grating

Suppose we have a Bragg grating fiber of length L_o and refractive index (we use this notation to avoid confusion with the number of unit cells, n, in the fiber Bragg grating):

$$r(z) = r_o\left(1 + 2V_o \cos\frac{2\pi z}{\Lambda}\right), \quad (1)$$

where Λ is the grating period ($\Lambda = \lambda/2r_o$), r_o is the effective refractive index, λ is the vacuum wave length, and V_o is the modulation amplitude. This is a typical periodic system. The space-time evolution of electromagnetic fields in these systems is governed by the wave equation.

$$\frac{\partial^2 E}{\partial z^2} = \frac{r^2(z)}{c^2}\frac{\partial^2 E}{\partial t^2}. \quad (2)$$

To study the space-time evolution of electromagnetic waves, we shall, first, recall the one coupled mode approximation and obtain the transmission coefficient and the transmission time as functions of the frequency. We will then obtain the same quantities using the Mathieu functions and, at the end, using the theory of finite periodic systems.

2.1. The FBG in the One Coupled Mode Approximation. In the one coupled mode approximation and following Erdogan's notation, the system of equations for the left and right moving field amplitudes, dominant at the Brag reflection frequency, is written as

$$\begin{aligned}\frac{\partial u}{\partial z} &= \eta u + \kappa v,\\ \frac{\partial v}{\partial z} &= -\eta v - \kappa u.\end{aligned} \quad (3)$$

If we define the wave vector

$$f(z) = \begin{pmatrix} u(z) \\ v(z) \end{pmatrix}, \quad (4)$$

the system of (3) can be written as

$$\frac{d}{dz} f = \begin{pmatrix} \eta & \kappa \\ -\kappa & -\eta \end{pmatrix} f, \quad (5)$$

whose solution, for η and κ independent of z, is

$$f(z) = e^{\left(\begin{smallmatrix} \eta & \kappa \\ -\kappa & -\eta \end{smallmatrix}\right) z}. \quad (6)$$

It is easy to verify that the transfer matrix

$$M(z_2, z_1) = e^{(\eta\sigma_z - i\kappa\sigma_y)(z_2 - z_1)}, \quad (7)$$

with σ_z and σ_y being the Pauli matrices, satisfies the relation

$$\begin{pmatrix} u(z_2) \\ v(z_2) \end{pmatrix} = M(z_2, z_1)\begin{pmatrix} u(z_1) \\ v(z_1) \end{pmatrix}. \quad (8)$$

Expanding the exponential function, the transfer matrix in (7) becomes [16]

$$M(z_2, z_1) = \begin{pmatrix} \alpha_G & \beta_G \\ -\beta_G & \delta_G \end{pmatrix}, \quad (9)$$

where

$$\begin{aligned}\alpha_G &= \cosh K(z_2 - z_1) + \frac{\eta}{K}\sinh K(z_2 - z_1),\\ \beta_G &= \frac{\kappa}{K}\sinh K(z_2 - z_1),\\ \delta_G &= \cosh K(z_2 - z_1) - \frac{\eta}{K}\sinh K(z_2 - z_1).\end{aligned} \quad (10)$$

Here $K = \sqrt{\eta^2 - \kappa^2}$. If we consider the parameters

$$\begin{aligned}\eta &= i\frac{r_o}{c}(\omega - \omega_B) = i\frac{r_o}{c}\delta,\\ \kappa &= ik_B V_o = i\frac{r_o}{c}\omega_B V_o = i\frac{r_o}{c}\omega_o,\end{aligned} \quad (11)$$

the transfer matrix of a single Bragg grating of length $L_o = z_2 - z_1$ becomes

$$M_{BGa} = \begin{pmatrix} \cosh KL_o + i\frac{\delta}{\Omega}\sinh KL_o & i\frac{\omega_o}{\Omega}\sinh KL_o \\ -i\frac{\omega_o}{\Omega}\sinh KL_o & \cosh KL_o - \frac{i\delta}{\Omega}\sinh KL_o \end{pmatrix}, \quad (12)$$

with $\Omega = \sqrt{\omega_o^2 - \delta^2} = \sqrt{\omega_o^2 - (\omega - \omega_B)^2} = cK/r_o$. The frequency $\delta = \omega - \omega_B$ is known as the detuning frequency. It is worth noticing that this matrix lacks the information of the air-BG and the BG-air interfaces. To describe transmission through a finite BG, bounded by air or some other media, one needs in principle to multiply by

$$\begin{aligned}M_{BGa} &= \frac{1}{2r_a}\begin{pmatrix} r_a + r & r_a - r \\ r_a - r & r_a + r \end{pmatrix},\\ M_{aBG} &= \frac{1}{2r}\begin{pmatrix} r + r_a & r - r_a \\ r - r_a & r + r_a \end{pmatrix},\end{aligned} \quad (13)$$

on the left and right hand sides, respectively. For BG bounded by air $r_a = 1$ and $r \simeq r_o$. It we take into account these matrices, the effect on the physical quantities is negligible. For this reason one can keep the manageable form (12). Given the simple relation

$$t_G = \frac{1}{\alpha_G^*} \quad (14)$$

between the transmission amplitude and the transfer matrix elements, the transmission coefficient and phase time of a FBG in the OCMA become

$$\begin{aligned}T_G &= \frac{1}{\cosh^2 KL_o + (\delta^2/\Omega^2)\sinh^2 KL_o},\\ \tau_G &= \frac{n_o L_o T_G}{c}\frac{\delta^2}{\Omega^2}\left(\sinh^2 KL_o - \cosh^2 KL_o\right)\\ &\quad + \frac{n_o T_G}{c\Omega}\left(1 + \frac{\delta^2}{\Omega^2}\right)\cosh KL_o \sinh KL_o.\end{aligned} \quad (15)$$

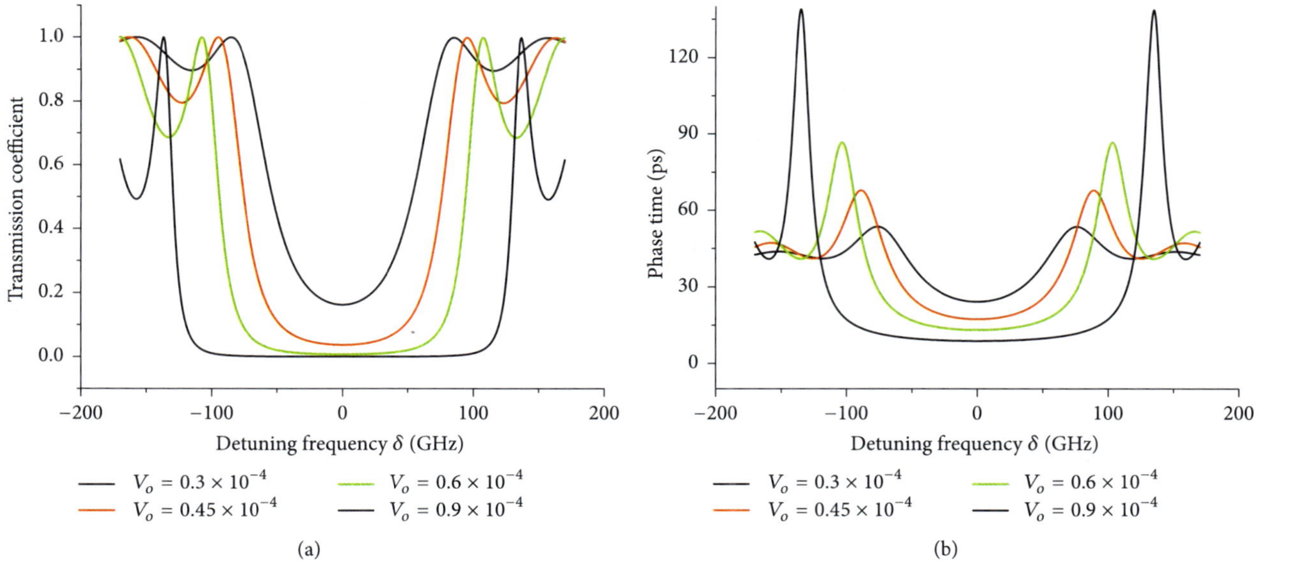

FIGURE 1: The transmission coefficient (a) and the phase time (b) in (15), as functions of the detuning frequency $\delta = \omega - \omega_B$, for different values of the modulation amplitude V_o with $w_B = 1.261 \times 10^{15}$ Hz, $r_0 = 1.452$, and $L_o = 8.5$ mm.

In the particular case of $\delta = w - w_B = 0$ we have

$$T_s \equiv T_G\big|_{\delta=0} = \frac{1}{\cosh^2(k_B V_o L_o)},$$
$$\tau_s \equiv \tau_G\big|_{\delta=0} = \frac{r_o}{c_o k_B V_o}\tanh(k_B V_o L_o). \quad (16)$$

These expressions, written here with the notation of [24], are characteristic from the one coupled mode approach [16]. In Figures 1(a) and 1(b) we plot the transmission coefficient T_G and the tunneling time τ_G, as functions of the detuning frequency δ, for different values of the modulation amplitude V_o. For these graphs we considered $L_o = 8.5$ mm, $\omega = 1.261 \times 10^{15}$ Hz, and $r_o = 1.452$. Increasing the amplitude V_o, the optical gap becomes deeper and wider. At the same time, the transmission time in the gap becomes smaller. In the next sections we will present the Mathieu functions approach and the theory of finite periodic systems and, with the purpose of comparing with the results shown in Figure 1, we will calculate the same physical quantities for a system similar to that considered here.

2.2. The FBG Described by Mathieu Functions. If we come back to (2) and write the refractive index r^2, for $2V_o \ll 1$, as

$$r^2(z) = r_o^2\left(1 + 2V_o\cos\frac{2\pi z}{\Lambda}\right)^2$$
$$\simeq r_o^2\left(1 + 4V_o\cos\frac{2\pi z}{\Lambda}\right), \quad (17)$$

the wave equation becomes the Mathieu differential equation

$$\frac{d^2E(z)}{dz^2} + \left(a_E - 2q_E\cos\frac{2\pi z}{\Lambda}\right)E(z) = 0, \quad (18)$$

whose solutions $E_1(z) = S_e(a_E, q_E; k_B z + \pi/2)$ and $E_2(z) = S_o(a_E, q_E; k_B z + \pi/2)$ are the well known even and odd Mathieu functions [25], with $k = \omega/c$, $a_E = k^2 r_o^2$, $q_E = 2k^2 r_o^2 V_o$, and $z \to z \pm \Lambda/2$. Defining the wave vector

$$F(z) = \begin{pmatrix} E(z) \\ E'(z) \end{pmatrix} = \begin{pmatrix} AS_e + BS_o \\ AS_e' + BS_o' \end{pmatrix}, \quad (19)$$

it is possible to show that the transfer matrix

$$W(z_2, z_1) = \begin{pmatrix} \theta & \mu \\ \nu & \chi \end{pmatrix} \quad (20)$$

with

$$\theta(z_1, z_2) = \frac{S_o'(z_1)S_e(z_2) - S_e'(z_1)S_o(z_2)}{W_r}, \quad (21a)$$

$$\mu(z_1, z_2) = \frac{S_e(z_1)S_o(z_2) - S_o(z_1)S_e(z_2)}{W_r}, \quad (21b)$$

$$\nu(z_1, z_2) = \frac{S_o'(z_1)S_e'(z_2) - S_e'(z_1)S_o'(z_2)}{W_r}, \quad (21c)$$

$$\chi(z_1, z_2) = \frac{S_e(z_1)S_o'(z_2) - S_o(z_1)S_e'(z_2)}{W_r}, \quad (21d)$$

$$W_r = S_e(z_1)S_o'(z_1) - S_o(z_1)S_e'(z_1), \quad (22)$$

satisfies the relation

$$F(z_2) = W(z_2, z_1)F(z_1). \quad (23)$$

The transfer matrix $W(z_2, z_1)$ connects the wave vectors of Mathieu functions and their derivatives at any two points, z_2 and z_1, of an infinite periodic system. The actual Bragg gratings are finite with thousands of unit cells. In this case, like in the standard approach of infinite periodic systems, we assume a kind of Born-von Kármán approximation, but for the transfer matrix. More precise results will be obtained using the theory of finite periodic functions. To evaluate transmission coefficients and phase times of a BG whose length is L_o, we need to transform the matrix $W(L_o, 0)$ into the transfer matrix $M(L_o, 0)$ that connects wave vectors of propagating functions at two points just outside the BG. When the wave number of the propagating functions in vacuum is k, the relation between these matrices is [26]

$$M_G(L_o, 0) = \frac{1}{2}\begin{pmatrix} 1 & -ik^{-1} \\ 1 & ik^{-1} \end{pmatrix} W(L_o, 0) \begin{pmatrix} 1 & 1 \\ ik & -ik \end{pmatrix}. \quad (24)$$

After multiplying, the transfer matrix that we obtain for a single Bragg grating is

$$M_G(L_o, 0) = \begin{pmatrix} \alpha_G & \beta_G \\ \beta_G^* & \alpha_{G^*} \end{pmatrix} \quad (25)$$

with

$$\alpha_G = \frac{\theta(0, L_o) + \chi(0, L_o)}{2} + i\frac{k^2\mu(0, L_o) - \nu(0, L_o)}{2k},$$
$$\beta_G = \frac{\theta(0, L_o) - \chi(0, L_o)}{2} - i\frac{k^2\mu(0, L_o) + \nu(0, L_o)}{2k}. \quad (26)$$

The transmission coefficients and the phase time are therefore given by

$$T_G = \frac{1}{|\alpha_G|^2} = \frac{4k^2}{k^2(\theta + \chi)^2 + (k^2\mu - \nu)^2}, \quad (27)$$

$$\tau_G = T\left(\alpha_{Gr}\frac{\partial \alpha_{Gi}}{\partial \omega} - \alpha_{Gi}\frac{\partial \alpha_{Gr}}{\partial \omega}\right). \quad (28)$$

In Figure 2 we plot the transmission coefficient T_G, for the same parameter values used in Figure 1. The qualitative and quantitative agreements are good. We do not present the phase time of (28) because the evaluation, using Mathematica code, diverges for Bragg-grating lengths of the order of 10μ. To overcome this difficulty one can combine this Mathieu functions approach with the theory of finite periodic systems. The results of this combined approach will be shown after discussing the results obtained in the TFPS.

2.3. The FBG in the TFPS. To simplify the calculation we shall assume that the refractive index is $r(z) = r_o(1 + q(z))$ with $q(z)$, a sectionally constant periodic function defined as

$$q(z) = \begin{cases} \dfrac{V_o}{2} & 0 < z \bmod (\Lambda) \le \dfrac{\Lambda}{2} \\ \dfrac{-V_o}{2} & \dfrac{\Lambda}{2} < z \bmod (\Lambda) \le \Lambda. \end{cases} \quad (29)$$

In this case the unit cell transfer matrix is

$$M(\Lambda, 0) = \begin{pmatrix} \alpha & \beta \\ \beta^* & \alpha^* \end{pmatrix} \quad (30)$$

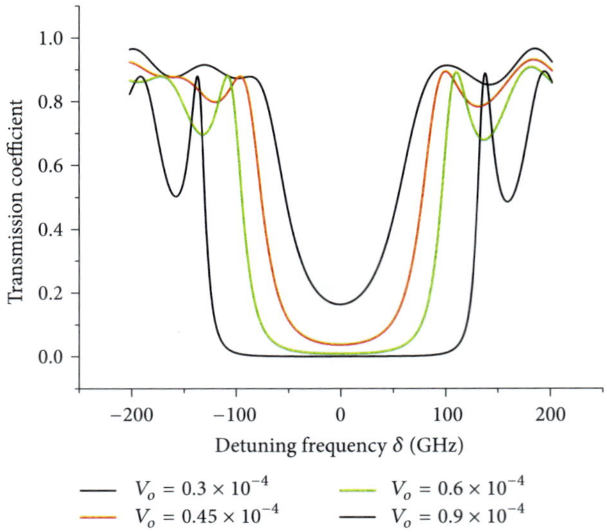

FIGURE 2: The transmission coefficient as a function of the detuning frequency, evaluated with (27) and the Mathieu functions, for different values of the modulation amplitude V_o and the same parameter values used in Figure 1. The behavior compares quite well with Figure 1; however the phase time using a code like Mathematica diverges for BG length above $L_o \simeq 10\mu$.

with the real and imaginary parts of α and β given by

$$\alpha_R = \cos(kr_1a)\cos(kr_2b) - \frac{r_1^2 + r_2^2}{2r_1r_2}\sin(kr_1a)\sin(kr_2b),$$
$$\alpha_I = \sin(kr_1a)\cos(kr_2b) + \frac{r_1^2 + r_2^2}{2r_1r_2}\cos(kr_1a)\sin(kr_2b),$$
$$\beta_I = -\frac{r_1^2 - r_2^2}{2r_1r_2}\sin(kr_2b),$$
$$\beta_R = 0. \quad (31)$$

Here $a = b = \Lambda/2$ are the half-cell lengths with refractive indices $r_1 = r_0 - V_o/2$ and $r_2 = r_0 + V_o/2$, respectively. In the theory of finite periodic systems, the transfer matrix of the whole n-cells Bragg grating is [21]

$$M(n\Lambda, 0) = \begin{pmatrix} \alpha_n & \beta_n \\ \beta_n^* & \alpha_n^* \end{pmatrix} \quad (32)$$

with

$$\alpha_n = U_n(\alpha_R) - \alpha^* U_{n-1}(\alpha_R).$$
$$\beta_n = \beta U_{n-1}(\alpha_R), \quad (33)$$

where $U_n(\alpha_R)$ is the Chebyshev polynomial of the second kind and order n, evaluated at α_R. The transmission

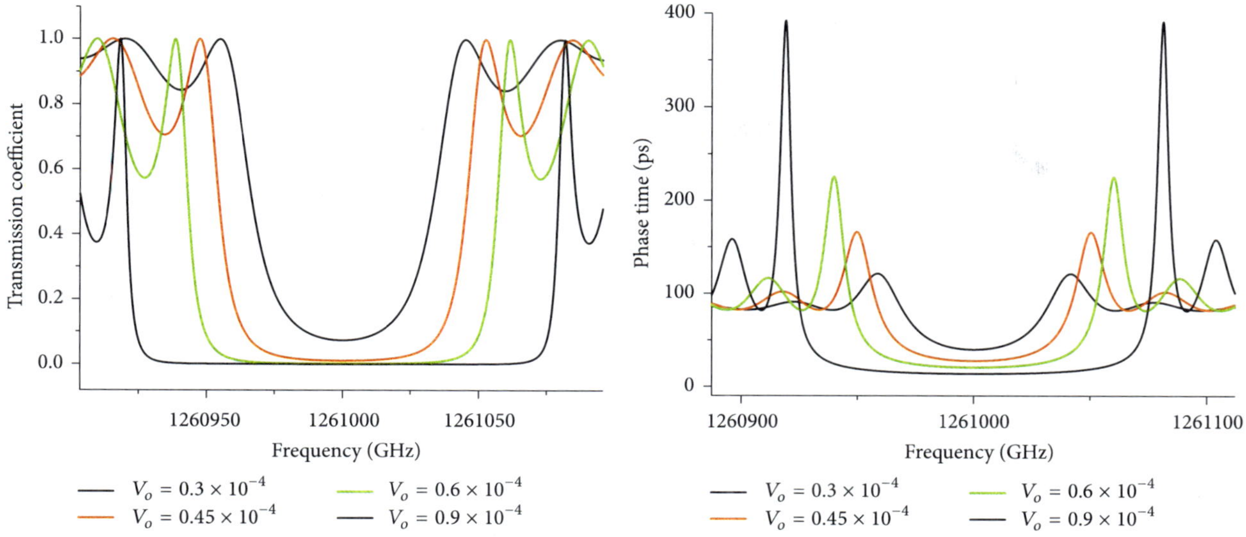

FIGURE 3: The transmission coefficient (a) and the phase time (b) from the TFPS, as functions of the frequency, for different values of the "modulation amplitude" V_o with $w_B = 1.261 \times 10^{15}$ Hz, $r_0 = 1.452$, and $L_o = 8.5$ mm.

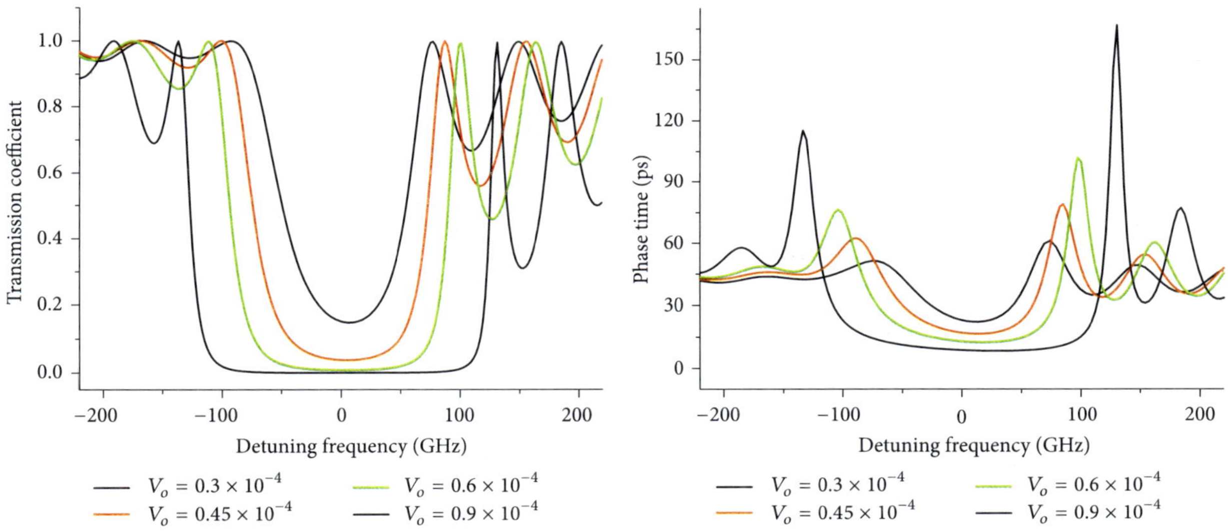

FIGURE 4: The transmission coefficient (a) and the phase time (b) from a combined Mathieu functions and FPS theory, as functions of the detuning frequency, for different values of the "modulation amplitude" V_o with $w_B = 1.261 \times 10^{15}$ Hz, $r_0 = 1.452$, $n = 3445$, and $L_o = 8.5$ mm.

coefficient and the phase time of the n-cells Bragg grating can be evaluated from

$$T_n = \frac{1}{|\alpha_n|^2} = \frac{1}{|U_n - \alpha^* U_{n-1}|^2},$$

$$\tau_n(E) = -a\hbar \frac{dk}{dE} + \frac{\hbar}{|\alpha_n|^2}\left\{\frac{1}{2}U_{2n-1}(\alpha_R)\frac{d\alpha_I}{dE} - \frac{\alpha_I}{(1-\alpha_R^2)}\left[n - \frac{\alpha_R}{2}U_{2n-1}(\alpha_R)\right]\frac{d\alpha_R}{dE}\right\}. \quad (34)$$

In Figure 3 we plot these quantities for the same parameters as in Figure 1. There is also a very good qualitative and quantitative agreement with the results in Figure 1.

2.4. The FBG Combining Mathieu Functions and the TFPS. To avoid divergencies when standard codes like Mathematica are used to evaluate Mathieu functions and their derivatives, one can divide the Bragg grating of length L_o by some natural number $n \gtrsim 1000$ such that the Mathieu functions evaluated at the new BG length $l_o = L_o/n$ is congruent with Λ. The smaller Bragg gratings can now be taken as the unit cell in the TFPS. For n a multiple of 689 with $w_B = \pi c/r_o\Lambda =$

1.2618710^{15} Hz and $r_o = 1.452$ we obtain the transmission coefficients and phase times shown in Figure 4, for the same values of V_o as in Figures 1, 2 and 3. The qualitative and quantitative agreement is really good.

The theory of finite periodic systems and the Mathieu functions approach were also applied to study the double Bragg grating, particularly to understand the effect of the Bragg gratings separation in the tunneling time behavior and to compare with other approaches' predictions [27].

2.5. Conclusions. We have shown that the transmission of electromagnetic waves, through fiber Bragg gratings, can be faithfully studied using the theory of finite periodic system either alone or combined with the Mathieu functions.

Competing Interests

The author declares that there is no conflict of interests regarding the publication of this paper.

References

[1] A. Othonos and K. Kalli, *Fiber Bragg Gratings: Fundamentals and Applications in Telecommunications and Sensing*, Artech House, 1999.

[2] P. Ferraro and G. De Natale, "On the possible use of optical fiber Bragg gratings as strain sensors for geodynamical monitoring," *Optics and Lasers in Engineering*, vol. 37, no. 2-3, pp. 115–130, 2002.

[3] T. Guo, X. Qiao, Z. Jia, Q. Zhao, and X. Dong, "Simultaneous measurement of temperature and pressure by a single fiber Bragg grating with a broadened reflection spectrum," *Applied Optics*, vol. 45, no. 13, pp. 2935–2939, 2006.

[4] V. V. Spirin, M. G. Shlyagin, S. V. Miridonov, and I. Marquez, "Temperature-insensitive strain measurement using differential double Bragg grating technique," *Optics and Laser Technology*, vol. 33, no. 1, pp. 43–46, 2001.

[5] N. Tanaka, Y. Okabe, and N. Takeda, "Temperature-compensated strain measurement using fiber Bragg grating sensors embedded in composite laminates," *Smart Materials and Structures*, vol. 12, no. 6, pp. 940–946, 2003.

[6] A. Sun, X. G. Qiao, Z. A. Jia, M. Li, and D. Z. Zhao, "Study of simultaneous measurement of temperature and pressure using double fiber Bragg gratings with polymer package," *Optical Engineering*, vol. 44, no. 3, Article ID 034402, 2005.

[7] O. Frazão, L. A. Ferreira, F. M. Araújo, and J. L. Santos, "Applications of fiber optic grating technology to multi-parameter measurement," *Fiber and Integrated Optics*, vol. 24, no. 3-4, pp. 227–244, 2005.

[8] M. Attygalle, C. Lim, G. J. Pendock, A. Nirmalathas, and G. Edvell, "Transmission improvement in fiber wireless links using fiber Bragg gratings," *IEEE Photonics Technology Letters*, vol. 17, no. 1, pp. 190–192, 2005.

[9] H. Kogelnik, "Coupled wave theory for thick hologram gratings," *Bell System Technical Journal*, vol. 48, no. 9, pp. 2909–2947, 1969.

[10] A. Yariv, "Coupled-mode theory for guided-wave optics," *IEEE Journal of Quantum Electronics*, vol. 9, no. 9, pp. 919–933, 1973.

[11] A. W. Snyder, "Coupled-mode theory for optical fibers," *Journal of the Optical Society of America*, vol. 62, no. 11, pp. 1267–1277, 1972.

[12] D. Marcuse, "Coupled mode theory of round optical fibers," *The Bell System Technical Journal*, vol. 52, no. 6, pp. 817–842, 1973.

[13] W. Streifer, D. Scifres, and R. Burnham, "TM-mode coupling coefficients in guided-wave distributed feedback lasers," *IEEE Journal of Quantum Electronics*, vol. 12, no. 2, pp. 74–78, 1976.

[14] H. Kogelnik and C. V. Shank, "Coupled–wave theory of distributed feedback lasers," *Journal of Applied Physics*, vol. 43, no. 5, pp. 2327–2335, 1972.

[15] H. Kogelnik, E. Garmire, and J. M. Hammer, *Integrated Optics*, Spriger, Berlin, Germany, 1st edition, 1975.

[16] T. Erdogan, "Fiber grating spectra," *Journal of Lightwave Technology*, vol. 15, no. 8, pp. 1277–1294, 1997.

[17] F. Abeles, "Sur la propagation des ondes electromagnetiques dans les milieux stratifies," *Annales de Physique (Paris)*, vol. 5, pp. 504–520, 1950.

[18] M. Born and E. Wolf, *Principles of Optics*, Dover, Pergamon, Oxford, UK, 1980.

[19] M. Pacheco and F. Claro, "Simple results for one–dimensional periodic potentials," *Physica Status Solidi (B)*, vol. 114, no. 2, pp. 399–403, 1982.

[20] D. J. Griffiths and N. F. Taussig, "Scattering from a locally periodic potential," *American Journal of Physics*, vol. 60, no. 10, pp. 883–888, 1992.

[21] P. Pereyra, "Resonant tunneling and band mixing in multichannel superlattices," *Physical Review Letters*, vol. 80, no. 12, pp. 2677–2680, 1998.

[22] P. Pereyra and E. Castillo, "Theory of finite periodic systems: general expressions and various simple and illustrative examples," *Physical Review B*, vol. 65, no. 20, Article ID 205120, 2002.

[23] P. Pereyra and H. P. Simanjuntak, "Time evolution of electromagnetic wave packets through superlattices: evidence for superluminal velocities," *Physical Review E*, vol. 75, no. 5, Article ID 056604, 2007.

[24] S. Longhi, P. Laporta, M. Belmonte, and E. Recami Phys, "Critical number of atoms for attractive Bose-Einstein condensates with cylindrically symmetrical traps," *Physical Review E*, vol. 64, no. 5, Article ID 055602, 2001.

[25] P. M. Morse and H. Feshbach, *Methods of Theoretical Physics, Part I*, McGraw Hill, New York, NY, USA, 1953.

[26] A. Anzaldo-Meneses and P. Pereyra, "Sylvester theorem and the multichannel transfer matrix method for arbitrary transverse potential profile inside a wave guide," *Annals of Physics*, vol. 322, no. 9, pp. 2114–2128, 2007.

[27] H. P. Simanjuntak and P. Pereyra, "On the generalized Hartman effect presumption in semiconductors and photonic structures," *Nanoscale Research Letters*, vol. 8, article no. 145, 2013.

3

Light Polarization using Ferrofluids and Magnetic Fields

Alberto Tufaile,[1] Timm A. Vanderelli,[2] and Adriana Pedrosa Biscaia Tufaile[1]

[1] *Soft Matter Laboratory, Escola de Artes, Ciências e Humanidades, Universidade de São Paulo, 03828-000 São Paulo, SP, Brazil*
[2] *Ferrocell USA, 739 Route 259, Ligonier, PA 15658, USA*

Correspondence should be addressed to Alberto Tufaile; tufaile@usp.br

Academic Editor: Charles Rosenblatt

We are presenting an experimental setup based on polarized light, enabling the visualization of the magnetic field of magnetic assemblies using a Hele-Shaw cell filled with ferrofluids. We have simulated the observed patterns with hypergeometric polynomials.

1. Introduction

When a strong magnet is brought close to a Hele-Shaw cell containing a ferrofluid, this magnetic fluid undergoes structural transitions due to the magnetic field, forming linear chains or rods of nanoparticles along the field direction. This is a magnetooptic effect in which an electromagnetic wave propagates through the ferrofluid that has been altered by the presence of a quasi-static magnetic field, forming some patterns. One way to observe this pattern formation is to use polarized light [1–4].

Some interesting applications of this type of device include optical switches, or the simulation of atmospheric optics known as jumping laser dogs [5], just to cite few of them. Another potential application is characterizing and controlling the quality of compound magnets observing the patterns formed in this type of Hele-Shaw cell filled with ferrofluids, known as Ferrocell.

In this paper, we present how to use and characterize the magnetic field of compound magnets using the Ferrocell, analyzing some aspects of the pattern formation and their relationship with some optical properties obtained from the presence of different distribution of the magnetic field, obtaining a map of the magnetic field.

2. Materials and Methods

The diagram of the experiment is shown in Figure 1(a). The Ferrocell is placed between two crossed polarizers illuminated with white light. The white light source was generated using a LED panel light (12 W), illuminating over the complete surface of a square Ferrocell (22 mm × 22 mm) homogeneously, with a methacrylate diffuser and backlighting technique. The glass plates used in the Ferrocell in this case are microscope slides. The image is obtained directly from the system polaroids/Ferrocell, photographing the Ferrocell with a digital camera.

The ferrofluid is the EFH1, a stable colloidal dispersion using light mineral oil. The response time is of the order of 200 ms, and any pattern disappears almost instantly after removing the magnetic field. In very low fields, the ferrofluid is isotropic, and the polarized light passes through the Ferrocell. When the magnetic field is increased, we have the formation of needlelike structures of the ferrofluid. These structures are shown in Figure 1(b), using an optical microscope (200x magnification). These structures are formed by the ferrofluid in the presence of the magnetic field, ranging from 100 G to 500 G, creating some elongated structures, known as needlelike ferrofluid structures.

The magnets are placed at the top of the analyzer, as it is shown in Figure 1(a), and they can create different magnetic field configurations, depending on their orientation, measuring the magnetic field intensity with a Hall sensor, as it is shown in Figure 2(a). For example, a cylindrical magnet with the north pole facing the center system formed by the analyzer/Ferrocell/polarizer in vertical position can create the unipolar field configuration shown in the first plot of Figure 2(b). Another possibility is to place the same magnet in the horizontal position with the north pole and south pole in direct contact with the analyzer/Ferrocell/polarizer

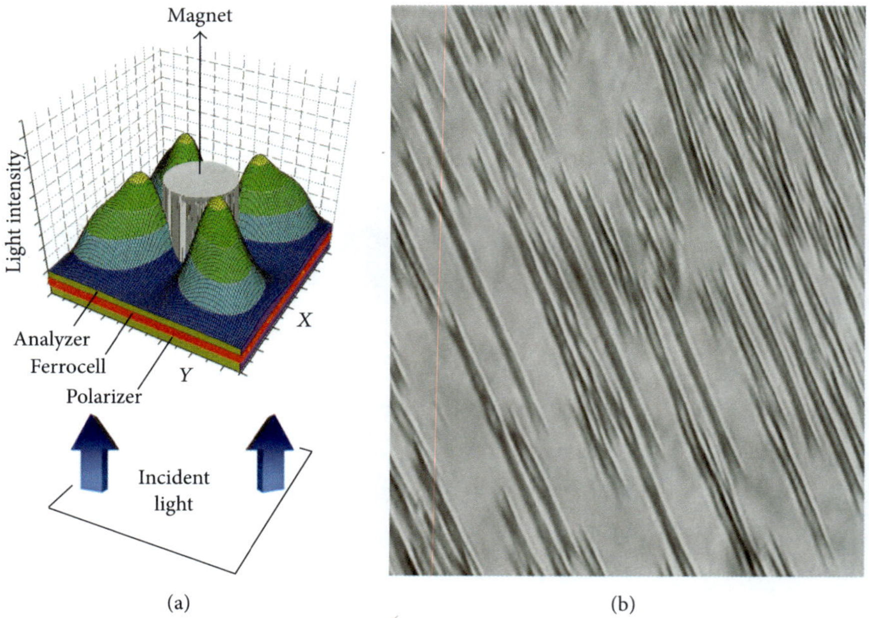

FIGURE 1: Experimental apparatus in (a), with the surface plot of the light intensity. Optical microscopic view (200x) of the structures formed by a ferrofluid in the presence of a magnetic field (from 100 G to 500 G) creating some needlelike structures in (b).

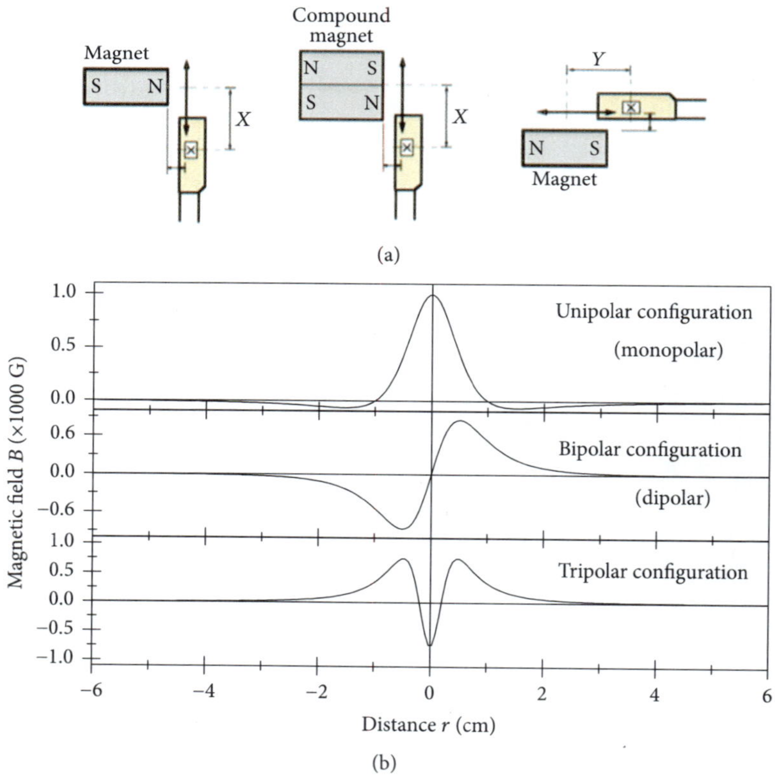

FIGURE 2: (a) The diagram of the measurements using the Hall sensor for the magnetic field of the compound magnets, in order to carry out a comparison with the light patterns observed in the Ferrocell. In (b), there are plots of the magnetic fields for different compound magnets. The coordinate r represents a generic coordinate, which can be replaced by x and y, or it represents the polar coordinate, depending on the magnetic field configuration.

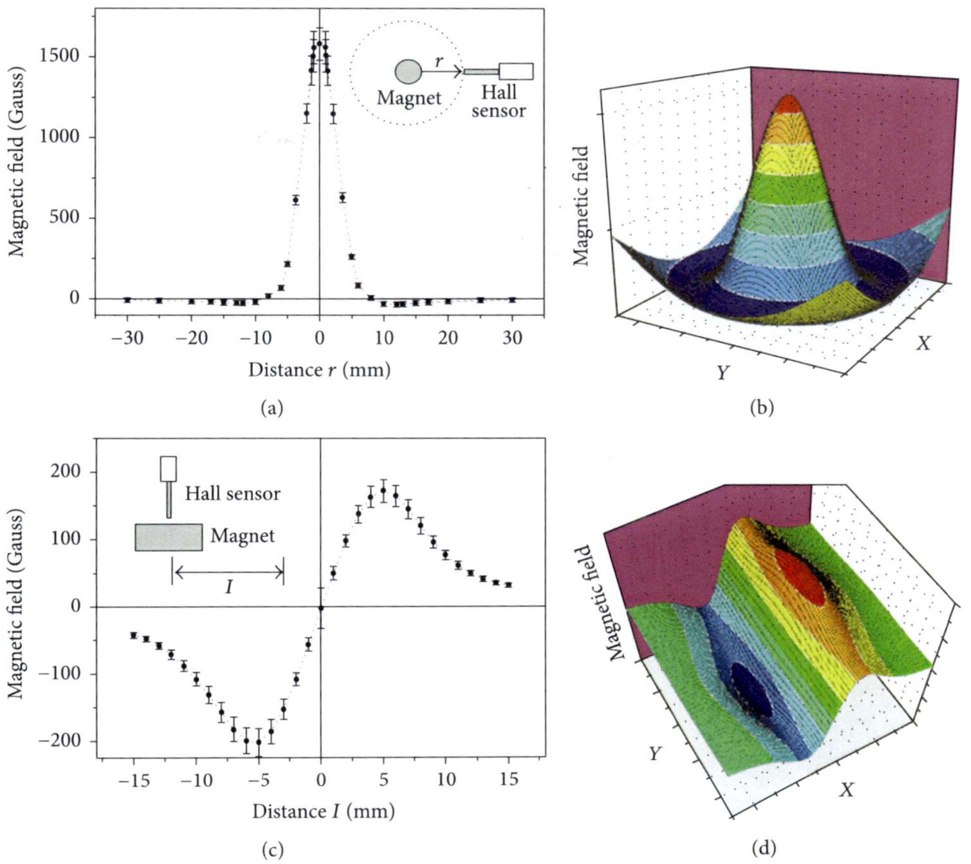

FIGURE 3: Plots representing the magnetic field using the Gaussmeter. (a) The plot of the magnetic field in the monopolar configuration (also known as unipolar configuration) and (b) the two-dimensional plot of the intensity of the magnetic field. (c) The plot of the magnetic field in the bipolar configuration and (d) the two-dimensional plot of the intensity of the magnetic field. When the magnet is placed at the top of the Ferrocell, the ferrofluid reacts to one of these configurations depending on its position.

system, creating the magnet field configuration of bipolar configuration of the second plot of Figure 2(b). Because the Hall sensor just measures the magnetic field at one point in the space at time, in order to reconstruct the magnetic field intensity in a similar plane of the Ferrocell, we have to slide it in certain directions. For example, the unipolar configuration of the diagram of the Hall sensor in Figure 2(a) and in the magnetic field intensity of Figure 2(b) is represented in Figure 3(a) with polar coordinates and with Cartesian coordinates in Figure 3(b). Placing the same magnet in the horizontal position at the center of the Ferrocell, the magnetic field in the plane of the Ferrocell of the bipolar configuration of Figure 2(b) is equivalent to the magnetic field represented in Figures 3(c) and 3(d).

We have used neodymium (Ne-Fe-B) magnets facing the center of the Ferrocell in order to change the light patterns. For example, the values of the magnetic field of some compound magnets were obtained with a Lakeshore Gaussmeter model 475 DSP in the plots of Figure 3, at room temperature of 27°C, with a Hall sensor. We can reconstruct the intensity of the magnetic field around the magnets using these data, as it is shown in Figure 3. The nanostructures inside the Ferrocell align to these local configurations of the magnetic field, whenever the magnet is placed at the top of the Ferrocell. In this way, the monopolar configuration is one of the magnetic poles facing directly the Ferrocell, and it is almost the half part of the bipolar configuration of Figure 3(d).

In order to simulate the polarized light patterns observed in the Ferrocell, we have used some routines written in commercial software (OriginLab), computing matrices with 1000 × 1000 elements.

3. Results and Discussion

The distinguishing basic feature of the Ferrocell in the presence of magnetic field, as far as optical properties are concerned, is the fact that the Ferrocell is generally magnetically anisotropic, such as the case of the Cotton-Mouton effect [5], in which the substance inside the Ferrocell behaves like a large group of small compasses, with the magnetic field defining the optic orientation. This means that the polarization in the Ferrocell by a given magnetic field is not just a simple scalar constant times the field but varies in a manner that depends on the direction of the applied field in relation to the plane of the Hele-Shaw cell. In this way, the polarization is attributed to the alignment of the nanoparticles by the magnetic field. In Figure 4, we present the light pattern obtained with a

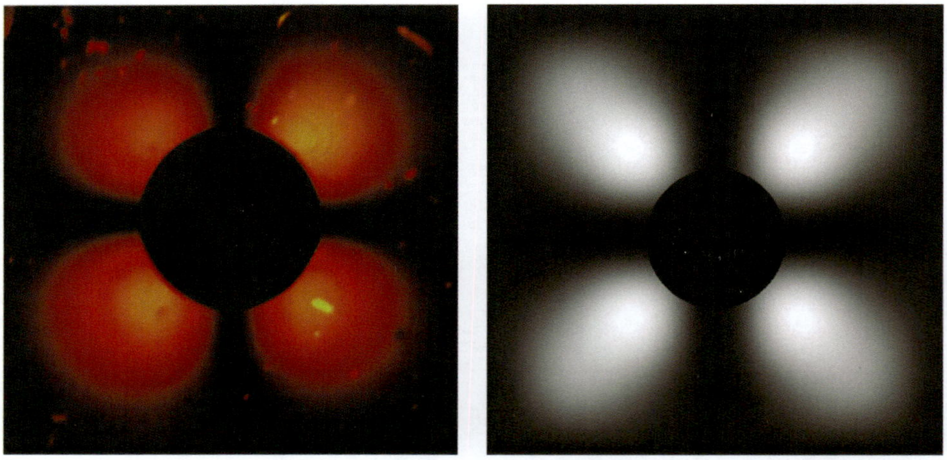

(a) (b)

FIGURE 4: (a) The pattern of light polarization from the Ferrocell with a cylindrical magnet placed in the monopolar configuration. The simulation of this configuration is presented in (b).

cylindrical magnet placed at the Ferrocell in the monopolar (or unipolar) configuration. This light pattern is associated with the magnetic field of Figure 3(b).

Using the information obtained from this pattern, we can explain the observed patterns with a phenomenological model. In order to obtain this model, we need to know the mathematical expression of the external magnetic field and the equation representing the needlelike structure. First, the calculation of the exact magnetic field of the magnet at any point in space has some mathematical complexities and involves the multiple properties of the magnetic materials. Things are simpler using direct measures of the magnetic field discussed in the previous section, such as the equation obtained for the horizontal component of the magnetic field B_x of the magnet in the monopolar configuration, as discussed in [4, 5], which is given by

$$B_x = \frac{D\left(1 - x^2\right)}{4\pi\left(x^2 + D^2\right)^{5/2}}, \tag{1}$$

where D is a constant featured for this magnet and x is the distance.

The formation of the needlelike structures of Figure 1(a) is directly related to the magnetic field B around 200 G; the direction of the polarization of the needles inside the circular region, with radius R and width a, is given by the coordinates (x_i, y_i) and (x_{i+1}, y_{i+1}) for the angle θ:

$$\begin{pmatrix} x_i & x_{i+1} \\ y_i & y_{i+1} \end{pmatrix} = \begin{pmatrix} \cos\theta \\ \sin\theta \end{pmatrix} \left(R \quad (R + a)\right). \tag{2}$$

Using (1) and (2), we are proposing that the light intensity of the pattern observed in the Ferrocell for the monopolar configuration [4, 5] is given by

$$I(r, \theta) = \frac{D\left(1 - r^2\right)}{16\pi\left(r^2 + D^2\right)^{5/2}} \sin^2 2\theta, \tag{3}$$

where r ranges from 0 to infinity and θ ranges from 0° to 360°.

This pattern suggested empirically a resemblance of this system with the patterns involving the concept of spatial modes of electromagnetic radiation in an open-sided resonator. In order to explore some of the interesting patterns observed with the traditional solutions of paraxial wave equation [6], we have used a simple hint used in computational graphics: the pattern recognition. Based on the idea that any picture may be conceived as a mathematical object, lying on part of the Euclidean plane and each point having its own light intensity, we have explored the pattern formation using matrices with 1000 × 1000 elements, developing and testing several mathematical functions. Even though this phenomenological method is very empirical and takes a lot of time to find some useful images, many commercial software packages available have a wealth of built-in mathematical functions, facilitating the exploration of image formation.

Following the previous method, we have found that the mathematical problem of image formation involves as solutions for this kind of system the Hermite-Gauss and other hypergeometric polynomials [7]. We have approximated the intensity of the light polarization observed for the magnetic field generated by these magnets by

$$\begin{aligned} B(x, y, z) &= B_0 H_m(x) H_n(y) e^{-(x^2+y^2)/w^2(z)} e^{i\phi(z)}, \\ I(x, y, z) &= \left|B(x, y, z)\right|^2, \end{aligned} \tag{4}$$

in which B_0 is the maximal intensity of the magnetic field and $H_m(x)$ and $H_n(y)$ are the spatial derivatives of the functions of the magnetic field in the x-axis and y-axis. $I(x, y, z)$ is the light intensity observed. We also have used the expressions of the magnetic field in non-Cartesian coordinates in order to obtain different types of light patterns, besides Hermite-Gauss polynomials.

Changing the orientation of the magnetic field, we have obtained different configuration patterns. For example, in Figure 5(a), we present some patterns obtained for the case of the bipolar configuration. The case of the compound magnets

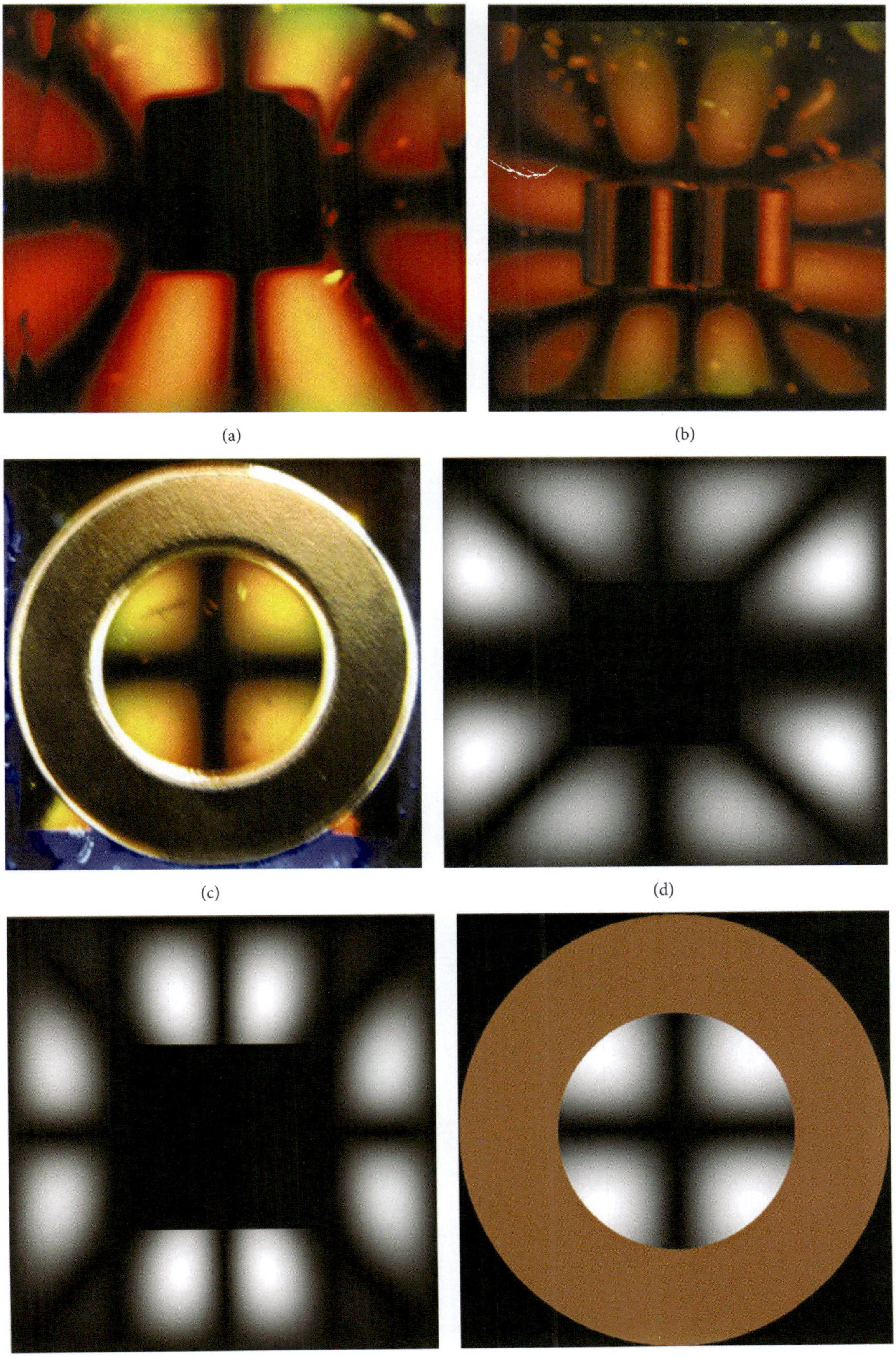

FIGURE 5: Patterns obtained with bipolar configuration (a); (b) tetrapolar configuration; (c) using a magnetic ring. Simulations for bipolar configuration (d); (e) tetrapolar configuration; (f) for the case of the magnetic ring.

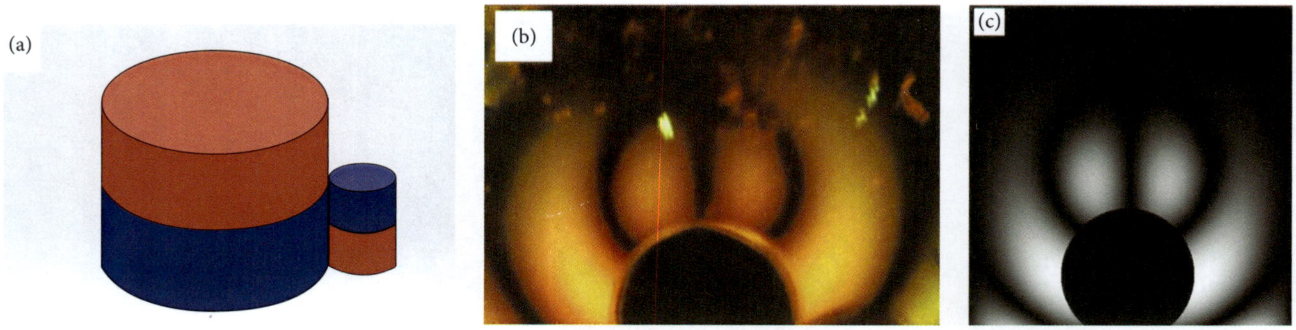

FIGURE 6: The magnetic field of two cylindrical magnets, placed side by side with poles in opposition, as it is shown in the diagram (a), one magnet with 900 Gauss and the small one with 200 Gauss. The pattern obtained with the Ferrocell is shown in (b) with the small magnet at the bottom. (c) presents the simulation for this compound magnet.

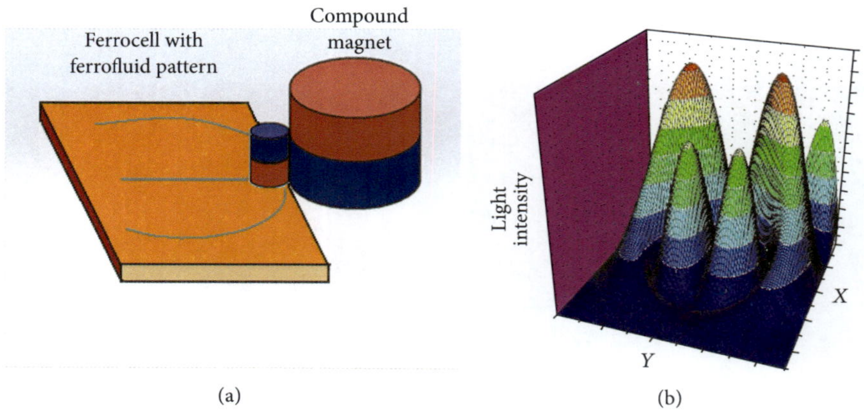

FIGURE 7: The diagram of the magnets of the previous case in (a) and the plot of the light intensity of two cylindrical magnets with different sizes and magnetic field intensities in (b), showing the result of the magnetic field of the small magnet being embraced by the magnetic field of the large magnet, resembling the interaction in the solar wind being deflected by the magnetic field of the earth.

in the tetrapolar configuration is shown in Figure 5(b), and this pattern is associated with the magnetic field of Figure 3(d). The pattern of light polarization of a magnetic ring is shown in Figure 5(c). The respective simulation of each case is presented in Figure 5(d) for bipolar configuration, in Figure 5(e) for the tetrapolar configuration, and in Figure 5(f) for the case of the magnetic ring.

These patterns indicate that the Ferrocell has the same role of a diffraction grating in the presence of the magnetic field, similar to the case of liquid crystal devices [1, 2].

An interesting case to report is the light pattern obtained for a compound magnet of Figure 6, with two cylindrical magnets placed side by side with poles in opposition, in which a more intense magnetic field "embraces" the magnetic field of the smaller cylindrical magnet. Due to the symmetry of this system, this pattern resembles one of the modes of the Laguerre-Gauss polynomials, with decay of the magnetic field pointing to the top of Figure 6(c). The diagram of this setup is represented in Figure 7(a), and the light intensity of this configuration is shown in Figure 7(b), without the representation of the small magnet.

One possible application of this system is with physics demonstrations, in order to introduce the students to more complex concepts observed in nature, because the reaction of the ferrofluid to the external magnetic field is very fast. For example, this setup can be used to construct demonstrations of the interaction between the extreme solar wind conditions stressing the earth magnetosphere, moving one of the magnets, while the small magnet is kept fixed.

As a final note, in order to show the complex behavior of the Ferrocell in the presence of magnetic field of a ring, we present in Figure 8 a different setup just using a round Ferrocell and two lasers crossing each other, in the same plane of the Ferrocell. There are no polarizers in this setup, and the magnetic ring is covered with an opaque sheet of paper, so the image formed is only due to the arrangement of the ferrofluid particles scattering the light. Although this setup is out of the scope of the present study, many important features of the Ferrocell used in the polarization study can be found in [8], in which the authors describe some techniques for the visualization of magnetic fields and explain in more detail the constitutive aspects of the Ferrocell.

4. Conclusions

We have presented an experimental setup with potential application in characterizing and controlling the quality of magnets, enabling the visualization and a fast analysis of the

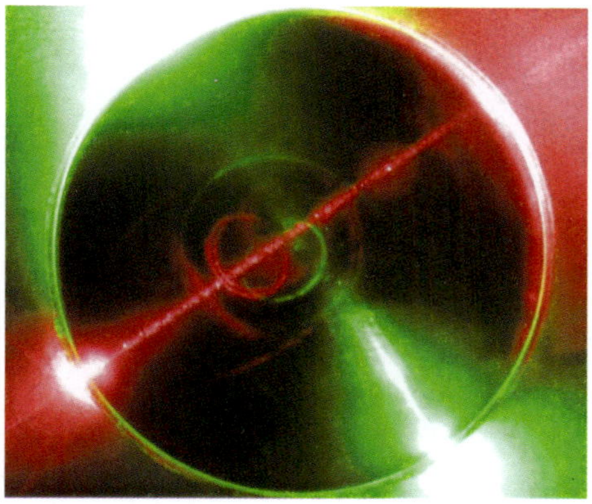

FIGURE 8: Two crossed laser beams (red and green) in a circular Ferrocell with a magnetic ring behind it. There is no polarizer in this setup. We can observe the ferrofluid particles scattering light forming curved structures at the center of the picture, which resembles some parts of the magnetic ring.

magnetic field of small multipolar magnets and magnetic assemblies. The orientation of the particles inside the ferrofluid controls the transmittance intensity.

The pattern formation curiously resembles the nodal lines of the transverse electric and magnetic modes observed in optical resonant cavities, associated with the hypergeometric polynomials, such as Hermite-Gauss, Laguerre-Gauss, and Ince-Gauss polynomials. In this analogy, the magnetic field has an equivalent role of the Gaussian irradiance distribution, and the light source is a very flat lamp normal to z, which nearly corresponds to a plane wave with constant flux set up to the z-axis, analogous to a standing wave in the optical cavity.

Conflicts of Interest

The authors declare that they have no conflicts of interest.

Acknowledgments

This work was partially supported by Conselho Nacional de Desenvolvimento Científico e Tecnológico (CNPq), Instituto Nacional de Ciência e Tecnologia de Fluidos Complexos (INCT-FCx), and by Fundação de Amparo à Pesquisa do Estado de São Paulo (FAPESP), FAPES/CNPq#573560/2008-0.

References

[1] Y.-H. Wu, Y.-H. Lin, H. Ren, X. Nie, J.-H. Lee, and S.-T. Wu, "Axially-symmetric sheared polymer network liquid crystals," *Optics Express*, vol. 13, no. 12, pp. 4638–4644, 2005.

[2] S. Nersisyan, N. Tabiryan, D. M. Steeves, and B. R. Kimball, "Fabrication of liquid crystal polymer axial waveplates for UV-IR wavelengths," *Optics Express*, vol. 17, no. 14, pp. 11926–11934, 2009.

[3] C. Penfold, R. T. Collins, A. P. B. Tufaile, and Y. Souche, "Transverse magneto-optical Kerr effect: the phase change of reflected light," *Journal of Magnetism and Magnetic Materials*, vol. 242–245, pp. 964–966, 2002.

[4] A. Tufaile, T. A. Vanderelli, and A. P. B. Tufaile, "Light polarization using ferrofluids and magnetic fields," in *Encontro de Física 2016*, Natal, Brazil, September 2016.

[5] A. Tufaile, T. A. Vanderelli, and A. P. Biscaia Tufaile, "Observing the jumping laser dogs," *Journal of Applied Mathematics and Physics*, vol. 04, no. 11, pp. 1977–1988, 2016.

[6] G. R. Fowles, *Introduction to Modern Optics*, Dover, New York, NY, USA, 2013.

[7] M. A. Bandres and J. C. Gutiérrez-Vega, "Ince-Gaussian modes of the paraxial wave equation and stable resonators," *Journal of the Optical Society of America A*, vol. 21, no. 5, pp. 873–880, 2004.

[8] M. Snyder and J. Frederick, "Photonic dipole contours of ferrofluid Hele-Shaw cell," https://arxiv.org/ftp/arxiv/papers/0805/0805.4364.pdf.

4

Thermotunable Terahertz Negative-Index Metamaterials with Dielectric Spheres Embedded in Semiconductor Host

Yu Chen[1] **Zhihao Chen**[1] **Binyi Qin**[2] **and Xiaoyu Dai**[1]

[1] *International Collaborative Laboratory of 2D Materials for Optoelectronics Science and Technology, Key Laboratory of Optoelectronic Devices and Systems of Ministry of Education and Guangdong Province, College of Optoelectronic Engineering, Shenzhen University, Shenzhen 518060, China*
[2] *School of Electronics and Communication Engineering, Yulin Normal University, Yulin 537000, Guangxi, China*

Correspondence should be addressed to Xiaoyu Dai; xiaoyudai@126.com

Academic Editor: Charles Rosenblatt

A possibility to realize thermotunable isotropic negative-index metamaterials up to terahertz regime is theoretically investigated. The proposed composite metamaterials consist of dielectric spheres embedded randomly in a semiconductor host medium. As the variation in the intrinsic carrier density in InSb due to a variation in temperature thus changes the plasma frequency, we show theoretically the provided composite metamaterials whose index of refraction is thermally tunable and potentially less propagation loss. Furthermore, the effects of the radius of the dielectric sphere on the modulation frequency and bandwidth are discussed. Finally, we find that the design parameters for the composites can be scaled for application in the higher frequency regions.

1. Introduction

Negative-index metamaterials (NIMs), with simultaneously negative permittivity and permeability, were introduced by Veselago nearly forty years ago [1] and have recently received much attention in the literature owing to great potential for applications, such as perfect lens [2], optical cloaking [3], and perfect absorber [4]. Recently, metamaterials have also been used to regulate the light field [5–7]. There has been much progress in the development of the metamaterials (MMs) from the microwave frequencies to infrared and optical frequencies. In the microwave regime, the MMs can be constructed as a combined arrays consisting of metal wires and split-ring resonators [8]. MMs at the infrared and optical frequencies were accomplished with pairs of metal rods [9] and for the inverted system of pairs of dielectric voids in metal [10]. Most of the MMs designs are constructed by the metallic elements, which incur many drawbacks such as high loss, very narrow bandwidth, and highly anisotropic. An alternative scheme was proposed based on the resonance in individual nonmagnetic dielectric scatterer with very high permittivity [11–14]. However, dielectrics with extremely high permittivity also suffer from large damping. Recently, Seo et al. have shown that isotropic NIMs up to optical frequencies can be realized with dielectric spheres embedded randomly in a negative permittivity host (NPH) medium [15]. As a result, the DS/NPH shows more robust characteristics over the DS/DH in terms of fabrication tolerance, bandwidth, and propagation loss.

Most provided NIMs usually display a narrow operational bandwidth, and the operational frequency is not tunable. The tunable MMs are obviously very highly desired and potential expectations for future optical devices are, for example, spatial light modulators and tunable optical filters. There are various strategies reported for tuning the magnetic resonance of the permeability include electrical control [16, 17], magnetically control [18], and temperature changing [19, 20]. Generally, tunability is introduced by ferromagnetic materials [21, 22], ferroelectric [13, 23], liquid crystal [20, 24, 25], etc. However, the abovementioned tunable NIMs mostly operated at the low frequencies or high losses. In this paper, we will propose the designs of the thermotunable and low loss isotropic negative-index metamaterials up to terahertz region, which consists of

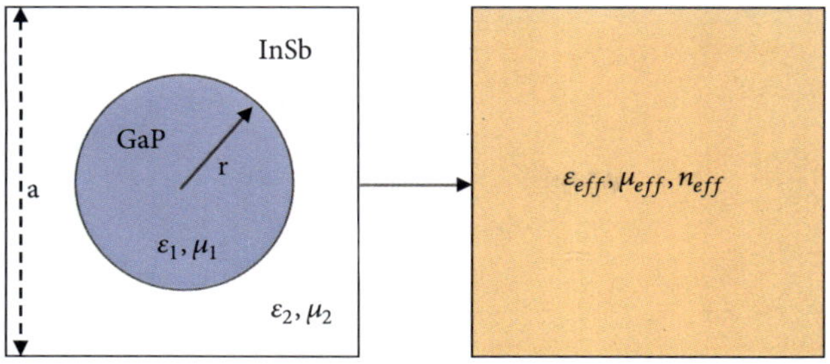

FIGURE 1: A schematic of a dielectric sphere is embedded into the semiconductor host. r is the radius of the dielectric sphere and a is the unit cell size.

dielectric spheres embedded randomly in a semiconductor host medium.

2. The Permittivity of the Semiconductor and the Effective Medium Theory

In the terahertz regime, the complex-valued relative permittivity of InSb is given by the simple Drude model [26]

$$\varepsilon(\omega) = \varepsilon_\infty - \frac{\omega_p^2}{(\omega^2 + i\gamma\omega)}, \quad (1)$$

where ω is the angular frequency, ε_∞ represents the high-frequency permittivity,and γ is the damping constant; ω_p is the plasma frequency and $\omega_p = \sqrt{Ne^2/\varepsilon_0 m^*}$ depends on the intrinsic carrier density N, the effective mass m^* of free carriers, the electronic charge e, and the free-space permittivity ε_0. Compared to metals, the plasma frequency ω_p of InSb depends strongly on the temperature T. The intrinsic carrier density N (in m^{-3}) in InSb obeys the relationship [27]

$$N = 5.76 \times 10^{20} T^{3/2} \exp\left(-\frac{0.26}{2k_B T}\right), \quad (2)$$

where k_B is the Boltzmann constant and the temperature is in Kelvin. A variation in N due to a variation in T thus changes ω_p. Consequently, in the far-infrared portion of the terahertz regime, $\varepsilon(\omega)$ of InSb is very sensitive to T. Hence, for NIMs comprising InSb host and dielectric sphere, we can expect that temperature variations can cause substantial variations in the optical response characteristics.

The composite material consisting of spherical scatters can be described as a homogeneous medium in the subwavelength limit and the effective electric permittivity and effective magnetic permeability are $\varepsilon_{eff}(\omega)$ and $\mu_{eff}(\omega)$, respectively. As shown in Figure 1, we assume that the dielectric spheres possessing a relative dielectric permittivity ε_1 and relative magnetic permeability μ_1 are embedded in a semiconductor host medium whose relative permittivity and relative permeability are $\varepsilon_2(\omega)$ and $\mu_2 = 1$, respectively. Assuming that the dielectric spheres are located in the host medium randomly and homogeneously, in the quasistatic limit, we can derive the relative effective permittivity $\varepsilon_{eff}(\omega)$ and the relative effective permeability $\mu_{eff}(\omega)$ from the extended Maxwell-Garnett theory (EMG), as shown in [14]

$$\varepsilon_{eff}(\omega) = \varepsilon_2 \frac{x^3 - 3ifT_1^E}{x^3 + (3/2)ifT_1^E}$$
$$\mu_{eff}(\omega) = \mu_2 \frac{x^3 - 3ifT_1^H}{x^3 + (3/2)ifT_1^H} \quad (3)$$

where f is the volume fraction representing the ratio of the volume of the dielectric spheres to the total volume and $f = (4/3)\pi r^3/a^3$, where r denotes the sphere radius and a is the unit cell size. T_1^E and T_1^H are the electric-dipole and magnetic-dipole components of the scattering T matrix of a single sphere, respectively. They can be written as

$$T_1^E(\omega) = \left[\frac{j_1(x_1)[x_2 j_1(x_2)]'\varepsilon_1 - j_1(x_2)[x_1 j_1(x_1)]'\varepsilon_2}{h_1^+(x_2)[x_1 j_1(x_1)]'\varepsilon_2 - j_1(x_1)[x_2 h_1^+(x_2)]'\varepsilon_1}\right], \quad (4)$$

$$T_1^H(\omega) = \left[\frac{j_1(x_1)[x_2 j_1(x_2)]'\mu_1 - j_1(x_2)[x_1 j_1(x_1)]'\mu_2}{h_1^+(x_2)[x_1 j_1(x_1)]'\mu_2 - j_1(x_1)[x_2 h_1^+(x_2)]'\mu_1}\right], \quad (5)$$

where j_1 is the spherical Bessel function and h_1^+ is the spherical Hankel function for $l = 1$, respectively. $[X(x)]' = dX/dx$. x_1 and x_2 stand for the sphere size parameter, $x_1 = (\omega/c)r\sqrt{\varepsilon_1\mu_1} = 2\pi r/\lambda_1$ and $x_2 = (\omega/c)r\sqrt{\varepsilon_2\mu_2} = 2\pi r/\lambda_2$,and λ_1 and λ_2 are the wavelengths in the sphere medium and host medium, respectively.

3. Numerical Results and Discussion

Figure 2 shows the simulation results of $\varepsilon_{eff}, \mu_{eff}, n_{eff}$ for $T = 220$ K, $f = 0.2$, and $r = 60$ um using [3–5]; here we have assumed that $\varepsilon_\infty = 15.68$, $m^* = 0.015 m_e$ kg, $m_e = 9.1 \times 10^{-31}$, and $\gamma = 2\pi \times 0.05$ THz [27], and the gallium phosphide (GaP) is assumed for the nonmagnetic dielectric sphere, and the permittivity $\varepsilon_1 = 12.25 \times (1 + i \times 10^{-3})$.

We can see from the figures that the real part of the effective permittivity ε_{eff} is negative from 0.58 THz to 0.9 THz, and the real part of the effective permeability μ_{eff} is negative from 0.65 THz to 1.0 THz; hence the double region is between about 0.65THz and 0.9 THz and the bandwidth is about 0.25 THz, as shown in the Figure 2(c). Also we can find that the value of ε_{eff} is far larger than μ_{eff}; however the imaginary parts of $\varepsilon_{eff}, \mu_{eff}, n_{eff}$ are all much smaller than the composite MMs constructed by the metallic sphere when the incident wave frequency is far away from the resonance regions.

In order to discuss the effects of the temperature T on the properties of $\varepsilon_{eff}, \mu_{eff}, n_{eff}$, in Figure 3, we have plotted the frequency-dependent refractive-index of composite MMs at various temperatures. We can find from Figure 3(a) that the resonance dips are almost kept fixed as the temperature increases; however, the value of the real part

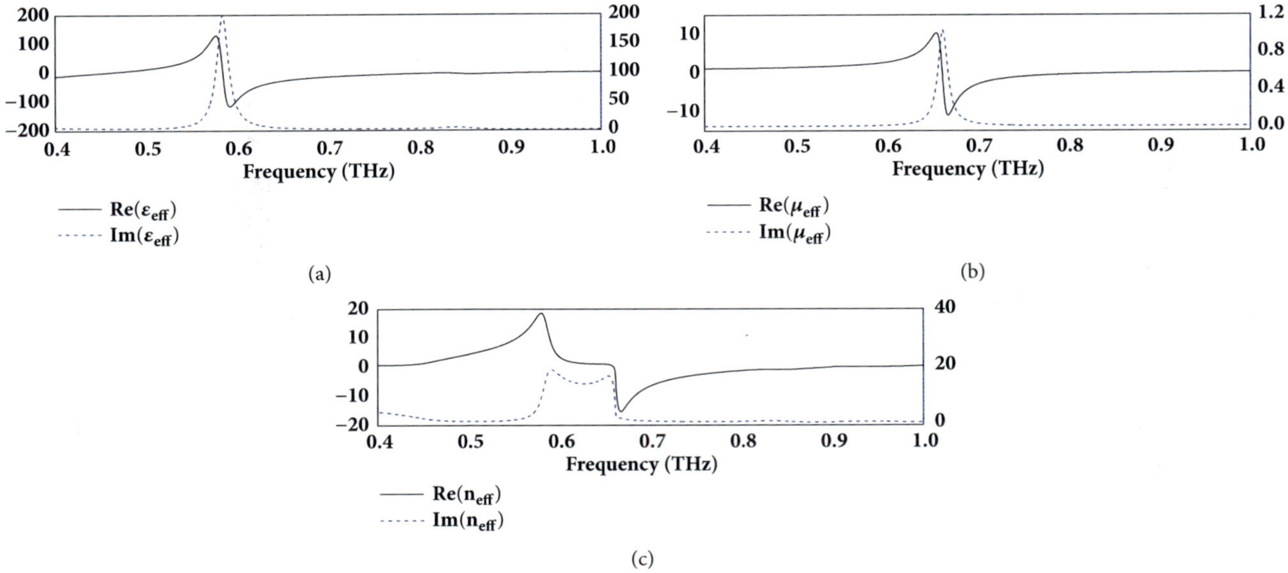

FIGURE 2: The numerical results of the real part (Re) and imaginary (Im) part of $\varepsilon_{eff}, \mu_{eff}, n_{eff}$ of the composite DS/NPH effective metamaterials.

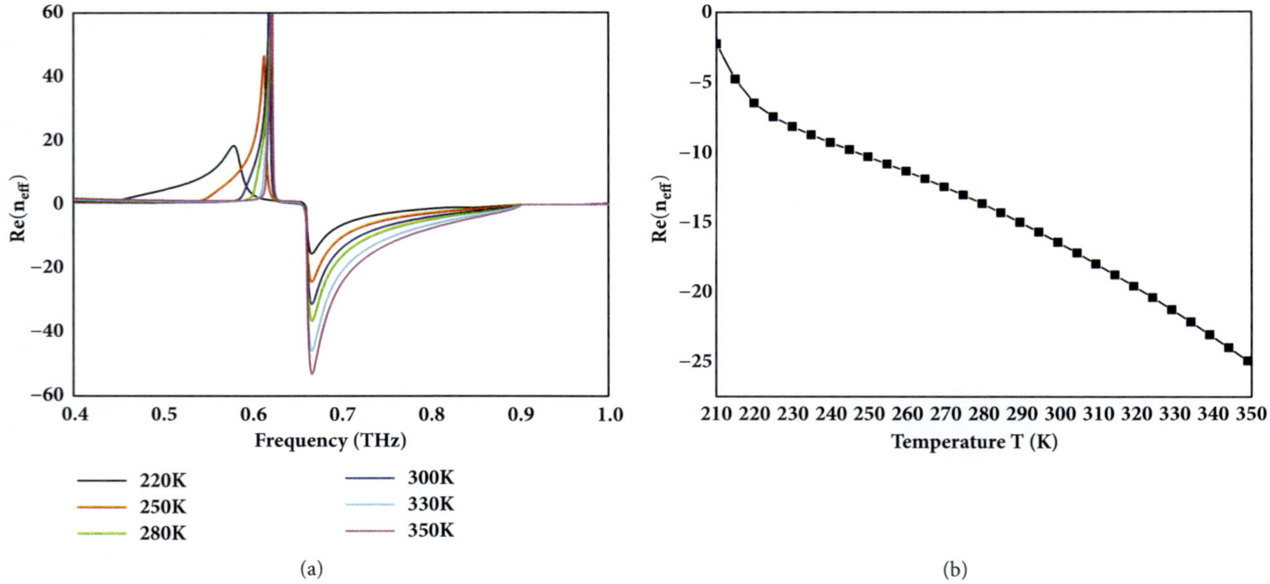

FIGURE 3: The thermotunable negative-index metamaterials. (a) The real part of the refractive-index n_{eff} at different temperatures ranging from 220 to 350 K. (b) The real part of the refractive-index n_{eff} at $\omega/2\pi = 0.7$THz for different temperatures ranging from 210 to 350 K.

of n_{eff} will become bigger and bigger as the temperature increases. Clearly, the optical properties of the MMs, such as $\varepsilon_{eff}, \mu_{eff}, n_{eff}$, all change significantly with temperature. For proving these properties clearly, in Figure 3(b), we have shown the real part of n_{eff} as a function of temperature from 210 K to 350 K. As expected, n_{eff} can be significantly enlarged by raising the temperature. The temperature-dependent optical properties of composite MMs can be attributed directly to the thermally sensitiveness of the plasma frequency for the permittivity of the semiconductor InSb.

Further, we find that the properties of $\varepsilon_{eff}, \mu_{eff}, n_{eff}$ in our composite structure depend on the radius of the dielectric sphere strongly; hence the spectral position of the resonance point can be tuned by varying the radius of the dielectric sphere. For proving this, we have shown the dependence of the real part of n_{eff} on the radius of the dielectric sphere in Figure 4. It is shown that when the radius of the sphere changes from 50 um to 80 um, the resonance frequency points of the negative refractive-index will move from 0.8 THz to 0.5 THz and the value of the resonance negative

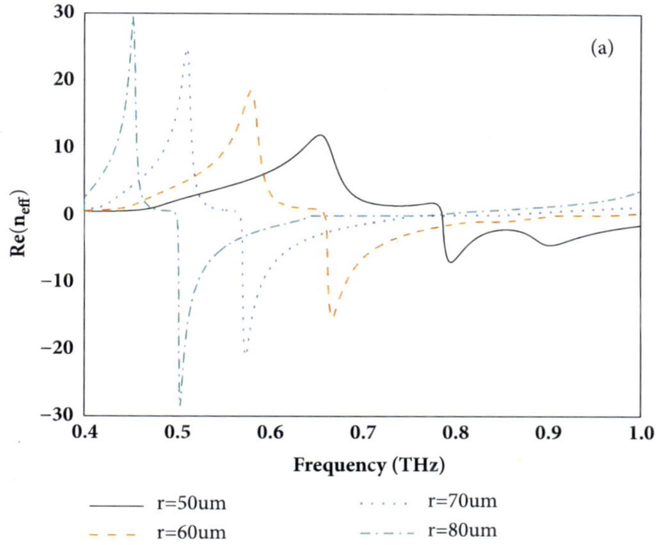

FIGURE 4: The real part of n_{eff} as a function of the radius of the dielectric sphere, where $f = 0.2$ and $T = 220$ K.

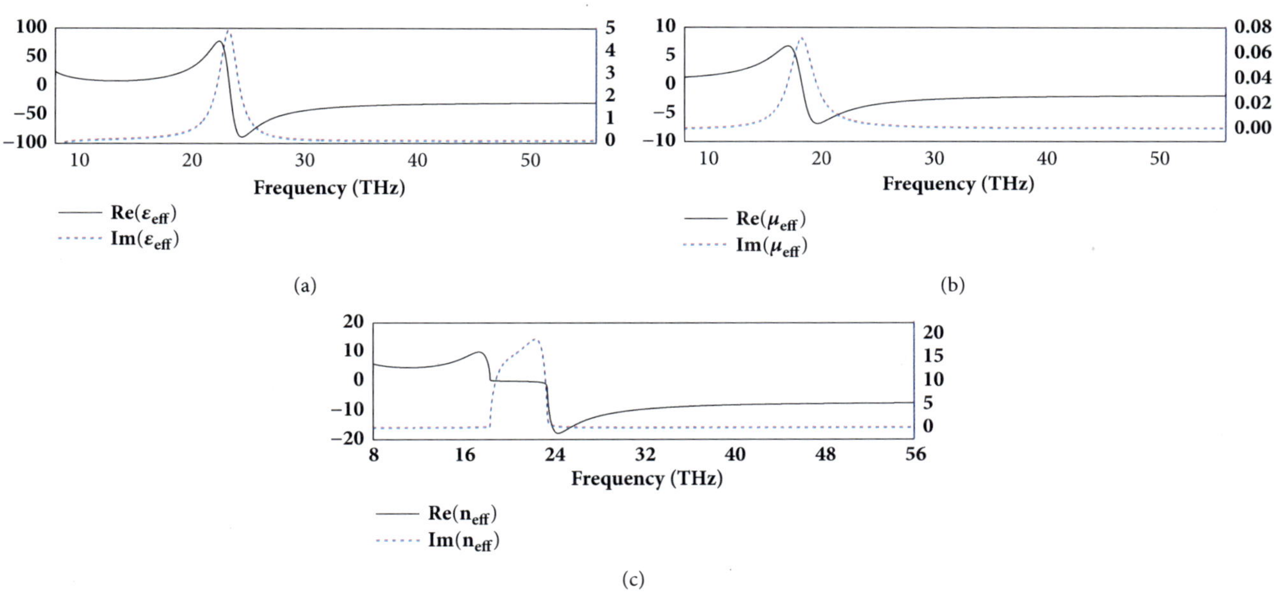

FIGURE 5: Calculated real and imaginary part of the complex permittivity, permeability, and refractive-index of the composite MMs in the terahertz region between 8 and 56 THz. Where $r = 1$ um and $T = 350$ K, the other parameters are same as in Figure 2.

refractive-index point will become larger and larger as the radius of the dielectric sphere increases.

We have shown that the low loss and thermotunable NIMs can be realized at the terahertz frequencies; however the design of the material is scalable within a large dynamic range of operating frequencies as long as we change the radius of the dielectric sphere and the temperature simultaneity. First, the intrinsic carrier density in InSb can be changed from $N = 1.977 \times 10^{21} \mathrm{m}^{-3}$ at 220 K to $N = 5.065 \times 10^{22}$ at 350 K, which will lead to the plasma frequency of the semiconductor host altering from $\omega_p = 2.046 \times 10^{13}$ at 220 K to $\omega_p = 1.036 \times 10^{14}$ at 350 K; therefore if only we adjust the scale of the radius of the dielectric sphere, we can obtain the double negative metamaterials at higher frequency. Figure 5, for example, shows the corresponding real part of the refractive-index from 8 THz to 56 THz, where $r = 1$ um and $T = 350$ K, the other parameters are the same as in Figure 2. Again, it clearly demonstrates that one can obtain a bulk material of double negative metamaterials by embedding the dielectric spheres in a semiconductor host medium randomly at the terahertz even higher frequency. Moreover, the frequency bandwidth also becomes very large to reach 32 THz.

4. Conclusions

In conclusion, we have provided a scheme to realize a thermotunable and less loss isotropic negative-index metamaterials up to terahertz regime by embedding the dielectric spheres in a semiconductor host medium randomly. The composite metamaterial is sensitive to the outer temperature, which makes the double negative frequency regions tunable. Furthermore, the radius of the dielectric sphere will also make the frequency ranges of the double regions tunable. Moreover, by scaling the size of the spheres and outer temperature, we can vary the operating frequency to higher frequency. Such reconfigurable and thermotunable metamaterials have very promising future for designing tunable devices, such as tunable optical cloaking, tunable spatial light modulators, and tunable novel microwave antenna.

Conflicts of Interest

The authors declare that they have no conflicts of interest.

Acknowledgments

This work is supported by the National Natural Science Foundation of China (Grants nos. 61505122, 61505124), the Natural Science Foundation of SZU (Grant nos. 2016031, 2017018), the program of Fundamental Research of Shenzhen Science and Technology Plan (Grant no. JCYJ20160422152152634), the project supported by Guangdong Natural Science Foundation (Grant no. 2016A030310065), the Natural Science Foundation Guangdong Education Department (2015KQNCX146), and Yulin Normal University Research Grant (Grant no. 2016YJKY06).

References

[1] V. G. Veselago, "The electrodynamics of substances with simultaneously negative values of and μ," *Soviet Physics Uspekhi*, vol. 10, no. 4, pp. 509–514, 1968.

[2] J. B. Pendry, "Negative refraction makes a perfect lens," *Physical Review Letters*, vol. 85, no. 18, pp. 3966–3969, 2000.

[3] D. Schurig, J. J. Mock, B. J. Justice et al., "Metamaterial electromagnetic cloak at microwave frequencies," *Science*, vol. 314, no. 5801, pp. 977–980, 2006.

[4] R. Alaee, M. Farhat, C. Rockstuhl, and F. Lederer, "A perfect absorber made of a graphene micro-ribbon metamaterial," *Optics Express*, vol. 20, no. 27, pp. 28017–28024, 2012.

[5] J. Li, T. Tang, L. Luo, N. Li, and P. Zhang, "Spin Hall effect of reflected light in dielectric magneto-optical thin film with a double-negative metamaterial substrate," *Optics Express*, vol. 25, no. 16, pp. 19117–19128, 2017.

[6] T. Tang, J. Li, Y. Zhang, C. Li, and L. Luo, "Spin Hall effect of transmitted light in a three-layer waveguide with lossy epsilon-near-zero metamaterial," *Optics Express*, vol. 24, no. 24, pp. 28113–28121, 2016.

[7] S. Liu, T. Cui, Q. Xu et al., "Anisotropic coding metamaterials and their powerful manipulation of differently polarized terahertz waves," *Light Science & Applications*, vol. 5, no. 5, Article ID e16076, 2016.

[8] R. A. Shelby, D. R. Smith, and S. Schultz, "Experimental verification of a negative index of refraction," *Science*, vol. 292, no. 5514, pp. 77–79, 2001.

[9] A. Moreau, C. Ciracì, J. J. Mock et al., "Controlled-reflectance surfaces with film-coupled colloidal nanoantennas," *Nature*, vol. 491, no. 7427, pp. 86–89, 2012.

[10] S. Yang, X. Ni, B. Kante et al., "Experimental Demonstration of Optical Metamaterials with Isotropic Negative Index," in *Proceedings of the Conference on Lasers and Electro-Optics*, pp. 1-2, San Jose, California, 2016.

[11] Y. Ma, H. Zhang, Y. Li, and Y. Wang, "Miniaturized and dual-band metamaterial absorber with fractal Sierpinski structure," *Journal of the Optical Society of America B: Optical Physics*, vol. 31, no. 2, pp. 325–331, 2014.

[12] W. Walasik, "Influence of the Filling Factor and Size of Coated Spheres on Effective Refractive Index in Nanosphere Dispersed Liquid Crystal Metamaterial," *Nonlinear Optics Quantum Optics*, vol. 44, no. 44, p. 129, 2012.

[13] M. V. Rybin, K. B. Samusev, and M. F. Limonov, "Multiscale modeling of all-dielectric metamaterials," in *Proceedings of the 2015 Days on Diffraction (DD)*, pp. 1–5, St.Petersburg, Russia, May 2015.

[14] J. Li, Y. Xuan, and Q. Li, "Analysis of influence factors of negative refraction phenomenon in particulate composite," *Science China Technological Sciences*, vol. 55, no. 6, pp. 1719–1724, 2012.

[15] B.-J. Seo, T. Ueda, T. Itoh, and H. Fetterman, "Isotropic left handed material at optical frequency with dielectric spheres embedded in negative permittivity medium," *Applied Physics Letters*, vol. 88, no. 16, 2006.

[16] S. Qiao, Y. Zhang, S. Liang et al., "Multi-band terahertz active device with complementary metamaterial," *Journal of Applied Physics*, vol. 118, no. 12, p. 123106, 2015.

[17] T. H. Hand and S. A. Cummer, "Frequency tunable electromagnetic metamaterial using ferroelectric loaded split rings," *Journal of Applied Physics*, vol. 103, no. 6, 2008.

[18] S.-C. Zhao, X.-F. Qian, Y.-P. Zhang, and Y.-A. Zhang, "Negative refraction with little loss manipulated by the voltage and pulsed laser in double quantum dots," *Progress of Theoretical and Experimental Physics*, vol. 128, no. 2, pp. 243–250, 2012.

[19] W. Lewandowski, M. Fruhnert, J. Mieczkowski, C. Rockstuhl, and E. Górecka, "Dynamically self-assembled silver nanoparticles as a thermally tunable metamaterial," *Nature Communications*, vol. 6, 2015.

[20] S. T. Bui, V. D. Nguyen, X. K. Bui et al., "Thermally tunable magnetic metamaterials at THz frequencies," *Journal of Optics*, vol. 15, no. 7, p. 075101, 2013.

[21] B. Du, Z. Xu, J. Wang, and S. Xia, "Magnetically tunable ferrite-dielectric left-handed metamaterial," *Progress in Electromagnetics Research C*, vol. 66, pp. 21–28, 2016.

[22] K.-M. Li, C. Sun, and J.-S. Li, "Magnetically tunable terahertz wave left-handed material," in *Proceedings of the Laser and Tera-Hertz Science and Technology, LTST 2012*, China, November 2012.

[23] M. Gil, C. Damm, A. Giere et al., "Electrically tunable split-ring resonators at microwave frequencies based on barium-strontium-titanate thick films," *IEEE Electronics Letters*, vol. 45, no. 8, pp. 417-418, 2009.

[24] M. Maasch, M. Roig, C. Damm, and R. Jakoby, "Planar metamaterial with separately voltage tunable electric and magnetic dispersion based on liquid crystal," in *Proceedings of the 2013 7th International Congress on Advanced Electromagnetic Materials in Microwaves and Optics, METAMATERIALS 2013*, pp. 556–559, France, September 2013.

[25] P. Yaghmaee, O. H. Karabey, B. Bates, C. Fumeaux, and R. Jakoby, "Electrically tuned microwave devices using liquid crystal technology," *International Journal of Antennas and Propagation*, vol. 2013, Article ID 824214, 9 pages, 2013.

[26] S. C. Howells and L. A. Schlie, "Transient terahertz reflection spectroscopy of undoped InSb from 0.1 to 1.1 THz," *Applied Physics Letters*, vol. 69, no. 4, pp. 550–552, 1996.

[27] P. Halevi and F. Ramos-Mendieta, "Tunable photonic crystals with semiconducting constituents," *Physical Review Letters*, vol. 85, no. 9, pp. 1875–1878, 2000.

5

Acoustic Signatures of the Phases and Phase Transitions in the Blume Capel Model with Random Crystal Field

Gul Gulpinar

Department of Physics, Dokuz Eylül University, 35160 Izmir, Turkey

Correspondence should be addressed to Gul Gulpinar; gul.gulpinar@deu.edu.tr

Academic Editor: Oleg Derzhko

Sound propagation in the Blume Capel model with quenched diluted single-ion anisotropy is investigated. The sound dispersion relation and an expression for the ultrasonic attenuation are derived with the aid of the method of thermodynamics of irreversible processes. A frequency-dependent dispersion minimum that is shifted to lower temperatures with rising frequency is observed in the ordered region. The thermal and sound frequency (ω) dependencies of the sound attenuation and effect of the Onsager rate coefficient are studied in low- and high-frequency regimes. The results showed that $\omega\tau \ll 1$ and $\omega\tau \gg 1$ are the conditions that describe low- and high-frequency regimes, where τ is the single relaxation time diverging in the vicinity of the critical temperature. In addition, assuming a linear coupling of sound wave with the order parameter fluctuations in the system and ε as the temperature distance from the critical point, we found that the sound attenuation follows the power laws $\alpha(\omega,\varepsilon) \sim \omega^2\varepsilon^{-1}$ and $\alpha(\omega,\varepsilon) \sim \omega^0\varepsilon^1$ in the low- and high-frequency regions, while $\varepsilon \to 0$. Finally, a comparison of the findings of this study with previous theoretical and experimental studies is presented and it is shown that a good agreement is found with our results.

1. Introduction

The attenuation studies of acoustic waves in the vicinity of a magnetic ordering transition points provide insight into the critical dynamics of spin systems. Resonant ultrasound spectroscopy study of CoF_2 has shown the existence of a peak in sound attenuation as the Néel point is approached [1]. Magnetic phase diagram of multiferroic $MnWO_4$ has been obtained by sound velocity and attenuation measurements [2]. Investigation of the critical dynamics of sound propagation near continuous phase transition points not only provides valuable information about phase change mechanisms but also enables the determination of the critical indices that characterize these transitions [3]. Investigation of nonequilibrium processes probed by ultrasound waves in the spin-ice materials such as $Yb_2Ti_2O_7$ and $Dy_2Ti_2O_7$ [4, 5] and the frequency-dependent anisotropy of sound velocity and attenuation of the acoustic wave propagation through the nematic liquid crystals [6–8] represent examples to studies that are related to ultrasonic propagation in systems that undergo phase transitions. Further, considerable attention has been focused on the investigation of sound attenuation in disordered conductors [9]: the behavior of sound propagation near the λ point of the confined liquid ^4He [10–13] and absorption of ultrasound near the critical mixing point of a binary liquid [14–16]. Finally, one should note that a great variety of magnetic materials including halides, metals, oxides, intermetallics, and sulphides exhibit anomalies in their elastic properties due to the fact that order-disorder transitions are typically accompanied by small lattice distortions [17].

The Blume Capel (BC) model formulated independently by Blume [18] and Capel [19] plays a fundamental role in the multicritical phenomena associated with various physical systems, such as liquid crystals [20, 21], metallic alloys [22], proteins [23], and polymeric systems [24, 25]. On the other hand, the introduction of randomness changes the critical behaviors of a spin model considerably; that is, random fields can altogether eliminate the phase transitions in low dimensions and affect the numerical values of the critical exponents in higher dimensions [26–28]. Since then investigation of the effect of crystal field disorder on the equilibrium phase diagram of spin-1 Ising models has been a research interest for many authors. The BC model with random single-ion anisotropy provides a microscopic model for phase transitions of ^{3}He-^{4}He mixtures in silica aerogel [29, 30] and relevant to the study of the phase separation in porous media in the vicinity of the superfluid transition [31, 32]. The interplay between quenched disorder provided by a random field and network connectivity in the BC model is investigated by using the replica method [33]. Moreover, the BC model with infinite-range ferromagnetic interactions and under the influence of a quenched disorder has been investigated and a classification of the phase diagrams in terms of their topology is presented in [34].

In this study, we investigate the critical dynamics of sound wave propagation in the BC model with bimodal crystal field by combining the statistical equilibrium theory and the thermodynamics of linear irreversible processes. This approach has been utilized to investigate the sound propagation in a great variety of model systems. Making use of the lowest approximation of the cluster variation method and linear response theory of irreversible processes, Erdem and Keskin performed the calculations of the sound attenuation near the critical point in the Blume-Emery-Griffiths (BEG) model with zero crystal field [35–37]. Gulpinar investigated the critical behavior of ultrasound wave absorption coefficient in the metamagnetic Ising model within the mean-field approximation [38]. Later, the absorption of sound in the spin-3/2 Ising model on the Bethe lattice is obtained and its temperature variance is analyzed near the phase transition points [39]. Recently, Cengiz and Albayrak studied the sound attenuation phenomena for a finite crystal field BEG model on the Bethe lattice in terms of the recursion relations by using the Onsager theory [40]. Due to mathematical complexity, the properties of critical sound propagation have not been studied in any spin system with random bond, random magnetic field, or random crystal field terms in the Hamiltonian expression. To the best of our knowledge, the critical dynamics of the sound propagation of the BC model with quenched diluted single-ion anisotropy have not been studied by the methods of irreversible thermodynamics. It is assumed in this manuscript that the sound wave is coupled to the order parameter fluctuations that decay mainly via order parameter relaxation process and the steady-state dynamics of the BC model with random diluted crystal field are formulated, while the system is under the effect of a propagating sound wave of frequency ω, which let us obtain the sound dispersion relation and sound attenuation coefficient for all temperatures and frequencies that contain effectively only one phenomenological rate coefficient. Temperature variance of sound dispersion and absorption has been investigated in the vicinity of the critical point. Finally, the frequency behavior of the attenuation coefficient for temperatures is presented in the vicinity of second-order transition from ordered to disordered phase.

The paper is organized as follows: the model and its static properties are presented briefly in Section 2. Next, the free energy production near the equilibrium is stated and the order parameter relaxation time is obtained in Section 3. The steady solution of the kinetic equation of the order parameter and expressions for the sound dispersion relation and ultrasonic attenuation coefficient are obtained in Section 4. Finally, frequency and temperature behaviors of the ultrasonic attenuation are analyzed and the discussion of the results is given in Section 5.

2. The Model and Its Equilibrium Properties

The BC model with random single-ion anisotropy in the presence of an external magnetic field H is described by the Hamiltonian

$$\widehat{H} = -J\sum_{\langle i,j\rangle}\sigma_i\sigma_j - H\sum_i \sigma_i + \sum_i \Delta_i\sigma_i^2, \tag{1}$$

with $\sigma_i = -1, 0, 1$. Here $J > 0$ is the exchange interaction due to ferromagnetic coupling and Δ_i is the crystal field acting on site i with a distribution function:

$$F(\Delta_i) = p\delta(\Delta_i - D) + (1-p)\delta(\Delta_i), \tag{2}$$

where p is concentration of the spins on the lattice which are influenced by a crystal field. The mean-field free energy for a nonvanishing external field is given by the following expression [41, 42]:

$$\begin{aligned}\Psi &= \Psi_0(V(a), T) + \frac{NzJ}{2}m^2 \\ &+ NT\int \ln\left(1 - \frac{2\cosh((Jzm+H)/T)}{2\cosh((Jzm+H)/T) + e^{\Delta_i/T}}\right)F(\Delta_i)\,d\Delta_i,\end{aligned} \tag{3}$$

where $\Psi_0(V(a), T)$ is the lattice-free energy that is independent of spin configuration. Here a, V, and $m =< \sigma_i >$ are the lattice constant, volume, and order parameter of the system, respectively. In addition, $< \ldots >$ denotes thermal expectation value. If one makes use of (2), the mean-field free energy becomes

$$\begin{aligned}\Psi &= \Psi_0(V(a), T) + NzJ\frac{m^2}{2} + Np\Delta \\ &- pNk_BT\ln\left(2\cosh\frac{Jzm+H}{T} + e^{\Delta/T}\right) \\ &+ (1-p)Nk_BT\ln\left(2\cosh\frac{Jzm+H}{T} + 1\right).\end{aligned} \tag{4}$$

For the sake of simplicity, we will take Boltzmann constant as unity ($k_B = 1$) from now on. The equilibrium conditions $(\partial\Psi/\partial m)|_{eq} = 0$ and $(\partial\Psi/\partial a)|_{eq} = 0$ give the following self-consistent equations for the magnetization and the lattice constant:

$$m = 2p\frac{\sinh((Jzm+H)/T)}{2\cosh((Jzm+H)/T)+e^{\Delta/T}} + 2(1-p) \cdot \frac{\sinh((Jzm+H)/T)}{2\cosh((Jzm+H)/T)+1}, \quad (5)$$

$$\left.\frac{\partial\Psi_0}{\partial a}\right|_{eq} = Nz\frac{\partial J}{\partial a}\left[-\frac{m^2}{2} + 2pzm\frac{\sinh((Jzm+H)/T)}{2\cosh((Jzm+H)/T)+e^{\Delta/T}} + 2(1-p)zm\frac{\sinh((Jzm+H)/T)}{2\cosh((Jzm+H)/T)+1}\right]. \quad (6)$$

The numerical solutions of the equation state given by (5) for vanishing external field have been performed and it has been reported that the BC model with quenched diluted single-ion anisotropy exhibits three distinct phase diagram topologies in the (T, Δ) plane depending on p [43, 44]. For $1 \geq p > 0.945$, the form of the phase diagram is identical to that of the pure BC model with homogenous crystal field, where the ferromagnetic phase is separated from the paramagnetic phase by a phase boundary that is of second order up to a tricritical point (TCP) at which the transition becomes first order. For $0.945 > p \geq 0.926$, TCP still exists but $\lambda-$ and first-order lines have reentrant parts. The phase diagram of the system changes dramatically in nature if one increases the dilutence further: For $0.926 > p \geq 8/9$, there exists discontinuous phase transitions between the ferromagnetic and paramagnetic phases at strong crystal fields and low temperatures. The transition becomes continuous at higher temperatures. Further, the λ-line displays reentrance. A portion of the second-order transition line is masked by a first-order transition line. This situation causes two distinct multicritical points to appear: critical endpoint (CEP) and double critical endpoint (DCP). Under a threshold value of the crystal field concentration ($p \leq 8/9$), the quenched disorder completely eliminates the first-order phase transitions, and the whole phase boundary is of second order.

3. Relaxation Dynamics of the BC Model with Quenched Diluted Single-Ion Anisotropy

One may assume the existence of a small uniform external field for a short while in order to be able to formulate the relaxation dynamics of the BC model with random single-ion anisotropy. In addition, the amplitude of the external field should be sufficiently small to allow the spin system to be in the neighborhood of equilibrium, where linear response theory can be utilized. In other words, we investigate the final stage of the approach to equilibrium. In the case of the existence of a small deviation of the magnetic field from its equilibrium value $\delta H = H - H_0$, the system will be removed slightly from equilibrium and a finite free energy production $\Delta\Psi$ will arise:

$$\Delta\Psi = \Psi(m, H, a) - \Psi_0(m_0, H_0, a_0), \quad (7)$$

where $\Delta\Psi$ corresponds to an increase in the corresponding thermodynamic potential related to the deviation of m, a, H from their equilibrium values and Ψ_0 is the equilibrium value of the free energy, while $m = m_0$, $a = a_0$, and $H = H_0$. In the neighborhood of equilibrium, the free energy production may be written as a Taylor-series expansion, in which deviations in the thermodynamic quantities are retained to second order:

$$\Delta\Psi = \frac{A}{2}(m-m_0)^2 - B(m-m_0)(H-H_0) + \frac{C}{2}(H-H_0)^2 + D(m-m_0)(a-a_0) + E(a-a_0)(H-H_0) + \frac{F}{2}(a-a_0)^2 + G(H-H_0). \quad (8)$$

Here, the coefficients A to G are the so-called free energy production coefficients and they are calculated by the following second-order derivatives:

$$A = \left.\frac{\partial^2\Psi}{\partial m^2}\right|_{eq},$$
$$B = -\left.\frac{\partial^2\Psi}{\partial m\partial H}\right|_{eq},$$
$$C = \left.\frac{\partial^2\Psi}{\partial H^2}\right|_{eq},$$
$$D = \left.\frac{\partial^2\Psi}{\partial m\partial a}\right|_{eq} = \left.\frac{\partial^2\Psi}{\partial m\partial J}\right|_{eq}\left.\frac{\partial J}{\partial a}\right|_{eq},$$
$$E = \left.\frac{\partial^2\Psi}{\partial H\partial a}\right|_{eq} = \left.\frac{\partial^2\Psi}{\partial H\partial J}\right|_{eq}\left.\frac{\partial J}{\partial a}\right|_{eq}, \quad (9)$$
$$F = \left.\frac{\partial^2\Psi}{\partial a^2}\right|_{eq} = \left.\frac{\partial^2\Psi_0}{\partial a^2}\right|_{eq} + \left.\frac{\partial^2\Psi}{\partial J^2}\right|_{eq}\left(\left.\frac{\partial J}{\partial a}\right|_{eq}\right)^2 + \left.\frac{\partial\Psi}{\partial J}\right|_{eq}\left.\frac{\partial^2 J}{\partial a^2}\right|_{eq},$$
$$G = \left.\frac{\partial\Psi}{\partial H}\right|_{eq},$$

where the subscript "eq" denotes the thermal equilibrium; thus the derivatives are evaluated for $m = m_0$, $a = a_0$, and $H = H_0$. The explicit expressions for the above-mentioned free energy production coefficients are given in the Appendix.

If the system is shifted from its equilibrium by the deviation of the external field from its equilibrium value ($\delta H \neq 0$) and/or by the volume change of the crystal which is proportional to $a - a_0$, the generalized force X_m arises, which can be regarded as the force that brings magnetization back to its equilibrium value. The generalized force conjugate to generalized current may be obtained by differentiating the free energy production with respect to $(m - m_0)$:

$$\begin{aligned} X_m &= -\frac{\partial \Delta \Psi}{\partial (m - m_0)} \\ &= -\left[A(m - m_0) - B(H - H_0) + D(a - a_0)\right]. \end{aligned} \tag{10}$$

In the realm of theory of irreversible thermodynamics, the time derivative of magnetization $(\dot{m})$ is treated as the generalized current conjugate to X_m:

$$\dot{m} = \frac{d(m - m_0)}{dt}. \tag{11}$$

If the deviation from the equilibrium condition is small, one can write a linear relation between the current and the force: $\dot{m} = LX_m$. Thus, the dynamics of the order parameter for the BC model with quenched diluted crystal field are ruled by the following rate equation:

$$\dot{m} = -L\left[A(m - m_0) - B(H - H_0) + D(a - a_0)\right], \tag{12}$$

where L is the order parameter Onsager coefficient. We should note that, in this study, the most simple temperature dependence is assumed for L, which must be found either in principle by a more powerful theory such as path probability method [45, 46] or in practice by fit with the experimental findings.

For the case of vanishing external stimulation ($H = H_0$, $a = a_0$), one obtains the relation that describes the rate of change in the order parameter relaxing to its equilibrium state as follows:

$$\dot{m} = -LA(m - m_0). \tag{13}$$

The solution of the kinetic equation given by (13) is of the form ($m - m_0 \simeq e^{-t/\tau}$), where τ is the relaxation time of the order parameter. Thus, the relaxation time corresponds to

$$\tau = \frac{1}{LA}. \tag{14}$$

The temperature behavior of the order parameter relaxation time near the phase transition points of the BC model with bimodal crystal field has been investigated in detail in [44] and a rapid increase in the single relaxation time is observed when the temperature approaches the critical and multicritical phase transition temperatures. In addition, τ presents a scaling relation $\tau \sim |T - T_C|^{-1}$, which corresponds to well-known phenomena of critical slowing down. Finally, as a signature of the first-order phase transition, a jump-discontinuity has been observed in the relaxation time near a first-order phase transition.

4. Derivation of the Sound Attenuation Coefficient

In this section, we will study the transport properties for the BC model with bimodal crystal field near its magnetic phase transition points. With this aim, we will consider the case in which the lattice is stimulated by the sound wave of frequency ω. Due to nature of the linear response theory, if you perturb the system at a frequency ω, the response will take place at same frequency. Thus, one can find the steady-state solution of (12) with an oscillating external force $a - a_0 = a_1 e^{i\omega t}$ as follows:

$$m - m_0 = m_1 e^{i\omega t}. \tag{15}$$

Introduce this expression into (12).

And assuming that $H = H_0$, one obtains the following nonhomogenous equation for m_1:

$$i\omega m_1 e^{i\omega t} = -LAm_1 e^{i\omega t} - LDa_1 e^{i\omega t}. \tag{16}$$

Solving (16) for m_1/a_1 gives

$$\frac{m_1}{a_1} = -\frac{LD}{i\omega + LA}. \tag{17}$$

In addition, if one makes use of (14) for the relaxation time of the BC model with a random crystal field, (17) becomes

$$\frac{m_1}{a_1} = -\frac{LD\tau}{1 + i\omega\tau}. \tag{18}$$

The response in the pressure $p - p_0$ is obtained by differentiating the minimum work with respect to $V - V_0$:

$$p - p_0 = \frac{\partial \Delta \Psi}{\partial (V - V_0)} = -\frac{a}{3V_0}\frac{\partial \Delta \Psi}{\partial (a - a_0)}. \tag{19}$$

Then, using (8), one obtains

$$p - p_0 = -\frac{a_0}{3V_0}\left[D(m - m_0) + F(a - a_0)\right]. \tag{20}$$

On the other hand, the derivative of pressure with respect to volume gives

$$\left(\frac{\partial p}{\partial V}\right)_{sound} = -\frac{a_0^2}{9V_0^2}\left[D\frac{m_1}{a_1} + F\right], \tag{21}$$

where D and F are the free energy production coefficients that are given in (9). If one introduces (18) and the density $\rho = M/V_0$ into (21),

$$\left(\frac{\partial p}{\partial \rho}\right)_{sound} = \frac{a_0^2}{9M}\left[F - \frac{LD^2\tau}{1 + i\omega\tau}\right]. \tag{22}$$

Finally, using the definition $c_{sound} = (\partial p/\partial \rho)_{sound}^{1/2}$ and the fact that the order parameter relaxation time tends to very large values in the vicinity of the continuous phase transition points, one obtains the complex effective elastic constant (complex velocity of sound) of the BC model with quenched diluted crystal field:

$$c_{sound} \cong \frac{a_0}{3}\sqrt{\frac{F}{M}}\left[1 - \frac{LD^2\tau}{2F(1 + i\omega\tau)}\right]. \tag{23}$$

It is a well-known fact that the lag between the oscillations of the pressure and the excitation of a given mode leads to the dissipation of energy and dispersion of the sound wave.

$$c(\omega) = \operatorname{Re}\{c_{sound}\} = \frac{a_0}{3\sqrt{M}}\sqrt{F}\frac{2F - LD^2\tau}{2F(1+\omega^2\tau^2)}. \tag{24}$$

Since we assume a linear coupling of sound wave with the order parameter fluctuations in the system, the dispersion which is the relative sound velocity change with frequency depends not on the sound wave amplitude but on frequency ω. Finally, if one makes use of the following definition for the sound attenuation constant,

$$\begin{aligned}\alpha(\omega) &= \operatorname{Im}\left\{\frac{-\omega}{c_{sound}}\right\} \\ &= \operatorname{Im}\left\{\left(\frac{a_0}{3}\sqrt{\frac{F}{M}}\right)^{-1}\left[\frac{-2\omega F(1+i\omega\tau)}{2F - LD^2\tau + i2F\omega\tau}\right]\right\}, \\ &= \frac{3\sqrt{M}}{2a_0}\frac{1}{(\sqrt{F})^3}\frac{LD^2\omega^2\tau^2}{\left(1-\dfrac{LD^2\tau}{2F}\right)+\omega^2\tau^2}.\end{aligned} \tag{25}$$

A finite attenuation above the critical temperature cannot be obtained due to the fact that the attenuation constant is proportional to free energy production constant D that vanishes above the critical temperature, where the order parameter is equal to zero (see (A.1)). This is entirely due to the mean-field approximation and one may expect a finite attenuation constant above the critical temperature for higher-order approximations (i.e., Bethe approximation).

Now, if one rewrites the free-energy production coefficient F given in (9), the following expressions are obtained for $\sqrt{F}$ and $(\sqrt{F})^{-3}$:

$$\begin{aligned}\sqrt{F} &\simeq \sqrt{\left.\frac{\partial^2\Psi_0}{\partial a^2}\right|_{eq}}\left\{1+\frac{1}{2}\left(\left.\frac{\partial^2\Psi_0}{\partial a^2}\right|_{eq}\right)^{-1}\right. \\ &\quad\left.\cdot\left[\left.\frac{\partial^2\Psi}{\partial J^2}\right|_{eq}\left(\left.\frac{\partial J}{\partial a}\right|_{eq}\right)^2+\left.\frac{\partial\Psi}{\partial J}\right|_{eq}\left.\frac{\partial^2 J}{\partial a^2}\right|_{eq}\right]\right\},\end{aligned} \tag{26}$$

$$\begin{aligned}(\sqrt{F})^{-3} &\simeq \left(\left.\frac{\partial^2\Psi_0}{\partial a^2}\right|_{eq}\right)^{-3/2}\left\{1-\frac{3}{2}\left(\left.\frac{\partial^2\Psi_0}{\partial a^2}\right|_{eq}\right)^{-1}\right. \\ &\quad\left.\cdot\left[\left.\frac{\partial^2\Psi}{\partial J^2}\right|_{eq}\left(\left.\frac{\partial J}{\partial a}\right|_{eq}\right)^2+\left.\frac{\partial\Psi}{\partial J}\right|_{eq}\left.\frac{\partial^2 J}{\partial a^2}\right|_{eq}\right]\right\}.\end{aligned} \tag{27}$$

In addition, if one uses the definition $c_\infty^2 = (a_0^2/9M)(\partial^2\Psi_0/\partial a^2)|_{eq}$, (24) and (25) become

$$\begin{aligned}c(\omega) &= c_\infty\left\{1+\frac{1}{2}\left(\left.\frac{\partial^2\Psi_0}{\partial a^2}\right|_{eq}\right)^{-1}\right. \\ &\quad\left.\cdot\left[\left.\frac{\partial^2\Psi}{\partial J^2}\right|_{eq}\left(\left.\frac{\partial J}{\partial a}\right|_{eq}\right)^2+\left.\frac{\partial\Psi}{\partial J}\right|_{eq}\left.\frac{\partial^2 J}{\partial a^2}\right|_{eq}\right]\right\} \\ &\quad\cdot\frac{2F - LD^2\tau}{2F(1+\omega^2\tau^2)},\end{aligned} \tag{28}$$

$$\begin{aligned}\alpha(\omega) &= \frac{a_0^2}{18M}\frac{1}{c_\infty^3}\left\{1-\frac{3}{2}\left(\left.\frac{\partial^2\Psi_0}{\partial a^2}\right|_{eq}\right)^{-1}\right. \\ &\quad\left.\cdot\left[\left.\frac{\partial^2\Psi}{\partial J^2}\right|_{eq}\left(\left.\frac{\partial J}{\partial a}\right|_{eq}\right)^2+\left.\frac{\partial\Psi}{\partial J}\right|_{eq}\left.\frac{\partial^2 J}{\partial a^2}\right|_{eq}\right]\right\} \\ &\quad\cdot\frac{LD^2\omega^2\tau^2}{(1-LD^2\tau/2F)+\omega^2\tau^2},\end{aligned} \tag{29}$$

where c_∞ is the velocity of sound for very high frequencies at which order parameter can no longer follow the sound wave motion.

5. Results And Discussion

The calculated data that provides information about the thermal variation of the sound velocity and attenuation coefficient in the ordered and disordered phases and the frequency dependence of isothermal attenuation coefficient in the ordered phase of the BC model with quenched diluted crystal field will be discussed in the following. In addition, the behavior of sound absorption in the neighborhood of critical temperature is analyzed according to various values of phenomenological rate coefficient. Finally, by making use of double logarithmic plots of the sound attenuation coefficient versus the distance from the critical temperature, the dynamical critical exponents of the sound absorption are obtained for low- and high-frequency regimes. Before starting to discuss the above-mentioned results, it is convenient to introduce the following reduced quantities:

$$\theta = \frac{T}{zJ}, \quad d = \frac{D}{zJ}, \quad h = \frac{H}{zJ}; \tag{30}$$

here, θ, d, and h are the reduced values of the temperature, crystal field, and magnetic field, respectively. One should stress that we focus on zero field case ($h = 0.0$) and c_∞^3 is taken as unity for the sake of simplicity throughout this section.

Figure 1 displays the temperature variation of the frequency-dependent sound velocity (dispersion) of the BC model with quenched diluted crystal field for several values of sound wave frequency, while $L = 0.02$, $p = 0.98$,

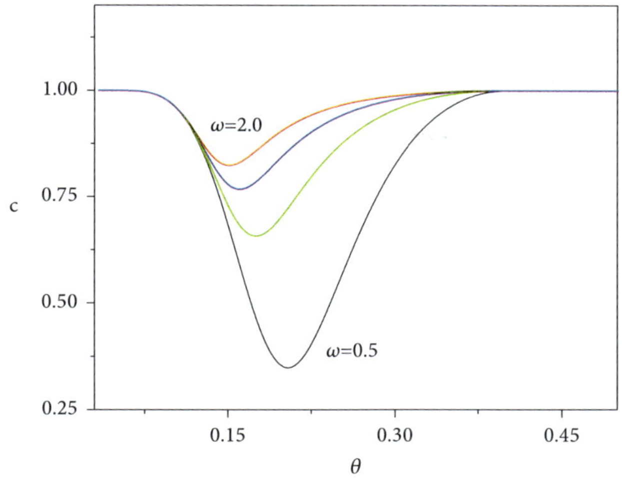

(a)

FIGURE 1: Sound dispersion at different frequencies of sound for $L = 0.02$, $p = 0.98$, and $d = d_C = 0.45$. The number accompanying each curve represents the value of the sound frequency ω.

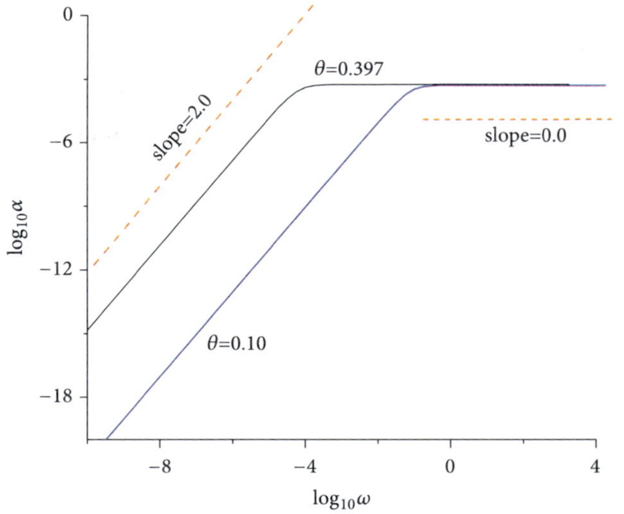

(a)

FIGURE 2: Double logarithmic plot of attenuation versus frequency for $L = 0.02$, $p = 0.98$, $d = d_C = 0.45$, and $\theta_C = 0.39747$. The number accompanying each curve represents the value of the reduced temperature.

and $d = 0.45$. The number accompanying each curve denotes the value of ω and one observes a characteristic sound velocity minimum that shifts to the lower temperatures with increasing frequency. The minima become deeper with decreasing ω and ultrasonic dispersion reaches the finite value of 1.0 at $\theta = \theta_C$ and remains temperature-independent after then. Similar temperature dependence of the sound dispersion has been reported near the order-disorder phase transition point of the BEG model [47]. Comparably, the existence of the minima of the sound velocity in the ordered phase and its shift to lower temperatures with increasing frequency have been observed in the classical liquid crystals [8], liquid ^{4}He confined in a microfluidic cavity [10], and the magnetic compound $RbMnF_3$ [48]. In addition, it can be seen from Figure 1 that the change in sound velocity becomes frequency-independent as one approaches to critical value of the reduced temperature and this behavior is in parallel with the findings of the molecular field theory [17, 49], dynamical renormalization group theory for the propagation of sound near the continuous structural phase transitions [50], and Brillouin scattering studies of A_2MX_6-crystals [49].

Figure 2 illustrates the double logarithmic plot of the isothermal α versus ω for two distinct values of the reduced temperature ($\theta = 0.10, 0.390 < \theta_C$) in the ferromagnetic phase, while $p = 0.98$, $L = 0.02$, $\theta_C = 0.39747$, and $d_C = 0.45$. One can observe the usual ω^2 dependence of the attenuation in the low-frequency region, where $\omega\tau << 1$. This result is in accordance with the theoretical and experimental findings for CoF_2 [1] and MnF_2 [51, 52]. The ultrasonic attenuation has different frequency dependence in the high-frequency regime at which order parameter can no longer follow sound wave motion: this part of the plot has zero slope; thus, ultrasonic attenuation behaves as $\alpha \propto \omega^0$, while $\omega\tau >> 1$. In addition, there is a crossover behavior in attenuation between low- and high-frequency regimes. The theoretical treatment of the sound propagation by mode-mode coupling method predicts that the sound attenuation coefficient scales as $\alpha \sim \omega^y$ for the sound frequencies that are quite larger than the characteristic frequency of the order parameter fluctuations ω_c and an extensive study of the frequency dispersion of the critical attenuation of $KMnF_3$ above 186.2 K measured $y = 0.13 \pm 0.05$ [53–55]. In addition, similar frequency dependence of the sound attenuation coefficient has been reported for various model systems such as spin-1 and spin-3/2 Ising models on the Bethe lattice [39, 40], zero crystal field BEG model [37], and metamagnetic Ising model [38] within mean-field theory.

Figures 3(a)-3(c) show the calculated sound attenuation coefficient as a function of the reduced temperature θ near the critical point for various values of the sound frequency whose value is denoted by the number accompanying each curve and the vertical dashed lines refer to critical temperature θ_c, while $p = 0.98$, $L = 0.02$, and $d = 0.45$. One can observe from these figures that the attenuation rises to a maximum in the ordered phase for all values of ω. But the frequency dependencies of the location form and amplitude of the absorption peak is different for low- and high-frequency regions. Figure 3(a) presents the temperature variance of α for two values of the sound frequency $\omega = 1.2 \times 10^{-5}, \omega = 1.8 \times 10^{-5}$ (blue and red curves), which satisfy the condition $\omega\tau \ll 1$ for the low-frequency regime. In the low-frequency region, the sound attenuation peak is rather sharp and located very near to the critical temperature θ_c and its amplitude increases with rising sound frequency.

Figure 3(b) presents the temperature dependence of α for higher values of the sound frequency ($\omega = 0.002, 0.005, 0.2, 0.5, 1.0$). It is observed from this figure that the absorption peak is rather broad and its amplitude grows with rising frequency for these ranges of the frequency. Further, the temperature at which attenuation maxima take

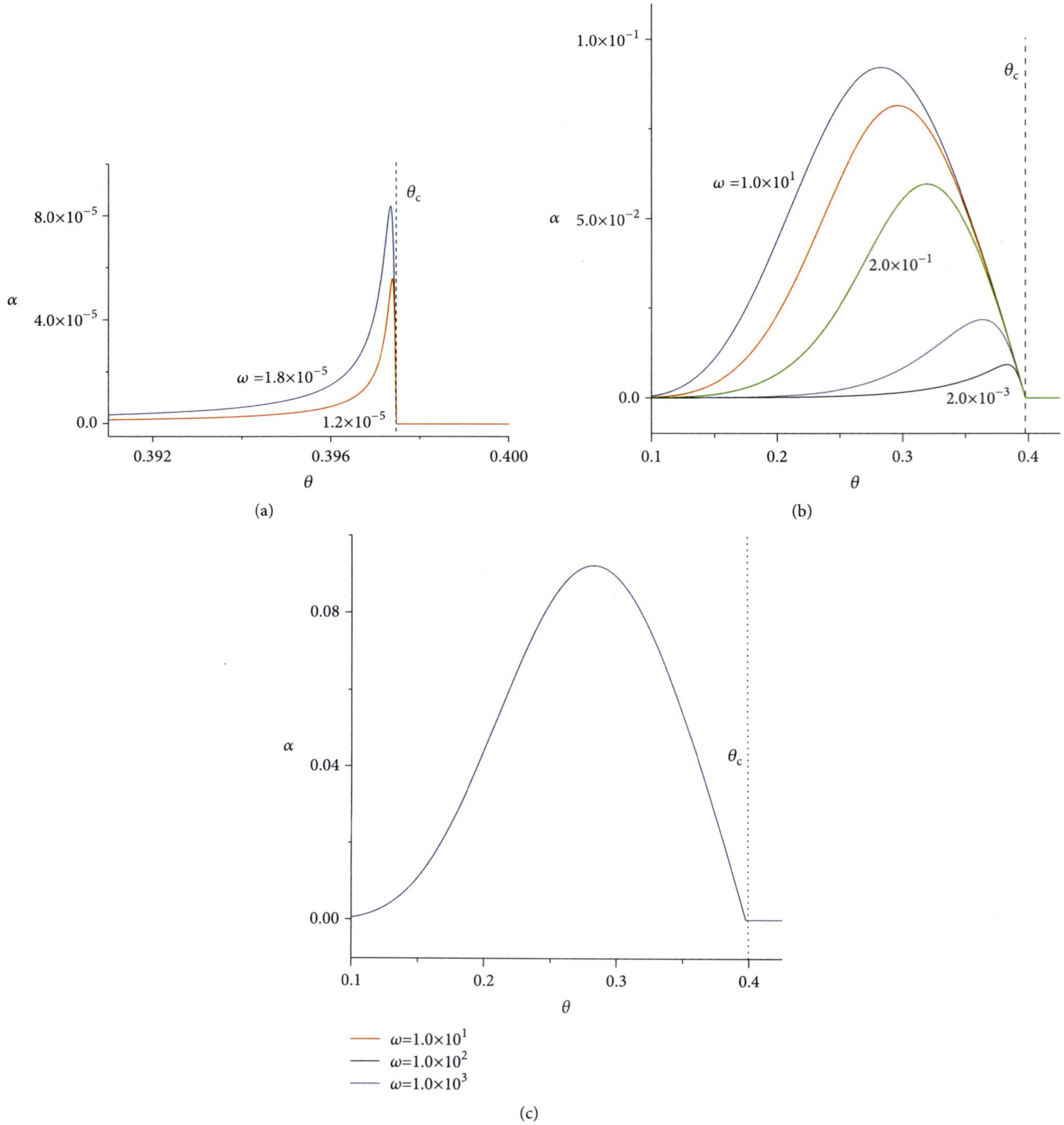

FIGURE 3: Sound attenuation α as a function of the reduced temperature at several sound frequency values ω in the neighborhood of a critical point for $p = 0.98$, $d = d_C = 0.45$, and $L = 0.02$. The number accompanying each curve denotes the value of the frequency ω and the vertical dotted line represents the critical temperature in the ordered phase.

place moves to lower temperatures with rising frequency. The frequency shift of the attenuation maxima is also observed in the experimental and theoretical studies of the ultrasonic investigation of liquid helium [56, 57] and Rb_2ZnCl_4 and K_2SeO_4 [58] as well as other Ising models [37–40] and elastically isotropic Ising system above the critical point on the basis of a complete stochastic model [59].

In Figure 3(c), we aim to illustrate the temperature dependence of the sound attenuation coefficient for three distinct values of the sound frequency ($\omega = 1, 10, 1000$) for which $\omega\tau >> 1$ and the order parameter can no longer follow sound wave motion. Both the location and the amplitude of the absorption peak do not change with the sound frequency and this is in accordance with the power law $\alpha \sim \omega^0$ observed

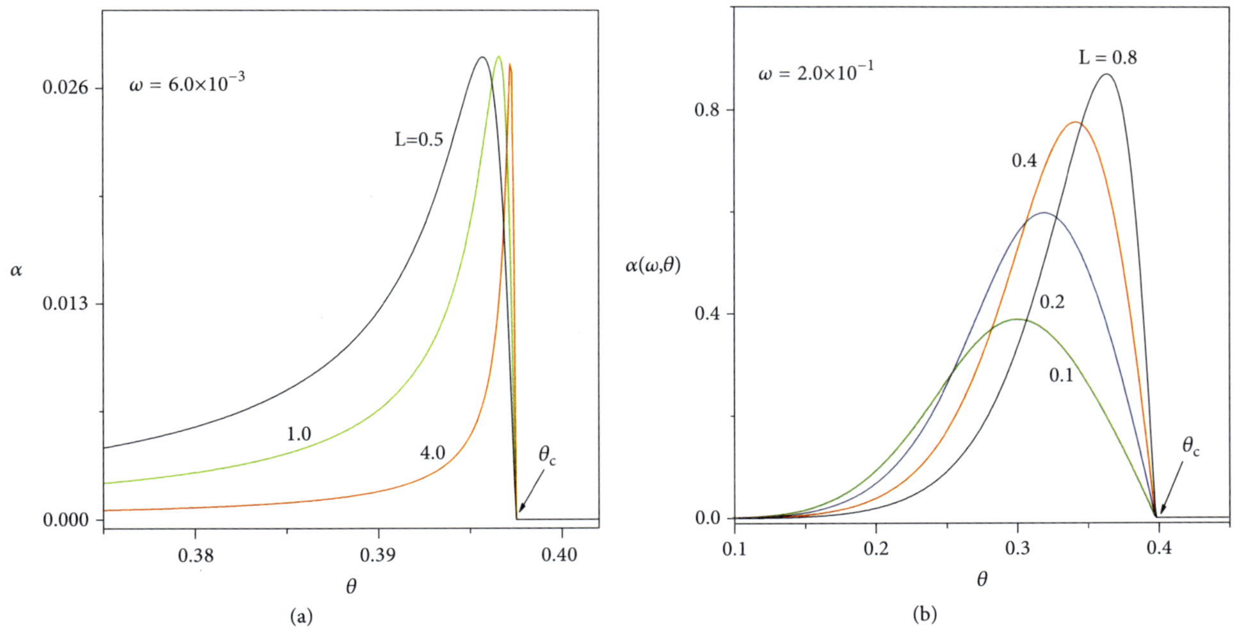

FIGURE 4: Ultrasonic attenuation as a function of reduced temperature for (a) $\omega = 0.006$ and (b) $\omega = 0.2$ while $p = 0.98$ and $d = d_C = 0.45$. The number accompanying each curve denotes the value of the order parameter Onsager coefficient (L) and the arrow denotes the critical temperature.

in Figure 2, while $\omega\tau >> 1$. Similar results are observed for metamagnetic Ising and BEG models in the high-frequency region. But since these models have two order parameters and their relaxation dynamics are governed by two relaxation times, their attenuation coefficient have two peaks for some frequency values in the low-frequency region as expected [37, 38].

It is clear from (14) and (28) that both the single relaxation time and sound attenuation coefficient depend on the phenomenological Onsager rate coefficient (L). Detailed investigation of the relaxation dynamics of the BC model with bimodal random crystal field revealed the fact that the relaxation time increases and the order parameter relaxation slows down with rising L [44]. Due to fact that thermal fluctuations become very large and the correlation distance extends over large number of spins, the coupling between sound waves and order parameter relaxation gives rise to an increase in the absorption of the sound waves as the temperature approaches its critical value. As the transition temperature of a solid is closely approached, the ordered parameter relaxes slow enough; the internal irreversible processes that tend to restore local equilibrium are switched on and they make the entropy increase, which results in energy dissipation in the system [17, 57]. Due to these facts, we found it interesting to study the effect of L on the temperature variance of the sound attenuation near the second-order phase transition point. Figures 4(a) and 4(b) illustrate the attenuation coefficient versus reduced temperature at several values of L for $p = 0.98$ and $d = 0.45$ at two distinct values of the sound wave frequency ω and the arrows refer to critical point temperature. It is apparent from Figure 4(a) which displays the calculated α for $\omega = 0.006$ that the maxima of α broaden and move to lower temperatures with decreasing L. Meanwhile the value of the sound attenuation maxima slightly increases with decreasing L. The attenuation peak broadens and its amplitude decreases and shifts to lower temperatures with decreasing L, while $\omega = 0.2$. We should note that, for systems that have two or more order parameters, the relation between the generalized forces and currents may be written in terms of a matrix of phenomenological rate coefficients and the effect of the nondiagonal Onsager coefficient has been investigated for BEG and metamagnetic Ising models in the mean-field approximation [37, 38] and spin-3/2 and spin-1 Ising models on the Bethe lattice [39, 40].

The critical behavior of the sound attenuation coefficient in the neighborhood of the second-order phase transition points is characterized by the dynamical sound attenuation critical exponent ρ. As discussed above, isothermal sound attenuation coefficient behaves as $\alpha \propto \omega^2$ for frequencies that satisfy the condition $\omega\tau << 1$ and $\alpha \propto \omega^0$ for $\omega\tau >> 1$. In addition, the comparison of Figures 3(a)–3(c) reveals the fact that the constant frequency temperature variance of α near the critical temperature is different in low- and high-frequency regions. Consequently, one may expect two distinct power law relations and two distinct exponents (ρ_l, ρ_h) for these two regions:

$$\alpha(\omega,\varepsilon) \approx \begin{cases} \omega^2\varepsilon^{-\rho_l} & \text{for } \omega\tau \ll 1, \\ \omega^0\varepsilon^{-\rho_h} & \text{for } \omega\tau \gg 1, \end{cases} \tag{31}$$

where $\varepsilon = \theta - \theta_C$ is the expansion parameter, which is a measure of the distance from the critical temperature.

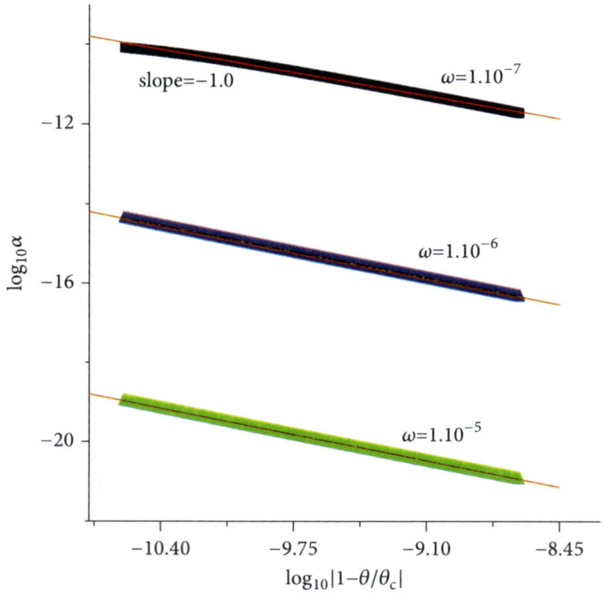

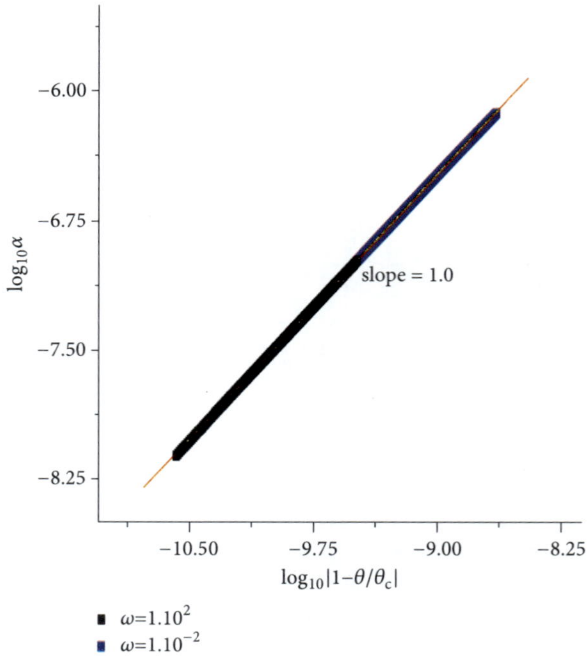

FIGURE 5: Double logarithmic plots of attenuation versus $|1-\theta/\theta_C|$ for (a) low-frequency and (b) high-frequency regimes in the neighborhood of a critical point for $p = 0.98$, $d = d_C = 0.45$, and $L = 0.02$.

The critical exponents for the function $\alpha(\omega,\varepsilon)$ are defined, respectively, as

$$\rho_l = -\lim_{\varepsilon\to 0}\frac{\ln|\alpha(\omega,\varepsilon)|}{\ln|\varepsilon|} \quad \text{for } \omega\tau \ll 1,$$
$$\rho_h = -\lim_{\varepsilon\to 0}\frac{\ln|\alpha(\omega,\varepsilon)|}{\ln|\varepsilon|} \quad \text{for } \omega\tau \gg 1. \tag{32}$$

In order to calculate the values of these exponents, one may sketch log – log plots of the sound attenuation coefficient versus $|1-\theta/\theta_C|$ and find the slopes of the linear parts of these curves, which will be equal to $-\rho_l$ and $-\rho_h$ for frequencies that satisfy the conditions $\omega\tau \ll 1$ and $\omega\tau \gg 1$, respectively. With this aim, Figures 5(a) and 5(b) illustrate the $\log_{10}\alpha - \log_{10}|1-\theta/\theta_C|$ plots for the constant values of sound frequencies in low- and high-frequency regions, respectively. Figure 5(a) displays three distinct curves for $\omega = 1.0\times10^{-7}$ (black curve), $\omega = 1.0\times10^{-6}$ (blue curve), and $\omega = 1.0\times10^{-5}$ (green curve), which obey the $\omega\tau \ll 1$ condition. As one can see from this figure, for all three frequencies in the low-frequency region, the slopes of the linear part of $\log_{10}\alpha - \log_{10}|1-\theta/\theta_C|$ curves are equal to −1; namely, $\rho_l = 1$. Utilizing (31) and (30), one obtains the power law $\alpha(\omega,\varepsilon) \sim \omega^2\varepsilon^{-1}$ of the sound attenuation coefficient for the BC model with bimodal random crystal field in the low-frequency regime. This result is in accordance with the scaling behavior of the sound attenuation coefficient α_{LK} obtained within the Landau-Khalatnikov theory, which applies to bilinear, quadratic coupling of strain-order parameter coupling [17, 57]: $\alpha_{LK} \sim \omega^2\tau$ and the order parameter relaxation time behaves as $\tau \sim \varepsilon^{-1}$ as $\varepsilon \to 0$ within the mean-field approximation [44]. A strong anomalous increase of the sound attenuation as the temperature approaches its critical value has been also observed for systems that undergo (elastic and distortive) structural phase transitions. The dynamical renormalization group calculations for distortive structural phase transitions report $\rho_\sigma = (z\nu + \alpha_\sigma)$, where $\alpha_\sigma = \alpha + 2(\phi-1)(1-\delta_{\sigma,1})$, where z, ν, ϕ, and are α are the dynamical, correlation length, heat capacity, and crossover critical exponents, respectively. Utilizing the same method, Schwabl showed that the sound attenuation coefficient scales as $\alpha \sim \varepsilon^{-3/2}\omega^2$ in the vicinity of the critical point of the elastic structural phase transitions [50]. Ultrasonic measurements reported small but positive attenuation exponents in isotropic garnet structured magnets such as $Y_3Fe_5O_{12}$ and $Gd_3Fe_5O_{12}$ [60]. Recently, resonant ultrasound spectroscopy studies of CoF_2 have shown that $\rho = 0.9$ to 1.1 for the resonance peaks near 487, 860, 1300 and 1882 kHz, which depend strongly on the shear modulus [1].

Finally, Figure 5(b) shows the linear part of the $\log\alpha$ versus $\log|1-\theta/\theta_C|$ curves for frequencies $\omega = 0.1$ (blue symbols) and $\omega = 100$ (black symbols), while $\omega\tau \gg 1$. One can see from this figure that the slope of the curves is +1; thus $\rho_h = -1$. If one again makes use of (30) and (31), the following power law can be written for the sound absorption: $\alpha(\omega,\varepsilon) \sim \omega^0\varepsilon^1$, while $\omega\tau \gg 1$ and verifies the convergence of the sound absorption to zero as the reduced temperature approaches its critical value (see Figure 3(b)). We should note that similar behavior is observed in the high-frequency region of the sound attenuation for the BEG model [37]. It has been shown by the phenomenological theory that sound attenuation reaches the saturation value $\alpha_{phe} \to B\tau^{-1}$ for very high frequencies, where B is independent of temperature and frequency for a system with a single relaxation process [3]. If one keeps in mind that $\tau \sim \varepsilon^{-1}$ as $\varepsilon \to 0$, this corresponds to $\alpha_{phe}(\omega, \varepsilon) \approx \omega^0 \varepsilon^1$, which is in parallel with our result.

6. Concluding Remarks

The present study is devoted to critical dynamics of sound propagation in BC model with a random single anisotropy by using a method that combines the statistical equilibrium theory of phase transitions with the linear response theory of irreversible thermodynamics. With this aim, it is supposed that the system is stimulated by a sound wave oscillating at an angular frequency ω and the complex effective elastic constant and sound velocity and sound attenuation coefficient are obtained for all temperatures and frequencies by studying the steady-state solutions. We have studied the temperature behavior of the sound velocity for various values of the sound frequency and observed a characteristic minimum that shifts to the lower temperatures with rising frequency and a frequency-independent sound velocity in the vicinity of critical temperature. Then, we have investigated the temperature, frequency, and order parameter Onsager rate coefficient dependencies of the sound attenuation near the second transition temperatures. Frequency-dependent shifts in attenuation maxima to lower temperatures with increasing frequency for temperatures less than the second-order phase transition temperatures are obtained and possible mechanisms for these phenomena are discussed. Our investigation of the frequency variance of the isothermal sound attenuation revealed the fact that sound attenuation follows the power law $\alpha \sim \omega^2$ for frequencies that satisfy the condition $\omega\tau \ll 1$ and $\alpha \sim \omega^0$, while $\omega\tau \gg 1$. Consequently, $\omega\tau$ is the quantity that determines the separation of the low- and high-frequency regimes. By making use of this information and obtaining the slope of the $\log \alpha$ versus $\log |1 - \theta/\theta_c|$ plots, we have obtained that sound absorption follows the scaling relation $\alpha(\omega, \varepsilon) \sim \omega^2 \varepsilon^{-1}$ in the low-frequency regime and the power law $\alpha(\omega, \varepsilon) \sim \omega^0 \varepsilon^1$ for frequencies that obey the condition $\omega\tau \gg 1$. Finally, we should note that the high-frequency behavior provides information about the damping mechanism.

In this study, the sound attenuation related to the relaxation process of the order parameter for BC model with a random single anisotropy is studied. We would like to emphasize that the above-mentioned theory is limited neither to the systems with only scalar spin nor to the magnetic phase transitions. It could also apply to a crystalline solid and other systems whenever the order parameter dynamics are relaxation dynamics. We should note that there are additional possible reasons for attenuation peaks in magnetic systems which are not taken into account in the current analysis, such as coupling of the sound waves to modes, domain effects, and nuclear acoustic resonance [17]. In addition, in the case of real crystals with many structural defects, the presence of disorder will affect the critical behavior of the sound attenuation in solid media. The field-theoretic description of dynamic critical effects of the disorder on acoustic anomalies near the temperature of second-order phase transition for three-dimensional Ising-like systems has shown that the presence of impurities brings additional couplings changing the value of the critical exponent of the sound attenuation and strongly affects the order parameter shape function at the low-frequency region in the vicinity of the critical point [61].

Appendix

The free energy production coefficients A to G that are defined in (9) are given as follows:

$$A = NJz - \frac{NJ^2z^2}{T}\left[\frac{2p\cosh(Jzm_0/T)}{2\cosh(Jzm_0/T) + e^{\Delta/T}} - \left(\frac{2\sqrt{p}\sinh(Jzm_0/T)}{2\cosh(Jzm_0/T) + e^{\Delta/T}}\right)^2 + \frac{2(1-p)\cosh(Jzm_0/T)}{2\cosh(Jzm_0/T) + 1} - \left(\frac{2\sqrt{1-p}\sinh(Jzm_0/T)}{2\cosh(Jzm_0/T) + 1}\right)^2\right],$$

$$B = \frac{NJz}{T}\left[\frac{2p\cosh(Jzm_0/T)}{2\cosh(Jzm_0/T) + e^{\Delta/T}} - \left(\frac{2\sqrt{p}\sinh(Jzm_0/T)}{2\cosh(Jzm_0/T) + e^{\Delta/T}}\right)^2 + \frac{2(1-p)\cosh(Jzm_0/T)}{2\cosh(Jzm_0/T) + 1} - \left(\frac{2\sqrt{1-p}\sinh(Jzm_0/T)}{2\cosh(Jzm_0/T) + 1}\right)^2\right],$$

$$C = -\frac{N}{T}\left[\frac{2p\cosh(Jzm_0/T)}{2\cosh(Jzm_0/T) + e^{\Delta/T}} - \left(\frac{2\sqrt{p}\sinh(Jzm_0/T)}{2\cosh(Jzm_0/T) + e^{\Delta/T}}\right)^2 + \frac{2(1-p)\cosh(Jzm_0/T)}{2\cosh(Jzm_0/T) + 1} - \left(\frac{2\sqrt{1-p}\sinh(Jzm_0/T)}{2\cosh(Jzm_0/T) + e^{\Delta/T}}\right)^2\right],$$

$$D = Nz\left.\frac{\partial J}{\partial a}\right|_{eq}\left[m_0 - \frac{2Jzm_0p\cosh(Jzm_0/T)}{T\left[2\cosh(Jzm_0/T) + e^{\Delta/T}\right]} - \frac{2p\sinh(Jzm_0/T)}{2\cosh(Jzm_0/T) + e^{\Delta/T}}\right.$$

$$+\frac{4Jzm_0 p\sinh^2(Jzm_0/T)}{T\left[2\cosh(Jzm_0/T)+e^{\Delta/T}\right]^2}$$

$$-\frac{2Jzm_0(1-p)\cosh(Jzm_0/T)}{T\left[2\cosh(Jzm_0/T)+1\right]}$$

$$-\frac{2(1-p)\sinh(Jzm_0/T)}{2\cosh(Jzm_0/T)+1}$$

$$\left.+\frac{4Jzm_0(1-p)\sinh^2(Jzm_0/T)}{T\left[2\cosh(Jzm_0/T)+1\right]}\right],$$

$$E=-\frac{Nzm_0}{T}\left[\frac{2p\cosh(Jzm_0/T)}{2\cosh(Jzm_0/T)+e^{\Delta/T}}\right.$$

$$-\left(\frac{2\sqrt{p}\sinh(Jzm_0/T)}{2\cosh(Jzm_0/T)+e^{\Delta/T}}\right)^2$$

$$+\frac{2(1-p)\cosh(Jzm_0/T)}{2\cosh(Jzm_0/T)+1}$$

$$\left.-\left(\frac{2\sqrt{1-p}\sinh(Jzm_0/T)}{2\cosh(Jzm_0/T)+1}\right)^2\right],$$

$$F=\left.\frac{\partial^2\Psi_0}{\partial a^2}\right|_{eq}+Nzm_0\left.\frac{\partial^2 J}{\partial a^2}\right|_{eq}\left[\frac{m_0}{2}\right. \tag{A.1}$$

$$-\frac{2(1-p)\cosh(Jzm_0/T)}{2\cosh(Jzm_0/T)+1}$$

$$\left.-\frac{2p\cosh(Jzm_0/T)}{2\cosh(Jzm_0/T)+e^{\Delta/T}}\right]+\frac{Nz^2m_0^2}{T}\left(\left.\frac{\partial J}{\partial a}\right|_{eq}\right)^2$$

$$\times\left[\frac{2p\cosh(Jzm_0/T)}{2\cosh(Jzm_0/T)+e^{\Delta/T}}\right.$$

$$-\left(\frac{2\sqrt{p}\sinh(Jzm_0/T)}{2\cosh(Jzm_0/T)+e^{\Delta/T}}\right)^2$$

$$+\frac{2(1-p)\cosh(Jzm_0/T)}{2\cosh(Jzm_0/T)+1}$$

$$\left.-\left(\frac{2\sqrt{1-p}\sinh(Jzm_0/T)}{2\cosh(Jzm_0/T)+1}\right)^2\right],$$

$$G=\frac{2N(1-p)\cosh(Jzm_0/T)}{2\cosh(Jzm_0/T)+1}$$

$$-\frac{2Np\sinh(Jzm_0/T)}{2\cosh(Jzm_0/T)+e^{\Delta/T}}.$$

Disclosure

The research and publication of this article was performed as a part of the employment of the author, Department of Physics, Dokuz Eylül University, Department of Physics, İzmir, Turkey.

Conflicts of Interest

The author declares that there are no conflicts of interest.

References

[1] R. I. Thomson, T. Chatterji, and M. A. Carpenter, "CoF_2: A model system for magnetoelastic coupling and elastic softening mechanisms associated with paramagnetic ↔ antiferromagnetic phase transitions," *Journal of Physics: Condensed Matter*, vol. 26, no. 14, Article ID 146001, 2014.

[2] V. Felea, P. Lemmens, S. Yasin et al., "Magnetic phase diagram of multiferroic $MnWO_4$ probed by ultrasound," *Journal of Physics: Condensed Matter*, vol. 23, no. 21, Article ID 216001, 2011.

[3] A. Pawlak, "Critical sound propagation in magnets," in *Horizons in World Physics*, L. Pedroza and M. Everett, Eds., vol. 268, pp. 1–69, Nova Science Publishers, Hauppauge, NY, USA, 2009.

[4] S. Bhattacharjee, S. Erfanifam, E. L. Green et al., " Acoustic signatures of the phases and phase transitions in ," *Physical Review B: Condensed Matter and Materials Physics*, vol. 93, no. 14, 2016.

[5] S. Erfanifam, S. Zherlitsyn, J. Wosnitza et al., "Intrinsic and extrinsic nonstationary field-driven processes in the spin-ice compound $Dy_2Ti_2O_7$," *Physical Review B: Condensed Matter and Materials Physics*, vol. 84, no. 22, Article ID 220404, 2011.

[6] S. S. Turzi, "Viscoelastic nematodynamics," *Physical Review E: Statistical, Nonlinear, and Soft Matter Physics*, vol. 94, no. 6, Article ID 062705, 2016.

[7] P. Biscari, A. DiCarlo, and S. S. Turzi, "Anisotropic wave propagation in nematic liquid crystals," *Soft Matter*, vol. 10, no. 41, pp. 8296–8307, 2014.

[8] C. Grammes, J. K. Krüger, K.-P. Bohn et al., "Universal relaxation behavior of classical liquid crystals at hypersonic frequencies," *Physical Review E: Statistical, Nonlinear, and Soft Matter Physics*, vol. 51, no. 1, pp. 430–440, 1995.

[9] A. Shtyk and M. Feigel'Man, "Ultrasonic attenuation via energy diffusion channel in disordered conductors," *Physical Review B: Condensed Matter and Materials Physics*, vol. 92, no. 19, Article ID 195101, 2015.

[10] X. Rojas, B. D. Hauer, A. J. MacDonald, P. Saberi, Y. Yang, and J. P. Davis, " Ultrasonic interferometer for first-sound measurements of confined liquid ," *Physical Review B: Condensed Matter and Materials Physics*, vol. 89, no. 17, 2014.

[11] E. Krotscheck and M. D. Miller, "Third sound and stability of thin ^{3}He-^{4}He films," *Physical Review B*, vol. 73, article 134514, pp. 1–14, 2006.

[12] J. M. Valles, R. M. Heinrichs, and R. B. Hallock, " ^{3}He-^{4}He Mixture Films: The He coverage dependence of the ^{3}He binding energy," *Physical Review Letters*, vol. 56, pp. 1704–1707, 1986.

[13] P. A. Sheldon and R. B. Hallock, "Third sound and energetics in," *Physical Review B: Condensed Matter and Materials Physics*, vol. 50, no. 21, pp. 16082–16085, 1994.

[14] D. B. Fenner, "Nonsingular absorption of ultrasound near the critical mixing point of a binary liquid," *Physical Review A:*

Atomic, Molecular and Optical Physics, vol. 23, no. 4, pp. 1931–1940, 1981.

[15] P. K. Khabibullaev, S. Z. Mirzaev, and U. Kaatze, "Critical fluctuations and noncritical relaxations of the nitrobenzene-isooctane system near its consolute point," *Chemical Physics Letters*, vol. 458, no. 1-3, pp. 76–80, 2008.

[16] S. Z. Mirzaev and U. Kaatze, "Scaling function of critical binary mixtures: Nitrobenzene-n-hexane data revisited," *Chemical Physics*, vol. 393, no. 1, pp. 129–134, 2012.

[17] B. Lüthi, *Physical Acoustics in the Solid State*, Springer, Berlin, Germany, 2005.

[18] M. Blume, "Theory of the first-order magnetic phase change in UO_2," *Physical Review A: Atomic, Molecular and Optical Physics*, vol. 141, no. 2, pp. 517–524, 1966.

[19] H. W. Capel, "On the possibility of first-order phase transitions in Ising systems of triplet ions with zero-field splitting," *Physica A: Statistical Mechanics and its Applications*, vol. 32, no. 5, pp. 966–988, 1966.

[20] P. E. Cladis, R. K. Bogardus, W. B. Daniels, and G. N. Taylor, "High-pressure investigation of the reentrant nematic - Bilayer-smectic-A transition," *Physical Review Letters*, vol. 39, no. 11, pp. 720–723, 1977.

[21] G. P. Johari, "The Tammann phase boundary, exothermic disordering and the entropy contribution change on phase transformation," *Physical Chemistry Chemical Physics*, vol. 3, no. 12, pp. 2483–2487, 2001.

[22] W. Sinkler, C. Michaelsen, R. Bormann, D. Spilsbury, and N. Cowlam, "Neutron-diffraction investigation of structural changes during inverse melting ofs," *Physical Review B: Condensed Matter and Materials Physics*, vol. 55, no. 5, pp. 2874–2881, 1997.

[23] M. I. Marqués, J. M. Borreguero, H. E. Stanley, and N. V. Dokholyan, "Possible mechanism for cold denaturation of proteins at high pressure," *Physical Review Letters*, vol. 91, no. 13, 2003.

[24] S. Rastogi, G. W. Höhne, and A. Keller, "Unusual Pressure-Induced Phase Behavior in Crystalline Poly(4-methylpentene-1): Calorimetric and Spectroscopic Results and Further Implications," *Macromolecules*, vol. 32, no. 26, pp. 8897–8909, 1999.

[25] A. L. Greer, "Too hot to melt," *Nature*, vol. 404, no. 6774, pp. 134-135, 2000.

[26] A. N. Berker, "Ordering under random fields: Renormalization-group arguments," *Physical Review B: Condensed Matter and Materials Physics*, vol. 29, no. 9, pp. 5243–5245, 1984.

[27] Y. Imry and S. Ma, "Random-field instability of the ordered state of continuous symmetry," *Physical Review Letters*, vol. 35, no. 21, pp. 1399–1401, 1975.

[28] G. Grinstein and S.-K. Ma, "Surface tension, roughening, and lower critical dimension in the random-field Ising model," *Physical Review B: Condensed Matter and Materials Physics*, vol. 28, no. 5, pp. 2588–2601, 1983.

[29] A. Falicov and A. N. Berker, "Correlated Random-Chemical-Potential Model for the Phase Transitions of Helium Mixtures in Porous Media," *Physical Review Letters*, vol. 74, no. 3, pp. 426–429, 1995.

[30] M. Paetkau and J. R. Beamish, "Tricritical point and superfluid transition in ^{3}He-^{4}He mixtures in silica aerogel," *Physical Review Letters*, vol. 80, no. 25, pp. 5591–5594, 1998.

[31] L. B. Lurio, N. Mulders, M. Paetkau, M. H. W. Chan, and S. G. J. Mochrie, "Small-angle x-ray scattering measurements of the microstructure of liquid helium mixtures adsorbed in aerogel," *Physical Review E: Statistical, Nonlinear, and Soft Matter Physics*, vol. 76, no. 1, Article ID 011506, 2007.

[32] T. R. Prisk, C. Pantalei, H. Kaiser, and P. E. Sokol, "Confinement-driven phase separation of quantum liquid mixtures," *Physical Review Letters*, vol. 109, no. 7, Article ID 075301, 2012.

[33] R. Erichsen, A. A. Lopes, and S. G. Magalhaes, "Multicritical points and topology-induced inverse transition in the random-field Blume-Capel model in a random network," *Physical Review E: Statistical, Nonlinear, and Soft Matter Physics*, vol. 95, no. 6, Article ID 062113, 2017.

[34] P. V. Santos, F. A. da Costa, and J. M. de Araujo, "Mean-field solution of the Blume-Capel model under a random crystal field," *Physics Letters A*, vol. 379, no. 22-23, pp. 1397–1401, 2015.

[35] M. Keskin and R. Erdem, "Critical behaviors of the sound attenuation in a spin-1 Ising model," *The Journal of Chemical Physics*, vol. 118, no. 13, pp. 5947–5954, 2003.

[36] R. Erdem and M. Keskin, "On the temperature dependence of the sound attenuation maximum as a function of frequency and magnetic field in a spin-1 Ising model near the critical region," *Physics Letters A*, vol. 326, no. 1-2, pp. 27–31, 2004.

[37] R. Erdem and M. Keskin, "Sound attenuation in a spin-1 ising system near the critical temperature," *Physics Letters A*, vol. 291, no. 2-3, pp. 159–164, 2001.

[38] G. Gulpinar, "Critical behavior of sound attenuation in a metamagnetic Ising system," *Physics Letters A*, vol. 372, no. 2, pp. 98–105, 2008.

[39] T. Cengiz and E. Albayrak, "The Bethe lattice treatment of sound attenuation for a spin- 3/2 Ising model," *Physica A: Statistical Mechanics and its Applications*, vol. 391, no. 10, pp. 2948–2956, 2012.

[40] T. Cengiz and E. Albayrak, "The crystal field effects on sound attenuation for a spin-1 Ising model on the Bethe lattice," *Journal of Statistical Mechanics: Theory and Experiment*, vol. 2012, no. 7, Article ID P07004, 2012.

[41] G. Gülpınar, R. Erdem, and M. Ağartıoğlu, "Critical and multicritical behaviors of static and complex magnetic susceptibilities for the mean-field Blume-Capel model with a random crystal field," *Journal of Magnetism and Magnetic Materials*, vol. 439, pp. 44–52, 2017.

[42] G. Gulpinar and R. Erdem, "High-frequency magnetic field on crystal field diluted S=1 Ising system: Magnetic relaxation near continuous phase transition points," *Canadian Journal of Physics*, 2018.

[43] A. Benyoussef, M. Saber, T. Biaz, and M. Touzani, "The spin-1 Ising model with a random crystal field: The mean-field solution," *Journal of Physics C: Solid State Physics*, vol. 20, no. 32, pp. 5349–5354, 1987.

[44] G. Gulpinar and F. Iyikanat, "Dynamics of the Blume-Capel model with quenched diluted single-ion anisotropy in the neighborhood of equilibrium states," *Physical Review E: Statistical, Nonlinear, and Soft Matter Physics*, vol. 83, no. 4, Article ID 041101, 2011.

[45] R. Kikuchi, "The path probability method," *Progress of Theoretical Physics Supplement*, vol. 35, pp. 1–64, 1966.

[46] A. Erdinç and M. Keskin, "Equilibrium and nonequilibrium behavior of the Blume–Emery–Griffiths model using the pair approximation and path probability method with the pair distribution," *International Journal of Modern Physics B*, vol. 18, no. 10n11, pp. 1603–1626, 2004.

[47] R. Erdem and M. Keskin, "Sound dispersion in a spin-1 Ising system near the second-order phase transition point," *Physics Letters A*, vol. 310, no. 1, pp. 74–79, 2003.

[48] T. J. Moran and B. Lüthi, "High-frequency sound propagation near magnetic phase transitions," *Physical Review B: Condensed Matter and Materials Physics*, vol. 4, no. 1, pp. 122–132, 1971.

[49] W. Henkel, J. Pelzl, K. H. Höck, and H. Thomas, "Elastic constants and softening of acoustic modes in A_2MX_6-crystals observed by Brillouin scattering," *Zeitschrift für Physik B Condensed Matter and Quanta*, vol. 37, no. 4, pp. 321–332, 1980.

[50] F. Schwabl, "Propagation of sound at continuous structural phase transitions," *Journal of Statistical Physics*, vol. 39, no. 5-6, pp. 719–737, 1985.

[51] A. Ikushima, "Anisotropy of the ultrasonic attenuation in MnF_2 near the Néel temperature," *Journal of Physics and Chemistry of Solids*, vol. 31, no. 5, pp. 939–946, 1970.

[52] A. Pawlak, "Critical sound propagation in MnF_2," *Physica Status Solidi (c)*, vol. 3, pp. 204–207, 2006.

[53] K. Kawasaki, "Sound Attenuation and Dispersion near the Liquid-Gas Critical Point," *Physical Review A*, vol. 1, pp. 1750–1757, 1970.

[54] Y. Shiwa and K. Kawasaki, "The Mode-Coupling Approach to Sound Propagation in a Critical Fluid. I: – Correction Due to Hydrodynamic Interactions –," *Progress of Theoretical and Experimental Physics*, vol. 66, no. 1, pp. 118–128, 1981.

[55] M. Suzuki, "Dynamical scaling and ultrasonic attenuation in $KMnF_3$ at the structural phase transition," *Journal of Physics C: Solid State Physics*, vol. 13, no. 4, article no. 013, pp. 549–560, 1980.

[56] D. T. Sprague, N. Alikacem, P. A. Sheldon, and R. B. Hallock, " ^{3}He binding energy in thin helium-mixture films," *Physical Review Letters*, vol. 72, no. 3, pp. 384–387, 1994.

[57] L. D. Landau, I. M. Khalatnikov,"On the anomalous absorption of sound near a second-order phase transition point," Dokl. Akad. Nauk SSSR 96, 469 (1954), reprinted in "Collected Papers of L. D. Landau", Ed. D. Ter Haar, Pergamon, London, UK, 1965.

[58] A. M. Schorgg and F. Schwabl, "Theory of ultrasonic attenuation at incommensurate phase transitions," *Physical Review B: Condensed Matter and Materials Physics*, vol. 49, no. 17, pp. 11682–11703, 1994.

[59] A. Pawlak, "Theory of critical sound attenuation in Ising-type magnets," *Acta Physica Polonica A*, vol. 98, no. 1-2, pp. 23–39, 2000.

[60] I. K. Kamilov and K. K. Aliev, "Ultrasonic studies of the critical dynamics of magnetically ordered crystals," *Physics-Uspekhi*, vol. 41, no. 9, pp. 865–884, 1998.

[61] P. V. Prudnikov and V. V. Prudnikov, "Critical sound attenuation of three-dimensional lsing systems," *Condensed Matter Physics*, vol. 9, no. 2, pp. 403–410, 2006.

6

Experimental and Simulation Research of AC Ripple Losses in a High Temperature Superconductor Tape

Hongwei Liu ,[1,2] Guosheng Song,[1,2] Chuanhui Zhang,[1,2] Jinwen Tan,[1,2] and Wengang Feng[1,2]

[1]*Key Laboratory of HV and EMC Beijing, Beijing 102206, China*
[2]*School of Electric and Electronic Engineering, North China Electric Power University, Beijing 102206, China*

Correspondence should be addressed to Hongwei Liu; hwliu@ncepu.edu.cn

Academic Editor: Gayanath Fernando

Current leads in superconducting magnets are widely adopted for heavy current transmission, and HTS DC cables have great advantages when used as current leads for such purposes. However, as an important parameter of HTS DC cables, AC loss has a strong impact on the stability and operation cost of current leads. In this paper, experiments were conducted to measure AC ripple losses of HTS tapes and HTS cables, and simulations of HTS tapes were carried out. The paper has reached conclusions on the relation between AC losses and power frequencies.

1. Introduction

Current leads are among the most important components in superconducting equipment, such as superconducting electric apparatus and superconducting magnets. Generally, high temperature superconductors (HTS) are widely used as current leads, as they may reduce heat leakage along current leads from room temperature to liquid nitrogen or liquid helium. Bi-2223-based hybrid current leads have been used extensively in accelerators and superconducting magnets [1]. Recently, 2G coated conductors have been reported for current leads (CL) application [2], demonstrating the feasibility of HTS tapes for CL application. Although DC current is always applied to the magnets and accelerators, AC ripple currents exist simultaneously with DC current during the transient processes of excitations and deexcitations, thus incurring AC Losses in the HTS tapes. AC loss is an important factor for HTS electric equipment in that it determines the capacity of the accompanying cryogenic cooling system [3]. References [3–7] have experimentally researched the AC loss behavior of superconducting tapes under AC ripple superimposed on a DC offset current, providing useful guidance for the evaluation of AC losses when the DC and AC ripple currents exist simultaneously. References [8–13] provide some measuring methods of AC losses, but the applied frequencies of AC ripple currents are not typical in actual power systems.

This paper aims to research experimentally AC ripple current losses of HTS tapes and the HTS cable for the common frequency range in the actual power system. Simulations of HTS tapes were also carried out under this study. The applied AC ripple current was the odd multiples of basic frequency. AC losses were measured under different amplitudes of AC currents and at different frequencies. Moreover, in the experiment, DC currents were the most part, while AC ripple currents were a small part of the total current, which is more consistent with superconducting electric DC equipment in reality. At the end of the paper, experiment data was provided and analysis was given about the causes for the phenomena.

2. Experimental Setup

2.1. Parameters of the YBCO Tape. The YBCO superconductors are made by Suzhou Advanced Materials Research Institute, and their critical current is 60 A at 77 K. The width is 2.5 mm, and the thickness is 0.18 mm. The specification of the superconductor was listed in Table 1. Although the critical current of the YBCO tape was a little lower, it was fit for the

Table 1: Specification of the superconductor.

Electric parameter	
Width	2.5 mm
Thickness	0.18 mm
Critical current	60 A

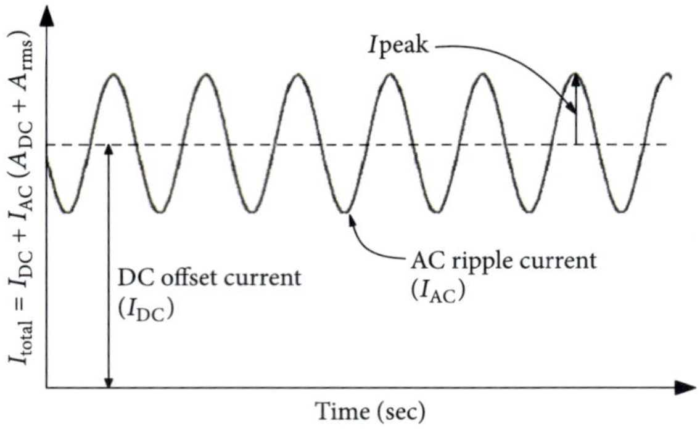

Figure 1: The schematic picture of the total current which includes the DC current and the AC ripple current.

experiments because of the capacity limit of the adjustable capacitor.

First, the critical characteristics of the HTS tape were measured, and the results were highly consistent with the critical current. Then we set up the experimental platform at which the AC losses were measured under the DC and AC ripple currents. The total current was shown in Figure 1, where we used the amplitude values for AC currents [5]. Thus, the total current is the sum of the DC current and the amplitude of the AC current.

2.2. Experimental Setup. In order to generate the hybrid source signal, a DC current source and an AC current source were paralleled. An adjustable capacitor was used in case the DC current might circulate to the branch of the AC current source. During the experiments, a fixed DC current was set to the HTS tape. Then, we changed the frequency of the AC current source to the set value. At the same time, the capacitor was adjusted because of the change of frequency. A lock-in amplifier was used to find the in-phase component of the voltage. In the experiment, we measured the frequency of 50 Hz, 150 Hz, 250 Hz, and 350 Hz. They are the odd multiples of the basic frequency and the most typical frequency components in the power system and the electric equipment. Every time the data acquisition system recorded the data for the following analysis. Figure 2 showed the schematic experimental platform and Figure 3 showed the real experiments photos. Since the critical current is 60 A, the highest total current we applied was 1.2 times I_c, that is, 72 A. The AC current was measured by the Rogowski coil and the total current by the shunt resistor.

3. Experimental Results

First, we fixed the DC current to 45 A. Then we increased the AC current step by step from 5 A to 27 A at a fixed frequency.

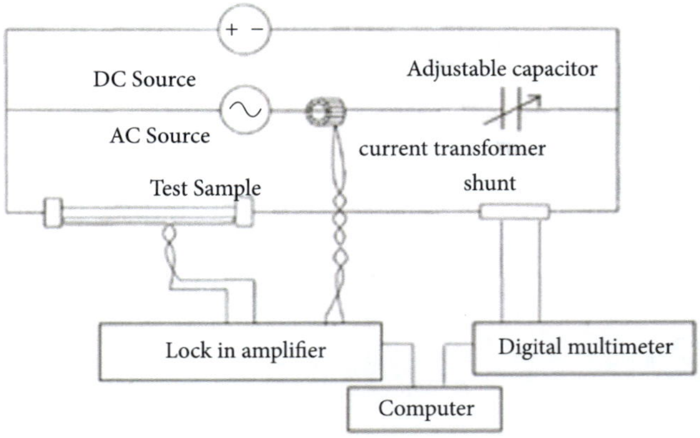

Figure 2: The schematic experimental platform.

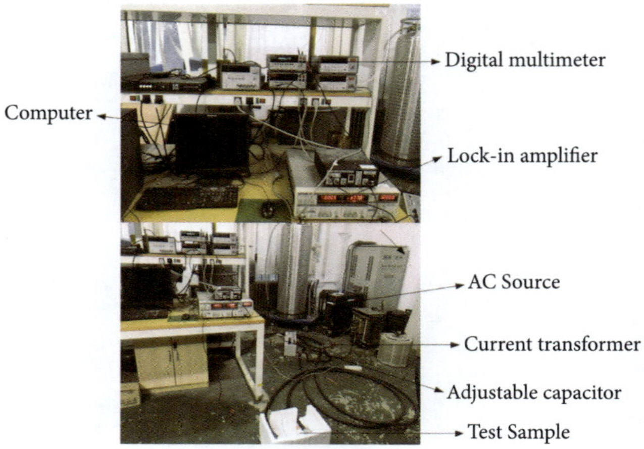

Figure 3: The real experimental photos.

Figure 4 showed the result of AC losses with the DC current of 45 A and the different amplitude of AC currents at the frequency of 250 Hz and 350 Hz. From the result, we could see that the AC losses increased with the amplitude of the AC current. When the total current approached 72 A, the losses increased greatly compared with that of 50 A. Figure 4 also showed that the AC loss was irrelevant to the frequency.

Figure 5 showed that, at a fixed AC current, the AC losses only slightly increased with the increase of the frequency. It signified that the AC losses were not much related to the frequency from 50 Hz to 350 Hz. On the other hand, the amplitude of the AC current had much more impact on the AC losses. With the increase of the amplitude of the AC current, the AC losses increased prominently.

Then, we fixed the DC current at 55 A. With the same steps as previously, we got the results as Figures 6 and 7 showed. We can see that the AC losses slightly increased with the frequency when the total current was 60 A. And the amplitude of the AC current had more effect on the AC losses. In order to research the effect of the DC current and the AC ripple current, we changed the different component of the total current while keeping the total current fixed. Figure 8 showed the experimental result. It showed that, at the same

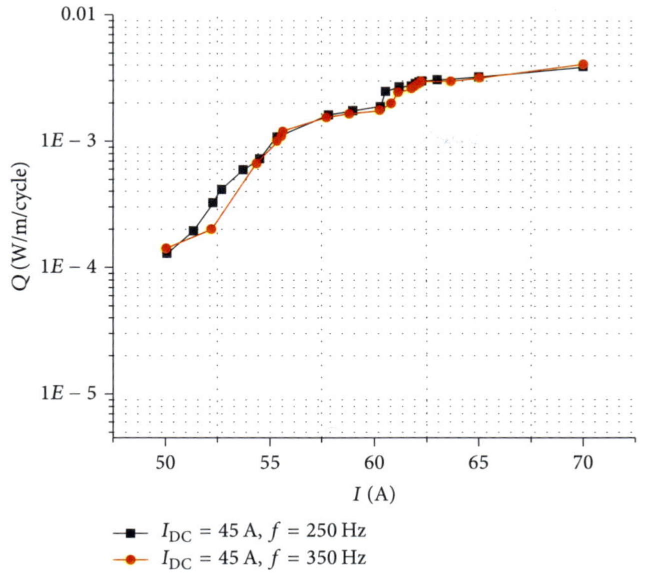

FIGURE 4: The AC losses at the DC current of 45 A at the frequency of 250 Hz and 350 Hz.

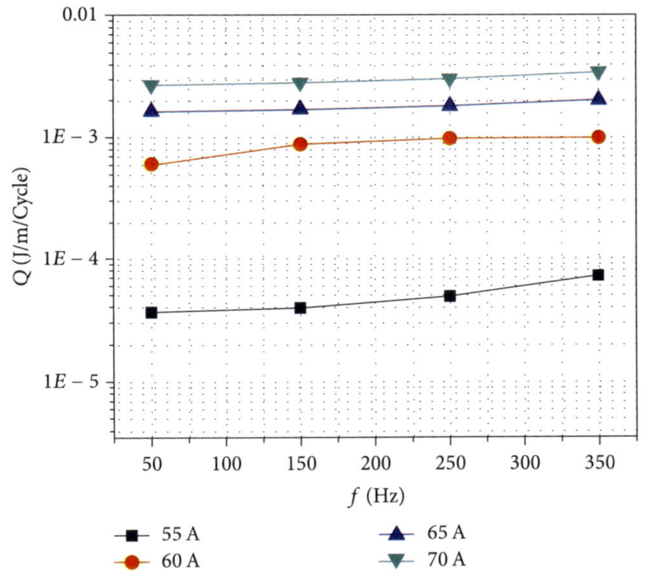

FIGURE 5: The AC losses at the DC current of 45 A at different amplitude and different frequencies.

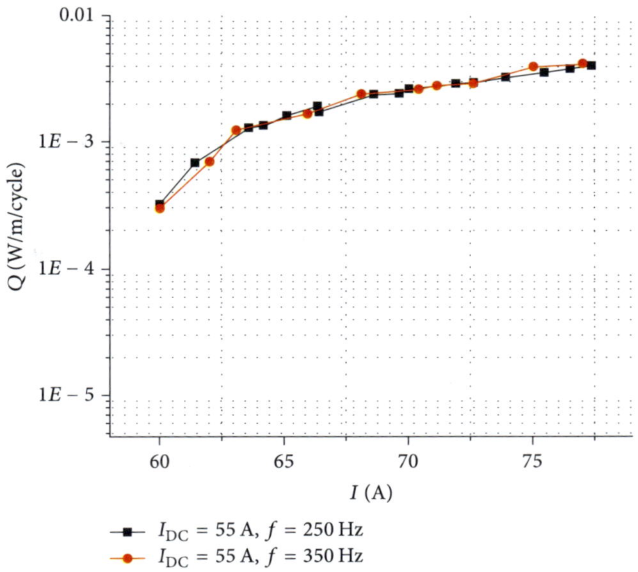

FIGURE 6: The AC losses at the DC current of 55 A at the frequency of 250 Hz and 350 Hz.

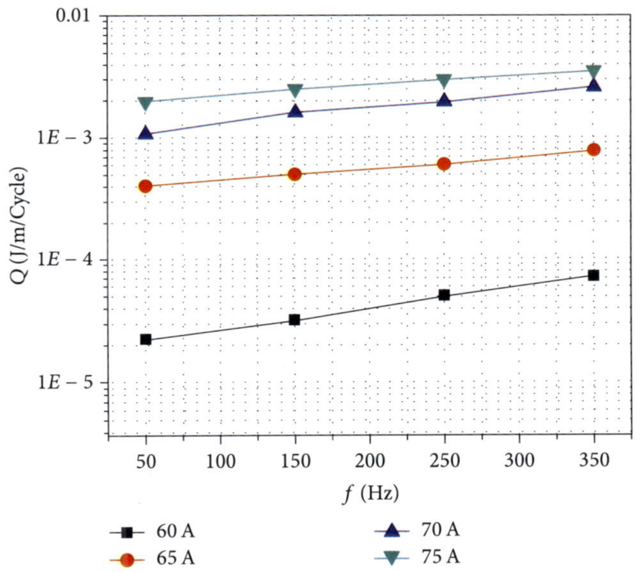

FIGURE 7: The AC losses at the DC current of 55 A at different amplitude and different frequencies.

total current of 65 A, the AC loss at which the DC current was 45 A and the AC current was 20 A was larger than that at which the DC current was 55 A and the AC current was 10 A. It signified that the component of the AC current had much more effect on the AC losses. Why did it happen like that? The current distribution was the starting point. With "ripples" of larger amplitudes, it is the edges of the HTS tape that make the predominant contribution to the loss [14].

Figure 9 showed the result of the AC losses of the 0.3 m HTS cable. The DC current was 200 A and the AC ripple current ranged from 5 A to 25 A. The AC losses were also insensitive to the frequency and increased with the AC ripple currents.

4. Simulation Results

To verify the measurement, a two-dimensional finite element method model based on H-formulation and the E-J power law was established. AC losses of HTS tapes were simulated, with the coexistence of DC and AC ripple currents.

The simulation result was shown in Figure 10. The comparison was made with the simulation result and the measurements. Experimental results had a little discrepancy

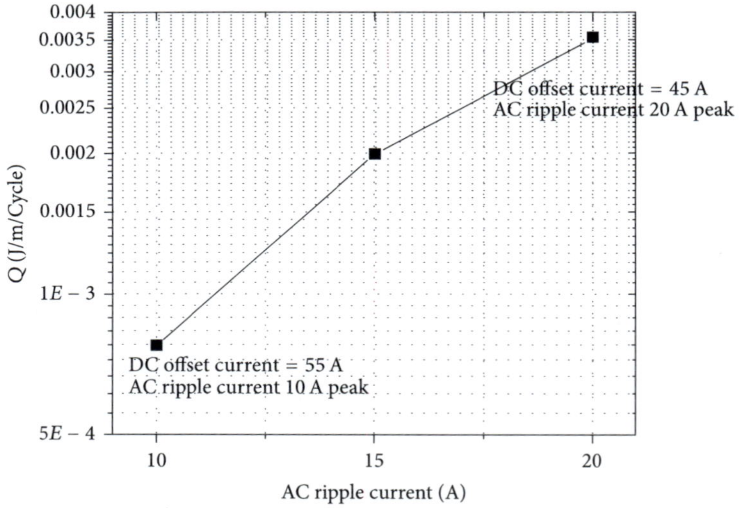

Figure 8: The AC losses at the same total current of 65 A but at the different DC current and the AC current.

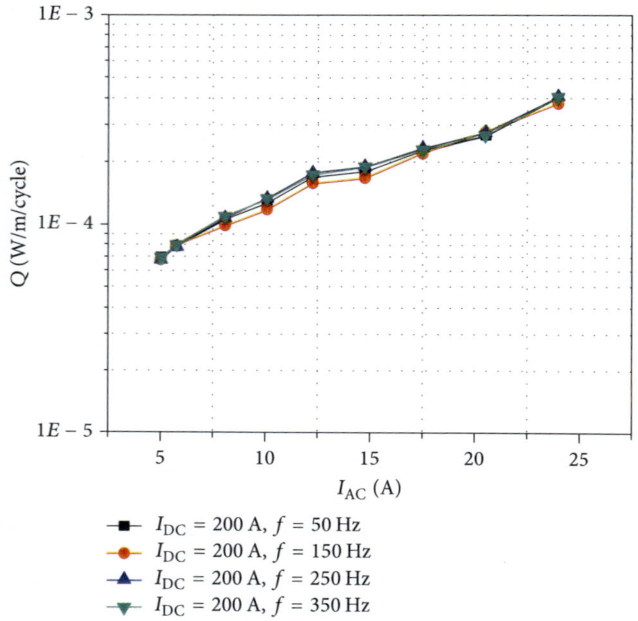

Figure 9: The AC losses of the 0.3 m HTS cable at the DC current of 200 A at different frequencies and different AC currents.

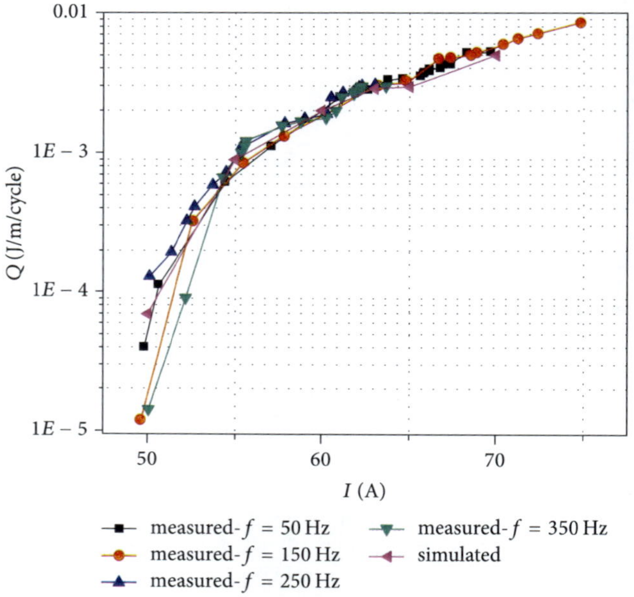

Figure 10: Comparisons of the measured results and the simulated results of the HTS tape.

compared to the simulated results when the AC current was small. This discrepancy was mostly caused by the background noise during measurement when the current was small. The trends of the measured curves and simulated curve were quite similar. Moreover, the losses all increase as the total currents increase towards the critical value.

5. Conclusion

This paper studies AC ripple losses of HTS tapes through experiments and simulations. We found the following in the experiments: (1) AC losses increased with the amplitude of AC currents, when holding the DC current constant. (2) When the sum of DC current and the amplitude of AC ripple current approaches the critical current, AC losses increased remarkably. (3) AC ripple currents had minor effects on the total AC loss when the ratio of AC to DC current was small.

Disclosure

A part of this article has been presented in the 25th International Conference on Magnet Technology.

Conflicts of Interest

The authors declare that there are no conflicts of interest.

Acknowledgments

This work was supported by Beijing Natural Science Foundation (3162032). This paper was developed further on the basis of the suggestions from experts and researchers in the 25th International Conference on Magnet Technology. The authors would like to thank them for their help in the suggestions.

References

[1] D. Uglietti, N. Bykovsky, R. Wesche, and P. Bruzzone, "Development of HTS Conductors for Fusion Magnets," *IEEE Transactions on Applied Superconductivity*, vol. 25, no. 3, 2015.

[2] I. A. Kovalev, M. I. Surin, A. V. Naumov et al., "Test results of 12/18 kA ReBCO coated conductor current leads," *Cryogenics*, vol. 85, pp. 71–77, 2017.

[3] L. Ying, J. Xu, J. Sheng et al., "Numerical and Experimental Analysis of AC Loss of YBCO Coated Conductor Carrying DC and AC Offset Transport Current," *IEEE Transactions on Applied Superconductivity*, vol. 23, no. 3, pp. 5900704-5900704, 2013.

[4] M.-C. Dinh, S.-K. Kim, J.-G. Kim, M. Park, I.-K. Yu, and B. Yang, "Loss characteristic analysis of an HTS dc model cable connected to a model VSC-HVDC System," *IEEE Transactions on Applied Superconductivity*, vol. 23, no. 3, 2013.

[5] J.-H. Kim and S. Pamidi, "Electrical characteristics of 2G HTS tapes under DC current with AC ripple," *IEEE Transactions on Applied Superconductivity*, vol. 22, no. 3, 2012.

[6] J. Ogawa, S. Fukui, T. Oka et al., "Experimental investigation of loss characteristics with AC and DC offset transport current in HTS double pancake coil," *IEEE Transactions on Applied Superconductivity*, vol. 22, no. 3, 2012.

[7] H. Zhang, Y. Wang, C. Kan, Y. Fu, and J. Xue, "Effects of HTS Tape Arrangements on AC Loss in Self-Shielding DC HTS Cable with AC Ripple Current," *IEEE Transactions on Applied Superconductivity*, vol. 27, no. 4, 2017.

[8] Z. Li, K. Ryu, S. Fukui, S. D. Hwang, and G. Cha, "AC loss measurement of a short HTS cable with shield by electrical method," *IEEE Transactions on Applied Superconductivity*, vol. 21, no. 3, pp. 1005–1008, 2011.

[9] B. C. Robert and H. S. Ruiz, "Magnetic characteristics and AC losses of DC type-II superconductors under oscillating magnetic fields," *Superconductor Science and Technology*, vol. 31, no. 3, p. 035006, 2018.

[10] J. P. Murphy, N. N. Gheorghiu, T. Bullard et al., "AC loss in YBCO coated conductors at high dB/dt measured using a spinning magnet calorimeter (stator testbed environment)," *Cryogenics*, vol. 86, pp. 57–69, 2017.

[11] B. J. de Bruyn, J. W. Jansen, and E. A. Lomonova, "AC losses in HTS coils for high-frequency and non-sinusoidal currents," *Superconductor Science and Technology*, vol. 30, no. 9, p. 095006, 2017.

[12] D. Hu, M. D. Ainslie, J. Kvitkovic et al., "Transport AC Loss Measurements of a Triangular Epoxy-Impregnated High-Temperature Superconducting Coil," *IEEE Transactions on Applied Superconductivity*, vol. 27, no. 4, pp. 1–6, 2017.

[13] J. Ogawa, S. Fukui, T. Oka, Y. Sano, M. Ogawa, and S. T. Ranecky, "Experimental Investigation of AC Loss Characteristics in HTS Tape With Rotating Magnetic Field," *IEEE Transactions on Applied Superconductivity*, vol. 27, no. 4, pp. 1–4, 2017.

[14] Z. Xu and F. Grilli, "Modelling ac ripple currents in HTS coated conductors," *Superconductor Science and Technology*, vol. 28, no. 10, Article ID 104002, 2015.

The Effect of Geometrical Parameters on Resonance Characteristics of Acoustic Metamaterials with Negative Effective Modulus

Xiao-Le Yan,[1] Li-Mei Hao,[1] Mei-Ling Men,[1] and Zhi Chen[2]

[1]*Department of Applied Physics, Xi'an University of Science and Technology, Xi'an 710054, China*
[2]*Department of Applied Physics, Northwestern Polytechnical University, Xi'an 710129, China*

Correspondence should be addressed to Li-Mei Hao; haolm@xust.edu.cn and Zhi Chen; c2002z@nwpu.edu.cn

Academic Editor: Liyuan Zhang

There has been an explosion of interest in acoustic metamaterial in the last ten years. The tunable negative acoustic metamaterial is an important issue for designing metamaterials. The acoustic metamaterial is restricted by the narrow bandgap of sound waves for the local resonance of metamaterial unit cells. By shifting the sizes of the unit cell, the acoustic metamaterial could potentially overcome the limit of the narrow resonance frequency. In this research, we focus on the resonant behavior of split hollow sphere (SHS) in the waveguide. Firstly, we analyze the resonance characteristics of SHS and get an analytical formula for the effect of geometrical parameters on the resonance frequency. Furthermore, the resonance frequency of SHS is verified with finite-element method analysis based on COMSOL Multiphysics simulation. The results are in good agreement with theory model. It is observed that there is a blue shift of the resonance frequency with the gradual increase of the neck radius of SHS, a V-type response curve with the increase of inner radius of SHS, and a red shift with the increase of outer radius of SHS. Using the method of estimate of resonance, we could get a precisely controllable unit structure with negative effective modulus and offer a way to optimize the realization of double negative acoustic metamaterial.

1. Introduction

Acoustic metamaterial is a kind of man-made material that exhibit superior properties, and it is applied to control the direction of sound wave. The idea of negative effective density of acoustic metamaterial using local resonance unit was first proposed by Liu et al. in 2000 [1]. In this way, more researchers pay attention to local resonant acoustic metamaterial with negative effective density mass [2–10]. At the year of 2006, Fang et al. [11] designed a one-dimensional array consisted of split hollow cavity with subwavelength which is a new class of acoustic metamaterial with negative effective modulus at the ultrasonic frequency. Later on, Lee et al. [12] made a kind of one-dimensional acoustic metamaterial with splitting hollow on the side of hollow tube and confirmed that the acoustic metamaterial has the property of negative effective modulus at the frequency range of 0~450 Hz. A simpler version of two-dimensional negative effective modulus based on SHS was prepared by Ding et al. [13]. By 2015, Jing et al. [14] proposed a new kind of monopole resonance characteristics of balloon soft resonator to realize negative effective modulus. However, the main disadvantage of the local resonant acoustic metamaterials is that the response frequencies of different units are different and the frequency bands are narrow. Therefore, similar to electromagnetic metamaterials, the tunable negative acoustic metamaterial is an important issue in the acoustic research. The tunable range of resonant frequency is one of the important parameters in the application of acoustic metamaterials.

At present, active and passive methods have been adopted to control the resonance frequency of metamaterials with negative modulus [15–17]. Akl and Baz [15] controlled the Helmholtz resonator with negative modulus with by coupling with piezoelectric film and applying the electric field. There are many factors that influence the resonance characteristics of acoustic metamaterial unit, such as the shape, size,

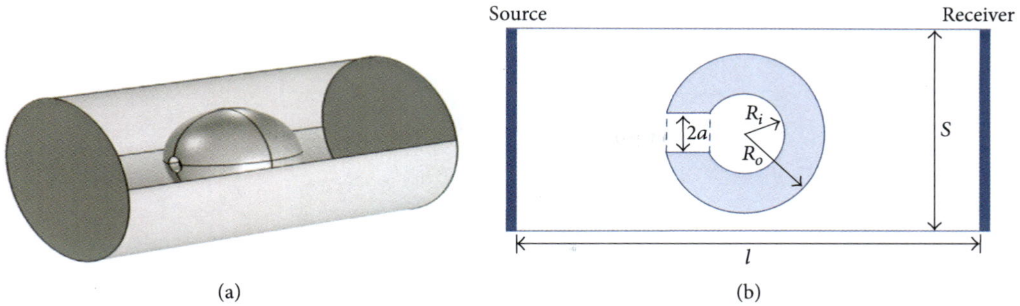

FIGURE 1: (a) The 3D scheme of the metamaterial unit cell. (b) The cross section of the cell.

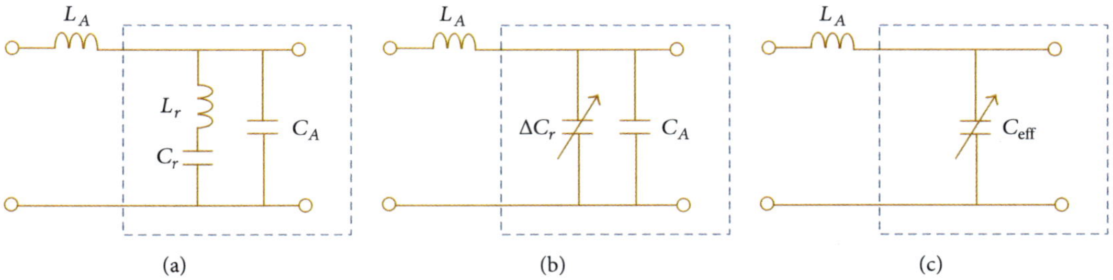

FIGURE 2: (a) An acoustic equivalent circuit using the impedance analogy. (b) The effective impedance circuit of SHS. (c) The effective impedance circuit of the two capacitors.

medium, and number of holes. Ding et al. [18] controlled the resonance frequency with negative modulus by shifting the hole size of SHS. Hao et al. [19] tuned the resonant frequency by changing the medium type of SHS. In addition, the resonant frequency can be controlled by multiple holes on the SHS [20]. At present, little attention has been paid to the precise relationship between the size of structure units and resonance frequency.

In this work, based on the Helmholtz theory and transmission line theory, the resonance characteristic of SHS was discussed. The effective bulk modulus and the amendment formula of resonance frequency were established. Furthermore, COMSOL Multiphysics with the finite-element method was used to design and simulate 3D structure of the acoustic metamaterials unit cell and to verify the effects of geometrical parameters on resonance characteristics consisting of size parameters of the neck radius, inner radius, and outer radius. This provides the basis and guidance for the design and fabrication of double negative acoustic metamaterials in arbitrary frequency by coupling structure units with negative mass density in the next step.

2. The Resonance Characteristics of SHS

2.1. Resonant Metamaterial Unit Cell. The Helmholtz resonators in [11] are arranged orthogonal to the propagation direction of the sound in the waveguide and the structure is successfully achieved negative effective bulk modulus. From transmission line concept, the equivalent circuit of a Helmholtz resonator is given by an LC circuit with no dissipation. And there is another LC section of the waveguide which is connected in parallel with the LC circuit. Based on this idea, we place a SHS in the middle of the cylindrical waveguide in the proposed metamaterial unit cell, as shown in Figure 1(a). The SHS is a type of Helmholtz resonator, including the sphere cavity and the short-tube formed by split hole, also called the neck. Here, it will be shown that the structure can be also used to support negative effective bulk modulus of acoustic metamaterial.

In Figure 1(b) a cross section of the metamaterial unit cell is shown. The cylindrical waveguide is driven at left end by a sound source and is measured at right end by a receiver. The cross section area of the waveguide is S, and the length is l. The SHS (a spherical cavity with the inner radius R_i and the outer one R_o, and a narrow neck with the radius a) is positioned at the source and levitated in the middle of the waveguide.

The equivalent circuit of the unit cell with lossless is given by an analogy circuit as Figure 2(a). In this model, the impedance of the waveguide is represented by the acoustic inertance L_A given by [21]

$$L_A = \frac{\rho l}{S} \tag{1}$$

and the acoustic compliance C_A, which is given by

$$C_A = \frac{V}{\rho v_s^2}, \tag{2}$$

where ρ is the density of air, v_s is the sound velocity, and V is the volume of the air in the waveguide except the SHS. Additionally, the SHS can be modeled with acoustic inertance

$$L_r = \frac{\rho L_p}{S_p} \tag{3}$$

and acoustic compliance

$$C_r = \frac{V_p}{\rho v_s^2}, \tag{4}$$

where L_p is the effective length of the neck, S_p is the cross-sectional area of the one, and V_p is the effective volume of cavity.

2.2. The Effective Bulk Modulus of the Unit Cell. It is well known that, according to the dynamical theory, the vibration system is characterized by its mechanical impedance. Similarly, acoustic impedance Z_a describes the relationship between the sound pressure and the particle velocity, which is given by

$$Z_a = R_a + jX_a, \tag{5}$$

where R_a is the acoustic resistance and X_a is the acoustic reactance. The acoustic resistance represents the energy losses and the acoustic reactance is a concentrated expression of the inertia and elasticity of the system.

Since the fluid medium is air in the waveguide, the shear viscosity and thermal conductivity of the gas are too small enough and can be neglected. Therefore, the acoustic resistance is equal to zero in the analog circuit.

In the case of the acoustic metamaterial unit cell, if the waveguide acts as a simple low-pass filter (LPF), the SHS can be viewed as an "internal heat" as shown in Figure 2(a). The resonance frequency of the SHS falls within the passband of the LPF. In the cylindrical waveguide, there is just one propagating acoustic mode below some cut-off frequency.

In order to simplify the question, the equivalent parameters method is used to the analysis of acoustic reactance. We use the effective acoustic compliance ΔC_r representing the quantity of the inertia and elasticity of SHS which is connected in parallel with C_A as shown in Figure 2(b). Consequently, the parallel combination of the two capacitors has one all effective acoustic compliance C_{eff}, Figure 2(c). C_{eff} could vary in a large number for local resonance and plays a major role in the circuit.

In the context of acoustics, the bulk modulus of unit cell B_A is an important parameter of sound propagation [11]. Suppose that sound propagation is an isentropic process, for which $v_s^2 = \partial p/\partial \rho = B_A/\rho$; bulk modulus $B_A = \rho v_s^2$ is reciprocal to the acoustic compliance from (2) and can be expressed as follows:

$$B_A = \frac{V}{C_A}. \tag{6}$$

Because the gas acts like a spring in the waveguide, the acoustic compliance is written as

$$C_A = \frac{1}{K_A} = \frac{S^2}{K}, \tag{7}$$

where K is the mechanical stiffness of the spring. Since the waveguide is the host frame of unit cell, the effective bulk modulus B_{eff} is defined as

$$B_{\text{eff}} = \frac{V}{C_{\text{eff}}}. \tag{8}$$

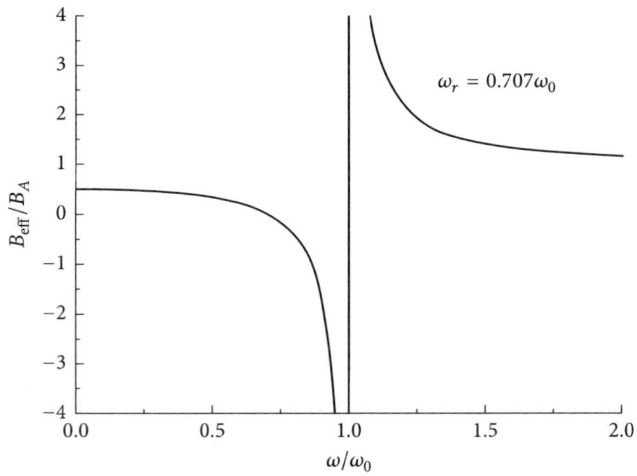

FIGURE 3: Effective bulk modulus of the unit cell obtained from the analytical expression.

Under the above assumption, acoustic impedance of the unit cell becomes

$$Z_a = j\omega L_A + \frac{1}{j\omega C_{\text{eff}}}. \tag{9}$$

Let us suppose that the waveguide has a larger cross section area and therefore the effective acoustic compliance C_{eff} is significantly greater than acoustic inertance L_A at near resonance frequency. Equation (9) is now

$$\begin{aligned} Z_a &\approx \frac{1}{j\omega C_{\text{eff}}} = \frac{(1/j\omega C_A)(j\omega L_r + 1/j\omega C_r)}{j\omega L_r + 1/j\omega C_A + 1/j\omega C_r} \\ &= \frac{1}{j\omega C_A}\frac{\omega^2 - \omega_r^2}{\omega^2 - \omega_0^2}, \end{aligned} \tag{10}$$

where $\omega_0^2 = \omega_A^2 + \omega_r^2$, $\omega_A = 1/\sqrt{L_r C_A}$, and $\omega_r = 1/\sqrt{L_r C_r}$. The effective acoustic compliance and effective bulk modulus are determined by the following expressions, respectively:

$$C_{\text{eff}} = C_A \frac{\omega^2 - \omega_0^2}{\omega^2 - \omega_r^2}, \tag{11}$$

$$B_{\text{eff}} = B_A \frac{\omega^2 - \omega_r^2}{\omega^2 - \omega_0^2}. \tag{12}$$

From (12), it can be seen that the inclusion of SHS in the waveguide results in the variation of effect bulk modulus of unit cell with frequency, as shown in Figure 3.

It can be seen that effective bulk modulus between frequencies $0 < \omega < \omega_r$ is greater than zero but is less than B_A. With the increasing of frequency, the effective bulk modulus is decreasing, because the acoustic inertance of SHS begins to take more effect. The frequency ω_r is the series resonance frequency of SHS unit. At this frequency, acoustic impedance is equal to zero which means the sound wave is by-passed and the effective bulk modulus is also zero.

In the frequencies range $\omega_r < \omega < \omega_0$, the inertia of SHS is so big that the effective bulk modulus is negative. The effective

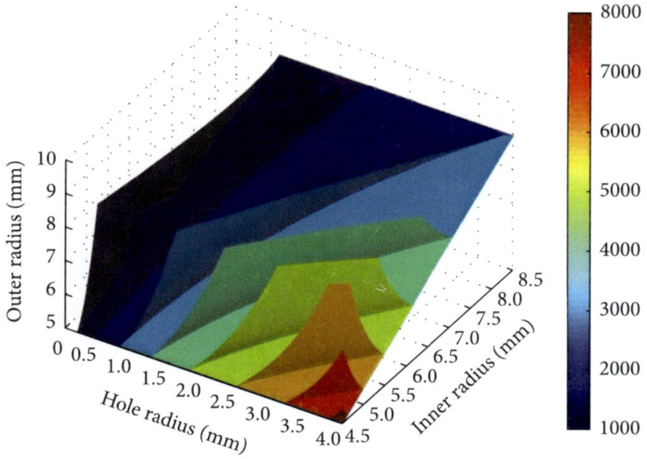

FIGURE 4: Relation between the parameters of SHS and the resonance frequency.

bulk modulus tends to minus infinity as frequency increases from ω_r to ω_0. It should be noted that the frequency ω_0 is the parallel resonance frequency of the ΔC_r and C_A and is also antiresonance frequency. At the frequency, the acoustic impedance is equal to minus infinity.

The effective bulk modulus is greater than zero in the frequency range above ω_0 and is greater than B_A. With the increasing of frequency, the effective bulk modulus is decreasing to B_A. It is also because of the more influence of the acoustic inertance of SHS.

2.3. The Resonance Frequency of SHS. From the above analysis, we can see that the resonance of SHS plays an important role in the negative effective bulk modulus of unit cell. Based on Helmholtz theory [21], the resonance frequency of SHS can be calculated by the similar LC circuit shown in Figure 1(a). The resonance frequency of SHS is

$$f_r = \frac{1}{2\pi\sqrt{L_r C_r}} = \frac{v_s}{2\pi}\sqrt{\frac{S_p}{L_p V_p}}. \quad (13)$$

The effective volume of cavity V_p, that is, the volume of inner sphere minus the volume of spherical crown as part of short-tube, is thus given by

$$V_p = \frac{4}{3}\pi R_i^3 - \pi\left(R_i - \sqrt{R_i^2 - a^2}\right)^2 \left(R_i - \frac{R_i - \sqrt{R_i^2 - a^2}}{3}\right), \quad (14)$$

where $S_p = \pi a^2$ is the area of cross section of the short-tube.

The geometric length of the short-tube is

$$l_p = \sqrt{R_0^2 - a^2} - \sqrt{R_i^2 - a^2}. \quad (15)$$

Since the effective length of air in short-tube is longer than geometric length, we define it as the effective length L_p. In this model, the short-tube has two end corrections, where the outside of short-tube is considered as nonflange and the correction length is $0.61a$; the inside is under the flange state and the correction length is $(8/3\pi)a$ [22]. Thus, the effective length of short-tube is

$$L_p = \sqrt{R_0^2 - a^2} - \sqrt{R_i^2 - a^2} + 0.61a + \frac{8}{3\pi}a. \quad (16)$$

From the results given by (14) and (16), the amendment formula of resonance frequency is therefore

$$f_r = \frac{v_s}{2\pi}\sqrt{\frac{\pi a^2}{\left(\sqrt{R_0^2 - a^2} - \sqrt{R_i^2 - a^2} + 0.61a + (8/3\pi)a\right)\left[(4/3)\pi R_i^3 - \pi\left(R_i - \sqrt{R_i^2 - a^2}\right)^2\left(R_i - \left(R_i - \sqrt{R_i^2 - a^2}\right)/3\right)\right]}}. \quad (17)$$

In order to analyze the influence of geometrical parameters of SHS on tunable frequency, the relation between the three parameters of SHS and the resonance frequency was shown in Figure 4 based on formula (17). It can be seen that there is a lower resonance frequency as the neck size is smaller, and there is a higher frequency as the neck size is larger. This provides us an important way to precisely control acoustic metamaterial with negative effective modulus.

2.4. Simulation Results. The acoustic field is simulated by COMSOL Multiphysics 4.3 and the pressure acoustic and solid mechanics are coupled in the acoustic model for the continuity. The model consists of a SHS unit and a waveguide. The plane sound wave was used as a sound source and the other open end of outlet waveguide was modeled with PML boundary condition. The remaining boundaries of waveguide are specified as sound hard condition for the sound waves reflecting from waveguide wall completely. The material of wall of SHS is polyethylene and the fluid domain is full of air. At standard temperature and pressure conditions, the bulk modulus B_o of air is 1.42×10^5 Pa for an isentropic process. Suppose that $l = 20.0$ mm, $S = 225\,\pi\text{mm}^2$, $R_o = 5.0$ mm, $R_i = 4.5$ mm, and $a = 0.5$ mm to 2.5 mm (Figure 1(b)).

The results of the transmission spectrum are got and shown in Figure 5. Except that the curve of the radius of 0 mm has no transmission dip, there is a dip in every transmission curve. Furthermore, the blue shift of transmission dip occurs as the neck radius of SHS becomes bigger. According to (3),

TABLE 1: Relation between the neck radius and resonance frequency, ($R_o = 5.0$ mm, $R_i = 4.5$ mm).

The radius of neck/mm	0.5	1.0	1.5	2.0	2.5
Theoretical frequency/Hz	2200	3502	4602	5602	6491
Simulate frequency/Hz	2190	3514	4607	5594	6546

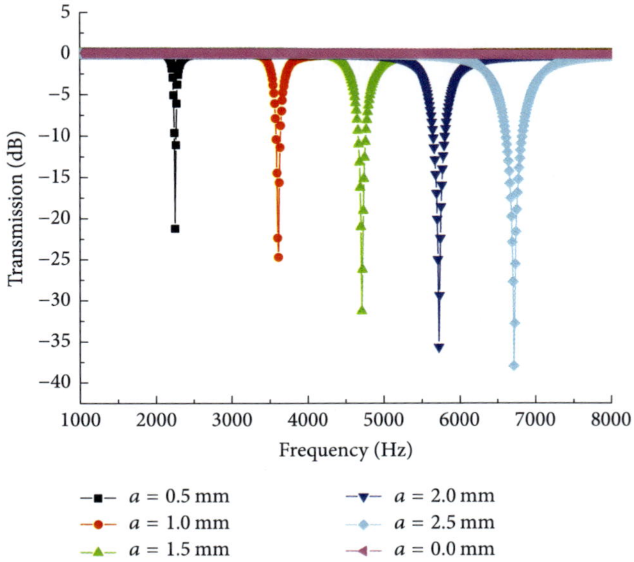

FIGURE 5: The transmission coefficients of Unit cell with different neck radius ($R_o = 5.0$ mm, $R_i = 4.5$ mm).

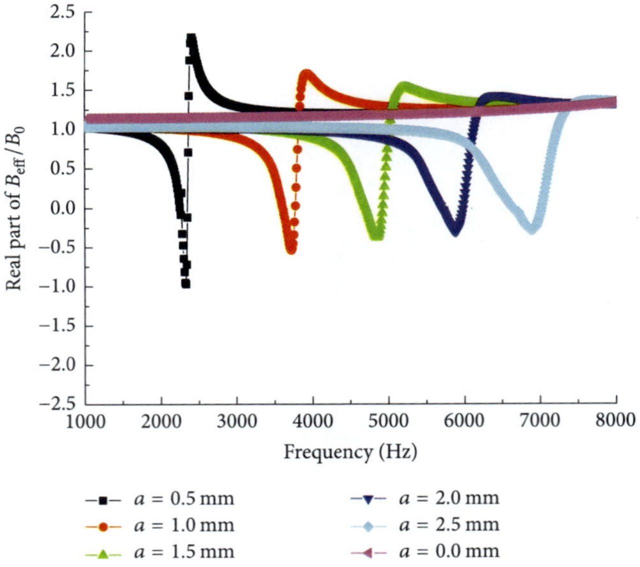

FIGURE 6: The effective modulus curves of unit cell with different neck radius ($R_o = 5.0$ mm, $R_i = 4.5$ mm).

the increased quality of the air in the neck enhances its acoustic inertance, while the volume value of cavity has decreased slightly. In terms of the serial inductor-capacitor circuit, if the radius of cavity is constant, the inductance of the SHS decreases with increasing neck radius. Although the inner radius of SHS has not been changed, some volume of them becomes a part of the short-tube. The ability of storing energy of SHS gets weak and makes the capacitance of the resonator smaller. Hence, resonance frequency goes up with the increase of the neck radius of SHS.

The values of resonance frequency are shown in Table 1. For example, the SHS with 0.5 mm neck radius has phase reverse near the frequency of 2190 Hz for the local resonance. As is shown in Figure 6, the real part of effective modulus is negative near the resonance frequency of SHS, and it represents the character of the negative dynamic response. Similarly, effective modulus with different neck radius was obtained, and negative effective modulus near the resonance frequency is achieved individually (see Figure 6). This result is consistent with that described in [11].

For verifying the validity of formula (17), we calculate the values of resonance frequency with COMSOL method and amendment formula (17), and they were listed in Table 1. The simulated value of resonance frequency is in good agreement with the analysis results of the equivalent circuit. Thus, using formula (17), we can obtain the precise value and then control resonance frequency of SHS accurately.

3. Influence of Geometrical Parameters of SHS on Resonance Frequency

Similarly, the effects of other parameters on the tunable frequency of SHS were analyzed with COMSOL Multiphysics and formula (17) in the section. Formula (17) suggests that the tunable parameters of SHS include the hole radius, outer radius, inner radius of sphere cavity, and so on. Because the inner radius and neck radius affect the cavity volume, the difference between outer and inner radius acts as the length of short-tube, and the neck radius affects the area of cross section of short-tube. Thus, three characteristic parameters of SHS have a simultaneous influence on the tunable frequency. In this section, the effects of the inner and outer radius on the resonance frequency were mainly discussed.

3.1. The Inner Radius of SHS. Supposing that $R_o = 9.0$ mm, $a = 0.5$ mm, and $R_i = 4.5$ mm to 8.5 mm, the transmission spectrum are measured and shown in Figure 7. The resonance frequency decreases firstly and then increases with increasing the inner radius gradually. In addition, the negative modulus can be obtained near their respective resonance frequency. The results of formula calculation and the simulation agree well, which suggests that we can obtain the precise resonance frequency by formula (17) and accurately control the AM with SHS.

There is a V-type character of the resonant frequency and the minimum frequency value is at the inner radius with 7.5 mm. This phenomenon explained that the acoustic inductance of SHS decreases with increasing the inner radius; on the other hand, the acoustic capacitance increases with increasing the volume of cavity. Under the influence of two opposite aspect, the resonance frequency decreases firstly and then increases with the increase of the inner radius.

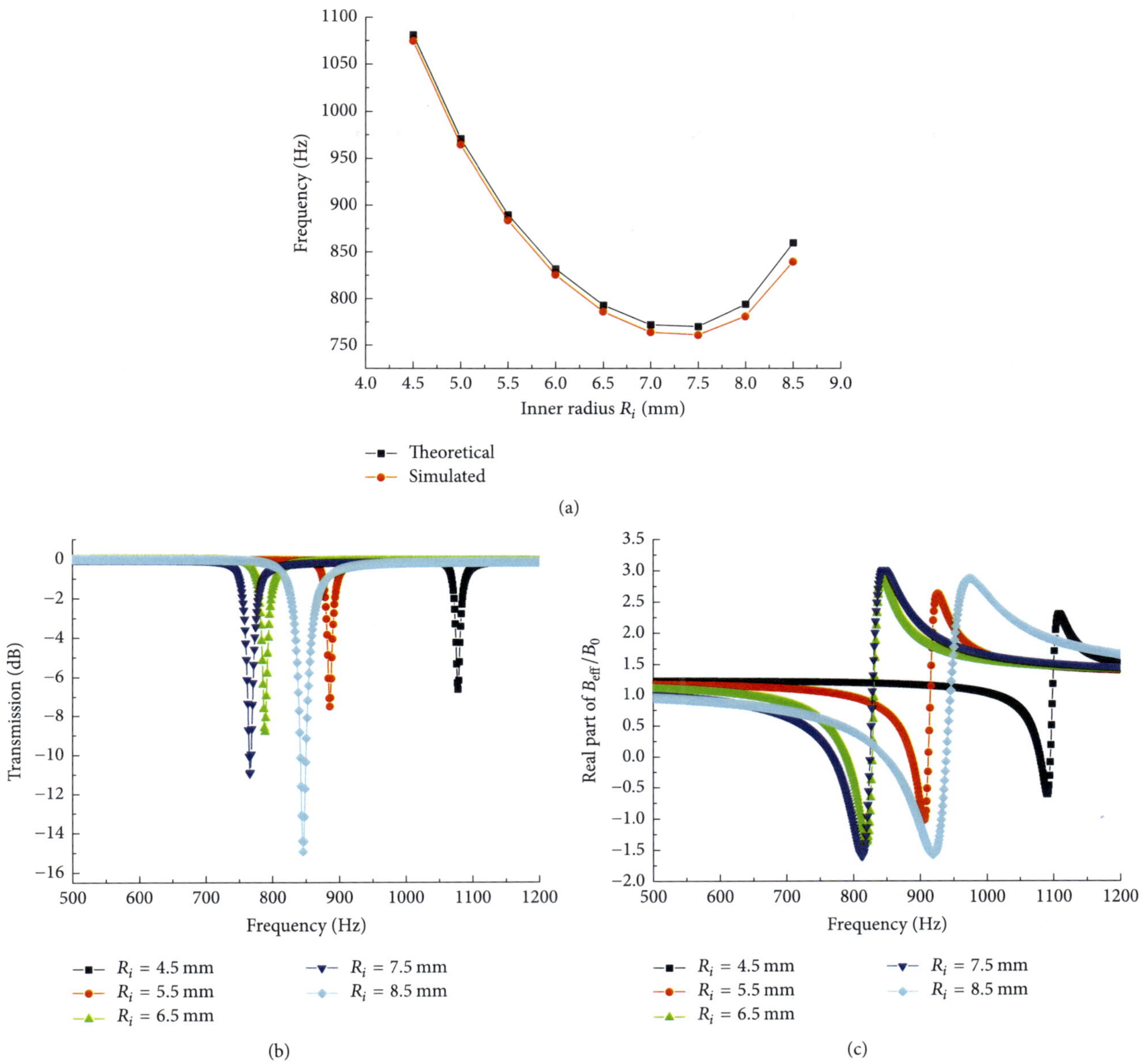

FIGURE 7: The relation curves present the relationship between the inner radius and related parameters (R_o = 9.0 mm, a = 0.5 mm). (a) The resonance frequency of SHS. (b) The transmission coefficient. (c) The effective modulus.

3.2. The Outer Radius of SHS. Supposing that R_i = 4.5 mm, a = 0.5 mm, and R_o = 5.0 mm to 9.0 mm, Figure 8 is the relation curves between the outer radius and related parameters, and the resonance frequency has a red shift with increasing the outer radius of SHS. This simulated result is in agreement with Helmholtz theory as well. In addition, the negative modulus can also be achieved near their respective resonance frequency.

The red shift explained that when the inner radius of SHS unchanged and the capacitance is constant, the sound energies of SHS unit with different outer radius are the same. However, the acoustic inductance of SHS increases with the increase of length of short-tube acted as inductor. Hence, the resonant frequency has a red shift.

The results of formula calculation and the simulation agree well, which suggests that we can obtain the precise resonance frequency by formula (17) and accurately control the acoustic metamaterial of SHS with negative modulus.

The results of formula calculation and the simulation agree well, which suggests that we can obtain the precise resonance frequency by formula (17).

4. Conclusion

Based on SHS unit, the tunable mechanism of structure size on controlling the resonance frequency of acoustic metamaterial with negative effective modulus was proposed from aspects of Helmholtz theory and the finite-element

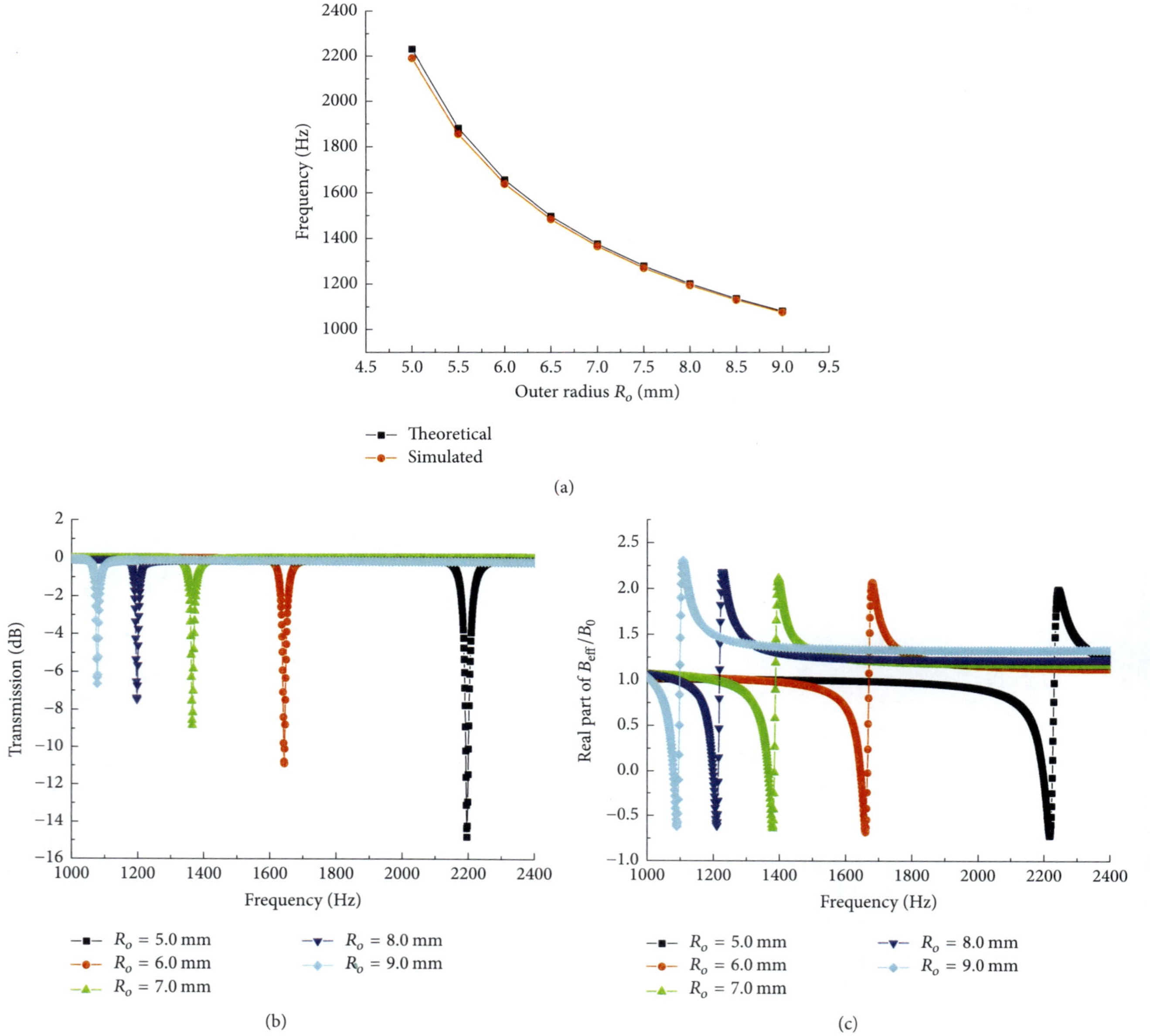

FIGURE 8: The relation curves present the relationship between the outer radius and related parameters ($R_i = 4.5$ mm, $a = 0.5$ mm). (a) The resonance frequency of SHS. (b) The transmission coefficient. (c) The effective modulus.

method. The amendment formula of resonance frequency of SHS was derived, and it can precisely describe the relation between resonance frequency and neck radius, inner radius and outer radius. Furthermore, COMSOL simulation was used to verify effectiveness of amendment formula, and it is revealed that simulated results of resonance frequency are in good agreement with theoretical results. The results indicate that the blue shift of resonance frequency occurs when the neck radius of SHS increases; the resonance frequency has a "V" change with the increase of inner radius of SHS; the resonance frequency has a red shift with increasing the outer radius of SHS. The method can precisely control the resonance frequency of acoustic metamaterial with negative effective modulus. There is a potential value on ultrasonic imaging, underwater acoustic, architectural acoustic, and sound absorbing material, and so on.

Conflicts of Interest

The authors declare that there are no conflicts of interest regarding the publication of this paper.

Acknowledgments

This work was supported by the National Natural Science Foundation of China (Grant nos. 11102164 and 11304243), the Natural Science Foundation of Shaanxi Province, China (Grant no. 2014JQ1039), the Scientific Research Program

Funded by Shaanxi Provincial Education Department (Grant no. 15JK1485), the Program of New Staff and Research Area Project of NPU (Grant no. 13GH014602), and the Cultivation Fund of Xi'an University of Science and Technology (Grant no. 2009029).

References

[1] Z. Liu, X. Zhang, Y. Mao et al., "Locally resonant sonic materials," *Science*, vol. 289, no. 5485, pp. 1734–1736, 2000.

[2] Z. Yang, J. Mei, M. Yang, N. H. Chan, and P. Sheng, "Membrane-type acoustic metamaterial with negative dynamic mass," *Physical Review Letters*, vol. 101, no. 20, Article ID 204301, 2008.

[3] C.-L. Ding and X.-P. Zhao, "Audible sound metamaterial," *Acta Physica Sinica*, vol. 58, no. 9, pp. 6351–6355, 2009.

[4] J. Pierre, B. Dollet, and V. Leroy, "Resonant acoustic propagation and negative density in liquid foams," *Physical Review Letters*, vol. 112, no. 14, Article ID 148307, 2014.

[5] M. Yang, G. Ma, Z. Yang, and P. Sheng, "Coupled membranes with doubly negative mass density and bulk modulus," *Physical Review Letters*, vol. 110, no. 13, Article ID 134301, 2013.

[6] J. J. Park, C. M. Park, K. J. B. Lee, and S. H. Lee, "Acoustic superlens using membrane-based metamaterials," *Applied Physics Letters*, vol. 106, no. 5, Article ID 051901, 2015.

[7] K. Song, S.-H. Lee, K. Kim, S. Hur, and J. Kim, "Emission enhancement of sound emitters using an acoustic metamaterial cavity," *Scientific Reports*, vol. 4, article no. 4165, 2014.

[8] J. Liu, Z. L. Hou, and X. J. Fu, "Mechanism for local resonant acoustic metamaterial," *Acta Physica Sinica*, vol. 64, no. 15, Article ID 154302, pp. 54302–154302, 2015.

[9] X. W. Yang, J. S. Lee, and Y. Y. Kim, "Effective mass density based topology optimization of locally resonant acoustic metamaterials for bandgap maximization," *Journal of Sound and Vibration*, vol. 383, no. 11, pp. 89–107, 2016.

[10] A. O. Krushynska, V. G. Kouznetsova, and M. G. Geers, "Visco-elastic effects on wave dispersion in three-phase acoustic metamaterials," *Journal of the Mechanics and Physics of Solids*, vol. 96, no. 11, pp. 29–47, 2014.

[11] N. Fang, D. Xi, J. Xu et al., "Ultrasonic metamaterials with negative modulus," *Nature Materials*, vol. 5, no. 6, pp. 452–456, 2006.

[12] S. H. Lee, C. M. Park, Y. M. Seo, Z. G. Wang, and C. K. Kim, "Acoustic metamaterial with negative modulus," *Journal of Physics: Condensed Matter*, vol. 21, no. 17, Article ID 175704, 2009.

[13] C. Ding, L. Hao, and X. Zhao, "Two-dimensional acoustic metamaterial with negative modulus," *Journal of Applied Physics*, vol. 108, no. 7, Article ID 074911, 2010.

[14] X. Jing, Y. Meng, and X. Sun, "Soft resonator of omnidirectional resonance for acoustic metamaterials with a negative bulk modulus," *Scientific Reports*, vol. 5, no. 11, Article ID 16110, 2015.

[15] W. Akl and A. Baz, "Multi-cell active acoustic metamaterial with programmable bulk modulus," *Journal of Intelligent Material Systems and Structures*, vol. 21, no. 3, pp. 541–556, 2010.

[16] X. Yu, Z. Lu, F. Cui, L. Cheng, and Y. Cui, "Tunable acoustic metamaterial with an array of resonators actuated by dielectric elastomer," *Extreme Mechanics Letters*, vol. 12, pp. 37–40, 2017.

[17] B. Xia, N. Chen, L. Xie, Y. Qin, and D. Yu, "Temperature-controlled tunable acoustic metamaterial with active band gap and negative bulk modulus," *Applied Acoustics*, vol. 112, no. 11, pp. 1–9, 2016.

[18] C. L. Ding, X. P. Zhao, L. M. Hao, and W. R. Zhu, "Acoustic metamaterial with split hollow spheres," *Acta Physica Sinica*, vol. 60, no. 4, Article ID 044301, 2011.

[19] L.-M. Hao, C.-L. Ding, and X.-P. Zhao, "Tunable acoustic metamaterial with negative modulus," *Applied Physics A: Materials Science & Processing*, vol. 106, no. 4, pp. 807–811, 2012.

[20] L.-M. Hao, C.-L. Ding, and X.-P. Zhao, "Design of a passive controllable negative modulus metamaterial with a split hollow sphere of multiple holes," *Journal of Vibration and Acoustics*, vol. 135, no. 4, Article ID 041008, 2013.

[21] G. H. Du, *Fundamentals of Acoustics*, Nanjing University Press, Nanjing, China, 2012.

[22] D. T. Blackstock, *Fundamentals of Physical Acoustics*, John Wiley & Sons, NJ, USA, 2000.

8

Intrinsic and Extrinsic Ferromagnetism in Co-Doped Indium Tin Oxide Revealed using X-Ray Magnetic Circular Dichroism

A. M. H. R. Hakimi,[1] F. Schoofs,[1] M. G. Blamire,[1] S. Langridge,[2] and S. S. Dhesi[3]

[1]Department of Materials Science and Metallurgy, University of Cambridge, 27 Charles Babbage Road, Cambridge CB3 0FS, UK
[2]Science and Technology Facilities Council, Rutherford Appleton Laboratory, Harwell Science and Innovation Campus, Didcot OX11 0QX, UK
[3]Diamond Light Source, Chilton, Didcot, Oxfordshire OX11 0DE, UK

Correspondence should be addressed to S. S. Dhesi; dhesi@diamond.ac.uk

Academic Editor: Ward Beyermann

The effects of high-temperature annealing on ferromagnetic Co-doped Indium Tin Oxide (ITO) thin films have been investigated using X-ray diffraction (XRD), magnetometry, and X-Ray Magnetic Circular Dichroism (XMCD). Following annealing, the magnetometry results indicate the formation of Co clusters with a significant increase in the saturation magnetization of the thin films arising from defects introduced during cluster formation. However, sum rule analysis of the element-specific XMCD results shows that the magnetic moment at the Co sites is reduced after annealing. The effects of annealing demonstrate that the ferromagnetism observed in the as-deposited Co-doped ITO films arises from intrinsic defects and cannot be related to the segregation of metallic Co clusters.

1. Introduction

Dilute magnetic oxides, formed by randomly substituting magnetic transition-metal (TM) ions into a host oxide, have attracted significant interest in recent years due to their potential application in spintronic devices [1, 2]. On the other hand room temperature ferromagnetism remains controversial in these materials because the classification of intrinsic or extrinsic ferromagnetism has proved particularly difficult [3–5]. Furthermore, ferromagnetism has been reported in oxides without TM ion doping [6, 7] calling into question the precise role of the TM impurities in establishing the ferromagnetic ordering.

A case in point is TM-doped Indium Tin Oxide (ITO) which exhibits room temperature ferromagnetism and has sparked considerable interest in its use as a transparent spin valve [8–12]. However, for optically transparent Co-doped ITO the Co ions have been observed to remain paramagnetic at low temperature ($T = 2\,\text{K}$) and hence do not contribute to the long range ferromagnetic ordering [13]. Instead, calculations indicate that ferromagnetism is related to defect concentration rather than TM ion doping [14]. In particular, oxygen vacancies may play an important role in the emergence of ferromagnetism in these materials [15, 16]. Baqiah et al. recently showed that the magnetic behavior of Co-doped ITO films made by a sol-gel process was sensitive to both the Co doping concentration and the density of oxygen vacancies as well as the estimated ratio of oxygen defects to Co ions [17]. Unfortunately, direct evidence for intrinsic ferromagnetism in this material remains elusive, but nowadays powerful element and site-specific probes such as X-Ray Magnetic Circular Dichroism (XMCD) and X-Ray Absorption Spectroscopy (XAS) are proving invaluable.

Here, we focus on ferromagnetic ordering in as-deposited and ultrahigh vacuum (UHV) annealed Co-doped ITO films using magnetometry together with XAS and XMCD. The annealing process is found to increase the saturation magnetization but decrease the magnetic moment per Co ion. This suggests that the ferromagnetism in the annealed films has a strong contribution from metallic Co clusters in contrast to the ferromagnetic response in the as-deposited

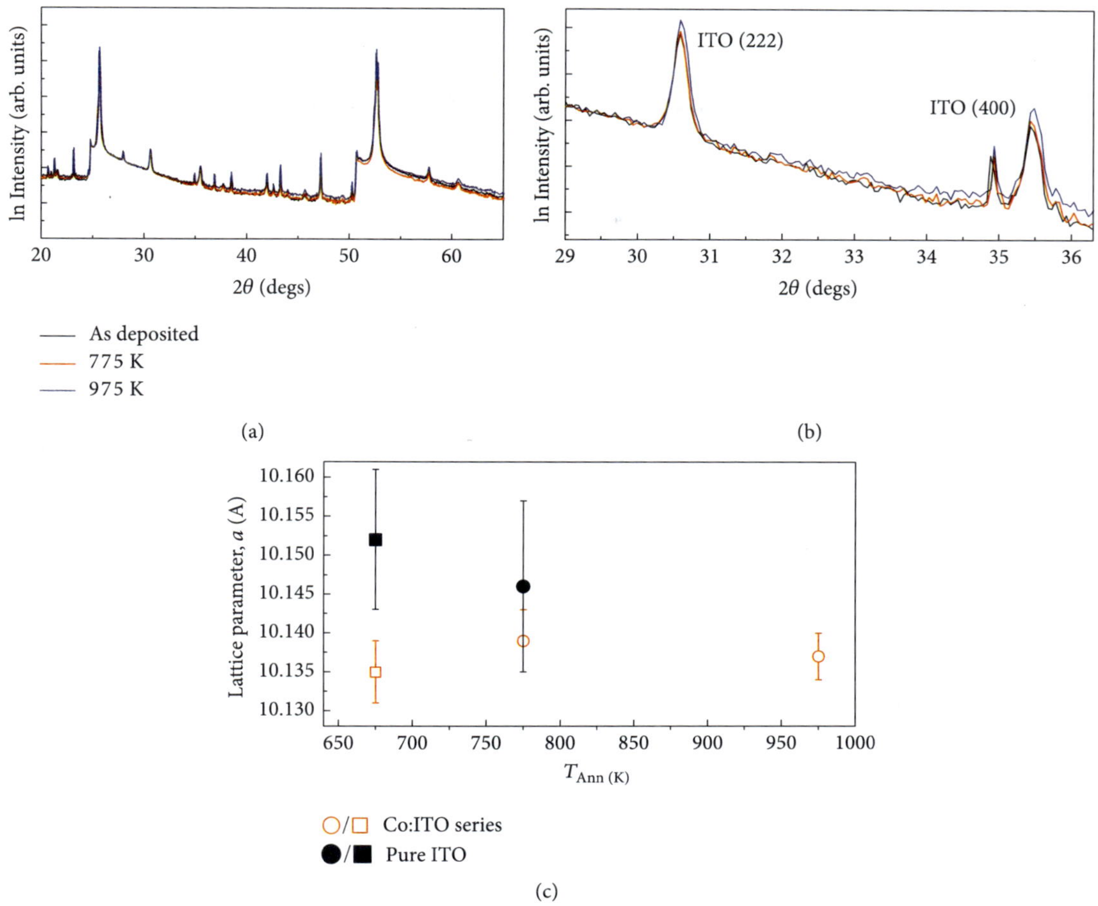

FIGURE 1: (a) θ-2θ XRD diffractograms of the as-deposited (solid black line) and annealed films (red and blue solid lines); (b) an expanded view of the same θ-2θ XRD diffractograms shown in (a) but focus is now put on the ITO (222) and ITO (400) peaks demonstrating the shift in the peak positions as the annealing temperature is increased to 975 K; (c) out-of-plane lattice parameter for as-deposited (squares) and annealed (circles) pure ITO (black) and Co-doped ITO (red) thin films.

films which we believe to be intrinsic and free from metallic clustering. In addition to this, magnetometry measurements reveal a ferromagnetic blocking temperature for the annealed films giving further evidence for metallic clustering.

2. Experimental Details

Co-doped (5.4 at.%) ITO thin films were codeposited using DC magnetron sputtering in a UHV system with a base pressure $<1 \times 10^{-6}$ Pa. The 250 nm thick films were grown at 675 K on *r*-plane sapphire substrates. A highly oxygen-deficient environment was used to create significant oxygen deficiency in the films, most likely in the form of oxygen vacancies. Films were sputtered in 1.2 ± 0.01 Pa of an Ar-O_2 (95 : 5) mix (0.06 Pa O_2). UHV annealing was performed for 2 hours in the deposition chamber at 975 K. Magnetic characterization was carried out with a superconducting quantum interference device (SQUID) over a temperature range 2–300 K using magnetic fields of up to 0.5 T. XAS and XMCD measurements were performed over the Co $L_{2,3}$ edges ($2p \rightarrow 3d$ dipole transitions) on beamline I06, Diamond Light Source, using the UHV compatible superconducting high-field magnet which provides fields of up to 6 T and operates in a temperature range of 2–370 K. The sample was mounted on a custom-designed pure Cu block using silver paste for good electrical contact. All SQUID magnetic characterization was performed before the XAS and XMCD measurements. The samples were carefully handled, that is, using nonmagnetic tweezers, to minimize the possibility of contamination leading to a false ferromagnetic effect.

3. Results and Discussion

3.1. Structural Properties. X-ray diffraction was performed on a Bruker D8 diffractometer in Bragg-Brentano geometry with Cu Kα radiation and a position sensitive detector. X-ray diffraction of the as-deposited and annealed films could be indexed assuming the cubic bixbyite structure of the pure ITO [18]. Lattice parameters were calculated based on profile-fitted cubic ITO reflections (211), (222), (400), (440), and (622), using the sapphire (001) reflections as an internal standard. The diffraction peaks were sharp, suggesting a high degree of crystallinity, with no detectable peaks corresponding to secondary Co oxides. Figure 1 shows θ-2θ

XRD scans of the as-deposited and that of the sample annealed at 975 K. To verify the absence of Co clusters further in our as-deposited samples, the integration time per point was raised significantly to 100 seconds. This still yielded spectra suggesting the absence of any metallic cluster formation. The same procedure was carried out on the annealed samples ending in the same outcome: within the detection limits of the instrument, no clusters could be detected even after high-temperature UHV annealing. Figure 1(b) shows an expanded view of the predominant ITO (222) peak and that of the ITO (400) peak for the as-deposited and two other films annealed at different temperatures.

There is no obvious significant shift in the ITO (222) peak position as the annealing temperature is changed. The only obvious change is the increased intensity of the ITO (222) peak at the highest annealing temperature (975 K), perhaps suggesting a marked increase in crystalline quality. However, there is a clear shift in the peak position of the ITO (400) peak towards higher values for 2θ as the annealing temperature is increased and would suggest a decrease in the out-of-plane lattice parameter, a. This is consistent with our own findings for pure ITO thin films which have been grown and annealed at different temperatures [19].

Figure 1(c) shows the shifts in a as a function of the annealing temperature for pure and Co-doped ITO thin films. It is clear that samples which have been doped with Co (red data points) have a lower a. This can be explained by the substitution of Co^{2+}, which has a smaller ionic radius, for In^{3+} within the ITO matrix. We have shown in a previous study that a will continue to decrease with increasing levels of Co doping [13]. Annealing the samples at 775 K resulted in a small increase in a and may correspond to the migration of Sn ions in the ITO matrix and the out-diffusion of Co^{2+} ions from In^{3+} lattice sites, resulting in the formation of interstitial metallic Co clusters. Increasing the annealing temperature further may compound this effect and also result in increased out-diffusion of oxygen from interstitial sites. This could result in the relaxation of the lattice and explain the apparent drop in a at higher 975 K. We note that these changes are very small and are within the specified experimental error. Therefore, from XRD alone we can only speculate such mechanisms upon annealing. Although there is no evidence for Co clustering from the XRD data, evidence that we present later suggests otherwise and again demonstrates the need for more powerful techniques to probe the true nature of magnetism in these materials.

3.2. Electrical Properties. The films were metallic with a room temperature carrier concentration (determined via Hall measurements using the van der Pauw configuration) of $8.5 \times 10^{20}\,cm^{-3}$ which increased to $9.5 \times 10^{20}\,cm^{-3}$ after annealing. This implies that the annealing process introduced additional oxygen vacancies into the film along with improved substitution of Sn ions for In ions. Both the improved incorporation of Sn ions into the matrix and the creation of oxygen vacancies provide extra carriers for conduction.

3.3. Magnetic Properties. In-plane magnetization (M) versus field ($\mu_0 H$) loops measured at room temperature for the as-deposited and annealed films are shown in Figure 2(a). The diamagnetic background of the substrate and sample holder has been subtracted from all the magnetization data shown here. The maximum saturation magnetization, M_S, for the as-deposited film is 1.02 emu cm^{-3} which increases to 2.55 emu cm^{-3} after annealing. In addition, the coercive field increases from ~8 mT to ~34 mT after annealing.

Figure 2(b) shows the temperature dependence of the zero-field cooled (ZFC) and field-cooled (FC) magnetization measured in a field of 10 mT for the annealed Co-doped ITO film. There is a clear separation between the ZFC and FC curves at ~272 K which represents the superparamagnetic blocking temperature, T_B, and indicates the presence of magnetic nanoclusters. The value of T_B can be used to estimate the cluster size using the relationship $T_B = K_{Co}V/k_B \ln(\tau_m/\tau_0)$, where K_{Co} is the magnetic anisotropy energy density for Co ($4.9 \times 10^5\,J\,m^{-3}$), V is the volume of the nanoparticle, k_B is the Boltzmann constant, τ_m is the experimental measurement time (taken to be 10 s), and τ_0 is the attempt period (1×10^{-10} s). Assuming spherically shaped clusters, the particle radius is estimated to be 3.8 ± 0.1 nm. The ZFC-FC magnetization curve for the as-deposited sample showed no evidence of a superparamagnetic T_B [13]. This clearly indicates that, aside from an increase in oxygen vacancy concentration as evidenced from the increase in carrier density, Co clusters are formed due to preferential reduction of Co^{2+} ions in the reducing environment.

Further evidence to support the role that oxygen vacancies have on these films is demonstrated in Figure 3 which shows a series of room temperature magnetization loops for 5.4 at.% Co-doped ITO thin films grown under varying oxygen partial pressures. The figure clearly shows the reduction in the strength of the magnetic response as the oxygen partial pressure during growth increases. The inset also shows a clear correlation between magnetization and the carrier concentration: increasing carrier concentration results in an increase in the magnetic response (as was also seen by Stankiewicz et al. for their Co-doped ITO films) [8]. This gives clear evidence that oxygen vacancies play a crucial role in mediating magnetism in these films. Coey et al. proposed a bound magnetic polaron (BMP) mechanism involving oxygen vacancies to explain the ferromagnetic behavior in these TM-doped oxide films [20]. These vacancies are capable of forming so-called *F*-centers with trapped electrons. The overlapping of *F*-center electron orbitals with the d orbitals of neighboring TM dopants to form TM-oxygen vacancy-TM groups is key to the observed ferromagnetism.

3.4. XAS and XMCD Measurements. XAS and XMCD measurements were performed over the Co $L_{2,3}$ edge at 2 K in an applied field of 4 T. The sample surface was oriented at 45° to the incident beam and the magnetic field was applied parallel to the direction of the incident X-ray beam. XAS spectra were recorded in total-electron yield (TEY) and fluorescence yield (FY) mode in order to measure the surface and bulk components, respectively. Figure 4 shows the TEY XAS spectra for a sample magnetization parallel (ρ_+) and antiparallel (ρ_-) to the photon helicity vector for a pure Co foil, the as-deposited film, and the annealed film. The XAS

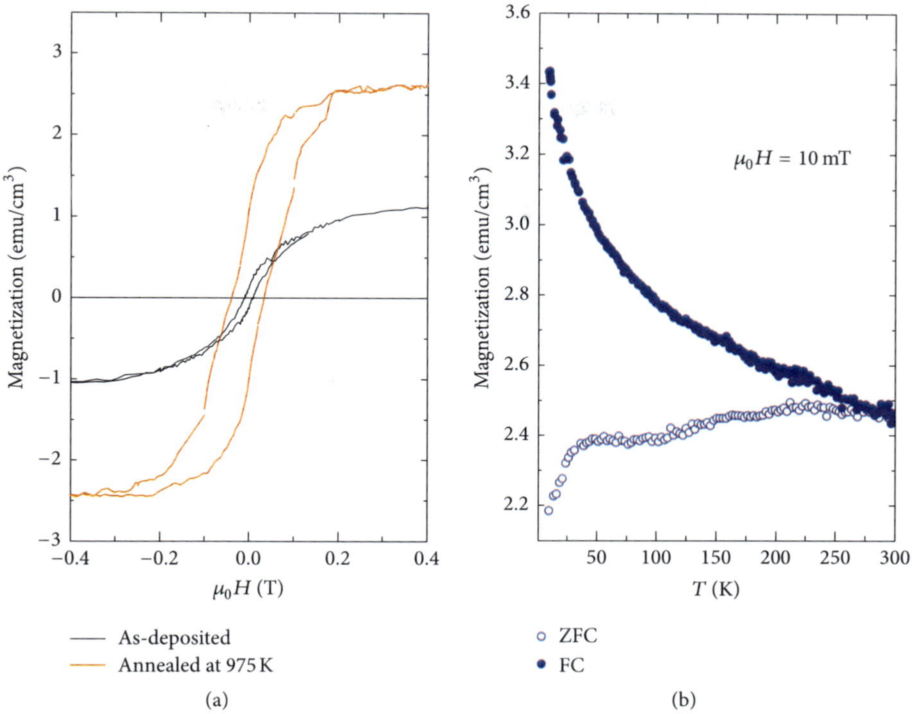

FIGURE 2: (a) Room temperature hysteresis loops for an as-deposited (black line) and annealed (red line) Co-doped ITO. (b) Zero-field cooled (open blue circles) and field-cooled (solid blue circles) magnetization versus temperature curves for the annealed Co-doped ITO film recorded in an applied magnetic field of 10 mT.

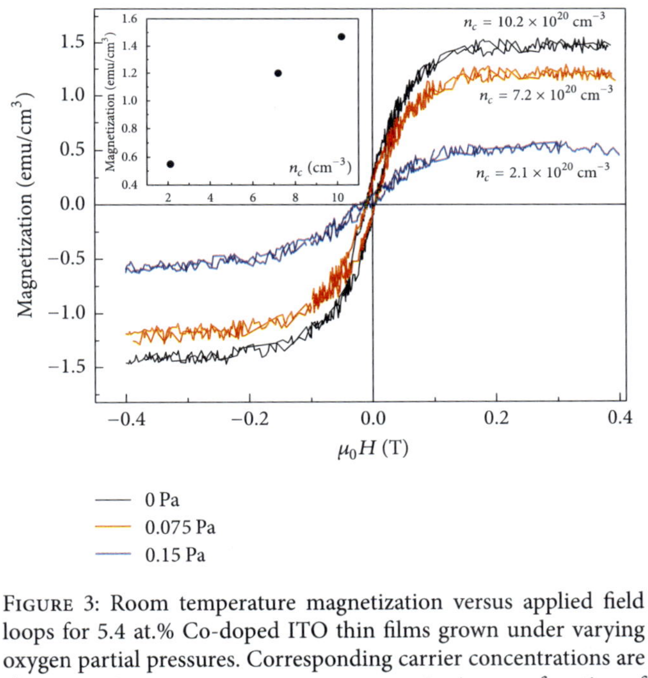

FIGURE 3: Room temperature magnetization versus applied field loops for 5.4 at.% Co-doped ITO thin films grown under varying oxygen partial pressures. Corresponding carrier concentrations are also given. Inset: room temperature magnetization as a function of carrier concentration for the same films.

and XMCD (ρ_{-}-ρ_{+}) from the Co foil shown in Figures 4(a) and 4(d), respectively, are representative of metallic Co [21]. On the other hand, the pronounced XAS multiplet structure at the Co $L_{2,3}$ edges for the as-deposited film (Figure 4(b)) arises from Coulomb and exchange interactions between $2p$ core hole and $3d$ valence states and is typical for a Co^{2+} ion. After annealing the XAS multiplet features at L_3 peak shown in Figure 4(c) are less pronounced, indicating an increased fraction of metallic Co. Figures 4(e) and 4(f) also show the corresponding XMCD spectra. The XMCD spectrum for the annealed sample (Figure 4(f)) exhibits features which are a mixture of the metallic and Co^{2+} spectra [13, 22] suggesting the presence of metallic and ionic Co^{2+} near the surface region after annealing.

The presence of metallic Co in the annealed Co-doped ITO sample was also confirmed by XAS and XMCD spectra measured in FY mode and is shown in Figure 5. Comparison of the XAS and XMCD spectra for the as-deposited and annealed films shows that the multiplet structure has been largely removed after annealing. The lineshape of the spectra is closer to that of metallic Co and suggests that the bulk of the film contains metallic Co clusters. This is consistent with the superparamagnetic blocking temperature measured in M-T (K) data shown in Figure 2(b). The arrows in Figure 5(a) indicate the emergence of two new features in the XAS spectrum which only appear after the annealing treatment. These new features may indicate hybridization of Co states to oxygen or In after the annealing process.

To probe the origins of the magnetism in the annealed films further, XAS and XMCD measurements were performed at 4 K and at remanence. The multiplet fine structure was still present in both TEY and FY XAS spectra, but no

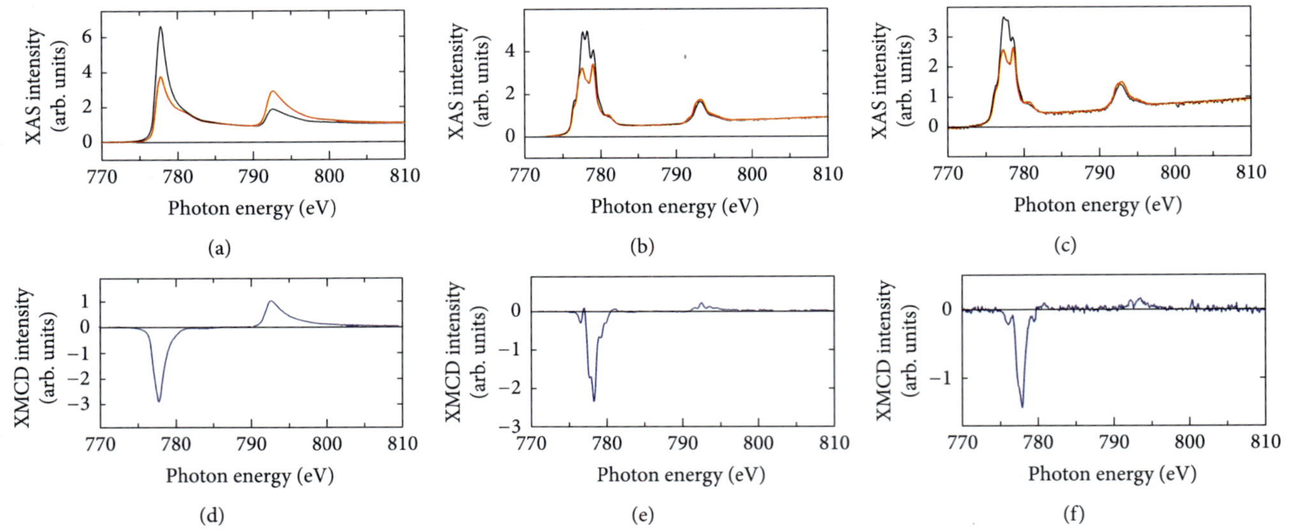

Figure 4: TEY XAS spectra taken with ρ_+ (black line) and ρ_- (red line) X-rays for (a) Co metal, (b) as-deposited Co-doped ITO, and (c) annealed Co-doped ITO at 2 K in an applied field of 4 T. Corresponding XMCD spectra for (d) Co metal, (e) as-deposited Co-doped ITO, and (f) annealed Co-doped ITO.

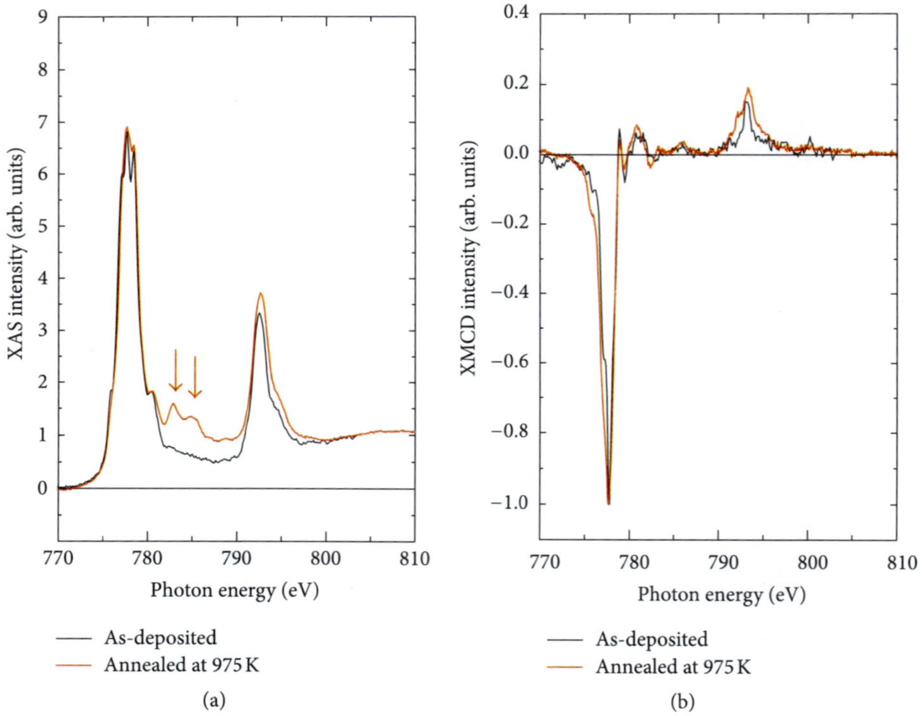

Figure 5: (a) Isotropic FY XAS spectra for the as-deposited (black line) and annealed (red line) Co-doped ITO film. Results recorded at 2 K in an applied magnetic field of 4 T. The arrows indicate spectral features that arise after annealing. (b) XMCD spectra for the as-deposited (black line) and annealed (red line) Co-doped ITO film. The new spectral features for the annealed sample are indicated with red arrows. The spectra have been normalized at L_3 edge.

XMCD was detected in TEY. A very weak dichroic effect was, however, observed in the FY XMCD spectra and is indicative of a remanent ferromagnetic ordering (see Figure 6). In the as-deposited film, no remanent XMCD in TEY or FY mode was detected [13] confirming the paramagnetic nature of the Co^{2+} ions.

Quantitative information regarding the orbital (μ_L) and spin (μ_S) components to the total magnetic moment per Co ion (μ_{TOT}) was calculated by applying sum rules [21, 23], where we have explicitly included the magnetic dipole term (μ_T) for the as-deposited and annealed films. For the Co foil the number of holes used for our calculations was 2.49 ([21]),

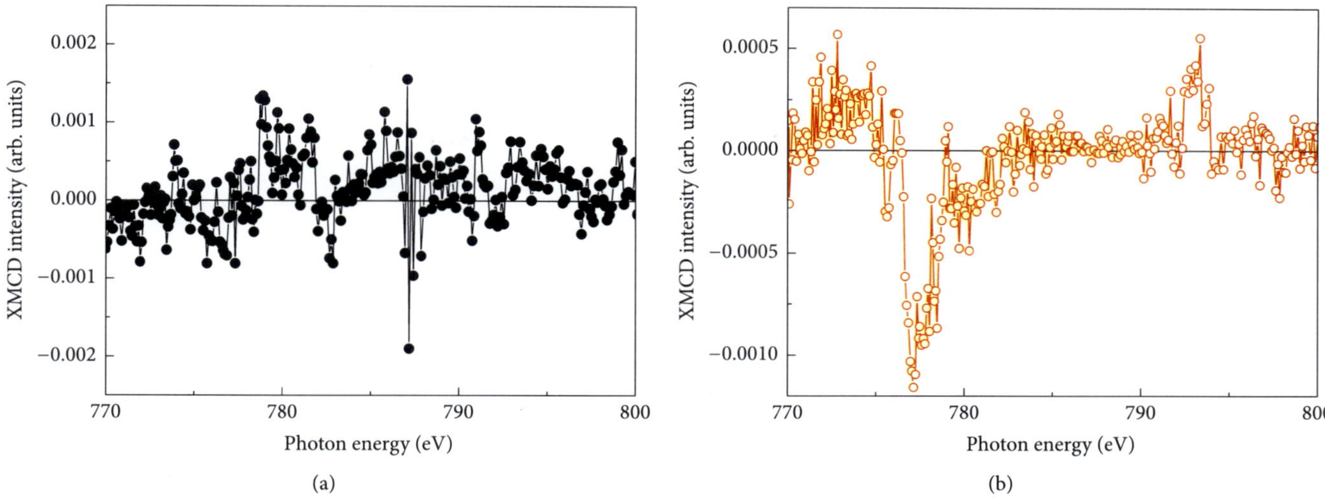

FIGURE 6: (a) TEY and (b) FY XMCD spectra at 2 K held at remanence for a Co-doped ITO film annealed at 975 K.

TABLE 1: Spin (μ_S) and orbital (μ_L) components of the total magnetic moment per Co ion (μ_{TOT}) at 4 K in a magnetic field of 4 T calculated using the XMCD sum rules.

Sample	$\mu_S(\mu_B/\text{Co})$	$\mu_L(\mu_B/\text{Co})$	$\mu_{TOT}(\mu_B/\text{Co})$
Co foil	1.55 ± 0.11	0.14 ± 0.04	1.69 ± 0.12
As-deposited	2.17 ± 0.09	0.56 ± 0.03	2.73 ± 0.10
Annealed	2.15 ± 0.12	0.41 ± 0.04	2.56 ± 0.13

but, in the case of the Co-doped samples, the number of holes was set to 3.00. The results of the analysis are shown in Table 1. The value for the Co foil total moment is in excellent agreement with the theoretical value for Co metal (1.72μ_B) [24]. Within the error it is evident that the as-deposited and annealed films have a similar μ_S, but there is a marked reduction in μ_L for the annealed film. The quenching of μ_L is most likely due to a lowering in the degeneracy of the states arising from a change in the crystal field at the Co sites due to the clustering of the Co dopants. The value of μ_{TOT} is lower than the theoretical value for the isolated Co^{2+} ion (~3.87 μ_B ([25])) with μ_{TOT} reduced further upon annealing and closer to that of pure Co metal. Rode et al. observed similar reductions in the sum rule calculated magnetic moments [24]. They suggested that the origins of the reduction may be due to a lack of magnetic saturation. However, this is unlikely to be the case for our study conducted in a field of 4 T. On the other hand, antiferromagnetic interactions between nearest neighbor Co^{2+} ions could also be responsible for the low value of the moment [24]. The rise in magnetization upon UHV annealing should therefore be largely ascribed to defects introduced during the formation of metallic Co clusters, as evidenced by changes in the XAS and XMCD spectra as well as the blocking temperature in the magnetization versus temperature response. The formation of oxygen vacancies during annealing provides a contribution to the overall ferromagnetic response by significantly enhancing the carrier density. Furthermore, these results demonstrate that the magnetism in the as-deposited Co-doped ITO film cannot be ascribed to cluster formation. We therefore class the ferromagnetic response in as-deposited Co-doped ITO as *intrinsic*.

4. Summary and Conclusions

In summary, we have shown that an increase in magnetization upon UHV annealing of Co-doped ITO at 975 K is due to the formation of ferromagnetic metallic Co clusters which have a lower magnetic moment per Co site than the Co impurities in the as-deposited film. Multiplet features in the Co $L_{2,3}$ edge XAS spectra measured in both TEY and FY from the as-deposited film indicate that the Co impurities are divalent. This is also the case for the annealed sample at the near surface region, but bulk sensitive FY TEY spectra reveal the presence of substantial Co metal clustering. The results demonstrate that clustering is not the origin of the ferromagnetism observed in the as-deposited Co-doped ITO film. This study should therefore inspire further work to understand the fundamental role of intrinsic lattice defects in dilute magnetic oxides.

Conflicts of Interest

The authors declare that they have no conflicts of interest regarding the publication of this paper.

Acknowledgments

This work was funded by the Engineering and Physical Sciences Research Council. The authors would like to thank Richard Mott for his support during the experiments at Diamond Light Source and the Science and Technology Facilities Council for access to the SQUID instrument.

References

[1] H. Ohno, "Making nonmagnetic semiconductors ferromagnetic," *Science*, vol. 281, no. 5379, pp. 951–956, 1998.

[2] S. A. Wolf, D. D. Awschalom, R. A. Buhrman et al., "Spintronics: a spin-based electronics vision for the future," *Science*, vol. 294, no. 5546, pp. 1488–1495, 2001.

[3] S. X. Zhang, W. Yu, S. B. Ogale et al., "Magnetism and anomalous Hall effect in Co- (La,Sr) TiO_3," *Physical Review B—Condensed Matter and Materials Physics*, vol. 76, no. 8, Article ID 085323, 2007.

[4] Y. J. Lee, M. P. de Jong, and R. Jansen, "Magnetism and heterogeneity of Co in anatase Co:TiO_2 magnetic semiconductor," *Applied Physics Letters*, vol. 96, no. 8, Article ID 082506, 2010.

[5] M. A. Garcia, E. Fernandez Pinel, J. De La Venta et al., "Sources of experimental errors in the observation of nanoscale magnetism," *Journal of Applied Physics*, vol. 105, no. 1, Article ID 013925, 2009.

[6] M. Venkatesan, C. B. Fitzgerald, and J. M. D. Coey, "Unexpected magnetism in a dielectric oxide," *Nature*, vol. 430, no. 7000, p. 630, 2004.

[7] N. H. Hong, J. Sakai, N. Poirot, and V. Brizé, "Room-temperature ferromagnetism observed in undoped semiconducting and insulating oxide thin films," *Physical Review B*, vol. 73, no. 13, Article ID 132404, 2006.

[8] J. Stankiewicz, F. Villuendas, and J. Bartolomé, "Magnetic behavior of sputtered Co-doped indium-tin oxide films," *Physical Review B*, vol. 75, no. 23, Article ID 235308, 2007.

[9] H. S. Kim, S. H. Ji, H. Kim et al., "Observation of ferromagnetism and anomalous Hall effect in laser-deposited chromium-doped indium tin oxide films," *Solid State Communications*, vol. 137, no. 1-2, pp. 41–43, 2006.

[10] J. Philip, N. Theodoropoulou, G. Berera, J. S. Moodera, and B. Satpati, "High-temperature ferromagnetism in manganese-doped indium–tin oxide films," *Applied Physics Letters*, vol. 85, no. 5, pp. 777–779, 2004.

[11] A. M. H. R. Hakimi, M. G. Blamire, S. M. Heald et al., "Donor-band ferromagnetism in cobalt-doped indium oxide," *Physical Review B*, vol. 84, no. 8, Article ID 085201, 2011.

[12] A. M. H. R. Hakimi, N. Banerjee, A. Aziz, J. W. A. Robinson, and M. G. Blamire, "Estimating the spin diffusion length of semiconducting Indium Tin Oxide using Co/Indium Tin Oxide/Co spin valve junctions," *Applied Physics Letters*, vol. 96, no. 10, Article ID 102514, 2010.

[13] A. M. H. R. Hakimi, F. Schoofs, R. Bali et al., "Origin of magnetism in cobalt-doped indium tin oxide thin films," *Physical Review B*, vol. 82, no. 14, Article ID 144429, 2010.

[14] J. M. D. Coey, P. Stamenov, R. D. Gunning, M. Venkatesan, and K. Paul, "Ferromagnetism in defect-ridden oxides and related materials," *New Journal of Physics*, vol. 12, Article ID 053025, 2010.

[15] R. P. Panguluri, P. Kharel, C. Sudakar et al., "Ferromagnetism and spin-polarized charge carriers in In_2O_3 thin films," *Physical Review B*, vol. 79, no. 16, Article ID 165208, 2009.

[16] V. N. Antonov, L. V. Bekenov, L. P. Germash, and N. A. Plotnikov, "X-ray magnetic dichroism in the cobalt-doped indium tin oxide from first principle calculations," *Journal of Solid State Physics*, vol. 2013, Article ID 213594, 10 pages, 2013.

[17] H. Baqiah, N. B. Ibrahim, S. A. Halim, M. H. Flaifel, and M. H. Abdi, "The role of cobalt doping on magnetic and optical properties of indium oxide nanostructured thin film prepared by sol-gel method," *Materials Research Bulletin*, vol. 63, pp. 147–154, 2015.

[18] F. S. Galasso, *Structure and Properties of Inorganic Solids*, Pergamon Press, Oxford, UK, 1970.

[19] A. M. H. R. Hakimi, *Magnetism and spin transport studies on indium tin oxide [Ph.D. thesis]*, University of Cambridge, Cambridge, UK, 2011.

[20] J. M. D. Coey, M. Venkatesan, and C. B. Fitzgerald, "Donor impurity band exchange in dilute ferromagnetic oxides," *Nature Materials*, vol. 4, no. 2, pp. 173–179, 2005.

[21] C. T. Chen, Y. U. Idzerda, H.-J. Lin et al., "Experimental confirmation of the X-ray magnetic circular dichroism sum rules for iron and cobalt," *Physical Review Letters*, vol. 75, no. 1, pp. 152–155, 1995.

[22] D. H. Kim, J. S. Yang, Y. S. Kim et al., "Effects of high-temperature postannealing on magnetic properties of Co-doped anatase TiO_2 thin films," *Physical Review B*, vol. 71, no. 1, Article ID 014440, 2005.

[23] B. T. Thole, P. Carra, F. Sette, and G. van der Laan, "X-ray circular dichroism as a probe of orbital magnetization," *Physical Review Letters*, vol. 68, no. 12, pp. 1943–1946, 1992.

[24] K. Rode, R. Mattana, A. Anane et al., "Magnetism of (Zn,Co)O thin films probed by x-ray absorption spectroscopies," *Applied Physics Letters*, vol. 92, no. 1, Article ID 012509, 2008.

[25] S. Blundell, *Magnetism in Condensed Matter*, Oxford University Press, 2001.

9

Low Temperature Conductivity in *n*-Type Noncompensated Silicon below Insulator-Metal Transition

A. L. Danilyuk,[1] A. G. Trafimenko,[1] A. K. Fedotov,[2] I. A. Svito,[2] and S. L. Prischepa[1,3]

[1]*Belarusian State University of Informatics and Radioelectronics, P. Browka 6, 220013 Minsk, Belarus*
[2]*Belarusian State University, Nezalezhnastsi Av. 4, 220030 Minsk, Belarus*
[3]*National Research Nuclear University (MEPHI), Kashirskoe Highway 31, Moscow 115409, Russia*

Correspondence should be addressed to S. L. Prischepa; prischepa@bsuir.by

Academic Editor: Da-Ren Hang

We investigate the transport properties of *n*-type noncompensated silicon below the insulator-metal transition by measuring the electrical and magnetoresistances as a function of temperature T for the interval 2–300 K. Experimental data are analyzed taking into account possible simple activation and hopping mechanisms of the conductivity in the presence of two impurity bands, the upper and lower Hubbard bands (UHB and LHB, resp.). We demonstrate that the charge transport develops with decreasing temperature from the band edge activation (110–300 K) to the simple activation with much less energy associated with the activation motion in the UHB (28–90 K). Then, the Mott-type variable range hopping (VRH) with spin dependent hops occurs (5–20 K). Finally, the VRH in the presence of the hard gap (HG) between LHB and UHB (2–4 K) takes place. We propose the empiric expression for the low T density of states which involves both the UHB and LHB and takes into account the crossover from the HG regime to the Mott-type VRH with increasing temperature. This allows us to fit the low T experimental data with high accuracy.

1. Introduction

Interest in studies of conductivity mechanisms in semiconductor materials, including traditional doped semiconductors near the metal-insulator transition (MIT), does not stop currently [1, 2]. This is due to both the fundamental problems of electron transport in the vicinity of such transition and applied aspects related to the development of highly sensitive sensors of magnetic and electric fields. In particular, it is not yet fully understood mechanisms of low temperature electrical conductivity in doped semiconductors involving multiply charged localized states, mechanisms of positive and negative magnetoresistance (MR), as well as mechanisms of localization and peculiarities of the energy band structures of impurity and localized states. Due to the above the detailed investigation of the conductivity of doped semiconductors near the MIT in a wide temperature range, influence on it the magnetic field is still relevant. Actually, the MIT in three dimensional (3D) system occurs when the Mott criterion $N_c^{1/3}a_B \approx 0.25$ is satisfied, where N_c is the critical concentration of the localized states for the MIT and a_B is the effective Bohr radius of an insulated defect center [3]. The validity of this criterion was confirmed in various experiments [4–6]. However, some ambiguity in correct understanding of the temperature dependence of the conductivity and MR near the critical concentration of the localized sites still does present. The main reason for this is related to the competition between various types of hopping conductivity, mechanisms of weak localization, percolation and metallic and/or impurity bands (Hubbard bands) conductivity. Therefore, a series of crossovers could be observed between different types of conductivity in a wide temperature range. This inevitably leads to the need for their very careful consideration during the processing of the experimental data. In particular, with decreasing temperature, the crossover from band edge activation (ϵ_1) to hopping (ϵ_3) conductivity is observed in doped semiconductors [6].

Hopping is one of the most likely mechanisms that determines the overall conductivity on the insulating side of the MIT. Among various types of hops, the variable range hopping (VRH) is one of the most relevant. The VRH mechanism, in turn, could be classified as the Mott mechanism, for which the density of states (DOS) on the Fermi level is constant, $g(E_F) = \text{const}$ [3] and the Efros-Shklovskii (ES) mechanism, which implies the soft Coulomb gap (CG), $g(E) \sim |E - E_F|^2$ [7, 8]. From a practical point of view, at a constant T, the Mott-type VRH occurs when the doping concentration N_d is far less than the N_c value, $N_d < 0.5N_c$, and one can neglect the Coulomb interaction [9]. The ES mechanism, in turn, dominates when $N_d > 0.6N_c$ and the CG has to be taken into account [10, 11]. From the literature a few models of crossovers between the Mott and ES hopping are known quite [12–14].

The temperature dependence of the resistivity for both mechanisms is described by the well-known expression

$$\rho(T) = \rho_0 \exp\left(\frac{T_h}{T}\right)^p, \quad (1)$$

where ρ_0 is the preexponential factor and T_h and p are the characteristic temperature and exponent, respectively. Last two quantities depend on the mechanism of hopping. In particular, for 3D systems and for the Mott mechanism, the exponent $p = 1/4$, while for the ES VRH $p = 1/2$ [3, 6]. The exponent p can be obtained from the experiment.

Actually, the concept of hopping between disorder-induced localized electronic states near the MIT is a universal feature of disordered Mott systems. In particular, it was successfully applied to describe transport in amorphous/nanocrystalline silicon hybrids [15], semiconductor nanocrystals [16], ruthenate [17], carbon nanotube fibers [18], as well as transport of single (few) donors in a silicon nanoscale transistor [19, 20]. Known models of crossover between different types of VRH are characterized by two different approaches to the calculation of the exponent p from (1). The first approach is based on the analysis of the percolation problem [12, 13, 21], while the second deals with the optimization of the exponent p using the interpolation expression for the DOS [14, 22, 23].

In general, the percolation approach can be applied when the spatial correlation length for the random potential is much larger than the phase coherence length [24, 25]. This is a very powerful tool for the description of charge transport in disordered systems with localized states, which occurs due to the electron hops from one site to another [26, 27]. It should be noted that the percolation theory was applied to explain the properties not only of doped semiconductors, but also granular metals [28], manganese oxides [29], quantum Hall plateau transistors [30], high-T_c cuprates [31], and so forth. Moreover, the percolation phase was directly observed in vanadium dioxide close to the Mott transition by means of nanoscale X-ray imaging [32].

The second approach is less stringent and leads to a noticeable overestimation of the width of the crossover region. General description of the crossover from the Mott to ES VRH on the basis of rather complex multivariable integral equation, which cannot be solved analytically and needs rather difficult numerical analysis, was proposed in [13]. In [14, 22, 23] the procedure based on the optimization of the exponent in the expression for the hopping probability and using the interpolation expression for the DOS, $g(E) \sim |E - E_F|^n$ (here n is integer which is equal to 0 for the Mott mechanism and to 2 for the ES VRH), has been proposed. This approach leads to quite simple analytical expressions but, as we mentioned above, is less stringent than the percolation task.

In addition to the crossover from the Mott to ES VRH, there is another, less studied low T crossover, between VRH and a simple activation dependence (SAD) [33]. For the SAD the temperature behavior of resistivity is described by (1) but with the exponent $p = 1$. Actually, the SAD may occur for different reasons. At high T it could indicate the nearest neighbor hopping (NNH) [23, 34–36], or a band conductivity [6, 37]. At low T, however, the probability of both the NNH and band conductivity becomes negligibly small. Thus, the SAD at low T is usually associated with a hard gap (HG) in the DOS [38], that is, in a certain energy range $g|E| \approx 0$. One of the reason of the HG in doped semiconductors could be the Coulomb interaction [39, 40]. The manifestation of these mechanisms depends not only on the concentration of the main impurities N_d, but also on the degree of compensation of the semiconductor, K.

Traditional Mott, ES, and NNH mechanisms are based on hops to the empty impurity centers. Therefore, the availability of sufficient number of empty donor cites is important for such kind of hops in n-type semiconductor. At low temperatures, this can only be achieved by compensation of semiconductor. Arisen due to compensation charged donors and acceptors create dispersion of the donor energy levels due to their chaotic potential. This dispersion exceeds significantly the exponentially small splitting of levels of neighbor donors caused by the overlap of the wave functions. The characteristic feature of noncompensated semiconductor ($K \ll 1$) is the rapid decrease of the empty (ionized) donors with decreasing temperature. In weakly compensated semiconductors, in a limited range of concentrations (close to the MIT), in addition to the band and hopping conductivity another activation mechanism develops in the temperature dependence of conductivity, ϵ_2 conductivity. This mechanism exhibits in the intermediate between band and hopping conductivity temperature interval [41–43]. It is believed that the ϵ_2 conductivity involves migration of electrons on the single occupied neutral donors (D^- states). They have large radius and, consequently, at the intermediate impurity concentrations are strongly overlapped. The result is a wide band of D^- states. This band is an analogue of the upper Hubbard band (UHB), formed in a disordered systems [6]. The decrease in donor impurity concentration leads to a strong narrowing of a D^- band. On the other hand, in the absence of compensation, when ϵ_3 conductivity is zero, there are most favorable conditions for ϵ_2 conductivity; that is, the concentration of neutral donors is the highest. The increase of the compensation improves significantly conditions for the ϵ_3 conductivity and worsens the ϵ_2 conductivity.

Electron hopping (ϵ_3 conductivity) may occur not only on free localized but also on occupied sites via the spin dependent transport [44]. In fact, for doped semiconductors, the spin degree of freedom could play a significant role in the electron hops. If, for example, the width of the distribution function of the energies of the localized sites overcomes the

Coulomb repulsion between electrons, double occupancy of the site becomes possible, as was first argued by Kurobe and Kamimura [44]. In this case, two types of hops contribute to the transport. The first occurs between singly occupied (SO) and unoccupied (UO) sites, while the second type of hops involves doubly occupied sites (DO) with opposite electron spins. As a result, the spin dependent charge transport could occur in this case. The UO site is considered as a singly ionized site with the elementary charge $+1e$. Consequently, SO site has charge $0e$, and DO site has charge $-1e$. Developing of a quantitative theory of the ϵ_2 conductivity faced with very great difficulties. The main difficulty is that in this concentration range the overlap of the wave functions play a role comparable to the Coulomb interaction of electrons with impurities and with each other. In addition, disorder in the distribution of impurities significantly complicates the analysis of experimental data. Therefore, many problems related to the interpretation of the experimental data for doped semiconductors, in which ϵ_2 conductivity is observed, are still unclear. In this regard, for doped semiconductors with a low degree of compensation, in which ϵ_2 conductivity is manifested, the study of the mechanisms of low temperature conductivity and crossover between ϵ_2 and hopping conductivity is a topical problem.

Therefore, the correct interpretation of the experimental $\rho(T)$ data measured in a wide T range can be significantly hampered, because, in order to compare the experimental values of the parameters with the calculated ones, it is important to identify and properly describe the transport mechanisms in different temperature intervals.

In this work we performed a thorough investigation of the low temperature ($2\,\mathrm{K} < T < 90\,\mathrm{K}$) conductivity of n-type noncompensated silicon close to the MIT taking into account possible VRH, SAD, and spin dependent mechanisms. We show that above 30 K the conductivity is of the activation type and is carried out in the UHB, while in the T range 5–25 K the Mott-type VRH dominates and is accompanied by the spin dependent charge transport. Below 5 K the crossover from the VRH to SAD occurs. We demonstrate that the reason for the SAD in this T range is the HG in the DOS. We developed a model of this crossover based on the percolation task, using the expression for DOS, which takes into account the evolution of the DOS from the HG at $T < 5\,\mathrm{K}$ to the Mott-type at $T > 5\,\mathrm{K}$. This approach allowed us to fit the experimental $\rho(T)$ dependence with high accuracy.

2. Samples and Experimental Details

Single crystalline n-type Si (100) grown by Czochralski method and doped with Sb was used in this work. Samples of rectangular shape with a width of 1 mm and a thickness of 0.5 mm were covered with 6 indium contacts as electric probes using ultrasonic soldering. Two contacts were for current supply, 2 were for Hall measurements, and, finally, last 2 contacts were for voltage measurements. All contacts were ohmic in the whole studied temperature range which was proved by the linear current-voltage characteristics. A dc current of 10 μA generated by multimeter Keithley 6430 was used to bias the sample during the resistivity measurements and to measure the voltage drop down to 5 μV. For low resistivity samples we used the two-channel nanovoltmeter Keithley 2182A. Samples were inserted into the cryogen free measuring system (Cryogenic Ltd., London) with the superconducting magnet. The system allowed performing measurements in the T range between 2 and 300 K in magnetic fields up to $H = 80\,\mathrm{kOe}$. Lakeshore controller allowed 0.1 K/min sweeping rate of temperature during the ρ versus T measurements and stabilizing temperature with the accuracy of ±0.005 K during the sweep of the magnetic field or current-voltage characteristics acquisition. Semiconducting GaAs thermometer was calibrated with an accuracy better than 0.1%. On the basis of measurements of the T dependence of the Hall effect the Sb concentration was estimated as $N_d = 1 \times 10^{18}\,\mathrm{cm}^{-3}$, which is 3 times less than the critical Mott concentration for Sb in Si [45, 46]. Therefore, our samples are on the insulating side of the MIT. Actually, a series of samples cut from different Si wafers belonging to the same technological set of fabrication were measured. All the data are in nice agreement with each other. Here we present typical results obtained on the investigated samples.

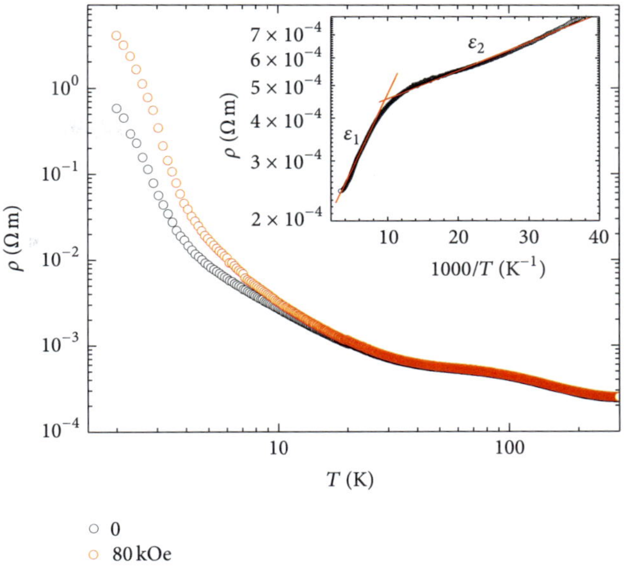

FIGURE 1: Temperature dependence of the resistivity at $H = 0$ (black circles) and $H = 80\,\mathrm{kOe}$ (red circles) in the range 2–300 K. Inset: resistivity versus T^{-1} (symbols). Experimental data are explained by two activation energies, $\epsilon_1 \approx 10.35\,\mathrm{meV}$ for 110–300 K and $\epsilon_2 \approx 1.73\,\mathrm{meV}$ for 28–90 K.

3. Results

In Figure 1 we show the resistivity versus temperature measured in the range 2–300 K at zero magnetic field and at the maximum applied field of 80 kOe. The strong change in ρ (4 orders of magnitude) is evident. The $\rho(T)$ dependence at $H = 0$ is linear in the Arrhenius coordinates in the temperature range 28–300 K, as is clearly seen from the inset to Figure 1. Actually, in this temperature interval there are two

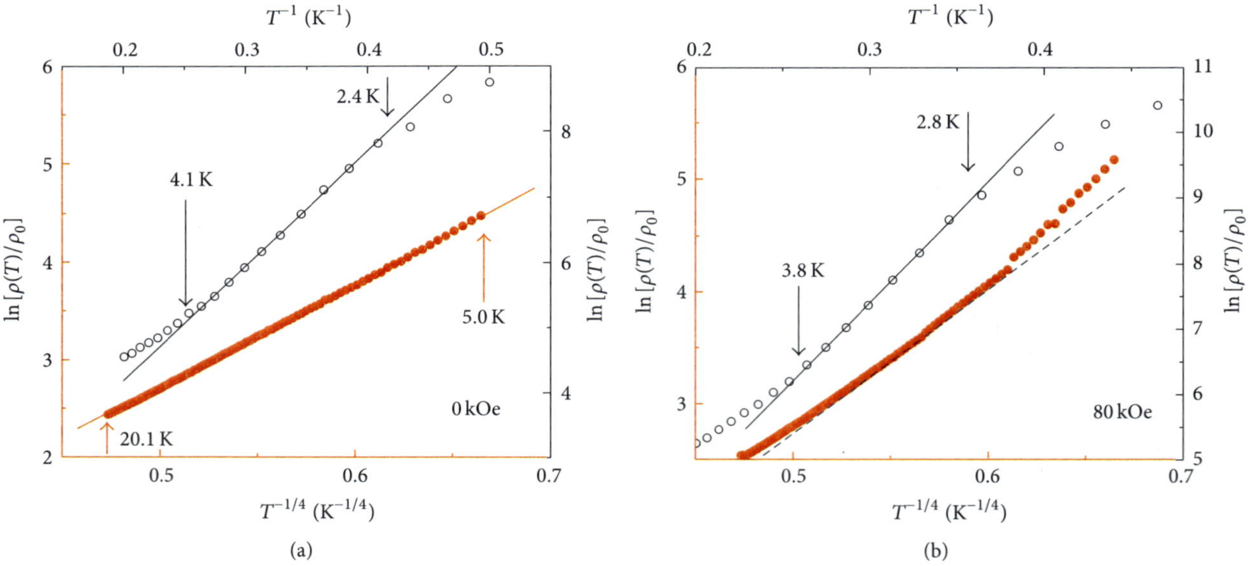

FIGURE 2: (a) $\ln\rho$ versus $T^{-1/4}$ (red symbols), left and bottom red axis. $\ln\rho$ versus T^{-1} (blue symbols), right and upper blue axis. Solid lines are the result of the best linear fit in these coordinates. Data are for $H = 0$. (b) $\ln\rho$ versus $T^{-1/4}$ (red symbols), left and bottom red axis. $\ln\rho$ versus T^{-1} (blue symbols), right and upper blue axis. Solid blue line is the result of the best linear fit in these coordinates. Dashed black line is tangent to the experimental data. Data are for $H = 80$ kOe.

significantly different activation energies: $\epsilon_1 \approx 10.35$ meV for the T range 110–300 K and $\epsilon_2 \approx 1.73$ meV for 28–90 K. Resistivity in the first T range is increased by two times, while in the second interval it is increased by only 1.2 times.

Below 20 K at $H = 0$ the SAD is changed to the Mott-type VRH. In Figure 2(a) we show the $\ln\rho(T^{-1/4})$ data for the temperature interval 5–20 K (red color). The obtained in these coordinates linear dependence unambiguously proves the Mott-type VRH. At $T < 5$ K the Mott-type VRH is no longer valid and crossover to the SAD with the exponent $p = 1$ and activation energy $\epsilon_4 \approx 1.48$ meV occurs (blue color).

Data at $H = 80$ kOe below 20 K reveal similar features as for zero field. However, only on the basis of the exponent p in (1) we cannot successfully fit experimental data in the T range 5–20 K. As it follows from Figure 2(b), red symbols, exponent $p = 1/4$ is not suitable to describe the experimental dependence. Other reasonable exponents in (1) also do not give satisfactory agreement with the experiment. Nevertheless, at lower T we can identify interval, where the exponent $p = 1$ is valid, see blue color in Figure 2(b). It turns out that low T SAD still exists even at high magnetic field.

For a more accurate identification of the observed low T SAD in Figure 3 by symbols we plot $\ln\rho$ versus T^{-p} dependence at zero field for two different p values, 1/2 (blue color, corresponds to SE VRH) and 1 (red color, corresponds to SAD). Solid lines in Figure 3 correspond to the result of the best linear fit. It follows from Figure 3 that exponent $p = 1/2$ gives worse agreement with the experiment. The same analysis for data measured in the presence of the magnetic field leads to similar result, namely, presence of SAD below $T = 4$ K.

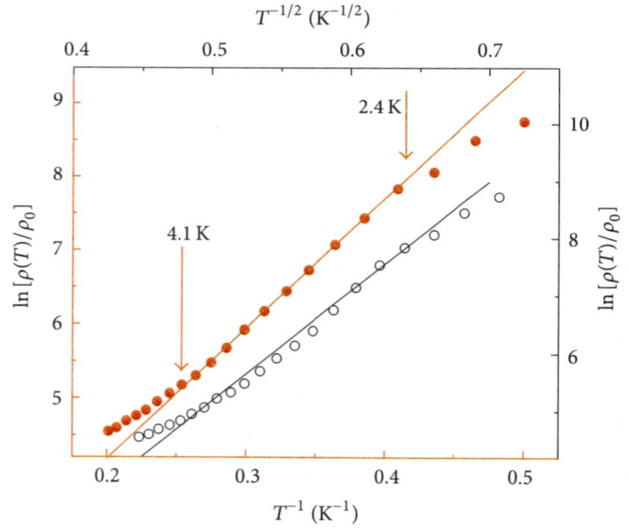

FIGURE 3: $\ln\rho$ versus T^{-1} (red symbols), left and bottom red axis. $\ln\rho$ versus $T^{-1/2}$ (blue symbols), right and upper blue axis. Solid lines are the result of the best linear fit in these coordinates. Data are for $H = 0$.

4. Discussion

4.1. Qualitative Analysis. The obtained experimental data revealed the following. In the T interval 110–300 K the conventional band edge activation caused by the transition from the Fermi level located in the tail of the lower Hubbard band (LHB) to the conduction band of silicon (the ϵ_1 conductivity [6]) takes place.

Another simple activation mechanism is manifested in the temperature dependence of the resistivity in the interval 28–90 K (ϵ_2 conductivity). It is believed that the ϵ_2 conductivity is associated with the activation motion of electrons along singly occupied neutral donors and corresponds to the states in the upper Hubbard band (UHB) [5, 6, 20, 37]. In this temperature interval the characteristic value of the thermal energy k_BT is greater than the activation energy ϵ_2 and all states are delocalized. Moreover, only slight increase in the resistivity as the temperature decreases or magnetic field increases indicates the extended UHB.

In the T range 5–20 K the Mott-type hopping ϵ_3 conductivity is observed in the absence of the magnetic field. Finally, in the temperature interval 2–4 K again the SAD with the activation energy $\epsilon_4 \approx 1.48$ meV is realized. At first glance, this is a reentrance of the previously observed within the temperature interval 28–90 K SAD with the activation energy $\epsilon_2 \approx 1.73$ meV. On the other hand, for the range 2–4 K the activation energy ϵ_4 is greater than the energy of thermal fluctuations, $\epsilon_4 \approx (3–7)k_BT$. It means that all states are localized; that is, the charge transport is hopping-like, but the mechanism of hops is different from usual VRH.

The reentrance of the SAD after VRH is well known for some dilute magnetic [38, 47, 48] and classical doped semiconductors, like Si:B [33] and Si:As [49] with doping concentration less than N_c, and for amorphous compounds like In/InO_x [50] or amorphous silicon after implantation of Au or Si ions [51]. For all these materials evidences of ϵ_2 conductivity before VRH were never reported. Typically, as T decreases first one observes the NNH, which is characterized by the exponent $p = 1$, then the VRH followed by SAD, again with the exponent $p = 1$. Moreover, for dilute magnetic semiconductors and for Si:B strong magnetic field removes the reentrance effect. But for In/InO_x the reentrance effect is insensitive to the magnetic field [50]. It is worth mentioning that for the electronic Si:As semiconductor the influence of the magnetic field on the reentrance effect was not studied. In our system, the reentrance effect in the presence of the magnetic field still does exist, see Figure 2(b). Moreover, the peculiar feature of the appearance of the SAD reentrance in our case is that not only the VRH, but also the ϵ_2 conductivity (not NNH) preceded it.

Thus, the performed brief qualitative analysis of the data disclosed that, in the range 28–90 K, the influence of both the temperature and magnetic field on the resistivity is very low. This indicates that the electron transport in this temperature range has apparently a band character, is realized in the relatively wide UHB and states are delocalized. In the range 5–20 K the electron transport is transformed to the hopping ϵ_3 conductivity. With further T decrease, below 4 K, the Mott-type VRH is transformed into more complicated hopping mechanism which formally corresponds to the SAD, but now states are localized.

Actually, the SAD at low T indicates the formation of the HG in the DOS [38]. The energy ϵ_4 characterizes the width of the HG, where the DOS is equal or close to zero. The magnetic field does not suppress the SAD. The reason of the HG appearance is hidden. In order to understand better the nature of the reentrance effect and the origin of the HG, we need to perform the quantitative analysis of the data.

TABLE 1: Parameters of the Mott-type VRH obtained within two different approaches.

Parameter	$\rho(T)$ data	MR data
a_M, nm	6.6	6.0–8.5
g_M, $eV^{-1} \times cm^{-3}$	5.408×10^{19}	4.7×10^{19}

4.2. Quantitative Analysis

4.2.1. Spin Dependent Hopping. For the T range 5–20 K the resistivity is described according to the Mott law, (1) [3, 6], in which $T_h = T_M$ and $p = 1/4$. Here $T_M = \beta_M[g_M a_M^3 k_B]^{-1}$ is the Mott parameter, $\beta_M = 21.2 \pm 1.2$ is the numerical factor, $g_M = \text{const}$ is DOS at the Fermi level, a_M is the localization length of the states around the Fermi level. We calculated the g_M and a_M values analyzing the experimental data $\rho(T)$ at zero magnetic field in the Mott coordinates ($\rho - T^{-1/4}$) and applying the evaluation expressions [3, 6, 52, 53]

$$g_M = \frac{N_d}{2k_B\left(T_M T_{vM}^3\right)^{1/4}}, \qquad a_M = \frac{\beta_M^{1/3}}{\left[k_B T_M g_M\right]^{1/3}}, \tag{2}$$

where T_{vM} is the temperature of the onset of the Mott-type VRH. Equation (2) gives approximate values of g_M and a_M. Note that on the basis of (1) and (2) the obtained g_M and a_M values must be temperature independent. On the other hand, the MR data allow checking whether these quantities are really T independent. Indeed, within the standard approach of the Mott-type VRH the positive MR which is usually related to the contraction of the wave function of the localized states in magnetic field is expressed as [6]

$$\ln\frac{\rho(H)}{\rho(0)} = t_3\frac{a_M^4}{l_H^4}\left(\frac{T_M}{T}\right)^{3/4} \sim H^2, \tag{3}$$

where numerical factor $t_3 = 5/2016$, $l_H = \sqrt{\hbar/e\mu_0 H}$ is the magnetic length, and $\rho(0)$ is the zero field resistivity.

On the basis of (3), knowing T_M, H, and MR, we calculated the a_M values at different T. Moreover, knowing already a_M, the value of g_M from results of the MR measurements can be calculated using the expression for the Mott parameter T_M.

The results of calculations of hopping parameters according to the $\rho(T)$ dependence and MR data at different T are summarized in Table 1. The Mott parameter was evaluated from (1) as $T_M = 1.524 \times 10^4$ K. The temperature T_{vM} was obtained from the $\rho(T)$ data as $T_{vM} \approx 20$ K [54].

It follows from Table 1 that the obtained within two different approaches a_M and g_M values are in good agreement with each other.

The surprising result, which we obtained, analyzing the a_M values from the MR measurements, is their T dependence.

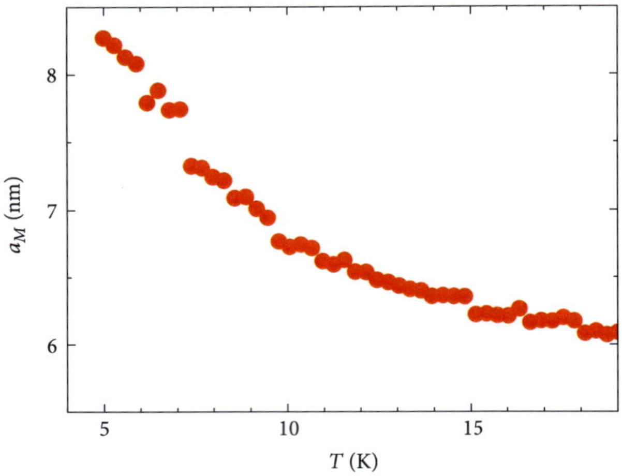

FIGURE 4: Temperature dependence of the localization length a_M. Data are for $H = 80$ kOe.

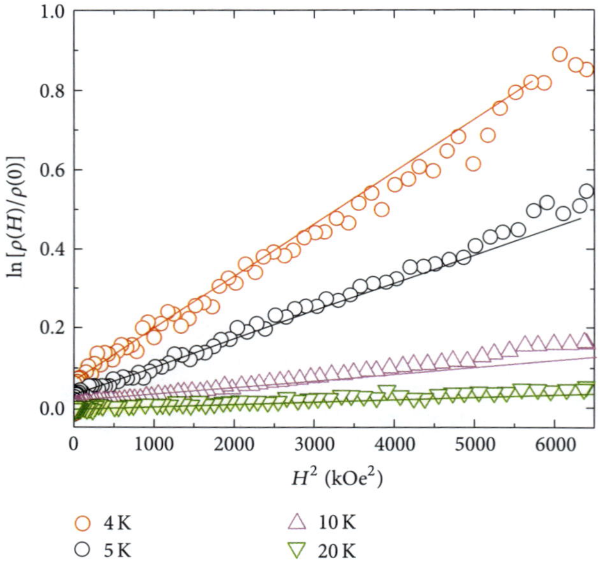

FIGURE 5: $\ln(\rho(H)/\rho(0))$ versus H^2 at different temperatures (symbols). Lines correspond to the results of the best linear fit in these coordinates.

This result for $H = 80$ kOe is presented in Figure 4, from which it follows that the $a_M(T)$ function is not constant. In particular, for the range 5–20 K the a_M values decrease from 8.3 to 6.0 nm. This means that the standard model of positive MR, which takes into account only the contraction of the electronic wave function in magnetic field and leads to the temperature independent localization length, cannot explain the obtained $a_M(T)$ result.

In Figure 5 we show the dependencies of $\ln(\rho(H)/\rho(0))$ versus H^2 in the temperature range 4–20 K together with the results of the best fit procedure according to (3). It is seen that the $\ln(\rho(H)/\rho(0)) \sim H^2$ law in the considered temperature range is valid only at $H < (50\text{–}60)$ kOe. From the results presented in Figures 2(b), 4 and 5 it follows that the transport mechanism at low T becomes more complicated

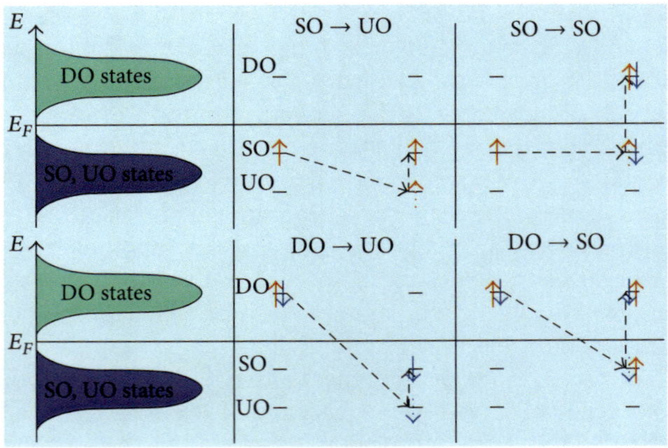

FIGURE 6: Possible electron hops considered for low temperature conductivity. For more details, see the text.

than conventional Mott-type VRH. To clarify the hopping mechanism, we examine other possible types of hops, to the occupied localized sites, which are accompanied by a spin flip [44].

The MR for a system of sites occupied by more than one electron was considered in detail some time ago by Kurobe and Kamimura [44], Matveev et al. [55], Meir [56], and Demishev and Pronin [57]. In particular, if the Coulomb repulsion between two electrons is smaller than the width of the Hubbard bands, 2 types of hops, on unoccupied (SO → UO, DO → UO) and occupied (SO → SO, DO → SO) sites are relevant. All these possibilities are shown in Figure 6, where we schematically draw the energy diagrams for each type of hops.

In general case, spin independent hops to unoccupied sites could be characterized by the localization length a_1 and DOS g_1, while spin dependent hops to occupied sites could be described, in turn, by a_2 and g_2. The quantity g_1 is assigned to the DOS in the LHB, and g_2 corresponds to the DOS in the UHB. The relation between g_1 and g_2 as well as between a_1 and a_2 is still an open question and different assumptions were put forward to simplify the task. Typically, it is assumed that the localization length for the DO state is more than for the SO, $a_2 > a_1$ [44]. In [56] the possibility of spin dependent transport was discussed considering inequality between two densities of states and two localization lengths. From a practical point of view, it seems problematic to separate these quantities in the framework of such general approach. It is necessary to introduce additional assumptions associated with the density of states or the localization length.

In [55] the spin polarized mechanism of MR was investigated in terms of the approach proposed by Kurobe and Kamimura [44]. It was supposed that both the states, UO and SO, are characterized by the same localization length, $a_1 = a_2 = a$, while densities of states differ from each other, $g_1 \neq g_2$.

We failed in the attempts to fit the experimental data of Figure 2(b) applying theoretical approaches of Refs. [55, 56]. We believe this is due to the fact that applied theoretical approaches do not take into account all complexity of the transport mechanisms. Further we investigate this problem in detail within the approach proposed by Demishev et al. [57, 58]. Authors [57], following Kurobe and Kamimura [44], considered only intrasite correlations and neglect the long range Coulomb repulsion. On the basis of [57] it is possible to determine which type of conductivity under the external magnetic field dominates: spin dependent or spin independent. In particular, the modified expression for the MR, which takes into account both types of hops, was deduced [58]:

$$\ln \frac{\rho(H)}{\rho(0)} = \left(\frac{T_M}{T}\right)^{1/4} \frac{1}{4} A_{\mathrm{eff}}(T) \left(\frac{\mu_0 \mu_B H}{k_B T}\right)^2, \quad (4)$$

where

$$A_{\mathrm{eff}}(T) = A + 4b\left(\frac{T}{T_M}\right)^{3/2}. \quad (5)$$

In (5) parameters $A = (g_2 a_2^3 - g_1 a_1^3)/(g_2 a_2^3 + g_1 a_1^3)$ and $b = t_3 a_2^4 (e^2 k_B^2 T_M^2 / \hbar^2 \mu_B^2)$. According to Demishev et al. [58], parameter A depends on the probability of hops, which involve both D^0 and D^- states, and is determined by the degree of polarization of the spin part of the wave function of these states in the magnetic field. Parameter b is determined by the contribution of the mechanism of the wave function contraction to the MR that acts on both types of hops.

Therefore, on the basis of the temperature dependence $A_{\mathrm{eff}}(T)$ it is possible to conclude which type of contribution determines the MR in the hopping region. If the spin dependent mechanism dominates, then $A_{\mathrm{eff}}(T) = \mathrm{const}$. Otherwise, $A_{\mathrm{eff}}(T)$ tends to 0 at $T \to 0$ according to the law $T^{3/2}$. When both types of contributions are comparable, the dependence of $A_{\mathrm{eff}}(T)$ is expected to be more complicated, it does not tend to 0 at $T \to 0$ [57, 58].

In Figure 7 we plot the $A_{\mathrm{eff}}(T)$ dependence obtained from the experiment applying (4). Data are for $H = 80$ kOe. As can be seen, the $A_{\mathrm{eff}}(T)$ dependence has two distinct regions. In one of them, in the T range 5–11 K, the A_{eff} value is temperature independent and the results of the fitting procedure give $A = 0.47 \pm 0.03$, $b = 0$ (blue line in Figure 7). Therefore, in this T range the spin dependent hopping prevails. In addition, the relation between DOS for two different types of hops can be estimated via the quantity $k_g = g_2 a_2^3 / g_1 a_1^3$. In this case $k_g \approx 2.8$.

In the T range 11–20 K the linear approximation $A_{\mathrm{eff}}(T) \sim T^{3/2}$ explains well the experimental data (red line in Figure 7), but A_{eff} does not tend to 0 at $T \to 0$. Consequently, the obtained result proves the absence of the dominant role of the spin independent hops in this temperature interval. From the fitting procedure we obtained coefficients $A = 0.29$ and $b = 0.96 \times 10^4$. That gives $a_2 = 5.5$ nm and $a_1 \approx a_2/4 = 1.38$ nm. The parameter k_g in this case depends on T. The obtained overall $k_g(T)$ dependence is shown in the inset to Figure 7. From k_g values we may estimate the g_1/g_2 relation, which is equal to 21.3 for T range 5–10 K and to 10.7 at $T = 20$ K.

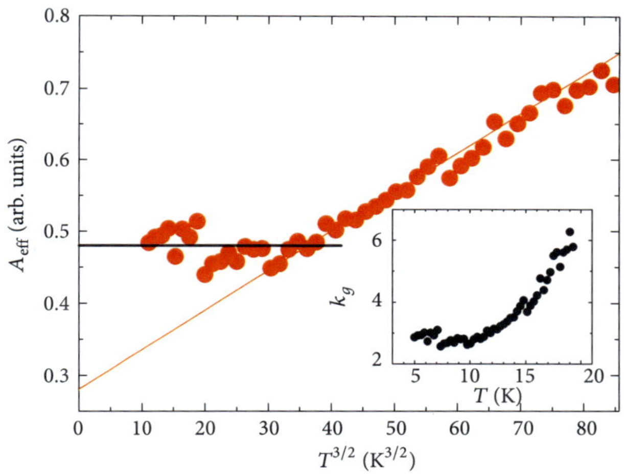

FIGURE 7: Evaluated temperature dependence $A_{\mathrm{eff}}(T)$ at $H = 80$ kOe (symbols). Lines are plotted according to (5). Inset: the temperature dependence of the calculated coefficient k_g.

4.2.2. Gap in the DOS. Thus, the performed estimations and calculations showed that, in the temperature range 5–20 K, on the whole, the Mott-type VRH is realized. Nevertheless, we obtained that two types of hops, traditional hops on UO sites and spin dependent hops on SO sites, have to be taken into account. Therefore, one needs to consider two localization lengths and two DOS. On the other hand, previously introduced g_1 and g_2 values are essentially model parameters that do not give information about the influence of the DOS on the mechanism of hopping conductivity. It should also be noted that neither approach [55] nor work [57] gives any information about the gap between two Hubbard bands.

Generally speaking, the origin of the gap could be twofold. First, it could be due to the CG; second, the HG between lower and upper Hubbard bands could be the reason. The soft CG at the energy minimum does not have a characteristic shelf, $g(E) \sim |E - E_F|^2$, while the true gap has the shelf of the width Δ. Moreover, the presence of the soft CG leads to the exponent $p = 1/2$, which we did not observe.

Below we propose the model expression for the DOS and evaluate the influence of the shape and parameters of the proposed model DOS on the T dependence of the hopping resistivity, which, in turn, is characterized by the critical values of the exponent $\xi(T) = \ln[\rho(T)/\rho_0]$ [21]. The relation between the critical exponent $\xi(T)$ and the exponent p is $\xi(T) \sim (T_h/T)^p$. Estimations are performed within the model of the low temperature crossover VRH-SAD.

To model this crossover let us consider the problem of the charge flow in the lattice of impurity atoms. In the theory of hopping conductivity, problems of charge flow in the system of random nodes, chaotically distributed in space, play the most important role. In this case the average number of nodes in a unit volume is supposed to be given and equal to the concentration of the impurities N_d. The average distance between nodes is $r_d = (3/4\pi N_d)^{1/3}$. For such a lattice, the percolation radius r_c is found from the condition of the connectivity of two and more nodes which form an infinite cluster [6], so that the value of r_c depends only on the nodes concentration N_d. Typically, the dimensionless threshold value B_c is used, having a meaning of the average number of bonds per node. In the 3D case [6]

$$B_c = \left(\frac{4\pi}{3}\right) r_c^3 N_d. \tag{6}$$

For random lattice the B_c value is typically determined numerically and varied in the range 2.65–2.76 [6]. Consequently, the percolation radius is given by $r_c = (0.865 \pm 0.015)N_d^{-1/3}$. In this case r_c is always larger than r_d. In particular, in our case $r_c = 8.64\,\text{nm} > r_d = 6.2\,\text{nm}$. In other words, the symmetry of the lattice is irrelevant and only the number of sites inside radius r_c determines the transport mechanism.

On the basis of the aforementioned estimations, we then developed a model of the VRH-SAD crossover; this model is based on a simplified procedure for solving the percolation problem using the empiric expression for the DOS valid for both types of the temperature dependent hopping. We start from the following expression for the dimensionless concentration of sites:

$$N(\xi) = 2\int_0^{E_{\max}} r_{\max}^3 g(E)\,dE, \tag{7}$$

which satisfies the condition of connectivity for the exponent of the hopping resistivity less than a certain value ξ_{min} [6, 13]. Here, $E_{\max}$ and $r_{\max}$ are maximal values of the energy and intersite spacing, respectively, which still allow the connectivity. The dimensionless concentration $N(\xi)$ in (7) is determined as a product of the concentration of sites with the energy $E < E_{\max}$ and the volume of these sites. This procedure for the temperature dependent DOS is not completely rigorous and the connectivity condition should be considered for each energy value separately, which leads to the complicated integral equation [13]. Here we will follow the procedure described in [21], which simplifies significantly the calculations.

Values of $r_{\max}$ and $E_{\max}$ are specified in the percolation theory by the relations [6] (see also Equation (4.30) in [59])

$$r_{\max} = \frac{a_2\xi}{2}, \qquad E_{\max} = k_B T\xi. \tag{8}$$

Substituting (8) into (7) and equating $N(\xi)$ to $2B_c$, we get the following expression:

$$2B_c = \left(\frac{a_2\xi}{2}\right)^3 \int_0^{k_B T\xi} g(E)\,dE, \tag{9}$$

which allows calculating the T dependence of ξ and compare it with the experimental data $\rho(T)$. The above procedure was carried out using three different types of expressions for the DOS.

First, we used the interpolation expression proposed in [22, 23]. Thus, the change of the DOS from constant value for the VRH Mott-type to the DOS $g(E) \sim |E - E_F|^n$, where for the SAD $n > 2$, was taken into account. In this case the

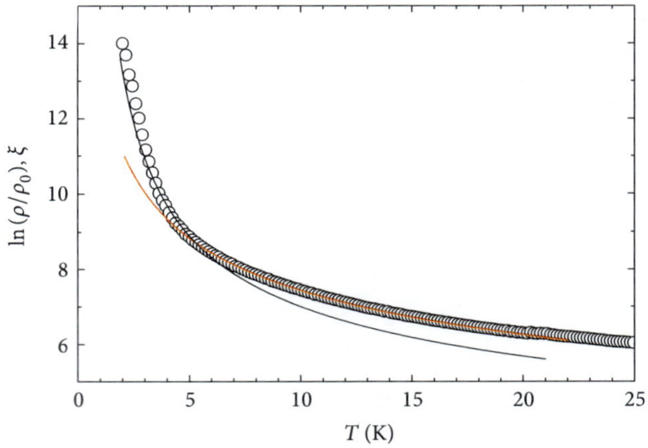

FIGURE 8: Experimental (symbols) $\ln[\rho(T)/\rho_0]$ and calculated (lines) according to (10) $\xi(T)$ dependencies. Red and blue lines correspond to $n = 0$ and 15, respectively.

DOS models, presumably, only the CG-*like*, but with steeper dependence on the energy. According to [22], the crossover at different exponent values of $p = (n+1)/(n+4)$ could be described applying the following expression for the DOS:

$$g(E,T) = g_0 \frac{\left[(E-E_F)/E_{sg}\right]^n}{1+\left[(E-E_F)/E_{sg}\right]^n}. \tag{10}$$

Parameter E_{sg} depends on the exponent n and, when it is fixed, is constant. Parameter g_0 is a constant related to g_M. For the case $n = 2$, (10) coincides with the expression used in [14] for the description of the crossover from the Mott to ES VRH.

Substituting (10) into (9) and using the previously obtained values, a_2 and $E_{sg} = \epsilon_4$, we calculated the dependencies $\xi(T)$ for the fixed values of the exponent n in (10), which were compared with the experimental data $\ln[\rho(T)/\rho_0]$. We obtained that theoretical results coincide with the experimental data only varying values of the exponent n with temperature [60]. For example, in the interval 5–25 K the agreement is observed only for $n = 0$ (the Mott-type VRH), while for 2–5 K to obtain the agreement between theory and experiment much greater value of n, $n > 6$, is required, Figure 8. Calculations of the $p(T)$ dependencies for different n values using the relation $p = -\partial \log\xi(T)/\partial \log T = -[T/\xi(T)][\partial\xi(T)/\partial T]$ [21] and evaluated the $\xi(T)$ dependence showed that, to achieve the value of the exponent p, close to unity, it is necessary to significantly increase n, that is, the variable n needs to tend to infinity, Figure 9. At $n = 0$ expressions for the crossover give the exact value $p = 1/4$. Thus, (10) for the DOS in this case has limited application because, firstly, it requires a change of the exponent n with T and, secondly, the value of n, which provides the exponent $p \to 1$, tends to infinity.

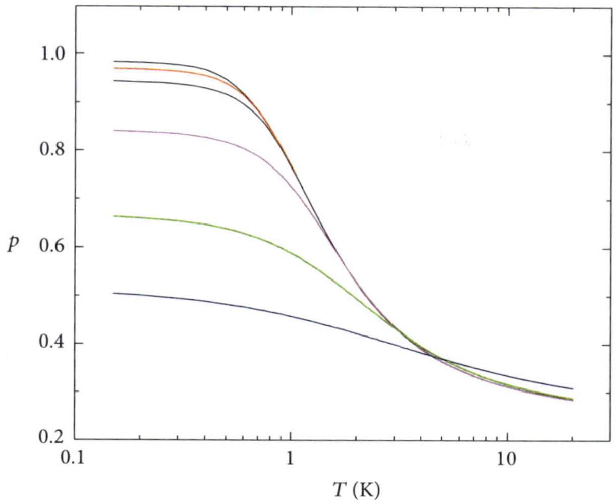

FIGURE 9: Temperature dependencies of the exponent p for different n values. From top to down, $n = 200, 100, 50, 15, 5$, and 2, respectively.

In the second attempt, we take into account the gap smearing with T increase [61],

$$g(E,T) = g_0\left[\gamma(T) + [1-\gamma(T)]\frac{|E-E_F|^n}{[\Delta(T)/2]^n + |E-E_F|^n}\right]. \quad (11)$$

The parameter γ describes a "residual" DOS at the Fermi level and Δ is the width of the gap.

Substitution of the modified expression (11) to (9) allowed comparing the value of the critical exponent ξ with the experimental data and finding the temperature dependencies $\gamma(T)$ and $\Delta(T)$, which adequately describe the experiment. This result is presented in Figure 10 for two n values, $n = 2.5$ and 3.0.

It follows from Figure 10 that γ varies in the range 0.1–1. At $T > 10$ K, $\gamma \to 1$, and the first constant term in (11) dominates. This corresponds to the Mott-type VRH. When $\gamma = 1$ (i.e., at $T > 15$ K), the result becomes insensitive to the Δ values. At low temperatures $\gamma \ll 1$ and the main contribution is due to the term $|E - E_F|^n$. The parameter Δ, which characterizes the width of the gap, decreases with increasing temperature, which is in contradiction with the result usually reported [61, 62]. Thus, the result of the fitting procedure found that the agreement with the experiment is achieved only taking into account the filling of the gap states between the Hubbard bands and suggested the narrowing of the gap with increasing temperature, during the crossover to the Mott-type DOS. In addition, insensitivity of the model to the value of n at $n > 3$ has been revealed. Therefore, it is obvious that there is a need to modify the formula for the DOS, in order to reflect the tendency both for smearing of the gap and its growth with increasing temperature. These are characteristic features of the HG. Moreover, the integer exponent n should be temperature independent.

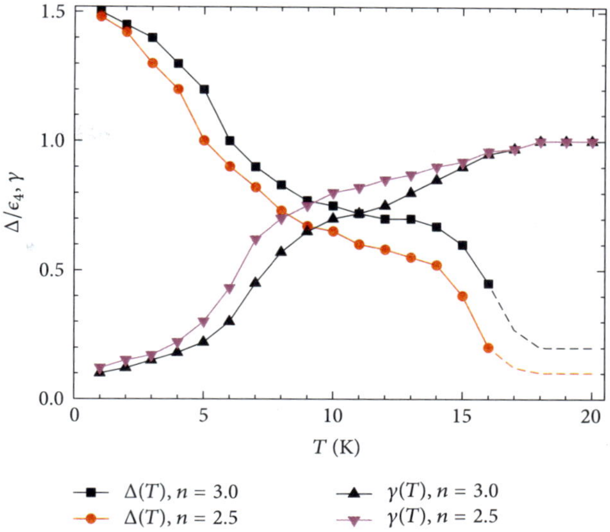

FIGURE 10: $\gamma(T)$ and $\Delta(T)$ dependencies as obtained using (11) for fitting the $\rho(T)$ data. Calculated curves correspond to the range of the exponent $n = 2.5$–3.0. The γ values are in arbitrary units. The Δ values are normalized to ϵ_4.

Within these assumptions the reasonable approximation for the DOS is of the following empiric type:

$$g(E,T) = g_0\left[\alpha(T)w^4(E) + \beta(T)w^2(E) + \gamma(T)\right] \times \exp\left(-\left[\frac{w(E)}{\sigma(T)}\right]^4\right), \quad (12)$$

where $w(E) = (E - E_F)/0.5\epsilon_4$ and α, β, γ, and σ are phenomenological *temperature dependent* coefficients.

The shape of the DOS in this case depends on T according to the α, β, γ and σ versus T dependencies. In particular, the first term corresponds to the HG, the second one is for the soft gap, the parameter γ is responsible for the filling of the HG with T and σ is responsible for the width of the Hubbard band. Indeed, the exponential term serves for smoothing of the overall DOS and for modelling of the Hubbard bands, in particular, their width. In other words, empirical expression (12) allows reproducing not only the HG, its filling, but also the width of the Hubbard bands. In this work, we assumed the identical widths of the upper and lower Hubbard bands. Widening of the UHB was taken into account by filling the HG. Function (12) reflects the change of the DOS with T, from HG at low T (with the dip between Hubbard bands) to the constant DOS close to the Fermi level at high T. This was enough for the experimental data interpretation. The fitting procedure was as for previous expressions for the DOS; that is, we substitute (12) into (9) and looked for such ξ values which adequately describe the $\rho(T)$ data. The result of the fitting procedure is shown in Figure 11. It follows that we were able to describe experimental data with very high accuracy in the T range 2–20 K. On the basis of the obtained $\xi(T)$ dependence it is possible to evaluate r_{max}, the maximal hopping length which still allows the connectivity. For the

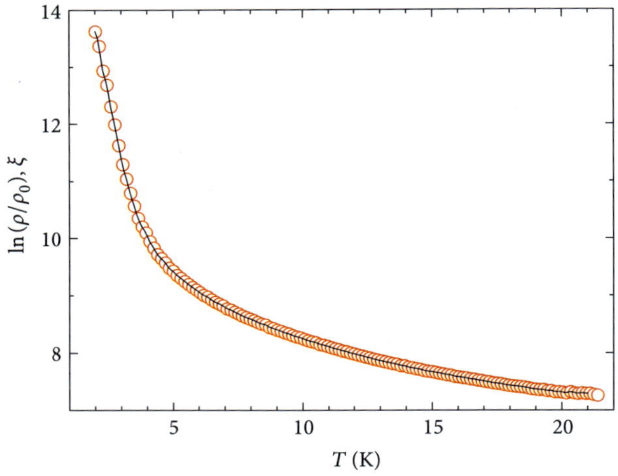

FIGURE 11: Experimental (symbols) $\ln[\rho(T)/\rho_0]$ and calculated (lines) according to (12) $\xi(T)$ dependencies.

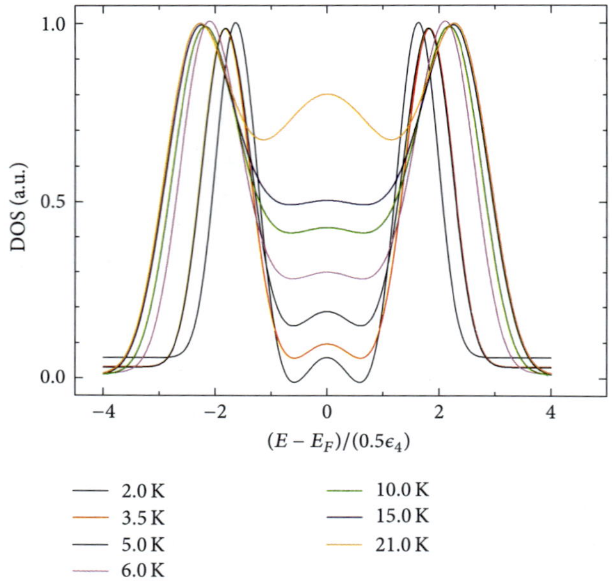

FIGURE 12: DOS at different temperatures as obtained from the experimental data within the proposed model (see (12)).

range 5–20 K we get $r_{\max}$ = 17.6–22.0 nm, while for the interval 2–4 K the result is $r_{\max}$ = 27.5–38.5 nm.

On the basis of the obtained model parameters, it is possible to reproduce the DOS at different temperatures. This result is shown in Figure 12. As it follows, the DOS evolves from HG at 2 K to almost filled states between Hubbard bands at 15 K. At that the widening of the Hubbard bands also occurs (within our approach, UHB and LHB are symmetrical; actually, they may differ in width and height, but this issue requires additional investigation and is outside the objectives of this work). Parameters of the model are changed as follows. The parameter α, which corresponds to the HG width is almost constant and is equal to 1 for the whole T range. The parameter β (characteristic of the soft gap) is almost constant with temperature, $\beta = -0.7$ except the region 15–21 K, where $\beta = -2$. The parameter γ, which is responsible for the HG filling, is changed from 0 at 2 K to 6 at 21 K. Finally, the parameter σ, the characteristic of the width of the Hubbard bands, also increases with temperature, from 1.5 to 2.3.

At the end of this section we summarize that, the conductivity in the T interval 5–20 K is of the hopping-type. On the whole, it obeys the Mott law and is determined by two type of hops: to UO neutral and to SO impurity sites. In the range 5–11 K the contribution of the spin dependent hopping dominates. This is valid also in the range 2–4 K. Simulations of the DOS revealed that in this T range hops are regulated by the HG. The presence of nonsmeared HG between Hubbard bands in the T range 2–4 K together with the spin dependent hops in the UHB provides reentrance of the SAD. *The latter provides a nonlinearity of the current-voltage characteristics at small bias voltage, which leads to negative differential resistance at electrical fields smaller than 1 V/cm [63]. It is unlikely that this effect may be due to the ES VRH, for which the delocalization occurs at significantly higher electrical fields.*

5. Conclusions

In conclusion, we performed the systematic study of the transport mechanisms in n-type noncompensated Si in the temperature range 2–300 K. Samples were grown by the Czochralski method, containing donor impurities of Sb with a concentration of 1×10^{18} cm^{-3}, It was found that, depending on the temperature range, it is possible to identify 4 different mechanisms of the electron transport. All of them include the upper and lower Hubbard bands.

In the T range 110–300 K the usual activation to the edge of the conduction band occurs.

At lower T, in the range 2–90 K, the conductivity of silicon is mainly due to the impurities and is caused by the hopping and band electron transport involving two Hubbard bands. In the high T range of this interval (28–90 K) electronic states are delocalized and charge transport occurs via the SO neutral donors, while in the low T range (2–20 K) electronic states are localized and the hopping prevails. The hopping region is characterized by two types of hops, on UO and SO sites, as well as by two modes, the Mott-type and SAD. The Mott regime, in turn, is characterized by two types of hops and two mechanisms of the MR, mainly the spin polarized mechanism in the T range 5–11 K and the mechanism, for which contribution of the spin polarization is approximately equal to the contribution of the wave function contraction (11–20 K). In the T interval 2–4 K the conductivity is governed by the HG and the charge transport is associated mainly with the spin dependent hopping in the UHB. *To improve the identification of the observed Mott-type VRH-SAD crossover and enhance the temperature interval of SAD the energy range of the impurity states as well as the level of the injection of the electrons should be increased [63].*

Simulation of the DOS within the model of crossover between the Mott-type VRH and SAD were carried out for two types of DOS, soft gap and HG. We showed that, the use of the interpolation expression for the CG-like DOS, but with the exponent n greater than 2 and neglecting the gap smearing, leads to the necessity of noticeable increase of the exponent n to get the exponent $p = 1$ at $T < 4$ K. Simulation of the CG-like DOS taking into account the width of the gap and smearing of the region between Hubbard bands showed that, with increasing T both smearing and decreasing of the gap occurs that does not quite correspond to the T dependence of the soft gap known from literature. Finally, simulation within the HG approach considering smearing, growth of the HG, and widening of the Hubbard bands with increasing T allowed describing with high accuracy the experimental $\rho(T)$ dependence in the low T range 2–20 K. The latter confirms the adequacy of the proposed HG model for interpretation of the conductivity in the range 2–4 K, although more detailed experimental and theoretical studies are required to clarify better the impact of the HG on the low T electron transport. *In particular, direct spectroscopic measurements of the density of states or conductivity measurements for n-type Si with impurities of various energy levels and different compensation degree could be potentially useful for these purposes.*

Competing Interests

The authors declare that there is no conflict of interests regarding the publication of this paper.

Acknowledgments

This work is partially supported by the Belarusian State Scientific Program "Physical Science, New Materials and Technologies," Project no. 2.44.

References

[1] H. V. Löhneysen, "Metal-insulator transition in heavily doped semiconductors," *Current Opinion in Solid State and Materials Science*, vol. 3, no. 1, pp. 5–15, 1998.

[2] V. Dobrosavljevic, N. Trivedi, and J. M. Jr. Valles, *Conductor-Insulator Quantum Phase Transitions*, Oxford University Press, 2012.

[3] N. F. Mott, *Metal-Insulator Transitions*, Taylor & Francis, London, UK, 2nd edition, 1990.

[4] P. P. Edwards and M. J. Sienko, "Universality aspects of the metal-nonmetal transition in condensed media," *Physical Review B*, vol. 17, no. 6, pp. 2575–2581, 1978.

[5] H. Fritzche, *The Metal Non-Metal Transition in Disordered Systems*, Edited by L. F. Friedman and D. P. Tunstall, University of St. Andrews, St. Andrews, UK, 1979.

[6] B. I. Shklovskii and A. L. Efros, *Electronic Properties of Doped Semiconductors*, Springer, Berlin, Germany, 1984.

[7] A. L. Efros and B. I. Shklovskii, "Coulomb gap and low temperature conductivity of disordered systems," *Journal of Physics C: Solid State Physics*, vol. 8, no. 4, pp. L49–L51, 1975.

[8] A. L. Efros, "Coulomb gap in disordered systems," *Journal of Physics C*, vol. 9, no. 11, pp. 2021–2030, 1976.

[9] J. R. Friedman, Y. Zhang, P. Dai, and M. P. Sarachik, "Magnetic-field-induced crossover from Mott variable-range hopping to weakly insulating behavior," *Physical Review B*, vol. 53, no. 15, pp. 9528–9531, 1996.

[10] J. Zhang, W. Cui, M. Juda et al., "Hopping conduction in partially compensated doped silicon," *Physical Review B*, vol. 48, no. 4, pp. 2312–2319, 1993.

[11] A. Fujimoto, H. Kobori, T. Ohyama et al., "Crossover from positive to negative magnetoresistance by the rise of electron temperature for Si:Sb in the variable-range hopping regime," *Physica B: Condensed Matter*, vol. 324, no. 1–4, pp. 1–8, 2002.

[12] A. Aharony, Y. Zhang, and M. P. Sarachik, "Universal crossover in variable range hopping with Coulomb interactions," *Physical Review Letters*, vol. 68, no. 26, pp. 3900–3903, 1992.

[13] Y. Meir, "Universal crossover between Efros-Shklovskii and mott variable-range-hopping regimes," *Physical Review Letters*, vol. 77, no. 26, pp. 5265–5267, 1996.

[14] R. Rosenbaum, N. V. Lien, M. R. Graham, and M. Witcomb, "A useful Mott-Efros-Shklovskii resistivity crossover formulation for three-dimensional films," *Journal of Physics Condensed Matter*, vol. 9, no. 29, pp. 6247–6256, 1997.

[15] L. R. Wienkes, C. Blackwell, and J. Kakalios, "Electronic transport in doped mixed-phase hydrogenated amorphous/nanocrystalline silicon thin films," *Applied Physics Letters*, vol. 100, no. 7, Article ID 072105, 2012.

[16] B. Skinner, T. Chen, and B. I. Shklovskii, "Theory of hopping conduction in arrays of doped semiconductor nanocrystals," *Physical Review B—Condensed Matter and Materials Physics*, vol. 85, no. 20, Article ID 205316, 2012.

[17] S. Nakatsuji, V. Dobrosavljević, D. Tanasković, M. Minakata, H. Fukazawa, and Y. Maeno, "Mechanism of hopping transport in disordered mott insulators," *Physical Review Letters*, vol. 93, no. 14, Article ID 146401, 2004.

[18] M. Salvato, M. Lucci, I. Ottaviani et al., "Transport mechanism in granular Ni deposited on carbon nanotubes fibers," *Physical Review B*, vol. 86, no. 11, Article ID 115117, 2012.

[19] F. A. Zwanenburg, A. S. Dzurak, A. Morello et al., "Silicon quantum electronics," *Reviews of Modern Physics*, vol. 85, no. 3, pp. 961–1019, 2013.

[20] E. Prati, K. Kumagai, M. Hori, and T. Shinada, "Band transport across a chain of dopant sites in silicon over micron distances and high temperatures," *Scientific Reports*, vol. 6, Article ID 19704, 2016.

[21] N. V. Agrinskaya and V. I. Kozub, "Universal description of crossover between the Mott regime and the Coloumb-gap regime in hopping conductivity: application to compensated CdTe," *Journal of Experimental and Theoretical Physics*, vol. 89, no. 6, pp. 1125–1129, 1999.

[22] N. V. Lien and R. Rosenbaum, "General crossovers from two-dimensional Mott $T^{-1/3}$ to soft-gap $T^{-\nu}$ variable-range hopping," *Physical Review B—Condensed Matter and Materials Physics*, vol. 56, no. 23, pp. 14960–14963, 1997.

[23] N. Van Lien and R. Rosenbaum, "General resistance crossover expressions for three-dimensional variable-range hopping," *Journal of Physics Condensed Matter*, vol. 10, no. 27, pp. 6083–6090, 1998.

[24] P. A. Lee and T. V. Ramakrishnan, "Disordered electronic systems," *Reviews of Modern Physics*, vol. 57, no. 2, pp. 287–337, 1985.

[25] A. Aharoni and D. Stauffer, *Introduction to Percolation Theory*, Taylor & Francis, London, UK, 2nd edition, 1994.

[26] B. I. Shklovskii and A. L. Efros, "Impurity band and conductivity of compensated semiconductors," *Soviet Physics—Journal of Experimental and Theoretical Physics*, vol. 33, no. 2, pp. 468–474, 1971.

[27] V. Ambegaokar, B. I. Halperin, and J. S. Langer, "Hopping conductivity in disordered systems," *Physical Review B*, vol. 4, no. 8, pp. 2612–2620, 1971.

[28] I. S. Beloborodov, A. V. Lopatin, V. M. Vinokur, and K. B. Efetov, "Granular electronic systems," *Reviews of Modern Physics*, vol. 79, no. 2, pp. 469–518, 2007.

[29] E. Dagotto, *Nanoscale Phase Separation and Colossal Magnetoresistance*, Springer, Berlin, Germany, 2002.

[30] D. Shahar, D. C. Tsui, M. Shayegan, E. Shimshoni, and S. L. Sondhi, "A different view of the quantum hall plateau-to-plateau transitions," *Physical Review Letters*, vol. 79, no. 3, pp. 479–482, 1997.

[31] S. H. Pan, J. P. ONeal, R. L. Badzey et al., "Microscopic electronic inhomogeneity in the high-T superconductor Bi Sr CaCu O," *Nature (London)*, vol. 413, no. 6853, pp. 282–285, 2001.

[32] M. M. Qazilbash, A. Tripathi, A. A. Schafgans et al., "Nanoscale imaging of the electronic and structural transitions in vanadium dioxide," *Physical Review B*, vol. 83, no. 16, Article ID 165108, 2011.

[33] P. Dai, Y. Zhang, and M. P. Sarachik, "Low-temperature transport in the hopping regime: evidence for correlations due to exchange," *Physical Review Letters*, vol. 69, no. 12, pp. 1804–1806, 1992.

[34] H. Fritzsche and M. Cuevas, "Impurity conduction in transmutation-doped p-type germanium," *Physical Review*, vol. 119, no. 4, pp. 1238–1245, 1960.

[35] R. K. Ray and H. Y. Fan, "Impurity conduction in silicon," *Physical Review*, vol. 121, no. 3, pp. 768–779, 1961.

[36] M. Inada, H. Yamamoto, M. Gibo et al., "Crossover from Efros-Shklovskii variable range hopping to nearest-neighbor hopping in silicon nanocrystal random network," *Applied Physics Express*, vol. 8, no. 10, Article ID 105001, 2015.

[37] H. Fritzsche, "Electrical properties of germanium semiconductors at low temperatures," *Physical Review*, vol. 99, no. 2, pp. 406–419, 1955.

[38] A. S. Ioselevich, "Spin polarons and variable range hopping in magnetically disordered systems," *Physical Review Letters*, vol. 71, no. 7, pp. 1067–1070, 1993.

[39] J. H. Davies, "The density of states in the coulomb gap," *Philosophical Magazine B*, vol. 52, no. 3, pp. 511–520, 1985.

[40] M. Pollak, "The Coulomb gap: a review, and new developments," *Philosophical Magazine Part B*, vol. 65, no. 4, pp. 657–667, 1992.

[41] C. Yamanouchi, "Hall coefficient and resistivity in the intermediate impurity conduction of n-type germanium," *Journal of the Physical Society of Japan*, vol. 20, no. 6, pp. 1029–1034, 1965.

[42] H. Nishimura, "Impurity conduction in the intermediate concentration region," *Physical Review*, vol. 138, no. 3A, pp. A815–A821, 1965.

[43] E. M. Gershenzon, A. P. Mel'nikov, R. I. Rabinovich, and N. A. Serebryakova, "H^--like impurity centers and molecular complexes created by them in semiconductors," *Soviet Physics—Uspekhi*, vol. 23, no. 10, pp. 684–698, 1980.

[44] A. Kurobe and H. Kamimura, "Correlation effects on variable range hopping conduction and the magnetoresistance," *Journal of the Physical Society of Japan*, vol. 51, no. 6, pp. 1904–1913, 1982.

[45] T. G. Castner, N. K. Lee, G. S. Cieloszyk, and G. L. Salinger, "Dielectric anomaly and the metal-insulator transition in *n*-type silicon," *Physical Review Letters*, vol. 34, no. 26, pp. 1627–1630, 1975.

[46] A. Ferreira Da Silva, "Metal-insulator transitions in doped silicon and germanium," *Physical Review B*, vol. 37, no. 9, pp. 4799–4800, 1988.

[47] I. Terry, T. Penney, S. Von Molnr, and P. Becla, "Low-temperature transport properties of Cd0.91Mn0.09Te:In and evidence for a magnetic hard gap in the density of states," *Physical Review Letters*, vol. 69, no. 12, pp. 1800–1803, 1992.

[48] H. Vinzelberg, A. Heinrich, C. Gladun, and D. Elefant, "Activated conduction in Amorphous Cr-SiOx Thin films," *Philosophical Magazine B*, vol. 65, no. 4, pp. 651–656, 1992.

[49] R. W. van der Heijden, G. Chen, A. T. A. M. de Waele, H. M. Gijsman, and F. P. B. Tielen, "Simple activated transport in ion-implanted Si:As at temperatures below 0.5 K," *Solid State Communications*, vol. 78, no. 1, pp. 5–8, 1991.

[50] J.-J. Kim and H. J. Lee, "Observation of a nonmagnetic hard gap in amorphous In/InO_x films in the hopping regime," *Physical Review Letters*, vol. 70, no. 18, pp. 2798–2801, 1993.

[51] V. Voegele, S. Kalbitzer, and K. Böhringer, "Observation of correlation effects in the hopping transport in amorphous silicon," *Philosophical Magazine B*, vol. 52, no. 2, pp. 153–168, 1985.

[52] K. G. Lisunov, E. K. Arushanov, C. Kloc, U. Malang, and E. Bucher, "Hopping conductivity in p-type β-FeSi2," *Physica Status Solidi (B) Basic Research*, vol. 195, no. 1, pp. 227–236, 1996.

[53] K. G. Lisunov, E. Arushanov, G. A. Thomas, E. Bucher, and J. H. Schön, "Variable-range hopping conductivity and magnetoresistance in n-CuGaSe2," *Journal of Applied Physics*, vol. 88, no. 7, pp. 4128–4134, 2000.

[54] A. Fedotov, S. Prischepa, A. Danilyuk, I. Svito, and P. Zukowski, "Spin-polarized and normal hopping magnetoresistance in heavily doped silicon," *Acta Physica Polonica A*, vol. 125, no. 6, pp. 1271–1274, 2014.

[55] K. A. Matveev, L. I. Glazman, P. Clarke, D. Ephron, and M. R. Beasley, "Theory of hopping magnetoresistance induced by Zeeman splitting," *Physical Review B*, vol. 52, no. 7, pp. 5289–5297, 1995.

[56] Y. Meir, "Universal spin-induced magnetoresistance in the variable-range hopping regime," *Europhysics Letters*, vol. 33, no. 6, pp. 471–476, 1996.

[57] S. V. Demishev and A. A. Pronin, "Magnetoresistance of carbon nanomaterials," *Physics of the Solid State*, vol. 48, no. 7, pp. 1363–1372, 2006.

[58] S. V. Demishev, A. D. Bozhko, V. V. Glushkov et al., "Scaling of magnetoresistance of carbon nanomaterials in Mott-type hopping conductivity region," *Physics of the Solid State*, vol. 50, no. 7, pp. 1386–1391, 2008.

[59] B. I. Shklovskiĭ and A. L. Éfros, "Percolation theory and conductivity of strongly inhomogeneous media," *Soviet Physics - Uspekhi*, vol. 18, no. 11, pp. 845–862, 1975.

[60] A. K. Fedotov, I. A. Svito, V. V. Fedotova, A. G. Trafimenko, A. L. Danilyuk, and S. L. Prischepa, "Low-temperature conductivity of silicon doped with antimony," *Semiconductors*, vol. 49, no. 6, pp. 705–711, 2015.

[61] B. Sandow, K. Gloos, R. Rentzsch, A. N. Ionov, and W. Schirmacher, "Electronic correlation effects and the Coulomb gap at finite temperature," *Physical Review Letters*, vol. 86, no. 9, pp. 1845–1848, 2001.

[62] I. Shlimak, M. Kaveh, R. Ussyshkin et al., "Temperature-induced smearing of the Coulomb gap: experiment and computer simulation," *Physical Review Letters*, vol. 75, no. 26, pp. 4764–4767, 1995.

[63] A. L. Danilyuk, A. G. Trafimenko, A. K. Fedotov, I. A. Svito, and S. L. Prischepa, "Negative differential resistance in n-type noncompensated silicon at low temperature," *Applied Physics Letters*, vol. 109, no. 22, Article ID 222104, 2016.

10

Phase Transitions and Magnetocaloric Properties in $MnCo_{1-x}Zr_xGe$ Compounds

Anil Aryal,[1] Abdiel Quetz,[1] Sudip Pandey,[1] Igor Dubenko,[1] Shane Stadler,[2] and Naushad Ali[1]

[1]Department of Physics, Southern Illinois University, Carbondale, IL 62901, USA
[2]Department of Physics & Astronomy, Louisiana State University, Baton Rouge, LA 70803, USA

Correspondence should be addressed to Anil Aryal; aryalanil@siu.edu

Academic Editor: Oleg Derzhko

The structural, magnetic, and magnetocaloric properties of $MnCo_{1-x}Zr_xGe$ ($0.01 \leq x \leq 0.04$) have been studied through X-ray diffraction, differential scanning calorimetry, and magnetization measurements. Results indicate that the partial substitution of Zr for Co in $MnCo_{1-x}Zr_xGe$ decreases the martensitic transition temperature (T_M). For $x = 0.02$, T_M was found to coincide with the ferromagnetic transition temperature (T_C) resulting in a first-order magnetostructural transition (MST). A further increase in zirconium concentration ($x = 0.04$) showed a single transition at T_C. The MST from the paramagnetic to ferromagnetic state results in magnetic entropy changes ($-\Delta S_M$) of 7.2 J/kgK for $\Delta H = 5$ T at 274 K for $x = 0.02$. The corresponding value of the relative cooling power (RCP) was found to be 266 J/kg for $\Delta H = 5$ T. The observed large value of MCE and RCP makes this system a promising material for magnetic cooling applications.

1. Introduction

The magnetocaloric effect (MCE) is a phenomenon in which a magnetic material heats up when a magnetic field is applied and cools down when the field is removed. In recent years, magnetic refrigeration based on the MCE has been considered as a possible ecofriendly and energy efficient cooling technology alternative to conventional vapor cycle refrigeration [1, 2]. Therefore, the developments of new materials that show large MCEs are essential. To date, the largest MCEs have been reported in materials such as $Gd_5Si_2Ge_2$ [3], MnFe(P,As) [4], $LaFe_{13-x}Si_x$ [5], NiMn(Ga,Sb,In) based Heusler alloys [6], and $MnAs_{1-x}Sb_x$ [7], which all show first-order crystallographic and magnetic phase transitions simultaneously, that is, magnetostructural transitions (MSTs). Hence, the exploration and study of the new materials that exhibit MSTs is important for applications and fundamental physics.

MnCoGe belongs to the family of ternary metallic compounds MM′X, where M and M′ are 3*d* transition metals and X is Si or Ge. Stoichiometric MnCoGe displays ferromagnetic (FM) properties below the Curie temperature ($T_C \sim 345$ K). In the paramagnetic (PM) region, the alloy transforms to a high-temperature, Ni_2In-type hexagonal structure (space group $P6_3/mmc$) from a low-temperature TiNiSi-type orthorhombic structure (space group Pnma) at $T_M \sim 650$ K [8]. If the structural transition temperature (T_M) shifts to coincide with T_C, a magnetostructural phase transition with a large change in magnetization can be expected and can lead to large MCE values. Previous studies indicate that changes in stoichiometry, chemical composition, or application of external pressure can result in concurrent magnetic and structural transitions and therefore may increase the magnitude of magnetic entropy change to show a large MCE [9–16]. In this work, we present results on the partial replacement of Co by Zr in the $MnCo_{1-x}Zr_xGe$ system, which exhibits a MST and a large MCE.

2. Experiment

Polycrystalline samples with compositions $0.01 \leq x \leq 0.04$ were prepared by arc-melting high purity elements (99.99%)

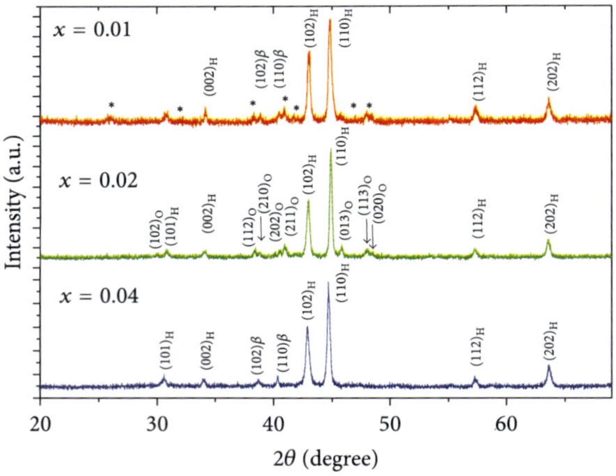

FIGURE 1: Room temperature XRD patterns of $MnCo_{1-x}Zr_xGe$ compounds with $0.01 \leq x \leq 0.04$. The orthorhombic and hexagonal peaks are indexed with *hkl* Miller indices with "O" and "H," respectively. The peaks indicated by "*" are unknown phase. The peaks labeled with *hkl* Miller indices with the symbol "β" are from the K_β wavelength, from the X-ray source.

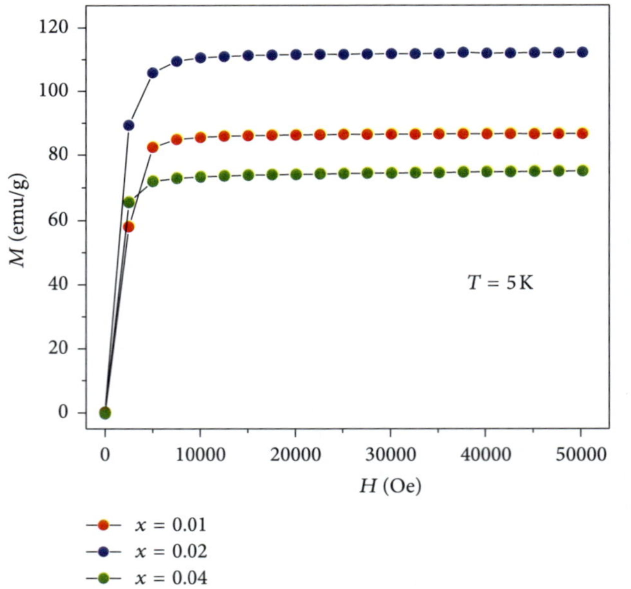

FIGURE 2: Field dependence of the magnetization $M(H)$ at $T = 5$ K.

in an ultrahigh purity argon atmosphere. The samples were turned over and remelted to ensure homogeneity. The crystal structures were determined by powder X-ray diffraction (XRD) using CuKα radiation at room temperature. A superconducting quantum interference device (SQUID by Quantum Design) magnetometer was used to measure the magnetic properties of the $MnCo_{1-x}Zr_xGe$ compounds in the temperature interval 5–380 K in applied magnetic fields up to 5 T. Differential scanning calorimetry (DSC) measurements were carried out using a DSC 8000 instrument (Perkin-Elmer) with a ramp rate of 30 K/min during heating and cooling in the temperature range 150–450 K to detect the temperature induced first-order transitions.

3. Results and Discussion

The room temperature XRD patterns of $MnCo_{1-x}Zr_xGe$ compounds are shown in Figure 1. The sample with $x = 0.01$ is in mixture state of the hexagonal (of about 80%) and unknown phase at room temperature. At Zr concentration, $x = 0.02$, the hexagonal (Ni_2In-type) structure stabilizes near room temperature with a small trace of orthorhombic phase (~15%). For $x = 0.04$, the compound crystallizes in a single-phase hexagonal Ni_2In-type structure. It has been reported that the orthorhombic phase of the MnCoGe system has smaller Co-Co separation [9]. Increasing the distance between the Co-atoms, either by creating Co vacancies or by substituting a larger element, stabilizes the hexagonal (Ni_2In-type) phase at lower temperature [9, 17]. Therefore, the partial substitution of Co ($R_{Co} = 1.252$ Å) by the larger Zr atoms ($R_{Zr} = 1.602$ Å) [18] in $MnCo_{1-x}Zr_xGe$ serves to stabilize the high-temperature austenite phase at lower temperatures.

Figure 2 shows the field dependence of the magnetization $M(H)$ at $T = 5$ K. The $M(H)$ curves at low temperature show that the compounds possess a ferromagnetic type of ordering in the ground state. The saturation magnetization (M_S) for the compound with $x = 0.02$ was found to be 3.72 μ_B/f.u., a value much larger than 2.87 μ_B/f.u. ($x = 0.01$) and 2.50 μ_B/f.u. ($x = 0.04$). It has been reported that the saturation magnetization of the parent MnCoGe in the orthorhombic structure ($M_S = 3.73\ \mu_B$/f.u.) is greater than that in the hexagonal structure (2.80 μ_B/f.u.) [19]. By comparing the saturation magnetization with that published in [19], and considering the results of the XRD and $M(T)$ measurements (Figures 1 and 3), we conclude that the compound with $x = 0.02$ is in the orthorhombic structure and that the samples with $x =$ 0.01 and 0.04 are in the hexagonal phase, respectively, in the ground state. Therefore, the increase in saturation magnetization for $x = 0.02$ is attributed to the low-temperature FM orthorhombic phase.

The temperature dependence of the magnetization $M(T)$ of the compounds in an applied field of 100 Oe is shown in Figure 3. The $M(T)$ curves were measured during heating and cooling cycles within 10 K–380 K. The values of the Curie temperatures (T_C) were calculated from the minimum of dM/dT in the magnetization curve during the heating cycle (see inset of Figure 3). The $M(T)$ graph shows that the compounds are ferromagnetic at low temperatures. With increasing temperature, a large jump in magnetization was observed, which is typical for a FM to PM transition at T_C. A thermal hysteresis between heating and cooling curves was observed for the compound with $x = 0.02$ which is a signature of a first-order structural transformation at T_M. The first-order transition seen in the $M(T)$ curves for $x = 0.02$ was further confirmed by the observed endothermic and exothermic peaks in the DSC data (see Figure 4). Thus, a first-order MST from a FM TiNiSi-type phase to a PM Ni_2In-type phase is observed in $x = 0.02$ due to the coincidence of T_C and T_M. For the compounds with $x = 0.01$ and 0.04, a second-order magnetic transition (SOT) from a FM to PM state was observed. The calculated values of T_M and T_C are shown in Table 1.

TABLE 1: Transition temperatures, Curie-Weiss temperatures, saturation magnetization (M_S), effective magnetic moment, $-\Delta S_M^{max}$, and RCP of $MnCo_{1-x}Zr_xGe$.

x	T_C (K)\$T_{M(DSC)}$ (K)	θ_P (K)	Ms (μ_B/f.u.) From $M(H)$	μ_{eff} (μ_B/f.u.)	Ms (μ_B/f.u.) From μ_{eff}	$-\Delta S_M^{max}$ (J/kgK) H = 1 T\2 T\3 T\4 T\5 T	RCP (J/kg) H = 1 T\2 T\3 T\4 T\5 T
0.01	300\—	308	2.87	4.51	3.92	0.9\1.7\2.4\3.0\3.5	32\81\134\193\251
0.02	274\286	311	3.72	4.31	3.42	1.6\3.2\4.5\6.0\7.2	42\99\149\210\266
0.04	262\—	280	2.50	4.28	3.40	0.8\1.4\1.9\2.4\2.8	27\67\108\145\192

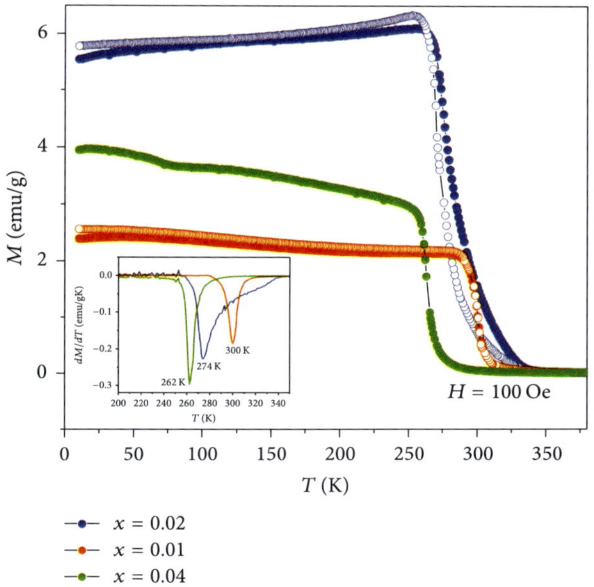

FIGURE 3: The temperature dependence of the magnetization $M(T)$ for $MnCo_{1-x}Zr_xGe$ with $0.01 \le x \le 0.04$ on heating and cooling in an applied field of 100 Oe. The solid and open symbols represent heating and cooling cycles, respectively. Inset: dM/dT versus T curves on heating.

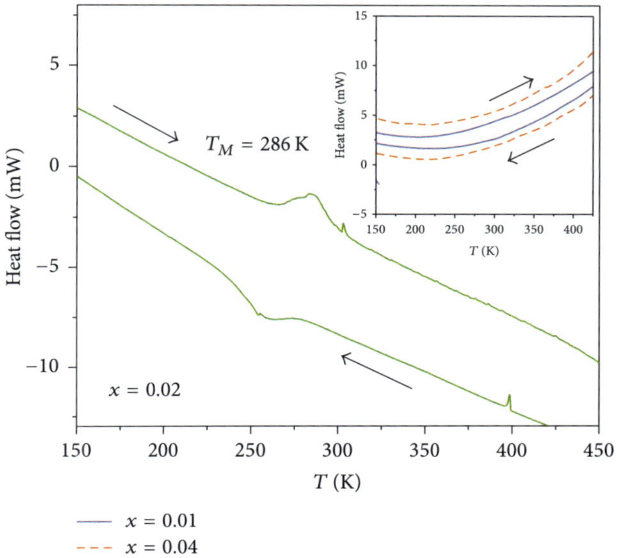

FIGURE 4: DSC heat flow curves (endothermic and exothermic peaks) as a function of temperature measured at the rate of 30 K/min during heating and cooling for the compound with $x = 0.02$. Inset: DSC heat flow curves for $x = 0.01$ and $x = 0.04$. The arrows indicate the heating and cooling cycles.

The temperature dependence of the inverse susceptibility $\chi^{-1}(T)$ for $MnCo_{1-x}Zr_xGe$ for H = 100 Oe is plotted in Figure 5. It is well known that the susceptibility (χ) follows the Curie-Weiss law, $\chi = C/(T - \theta_P)$, in the paramagnetic region, where C is the Curie constant and θ_P is the Curie-Weiss temperature. The values of the parameter θ_P were obtained by fitting the linear paramagnetic region and are given in Table 1. The observed values of θ_P are positive and larger than T_C, which indicates a ferromagnetic interaction between spins [20]. The effective magnetic moment (μ_{eff}) in the PM region has been calculated from $\chi^{-1}(T)$ data using the relation $\mu^2_{eff} = (3k_B/N\mu_B^2)C_m \approx 8\,C_m$, where $C_m = \chi T$ is the molar Curie constant [21]. The corresponding values of the saturation magnetization (M_S) were then calculated using $M_S = g \cdot J$ and $\mu_{eff} = g[J(J+1)]^{1/2}$ [21]. The calculated values of μ_{eff} and M_S are listed in Table 1. Comparing the values of the saturation magnetization obtained from $M(H)$ and $M(T)$ curves (see Table 1), one can conclude that the orientation of the magnetic moments in the ground state of the compound in the orthorhombic structure (x = 0.02) is close to a parallel ferromagnetic type compared to

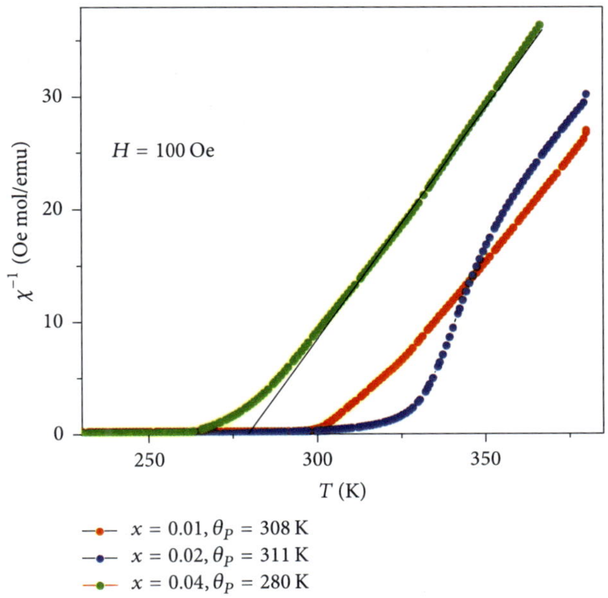

FIGURE 5: Temperature dependence of the inverse susceptibility in a field of 100 Oe. The solid line is the fitting according to the Curie-Weiss law.

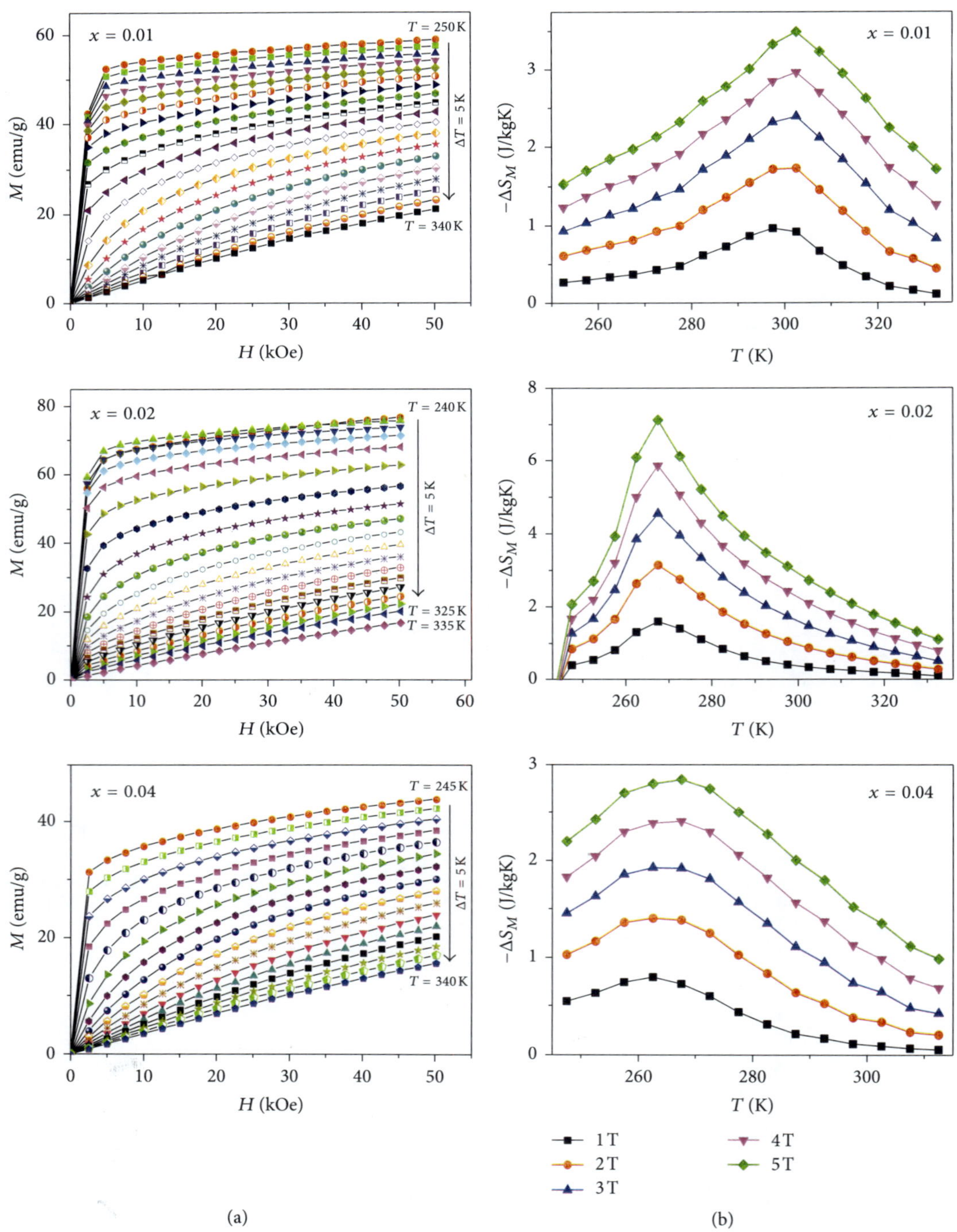

FIGURE 6: Isothermal magnetization $M(H)$ curves (a) and magnetic entropy changes, $-\Delta S_M(T)$ (b), with a magnetic field change $\Delta H = 5$ T for $MnCo_{1-x}Zr_xGe$.

a noncollinear ferromagnetic type in the compounds in the hexagonal structure ($x = 0.01, 0.04$).

The magnetic entropy change ($-\Delta S_M$) near the transition temperature for field changes up to 5 T is plotted in Figure 6. The $-\Delta S_M$ values were calculated from the isothermal magnetization curves $M(H)$ (shown in Figure 6) using the Maxwell relation, $\Delta S_M = \int_0^H (\partial M/\partial T)_H dH$ [1, 2]. The validity of the Maxwell relation to evaluate ΔS_M in the case of first-order transitions has been discussed in detail in [22]. A large $-\Delta S_M$ value of 7.2 J/kgK corresponding to a first-order magnetostructural transition was found for the compound with $x = 0.02$ for $\Delta H = 5$ T. The large value of $-\Delta S_M$ is attributed to the abrupt change in magnetization resulting from the coincidence of T_C and T_M brought forth by the Zr substitution for Co. For compounds with $x = 0.01$ and 0.04, smaller values of $-\Delta S_M$, 3.5 and 2.8 J/kgK, were found for $\Delta H = 5$ T, respectively.

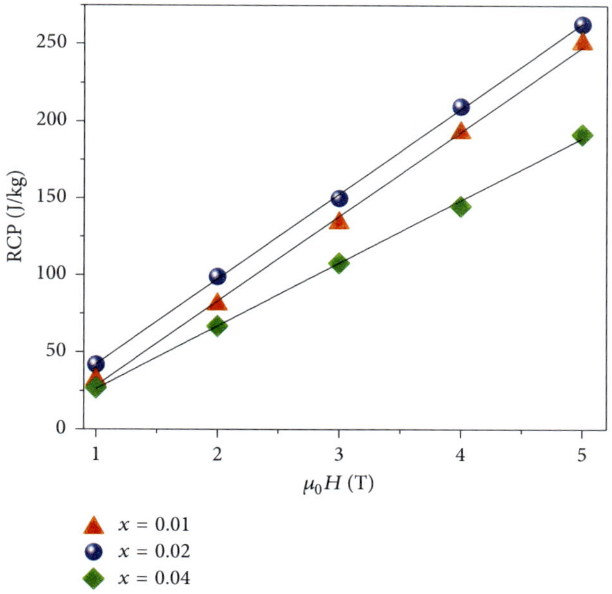

FIGURE 7: Field dependence of the RCP for $MnCo_{1-x}Zr_xGe$.

In addition to large magnetic entropy changes, large values of relative cooling power (RCP) are essential for magnetic refrigeration. RCP is an important parameter that estimates the usefulness of a material as a magnetic refrigerant. The RCP is a measure of amount of heat transferred between the hot and cold reservoirs in an ideal refrigeration cycle. It is defined as $RCP = -\Delta S_M\,(\max) \times \delta T_{FWHM}$, where δT_{FWHM} is the full width at half maximum of the $-\Delta S_M$ curve [23]. The field dependence of the RCP for $MnCo_{1-x}Zr_xGe$ is shown in Figure 7. Large values of the RCP 251, 266, and 192 J/kg with $\Delta H = 5$ T were found for $x = 0.01$, 0.02, and 0.04, respectively. The observed large values of $-\Delta S_M$ and RCP may be interesting for magnetic refrigeration applications.

4. Conclusions

The structural, magnetic, and magnetocaloric properties of $MnCo_{1-x}Zr_xGe$ compounds have been studied. Through the partial substitution of Zr for Co, the high-temperature austenite phase has been stabilized near room temperature. A first-order MST was observed for the composition with $x = 0.02$, resulting from the coincidence of T_C and T_M. Large values of $-\Delta S_M$ and RCP corresponding to first-order MSTs were observed for $x = 0.02$. Therefore, these large values of the magnetic entropy change and RCP and the nontoxic and less-expensive constituent elements make this system a promising material for magnetic cooling applications.

Conflicts of Interest

The authors declare no conflicts of interest in publication of this paper.

Acknowledgments

This work was supported by the US Department of Energy (DOE), Office of Science, Basic Energy Sciences (BES), under Award no. DE-FG02-06ER46291 (SIU) and DE-FG02-13ER46946 (LSU).

References

[1] K. A. Gschneidner Jr., V. K. Pecharsky, and A. O. Tsokol, "Recent developments in magnetocaloric materials," Reports on Progress in Physics 68, article 1479, 2005.

[2] A. M. Tishin and Y. I. Spichkin, *The Magnetocaloric Effect and Its Applications*, Great Britain Institute of Physics, 2003.

[3] V. K. Pecharsky and K. A. Gschneidner Jr., "Giant magnetocaloric effect in $Gd_5(Si_2Ge_2)$," *Physical Review Letters*, vol. 78, no. 23, article 4494, 1997.

[4] O. Tegus, E. Brück, K. H. J. Buschow, and F. R. de Boer, "Transition-metal-based magnetic refrigerants for room-temperature applications," *Nature*, vol. 415, pp. 150–152, 2002.

[5] A. Fujita, S. Fujieda, Y. Hasegawa, and K. Fukamichi, "Itinerant-electron metamagnetic transition and large magnetocaloric effects in $La(Fe_xSi_{1-x})_{13}$ compounds and their hydrides," *Physical Review B*, vol. 67, Article ID 104416, 2003.

[6] I. Dubenko, M. Khan, A. K. Pathak, B. R. Gautam, S. Stadler, and N. Ali, "Magnetocaloric effects in Ni-Mn-X based Heusler alloys with X=Ga, Sb, In," *Journal of Magnetism and Magnetic Materials*, vol. 321, no. 7, pp. 754–757, 2009.

[7] H. Wada and Y. Tanabe, "Giant magnetocaloric effect of $MnAs_{1-x}Sb_x$," *Applied Physics Letters*, vol. 79, no. 20, article 3302, 2001.

[8] T. Kanomata, H. Ishigaki, T. Suzuki, H. Yoshida, S. Abe, and T. Kaneko, "Magneto-volume effect of $MnCo_{1-x}Ge$ ($0 \leqslant x \leqslant 0.2$)," *Journal of Magnetism and Magnetic Materials*, vol. 140–144, no. 1, pp. 131-132, 1995.

[9] J.-T. Wang, D.-S. Wang, C. Chen et al., "Vacancy induced structural and magnetic transition in $MnCo_{1-x}Ge$," *Applied Physics Letters*, vol. 89, no. 26, Article ID 262504, 2006.

[10] N. T. Trung, L. Zhang, L. Caron, K. H. J. Buschow, and E. Bruck, "Giant magnetocaloric effects by tailoring the phase transitions," *Applied Physics Letters*, vol. 96, Article ID 172504, 2010.

[11] N. T. Trung, V. Biharie, L. Zhang, L. Caron, K. H. J. Buschow, and E. Bruck, "From single- to double-first-order magnetic phase transition in magnetocaloric $Mn_{1-x}Cr_xCoGe$ compounds," *Applied Physics Letters*, vol. 96, Article ID 162507, 2010.

[12] E. K. Liu, W. Zhu, L. Feng et al., "Vacancy-tuned paramagnetic/ferromagnetic martensitic transformation in Mn-poor $Mn_{1-x}CoGe$ alloys," *Europhysics Letters*, vol. 91, no. 1, article 17003, 2010.

[13] L. Caron, N. T. Trung, and E. Brück, "Pressure-tuned magnetocaloric effect in $Mn_{0.93}Cr_{0.07}CoGe$," *Physical Review B*, vol. 84, Article ID 20414, 2011.

[14] S. Niziol, R. Zach, J. P. Senateur, J. Beille, and J. Magn, "Pressure dependence of the magnetic transition temperature of the $CoMnGe_{1-x}Si_x$ system," *Journal of Magnetism and Magnetic Materials*, vol. 79, no. 3, pp. 333–337, 1989.

[15] V. Johnson, "Diffusionless orthorhombic to hexagonal transitions in ternary silicides and germanides," *Inorganic Chemistry*, vol. 14, no. 5, pp. 1117–1120, 1975.

[16] P. E. Markin, N. V. Mushnikov, V. I. Khrabrov, and M. A. Korotin, "Magnetic properties of the $Mn_{1.9-x}Co_xGe$ compounds with a hexagonal crystal structure," *Physics of Metals and Metallography*, vol. 106, no. 5, pp. 481–489, 2008.

[17] T. Samanta, I. Dubenko, Q. Abdiel, S. Stadler, and A. Naushad, "Large magnetocaloric effects over a wide temperature range in $MnCo_{1-x}Zn_xGe$," *Journal of Applied Physics*, vol. 113, article 17A922, 2013.

[18] W. B. Pearson, *The Crystal Chemistry and Physics of Metals and Alloys*, Wiley-interscience, New York, NY, USA, 1972.

[19] F. Guillou, F. Wilhelm, O. Tegus, and A. Rogalev, "Microscopic mechanism of the giant magnetocaloric effect in MnCoGe alloys probed by x-ray magnetic circular dichroism," *Applied Physics Letters*, vol. 108, Article ID 122405, 2016.

[20] A. Dhahri, M. Jemmali, M. Hussein, E. Dhahri, A. Koumina, and E. K. Hlil, "Critical behavior near the ferromagnetic to paramagnetic phase transition temperature in polycrystalline $La_{0.7}Ca_{0.2}Sr_{0.1}Mn_{1-x}Cr_xO_3$ ($x = 0.15$ and 0.2)," *Journal of Alloys and Compounds*, vol. 618, pp. 788–794, 2014.

[21] J. Crangle, *The Magnetic Properties of Solids*, Edward Arnold (publishers) Limited, London, UK, 1977.

[22] K. A. Gschneidner and V. K. Pecharsky, "Magnetocaloric materials," *Annual Review of Materials Science*, vol. 30, no. 1, pp. 387–429, 2000.

[23] A. Aryal, A. Quetz, S. Pandey et al., "Phase transitions and magnetocaloric and transport properties in off-stoichiometric $GdNi_2Mn_x$," *Journal of Applied Physics*, vol. 119, no. 4, Article ID 043905, 2016.

11

Photocatalytic Surface Modification of PI Film for Electroless Copper Plating

Wenxia Zhao,[1] Zenglin Wang,[2] Liang Qiao,[3] Shiwei Liu,[1] Hongjian Zhao,[1] and Yani Yan[1]

[1]*School of Chemistry and Chemical Engineering, Ningxia Normal University, Guyuan 756000, China*
[2]*Key Laboratory of Applied Surface and Colloid Chemistry, Ministry of Education, School of Chemistry and Chemical Engineering, Shaanxi Normal University, Xi'an 710119, China*
[3]*Department of Chemical and Biological Engineering, University at Buffalo (SUNY), Buffalo, NY 14260, USA*

Correspondence should be addressed to Wenxia Zhao; zwxchj2006@163.com

Academic Editor: Changlin Ding

This study investigated the surface modification of polyimide (PI) film through TiO_2 photocatalytic treatment. The effects of TiO_2 content, treatment duration, and UV power on the surface topography, surface contact angle, and adhesion strength of the surface-modified PI films were investigated. The results indicated that, after surface modification under the optimal photocatalytic conditions, the surface contact angle of the PI film decreased from 84.4° to 38.8°, and the adhesion strength between the PI film and the electroless copper film reached 0.78 kN/m. X-ray photoelectron spectroscopy analysis further demonstrated that carboxyl groups formed on the surface of the PI film after photocatalytic treatment. The surface hydrophilicity and adhesion strength of the surface-modified PI film were enhanced due to the numerous carboxyl groups formed on its surface. Therefore, the photocatalytic treatment is an environmentally friendly and effective method for the surface modification of PI films.

1. Introduction

Polyimides (PIs) possess many desirable traits, such as low dielectric constant, high breakdown voltage; good planarization, wear resistance, radiation resistance, inertness to solvents, and hydrolytic stability; low thermal expansion; long-term stability; and excellent mechanical properties. Given these attributes, PIs have been extensively used as substrates in the production of ultra-large-scale integrations, electronic packages, and circuit boards [1–3]. In several of these applications, the deposition of a metal layer on PI substrates is necessary to allow electrical conduction via interconnections [4]. However, poor adhesion strength between the copper film and the PI substrate is a serious problem in these applications.

Over the years, many studies have investigated surface modification methods for improving the adhesion of metals to PI film. The surfaces of PI films are commonly modified through plasma treatment [5–10], ion implantation [11, 12], chemical treatment [13–18], and UV/ozone treatment [19–21]. However, physical processes, such as plasma and ion implantation, require expensive equipment and are thus associated with high investment costs. Chemical treatments, such as oxidation, polymer grafting, and etching, require expensive waste disposal. The surfaces of polymers can be modified with photoirradiation, which occurs through the formation of an electron-hole pair in the semiconducting material when the photon energy exceeds the band gap [22]. Photocatalysis has been received increasing attention as an advanced oxidation process due to its low cost and environmental friendliness. In addition, the high chemical stability of TiO_2 and the potential use of sunlight as the irradiation source make photocatalysis an attractive method for the surface modification of PI films.

Thomas reported that base hydrolysis and acidification resulted in the formation of poly(amic acid) on the surface of PI film [23]. In this study, we introduced a method that combines the surface formation of hydrophilic species and the cleavage of imide rings to enhance the adhesion of copper to PI film. The effects of TiO_2 content (0.5, 1.0, 2.5, and 5.0 g/L), UV light power (100 and 300 W), and treatment duration (10, 20, 30, and 40 min) on the surface topography, surface roughness, adhesion strength, and surface chemistry of the

TABLE 1: The compositions of the electroless copper plating bath.

Composition	Content
Copper sulfate ($CuSO_4 \cdot 5H_2O$)	10 g/L
Ethylenediaminetetraacetic acid disodium salt (EDTA·2Na)	30 g/L
Formaldehyde (HCHO, 37%)	3 ml/L
2,2′-Dipyridyl	15 mg/L
Poly(ethylene glycol) (PEG 1000)	0.5 g/L

PI films were investigated. The surface topography and chemistry of the PI films were evaluated through scanning electron microscopy (SEM), atomic force microscopy (AFM), surface contact angle measurements, and X-ray photoelectron spectroscopy (XPS).

2. Experimental Section

PI films with a density of 1.42 g/cm^3 were procured from DuPont Chemical Co., Ltd., as Kapton HN in rolls that were 40 mm wide and 75 μm thick. The rolls were then cut into 40 mm × 25 mm samples for all experiments. Before photocatalytic treatment, the surfaces of the PI films were cleaned with acetone in an ultrasonic water bath for 30 min at room temperature and washed with deionized (DI) water. The photocatalysis experiments were conducted with a TiO_2 suspension over the PI surface. Then, 0.5–5.0 g of TiO_2 powder (JR05, Xuancheng Jingrui New Materials Co., Ltd., 5 nm), which was primarily in anatase phase, was dispersed in 1 L distilled water, respectively. Photocatalytic irradiation was conducted under a high-pressure mercury vapor lamp with a maximum wavelength of 365 nm. The distance between the PI film and the mercury lamp was fixed at 5 cm. After photocatalytic treatment, PI films were rinsed thrice with DI water.

The surface activation of the PI film was carried out in an activation solution (CATAPOSIT 44, purchased from Rohm and Haas Company) at 45°C for 5 min. Subsequently, the surface sensitization of the substrates was conducted by immersing the samples in 10% hydrochloric acid aqueous solution at 30°C for 1 min. The PI films were washed with distilled water after each step. The chemical composition of the electroless copper plating solution was the same as that in the literature [18], and the compositions are shown in Table 1. The pH of the solution was adjusted to 11.5 using NaOH, and the bath temperature was 70°C. After electroless plating for 40 min, the PIs were electroplated with copper at room temperature with a current density of 0.03 A/cm^2 for 1 h. The copper layer was deposited at a thickness of 20 μm. After annealing at 100°C for 120 min in an oven, the copper-coated PI films were cut into 10 mm × 40 mm strips. The adhesion strength of the PI films was measured through a 90° peel test at a peel rate of 25 mm/min [24]. The peel test was repeated four times, and the adhesion strength was reported as the average value of four measurements.

The surface topographies of the PI films were observed through SEM (Philips-FEI Quanta 200 electron microscope).

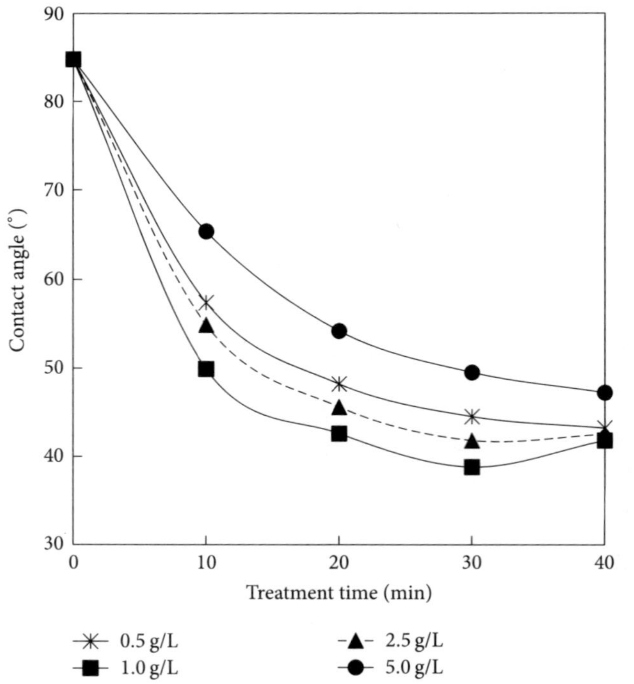

FIGURE 1: Dependence of the surface contact angle upon the TiO_2 content and the treatment time under 300 W UV light.

Moreover, the surface roughness of the PI films was measured through AFM (WETSPM-9500-J3, Shimazu Co., Ltd.). Surface roughness was reported as average roughness (R_a) and root-mean square roughness (R_{ms}). The surface contact angles of the PI films were determined as quickly as possible after different time photocatalytic treatment with a video-based contact angle measurement instrument. The volume of the water drop used in the measurements was 2 μL, and measurement values were taken at least five positions on the surface of the samples and averaged. The surface composition and chemistry groups of the PI films were determined by XPS measurements taken with a JPS90-MXV spectrometer from JEOL with a nonmonochromatized Mg Ka X-ray source. All binding energies were corrected on the basis of the Cls binding energy at 285 eV [25].

3. Results and Discussions

3.1. Effects of UV Power and Treatment Time on the Surface Contact Angle of PI Films. The wettability of the PI films was evaluated by measuring the surface contact angles of the film at room temperature. When the UV power was 300 W, effects of the TiO_2 content and the treatment duration upon the surface contact angle were measured, and the results are shown in Figure 1. As seen in the figure, the surface without photocatalytic treatment PI film was hydrophobic. The contact angle between distilled water and the untreated PI film was 84.8°. The surface contact angle of the PI film continuously decreased with increasing treatment duration when the PI films were treated with 0.5 g/L TiO_2 suspension under 300 W of UV light, suggesting that 0.5 g/L TiO_2 content

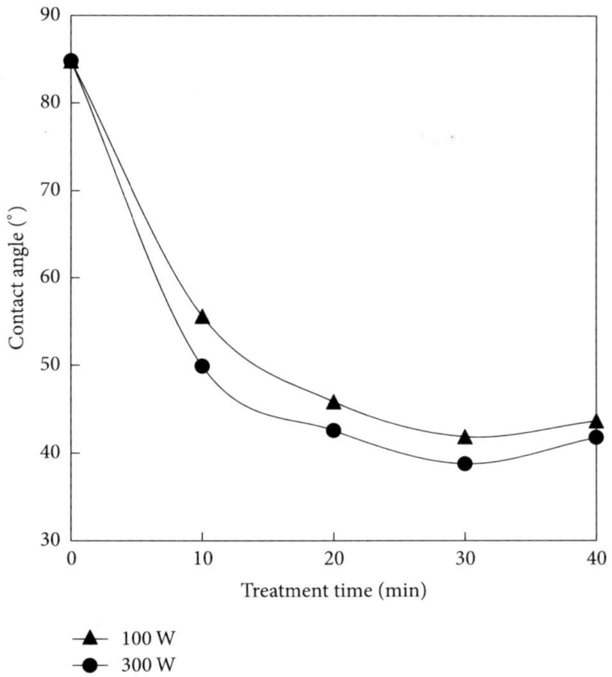

FIGURE 2: Dependence of the surface contact angle upon UV power and the treatment time at the TiO_2 content of 1.0 g/L.

was insufficient for inducing hydrophilicity. The surface contact angle of the PI film treated with 1.0 g/L TiO_2 suspension decreased with prolonged treatment duration and reached the minimum value with an irradiation time of 30 min; after that, the surface contact angle slowly increased with the further increase in the irradiation time. This result indicated that photocatalytic treatment induced the hydrophilicity of the PI film surface. Meanwhile, the surface contact angle of PI film treated with 1.0 g/L TiO_2 suspension was considerably lower than that of the PI film treated with 5.0 g/L TiO_2 suspension for the same treatment duration. This result was attributed to the agglomeration of TiO_2 particles in the saturated TiO_2 solution, because agglomeration decreases the oxidative capacity of generated $^{\bullet}OH$ radicals. Under 100 W of UV light, the change trend in the surface contact angle in response to TiO_2 content and treatment duration was similar to that under 300 W of UV light. The results indicated that 1.0 g/L TiO_2 suspension and 30 min of photocatalytic treatment are appropriate process parameters for the surface modification of PI film.

The effects of UV light power and treatment duration on the surface contact angle of PI films treated with 1.0 g/L TiO_2 solution are shown in Figure 2. Under 100 and 300 W of UV light, the surface contact angle of the PI film first decreased with prolonged treatment time and reached the minimum value after 30 min of treatment, and then, the surface contact angle increased when treatment time was continuously prolonged. Furthermore, the surface contact angle decreased with increased UV power at the same treatment time and fixed TiO_2 content. One potential reason for this result could be that the oxidation capacity of the photocatalytic system was strengthened by the increased intensity of UV light.

The roughness and hydrophilicity of the film surface affect the surface contact angle of the film. Specifically, the surface contact angle decreases with enhanced surface hydrophilicity and decreases with increasing surface roughness. Therefore, surface morphology and surface roughness observations were performed through SEM and AFM.

3.2. Effects of Surface Modification on the Surface Morphology and the Surface Roughness of PI Substrates. Figure 3 shows the changes in surface morphology before and after surface modification with 1.0 g/L TiO_2 suspension under 100 or 300 W UV light. The wet-chemical pretreatment may induce changes in morphology and the removal of weakly cohesive surface material. Thus, the surface morphology of the film is very rough and uneven after wet-chemical pretreatment [26, 27]. However, as shown in Figure 3, the surface morphology of the PI film treated with 1.0 g/L TiO_2 solution showed almost no change compared with that before photocatalytic treatment. Thus, the electroless copper film was more uniformly deposited through the photocatalytic reaction than through the wet-chemical pretreatment.

The AFM observations of the PI film morphology before and after pretreatment are shown in Figure 4. Before photocatalytic treatment, the PI film surface was flat and clean with R_a and R_{ms} of 13 and 17 nm, respectively. R_a and R_{ms} of the PI film modified under 100 W for 30 min were 14 and 19 nm, respectively. In addition, R_a and R_{ms} of the PI film modified under 300 W for 30 min were 15 and 21 nm, respectively. The results indicated that increasing UV power does not substantially change the surface topography of the PI films.

3.3. Effects of Surface Modification on the Adhesion Strength. The dependence of the adhesion strength on surface modification was investigated, and the results are shown in Figures 5 and 6. The adhesion strength between the electroless copper film and untreated PI film was zero. After photocatalytic treatment for 10 or 40 min, the adhesion strength between the electroless copper film and PI film increased with the increasing of TiO_2 content when the treatment time was fixed. And the adhesion strength between the electroless copper film and PI film reached the highest value after the PI film was treated with 1.0 g/L TiO_2 suspension. After that, the adhesion strength between the electroless copper film and PI film decreased when TiO_2 content was further increased. It can be seen from Figure 6, when PI film was treated with 1.0 g/L TiO_2 suspension, the adhesion strength between the electroless copper film and PI film increased with the increasing of UV power. Adhesion strength reached the maximum value when the PI film was treated for 30 min. This result was consistent with the changes in the surface contact angle of the PI film in response to TiO_2 content, treatment duration, and UV power. Given that hydrophilicity can improve the wettability of the film surface, the overall increased wettability of the PI film improved the intimate contact between the PI substrate and the electroless deposited copper film.

Adhesion strength is dependent not only on the surface roughness but also on surface hydrophilicity or density of

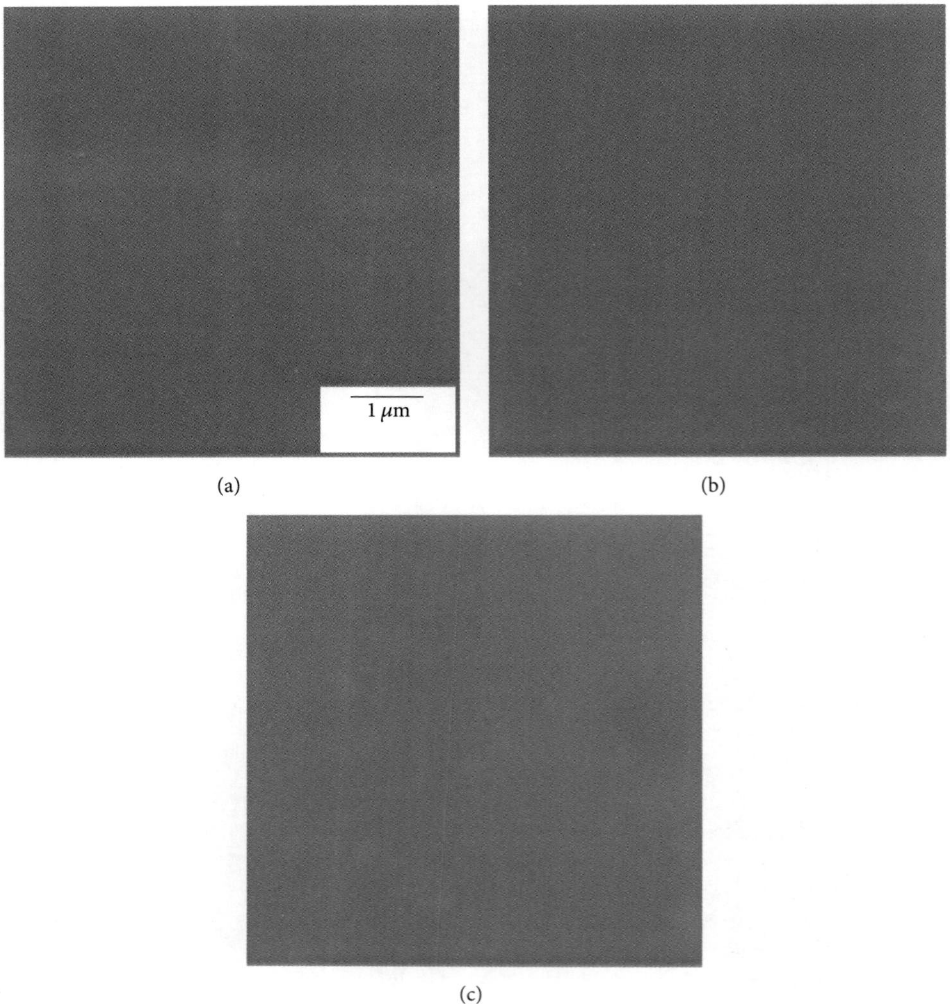

FIGURE 3: SEM images in surface morphology of PI films before and after surface modification with 1.0 g/L TiO_2 dispersed solution for 30 min, UV power: (a) 0 W, (b) 100 W, and (c) 300 W.

polar groups. Therefore, the effects of surface modification on the surface chemistry of PI films were investigated.

3.4. Surface Chemistry of the PI Film before and after TiO_2 Photocatalytic Treatment. The surface chemical properties of the PI films before and after photocatalytic treatment were characterized by analyzing the XPS spectra of the PI films. The surface element contents of the PI films were determined through XPS, and the results are shown in Table 2. It can be seen that the carbon content decreased with the increasing UV power and the oxygen content increased with the increasing UV power. The surface carbon and oxygen contents of the untreated PI film were 78.2% and 15.8%, respectively. When the UV power was 300 W, after 30 min of photocatalytic treatment, the carbon content of the PI film decreased from 78.2% to 75.4%, and the oxygen content of the PI film increased from 15.8% to 19.3%. This result was attributed to the increased density of hydrophilic groups under increased UV power. The trend in the variation of the aforementioned elemental contents corresponded with that of the surface contact angle.

TABLE 2: The element contents of PI surface before and after surface modification with 1.0 g/L TiO_2 dispersed solution under different UV power.

Treatment condition	C (%)	O (%)	N (%)
Before treatment	78.2	15.8	6.0
100 W, 30 min	77.0	17.5	5.5
300 W, 30 min	75.4	19.3	5.3

The C1s XPS spectra of PI films before and after treated with 1.0 g/L TiO_2 and UV power of 300 W were obtained and the results are shown in Figure 7. Four peaks were observed in the spectrum of the PI film before photocatalytic treatment. The peaks at 285.0, 286.2, 288.6, and 291 eV were attributed to C-H/C-C/C=C, C-O/C-N, $N(C{=}O)_2$, and π-π^*, respectively. A peak at 289.0 eV appeared in the spectrum of the PI film treated with 1.0 g/L TiO_2 for 30 min under 300 W UV light. This peak could be attributed to the -COOH group. The same phenomenon was observed when the PI film was treated with 1.0 g/L TiO_2 for 30 min under 300 W UV light. The

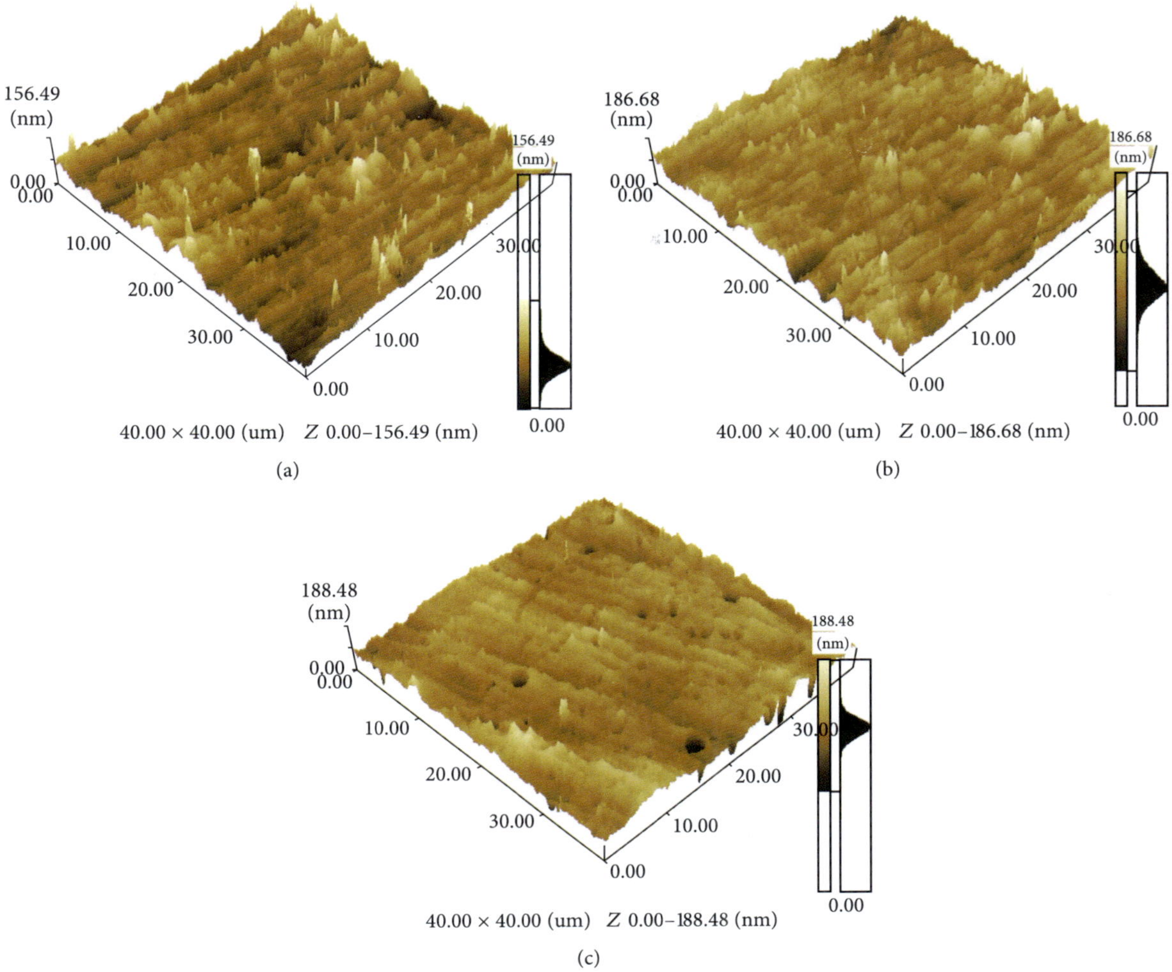

FIGURE 4: AFM images in surface morphology of PI films before and after surface modification with 1.0 g/L TiO_2 dispersed solution for 30 min, UV power: (a) 0 W, (b) 100 W, and (c) 300 W.

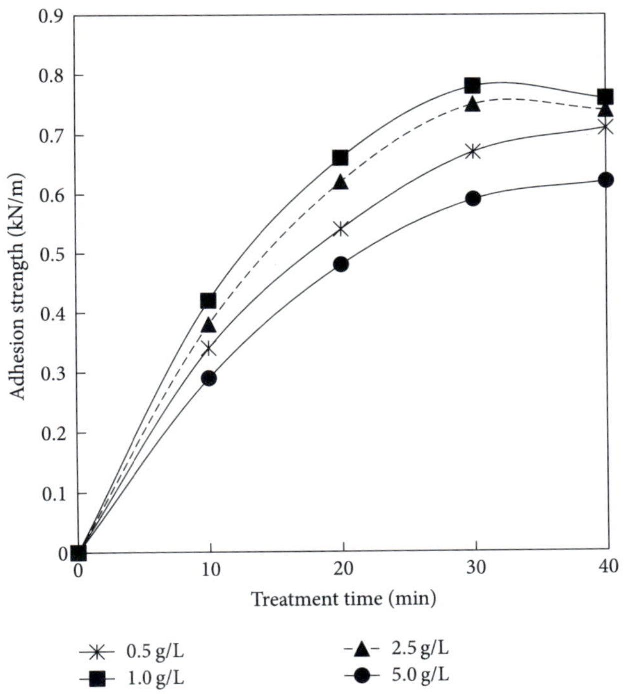

FIGURE 5: Dependence of the adhesion strength upon the TiO_2 content and the treatment time under 300 W UV light.

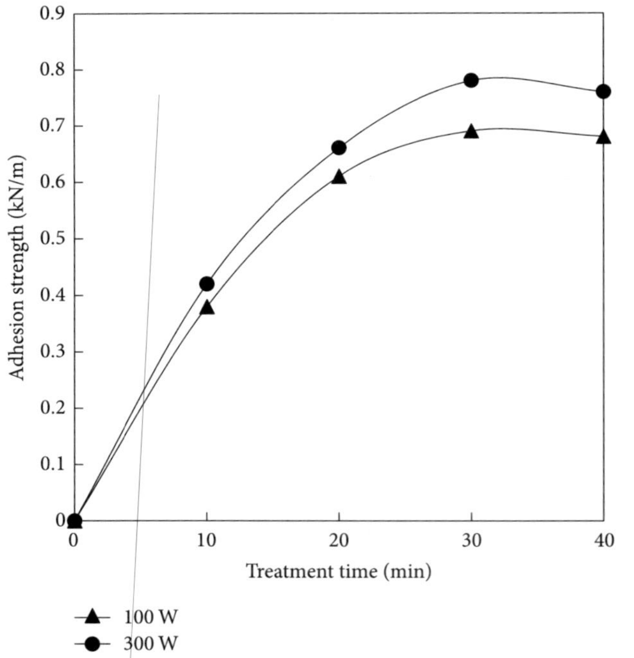

FIGURE 6: Dependence of the adhesion strength upon the UV power and the treatment time at the TiO_2 content of 1.0 g/L.

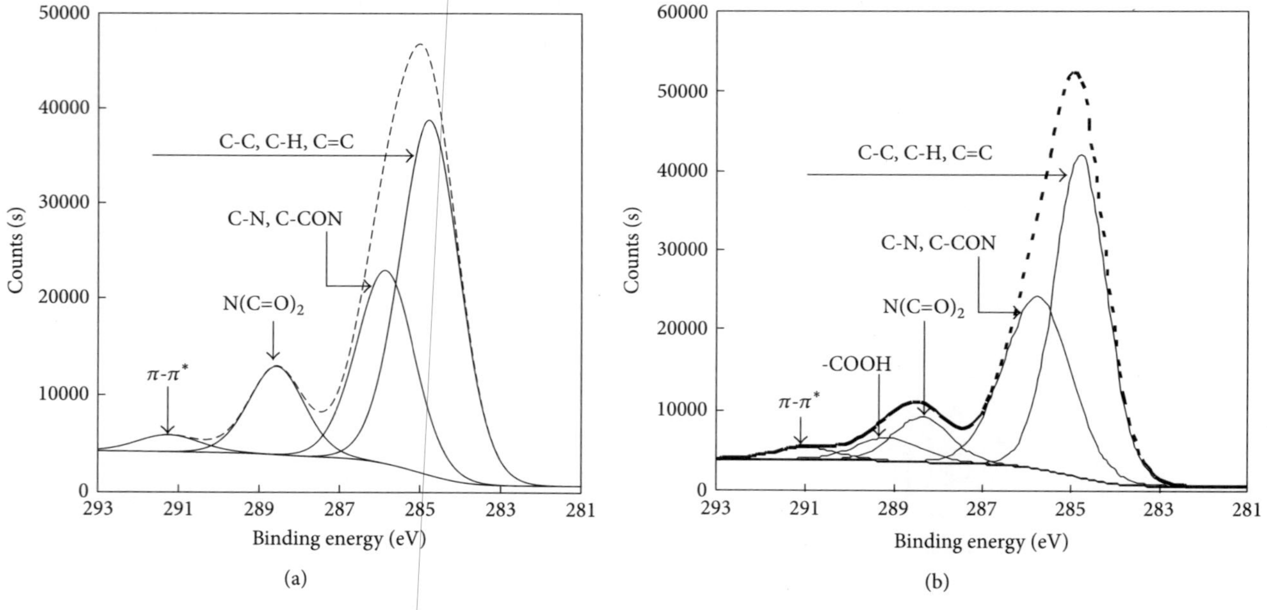

FIGURE 7: The Cls XPS spectra of the PI films before and after surface modification with 1.0 g/L TiO_2 dispersed solution for 30 min under 300 W UV light. (a) The Cls XPS spectra of the PI films before surface modification. (b) The Cls XPS spectra of the PI films after surface modification.

carbon group contents of the PI films before and after being treated for 30 min under different UV powers and 1.0 g/L TiO_2 content are shown in Table 3.

When the UV power increased from 0 W to 300 W, the content of hydrophobic groups (C-H/C-C/C=C) decreased from 53.7% to 46.5%, that of the $N(C{=}O)_2$ group decreased from 14.8% to 6.8%, and that of C-O/C-N and π-π^* groups showed almost no change. However, the content of hydrophilic groups (-COOH) increased from 0% to 6.1%, indicating that partial bonds of the PI film were cleaved and that amide and carboxyl groups formed on the surface of PI films after treatment with TiO_2. Given that carboxyl and amide groups can coordinate with the copper atom, their quantities can affect the adhesion strength of copper film to the PI film [28]. As a result, the high contents of -COOH groups enhanced the surface hydrophilicity of the PI film and

TABLE 3: The contents of carbon groups on the PI film surface before and after photocatalytic treatment (treatment time: 30 min, C_{TiO_2}: 1.0 g/L).

UV power	C-C, C-H, C=C (%)	C-O, C-N (%)	$N(C{=}O)_2$ (%)	π-π^* (%)	-COOH (%)
0	53.7	38.7	14.8	2.8	0
100 W	48.0	37.6	7.3	2.2	4.9
300 W	46.5	38.3	6.8	2.3	6.1

thus improved the adhesion strength between the electroless copper film and the PI film.

4. Conclusion

An environmentally friendly and effective photocatalytic treatment method was used to modify PI films. The effects of TiO_2 content, treatment duration, and UV power on the surface hydrophilicility and adhesion strength of the modified PI film were investigated through surface contact angle measurements, SEM observation, AFM, and XPS. The surface hydrophilicity and the adhesion strength of the photocatalytically treated PI film increased with increasing UV power and prolonged treatment duration. Surface modification with 1.0 g/L TiO_2 suspension was more effective than with other TiO_2 contents under 300 W of UV power. High TiO_2 content facilitated the aggregation of TiO_2 particles, thus decreasing the oxidative capacity of the hydroxyl radicals generated by the TiO_2 suspension. Under the optimal conditions of photocatalytic treatment, the adhesion strength between the electroless copper film and the PI film reached 0.78 kN/m, and the surface topography and the surface roughness of the PI film slightly changed. XPS results indicated that surface oxygen content increased with prolonged treatment duration and increased UV power. Meanwhile, the surface carbon content decreased with prolonged treatment duration because the content of -COOH polar group increased. Photocatalytic treatment enhanced the surface hydrophilicity of PI film, thus resulting in high adhesion strength between the PI film and electroless copper film.

Conflicts of Interest

The authors declare that there are no conflicts of interest regarding the publication of this article.

Authors' Contributions

Wenxia Zhao and Zenglin Wang contributed equally to this paper.

Acknowledgments

This work was partially supported by the Natural Science Foundation of China (Grants nos. 21561027 and 21273144), the Science and Technology Innovation Leading Talent Project of Ningxia Hui Autonomous Region (Grant no. Ning Ke Chuang Word [2016] 12), the Excellent Youth Teacher Training Fund Project for Institution of Higher Education of Ningxia (Grant no. NGY2016195), National Grade Academic Technical Leader of Ningxia Youth Top-Notch Talent (Office of Human Resources and Social Security of Ningxia Grant no. [2017] 787), Liupanshan Resources Engineering Technology Center (Grant no. HGZD17-03), and Key Disciplines of Inorganic Chemistry (Grant no. [2017] 83).

References

[1] K. S. Y. Lau, "High-Performance Polyimides and High Temperature Resistant Polymers," *Handbook of Thermoset Plastics*, pp. 297–424, 2013.

[2] I. A. Ronova, M. Bruma, and H.-W. Schmidt, "Conformational rigidity and dielectric properties of polyimides," *Structural Chemistry*, vol. 23, no. 1, pp. 219–226, 2012.

[3] D. Serbezeanu, I.-D. Carja, M. Bruma, and I. A. Ronova, "Correlation between physical properties and conformational rigidity of some aromatic polyimides having pendant phenolic groups," *Structural Chemistry*, vol. 27, no. 3, pp. 973–981, 2016.

[4] S. Joshi, A. v. Loon, A. Savov, and R. Dekker, "Adhesion Improvement of Polyimide/PDMS Interface by Polyimide Surface Modification," *MRS Advances*, vol. 1, no. 01, pp. 33–38, 2016.

[5] K. Fatyeyeva, A. Dahi, C. Chappey et al., "Effect of cold plasma treatment on surface properties and gas permeability of polyimide films," *RSC Advances*, vol. 4, no. 59, pp. 31036–31046, 2014.

[6] Y. Zhang, E. T. Kang, K. G. Neoh et al., "Surface plasma characterization of polyimide films for flexible electronics," *Advanced Materials Research*, vol. 970, pp. 132–135, 2014.

[7] A. B. Meddeb, Z. Ounaies, and M. Lanagan, "Enhancement of electrical properties of polyimide films by plasma treatment," *Chemical Physics Letters*, vol. 649, pp. 111–114, 2016.

[8] W. Kaczorowski, W. Szymanski, D. Batory, and P. Niedzielski, "Effect of plasma treatment on the surface properties of polydimethylsiloxane," *Journal of Applied Polymer Science*, vol. 132, no. 11, Article ID 41635, 2015.

[9] S.-H. Bang, "Improvement of NiMoNb to polyimide adhesion by inductively coupled nitrogen plasma treatment," *Applied Surface Science*, vol. 360, pp. 553–558, 2016.

[10] M. Akram, K. M. B. Jansen, L. J. Ernst, and S. Bhowmik, "Atmospheric plasma modification of polyimide sheet for joining to titanium with high temperature adhesive," *International Journal of Adhesion and Adhesives*, vol. 65, pp. 63–69, 2016.

[11] J. Won, M. H. Kim, Y. S. Kang et al., "Surface modification of polyimide and polysulfone membranes by ion beam for gas separation," *Journal of Applied Polymer Science*, vol. 75, no. 12, pp. 1554–1560, 2000.

[12] R. Mikšová, A. Macková, H. Pupikova, P. Malinský, P. Slepička, and V. Švorčík, "Compositional, structural, and optical changes of polyimide implanted by 1.0 MeV Ni + ions," *Nuclear Instruments and Methods in Physics Research Section B: Beam Interactions with Materials and Atoms*, vol. 406, pp. 199–204, 2017.

[13] F. J. Xu, J. P. Zhao, E. T. Kang, and K. G. Neoh, "Surface functionalization of polyimide films via chloromethylation and surface-initiated atom transfer radical polymerization," *Industrial & Engineering Chemistry Research*, vol. 46, no. 14, pp. 4866–4873, 2007.

[14] Z. Wu, D. Wu, W. Yang, and R. Jin, "Preparation of highly reflective and conductive metallized polyimide films through surface modification: Processing, morphology and properties," *Journal of Materials Chemistry*, vol. 16, no. 3, pp. 310–316, 2006.

[15] K. Akamatsu, S. Ikeda, H. Nawafune, and H. Yanagimoto, "Direct patterning of copper on polyimide using ion exchangeable surface templates generated by site-selective surface modification," *Journal of the American Chemical Society*, vol. 126, no. 35, pp. 10822-10823, 2004.

[16] L. Li, Y. Ma, J. Xie et al., "Metallization Process of a Polyimide Surface with Palladium-Free Activation for Electronic Field Applications," *Journal of Electronic Materials (JEM)*, vol. 44, no. 10, pp. 4042–4051, 2015.

[17] I. Ghosh, J. Konar, and A. K. Bhowmick, "Study of palladium catalyzation for electroless copper plating on polyimide film," *Journal of Nanoscience & Nanotechnology*, vol. 13, no. 1, pp. 517–522, 2013.

[18] Z. Wang, A. Furuya, K. Yasuda et al., "Adhesion improvement of electroless copper to a polyimide film substrate by combining surface microroughening and imide ring cleavage," *Journal of Adhesion Science and Technology*, vol. 16, no. 8, pp. 1027–1040, 2002.

[19] Y. I. Lee, Y. S. Goo, K. J. Lee, Y. G. Hwang, and Y. Byun, "Selective Cu patterning on polyimide using UV surface treatment and electroless plating," *Journal of Photopolymer Science & Technology*, vol. 28, no. 2, pp. 157–161, 2015.

[20] Y.-I. Lee, Y.-S. Goo, K.-J. Lee et al., "Effect of UV/ozone treatment on interactions between ink-jet printed Cu patterns and polyimide substrates," *Thin Solid Films*, vol. 519, no. 20, pp. 6853–6857, 2011.

[21] T. Kondo, R. Watanabe, Y. Shimoyama, K. Shinohe, S. A. Kulinich, and S. Iwamori, "Effect of reactive oxygen species generated with ultraviolet lamp and plasma on polyimide surface modification," *Surface and Interface Analysis*, vol. 49, no. 11, pp. 1069–1077, 2017.

[22] H. Schmidt, M. Naumann, T. S. Müller, and M. Akarsu, "Application of spray techniques for new photocatalytic gradient coatings on plastics," *Thin Solid Films*, vol. 502, no. 1-2, pp. 132–137, 2006.

[23] R. R. Thomas, "Wetting kinetics study of modified polyimide surfaces containing ionizable functional groups," *Langmuir*, vol. 19, no. 14, pp. 5763–5770, 2003.

[24] Z. Wang, O. Yaegashi, H. Sakaue, T. Takahagi, and S. Shingubara, "Suppression of native oxide growth in sputtered TaN films and its application to Cu electroless plating," *Journal of Applied Physics*, vol. 94, no. 7, pp. 4697–4701, 2003.

[25] L. Chen, B. Liao, J. Yu et al., "Study of adhesion strength of electroplated copper to polyimide using different types of tie layers," *Nanoscience and Nanotechnology Letters (NNL)*, vol. 7, no. 3, pp. 240–247, 2015.

[26] D. M. Brewis, "Electroless copper plating onto polyimide using polymer nanosheet as a nano-adhesive," *Polymer Journal*, vol. 39, no. 1, pp. 41–47, 2007.

[27] K.-S. Chou, X.-Y. Chen, and I.-H. Lai, "Surface Roughness on Adhesion of Silver Colloids to a Polyimide Substrate," *International Journal of Adhesion and Adhesives*, vol. 92, no. 6, pp. 429–439, 2016.

[28] T. Kondo, R. Watanabe, Y. Shimoyama, K. Shinohe, S. A. Kulinich, and S. Iwamori, "Effect of reactive oxygen species generated with ultraviolet lamp and plasma on polyimide surface modification," *Surface Interface Analysis*, vol. 49, no. 11, pp. 1069–1077, 2017.

12

Circular Polarization and Wavelength Selective Gratings based on Holographic Cholesteric Liquid Crystal Templates

Hsien-Kuo Chin,[1] Hui-Ying Kuo,[2,3] Yang-Chen Zheng,[4] Jia-De Lin,[4] and Chia-Rong Lee[4]

[1] *Division of Cardiovascular, Department of Surgery, Kaohsiung Armed Forces General Hospital, Kaohsiung 802, Taiwan*
[2] *Department of Optometry, Chung Shan Medical University, Taichung 402, Taiwan*
[3] *Department of Ophthalmology, Chung Shan Medical University Hospital, Taichung 402, Taiwan*
[4] *Department of Photonics, National Cheng Kung University, Tainan 701, Taiwan*

Correspondence should be addressed to Jia-De Lin; geman1218@yahoo.com.tw

Academic Editor: Jörg Fink

Refilling cholesteric liquid crystal (CLC) template gratings with circular polarization, wavelength selectivity, and dual operation mode are first demonstrated. The gratings with template and nontemplate regions are obtained through the two-beam interfering photo-polymerization, washing-out, and refilling procedure. The refilling CLC has different chiral handedness (left or right) and reflection band (at green and red regions). When the wavelength of a probe beam is within the refilled CLC reflection band of the nontemplate region, the device works as an amplitude grating since the sample in the template and nontemplate regions reflects distinct colour region. In addition, the diffraction intensity of the grating can be electrically controlled. Therefore, this CLC-based device can be potentially used in controllable diffraction elements in optics.

1. Introduction

Cholesteric liquid crystal (CLC) is a soft matter with a self-organized helical structure and can be regarded as a one-dimensional photonic crystal [1]. It has multiple responses under various external stimuli and can thus be applied on tunable photonic devices, such as tunable optical filter [2], diffraction grating [3], mirror-less laser [4], microlens [5], and eye protector [6]. CLCs can exhibit various textures, including planar (uniformly standing helix), fingerprint (uniformly lying helix), and focal conic, depending on the orientation of the helical axes in CLCs. Fingerprint CLCs exhibit modulation of refractive index in lateral direction and are suitable for use in developing optical gratings. By embedding the photosensitive moieties or doping some of the additives with photosensitivity into the fingerprint CLC, the helical pitch of the fingerprint CLC can be compressed or extended by light irradiation and can function as an optically tunable diffraction grating [7].

The optically tunable diffraction grating can be used for beam control and wide-range spectral scanning applications [7–9]. Fingerprint CLC can have the function of a tunable grating [10]. However, gratings based on fingerprint CLCs present no circular polarization selectivity, which is an attractive feature of CLCs. Therefore, gratings based on planar CLCs for diffracting circularly polarized light are developed in recent years [11–13]. Using holographic interference or a mask to generate a periodic distribution of light on photo-sensitive planar CLC-based materials leads to the periodic variations of CLC orientations and thus light diffraction [14, 15]. For example, a CLC grating can be fabricated via a spatial arrangement of the vertical and planar arrangement of the CLC molecules by two-beam interference [11]. A photo-induced surface relief phase grating can be realized on a photoresponsive CLC polymer film through the optical isomerization for reducing the molecular order parameters [12]. Polarization-selective CLC polymer gratings because of periodic phase modulation by phototunable helical pitch have also been demonstrated [13]. However, the planar CLC gratings based on photosensitive materials have stability issues. To improve the stability and functions of the planar CLC grating, a new type of CLC grating must be developed.

Table 1: Prescription of the refilled right-handed and left-handed CLC (RCLC and LCLC, respectively) materials. The indices of "red" and "green" behind RCLC or LCLC mean that the CLC materials reflect at red and green regions, respectively.

wt%	RCLC-red	RCLC-green	LCLC-red	LCLC-green
HTW114200-100	77	73	77	73
R811	23	27		
S811			23	27

In this work, optical gratings are fabricated on the basis of two-beam interference and fabrication technique of CLC template (washing-out/refilling) [16–18]. By refilling left- or right-handed CLC (LCLC and RCLC, respectively) into a right-handed CLC polymer template grating, the refilled CLC template grating can diffract light with specific wavelength. In addition, the refilled CLC template grating shows different diffraction efficiencies for left- and right-handed circular polarization (LCP and RCP, respectively) in reflection and transmission modes. The diffraction intensity can also be electrically controlled. The beam steering device exhibits electrical controllability and circular polarization and wavelength selectivity with dual operation modes.

2. Materials and Methods

The mixture used for fabricating CLC templates is composed of E7 (NLC, from Fusol-Material), R811 (right-handed chiral dopant, from Fusol-Material), R1011 (right-handed chiral dopant, from Fusol-Material), RMM691 (chiral monomer, from Merck), RMM257 (achiral diacrylate monomer, from Merck), and Irg184 (photoinitiator, from Pufeng). The weight ratio of these materials is 66.8:13:2:15.5:2.5:0.2. The uniform mixture was injected into an PVA-rubbed empty cell with a cell gap of 38 μm. Two UV laser beams from a He-Cd Laser (from Kimmon) were adjusted to construct an interference pattern on and cure the sample, resulting in a spatially periodic polymerized and nonpolymerized regions inside the sample. The cured sample was immersed in acetone for one day to wash-out the nonreactive materials. After evaporating the acetone, a dried grating with spatially periodic template and nontemplate regions was obtained. Last, another CLC material was then refilled into the grating to finish the fabrication of the refilling CLC template grating. In this paper, four types of refilled CLC materials composed of NLC (HTW114200-100, from Fusol-Material) and chiral dopants (S811 and R811, from Fusol-Material), as shown in Table 1, were prepared to refill into the sample. The four refilled CLCs are labelled as RCLC-red, RCLC-green, LCLC-red, and LCLC-green, in which "R" and "L" in front of CLC mean right- and left-handednesses of chirality, respectively, and "red" and "green" in the labels are the reflection bands at red and green regions, respectively.

As shown in Figure 1, one CW probe beam (λ = 532 nm or 633 nm) with right-handed or left-handed circular polarization was aligned to normally probe the refilling CLC template grating sample. Two photodetectors were used to measure the first-order diffraction efficiencies in the reflection and transmission modes of diffraction beams.

3. Results and Discussion

3.1. Refilled CLC Template Grating Samples with Various Refilled CLC Materials. As described in the previous section, the template grating was fabricated by refilling the CLC material (shown in Table 1) into the holographically cured sample with a grating spacing of about 7.3 μm. Two different regions are formed after refilling the CLC material into the sample, template, and nontemplate regions, which correspond to the constructive and destructive interference regions, respectively. Since the CLC cannot completely fill the nanopores of the template region, the reflection band of the refilled template region blue-shifts from its original state. The reflection band of the nontemplate region is attributed to the reflection of the refilled CLC. Therefore, the two regions can reflect at different colour regions. In other words, if a probe beam with a specific wavelength probes on the grating sample, the sample can be regarded as an amplitude grating. Given that the reflection of CLC is for a specific circular polarization, the refilled CLC template grating can be designed to reflect lights of a specific wavelength and circular polarization by refilling CLCs with different handednesses and reflection bands. For describing the circular polarization selectivity of the refilled template grating, a parameter called g-value is defined as [19]

$$g = \frac{2(\eta_L - \eta_R)}{(\eta_L + \eta_R)}, \quad (1)$$

where η_L and η_R represent left-handed and right-handed circularly polarized diffraction efficiencies, respectively. A positive g-value indicates that the diffraction efficiency with LCP is greater than that with RCP. Therefore, a large absolute g-value near 2 means a good circular polarization selectivity of the refilled CLC template grating.

Figure 2 shows the circular polarization selectivity of the template grating refilled with CLC material reflecting red light (RCLC-red and LCLC-red, reflection region: 590 nm–650 nm). When the refilled material is RCLC-red, the nontemplate region can reflect red RCP light but the template region reflects blue RCP light, which can be observed via the reflection spectrum shown in Figure 2(a). In the reflection mode, the diffraction efficiency of the RCP red light is greater than that of the LCP red light ($g = -1.14$) [Figure 2(a)]. On the contrary, the effect of the transmission mode is opposite to the reflection mode ($g = 1.38$) [Figure 2(b)]. If the wavelength of the probe beam is 532 nm, then the light cannot be reflected either by the template or the nontemplate region and thus should not experience an amplitude grating. However, the refractive index difference between the template

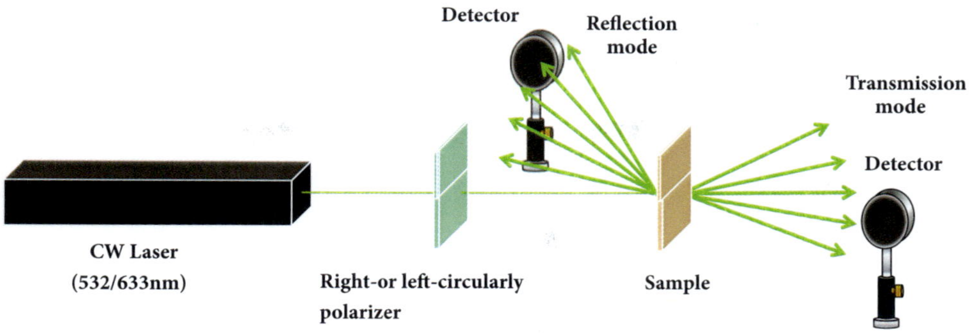

FIGURE 1: Setup for measuring the diffraction efficiencies of the refilled CLC template grating sample in reflection and transmission modes of diffraction beams.

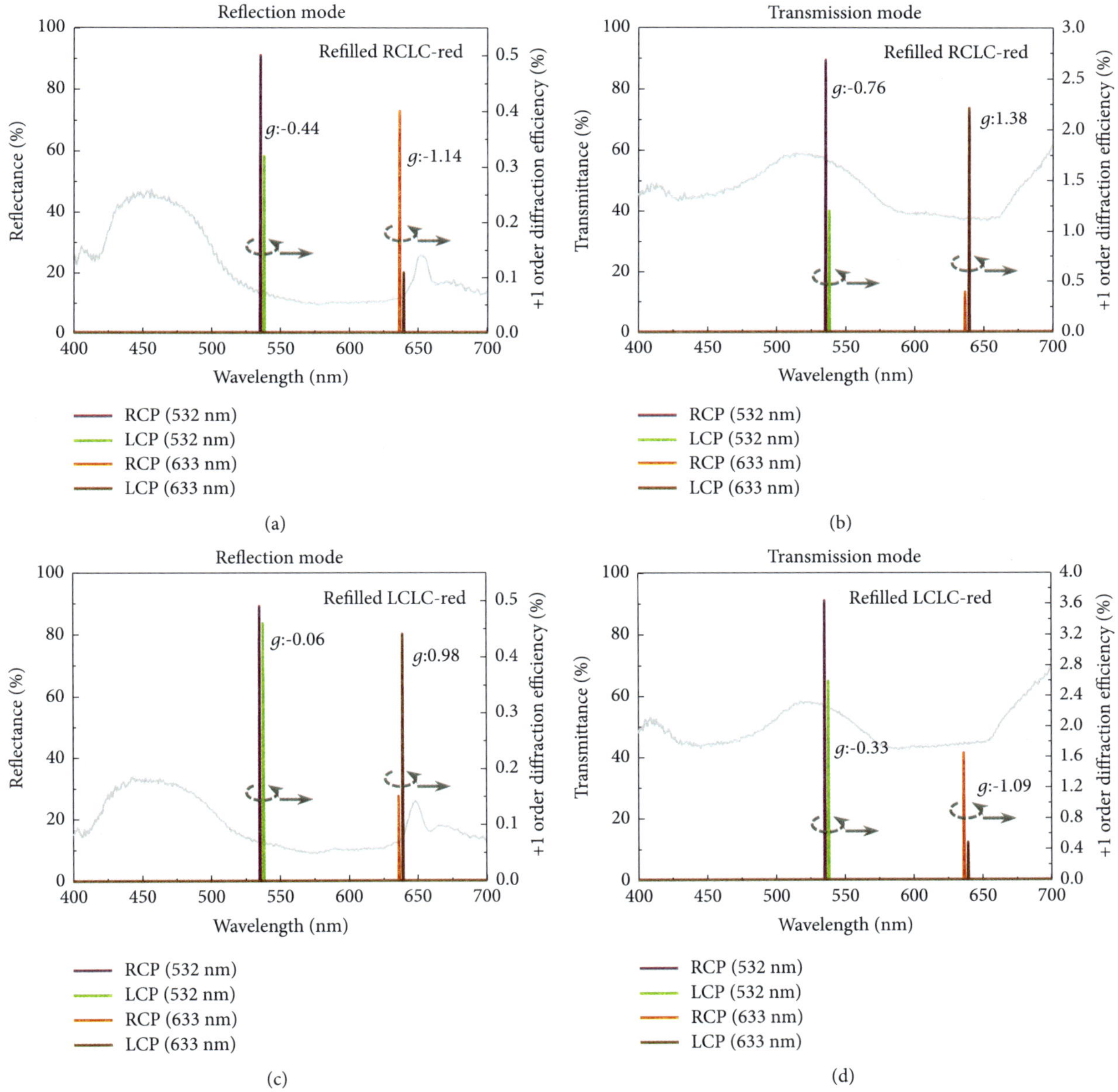

FIGURE 2: Measured +1 order diffraction efficiencies of the refilled CLC template grating in (a)/(c) reflection mode and (b)/(d) transmission mode by the RCP and LCP lights with the wavelengths of 532 and 632 nm, respectively. The refilled material is RCLC-red in (a)/(b) and LCLC-red in (c)/(d).

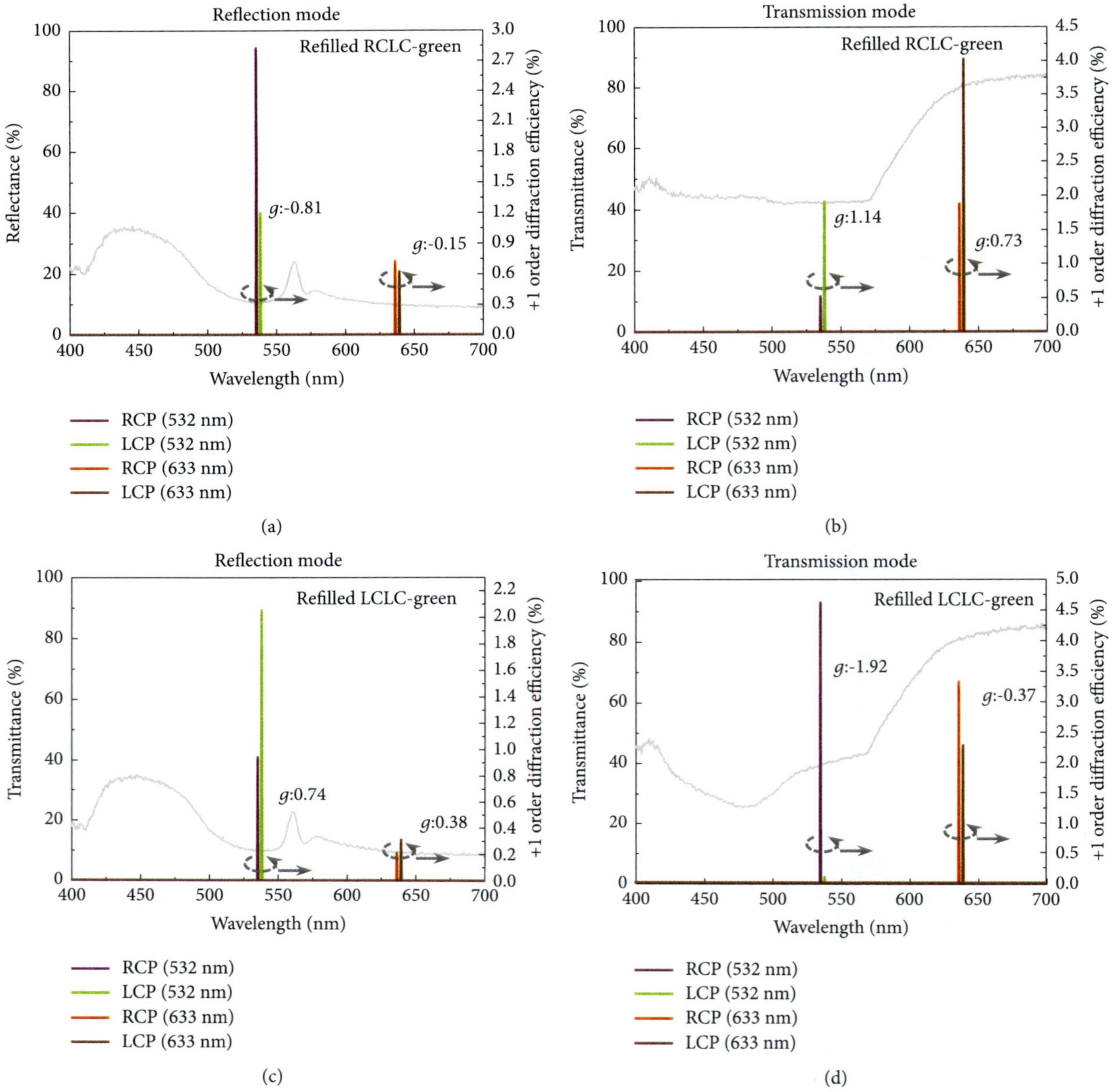

Figure 3: Measured +1 order diffraction efficiencies of the refilled CLC template grating in (a)/(c) reflection mode and (b)/(d) transmission mode by the RCP and LCP lights with the wavelengths of 532 and 632 nm, respectively. The refilled material is RCLC-green in (a)/(b) and LCLC-green in (c)/(d).

and nontemplate regions results in a slight effect of phase grating. Therefore, the green light can also be diffracted, but the absolute g-value is smaller than that of the red light. This condition represents no circular polarization selectivity for green light in this sample. When the LCLC-red is used as the refill material, the refilled CLC template grating sample exhibits two reflection bands in the blue and red regions, as shown in Figure 2(c). The properties of the refilled material are similar to RCLC-red, except for the opposite chirality. In the reflection mode, the diffraction efficiency of the RCP red light is smaller than that of the LCP red light ($g = 0.98$), as shown in Figure 2(c). On the contrary, the effect of the transmission mode is opposite to that of the reflection mode ($g = -1.09$), as shown in Figure 2(d). The absolute g-value in the green region is smaller than that in the red region, and no regular circular polarization selectivity occurs for green light in either the reflection or transmission mode. In other words, the polarization selectivity of the grating is only for red light when the refilled CLC reflects red light.

By refilling RCLC-green with reflection band in the green region (reflection region: 510–570 nm), the refilled CLC template grating causes the circular polarization selectivity of the green light. In the reflection mode, the diffraction efficiency of the RCP green light is greater than that of the LCP green light ($g = -0.81$), as shown in Figure 3(a). Conversely, the effect of the transmission mode is opposite

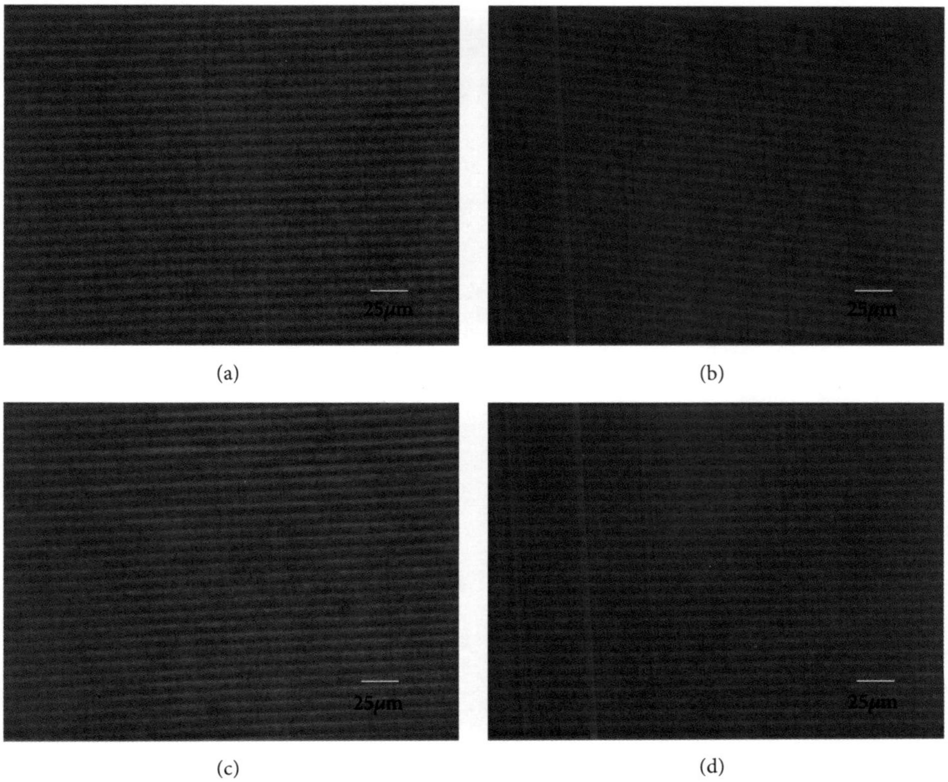

FIGURE 4: Reflected polarized optical microscopic image of the CLC polymer template gratings refilled with (a) RCLC-red, (b) LCLC-red, (c) RCLC-green, and (d) LCLC-green, respectively. The scale bar is 25 μm.

to that of the reflection mode ($g = 1.14$), as shown in Figure 3(b). When the refill material is LCLC-green, the refilled CLC template grating sample exhibits two reflection bands in the blue and green regions. The properties of the refilled material are similar to RCLC-green, except for the opposite chirality. In the reflection mode, the diffraction efficiency of the RCP green light is smaller than that of the LCP green light ($g = 0.74$), as shown in Figure 3(c). Conversely, the effect of the transmission mode is opposite to that of the reflection mode ($g = -1.92$), as shown in Figure 3(d). Figure 4 shows the reflected polarized optical microscopic images (with crossed polarizers) of the refilled template gratings discussed in Figures 2 and 3. In Figures 4(a) and 4(b), the red colour of the stripes is due to the reflection of the refilled red CLC materials, RCLC-red and LCLC-red. Similarly, the green stripes shown in Figures 4(c) and 4(d) are caused by reflection of the refilled green CLC materials.

Since the gratings in the work are fabricated by two-beam interference, the spatial distributions of refractive index and transmittance in the gratings can be expressed by a sinusoidal function. That is, the gratings can be regarded as sinusoidal transmittance gratings or sinusoidal phase gratings, depending on the experimental condition. If the incident light can be reflected by the CLC template or the refilled CLC, the sample is a sinusoidal transmittance grating. On the contrary, the sample can be regarded as a sinusoidal phase grating when the wavelength or polarization of the incident light does not match the reflection band or handedness of the CLC template or the refilled CLC. For a sinusoidal transmittance grating, the first-order diffraction efficiency varies as the square of the amount of swing of the sinusoidal transmittance modulation Δt, and is given by the formula $\eta_t = (\Delta t/2)^2$. Because the maximum value of Δt is 0.5, the maximum value of the first diffraction efficiency is 6.25% ideally [20]. The experimental values are lower than the maximum theoretical value because the transmittance modulations of the refilled CLC template gratings are smaller than 0.5. For a sinusoidal phase grating, the first-order diffraction efficiency can be expressed by a first-order Bessel function of the phase modulation of the grating $\Delta\varphi$, that is, $\eta_p = J_1^{\,2}(\Delta\varphi)$. It means the diffraction efficiency of a sinusoidal phase grating can be ranged from 0 to 33.8%, depending on the value $\Delta\varphi$, which is related to the variation of the refractive indices and thickness of the grating [20]. Since the variation of the refractive indices of the refilled CLC template gratings and thus $\Delta\varphi$ is very small, the diffraction efficiency is not high. In addition, the scattering of light from the refilled CLC template grating is another factor for the low diffraction efficiency.

3.2. Electrically Controlled Diffraction of Refilled CLC Template Grating Sample. Considering that the arrangement of the refilled CLC can be changed by applying an electric field, the diffraction of the refilled CLC template grating can be controlled by AC electric field (1 kHz). As displayed in Figure 5, when the electric field is applied, the CLC in the nontemplate region can be converted from planer to

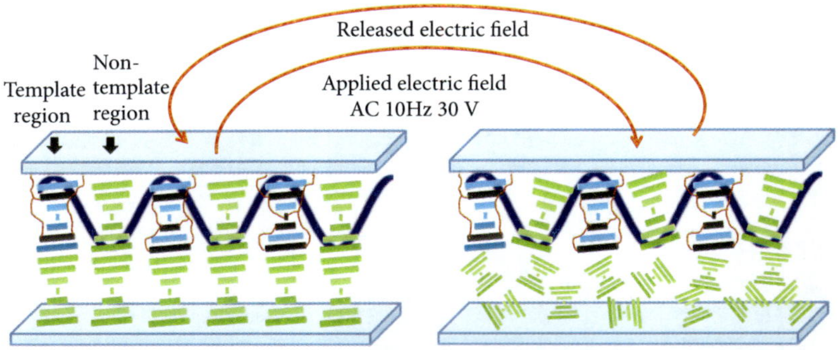

FIGURE 5: Schematic of conversion between planer and focal conic states in the refilled CLC template grating by the presence of an applied electric field.

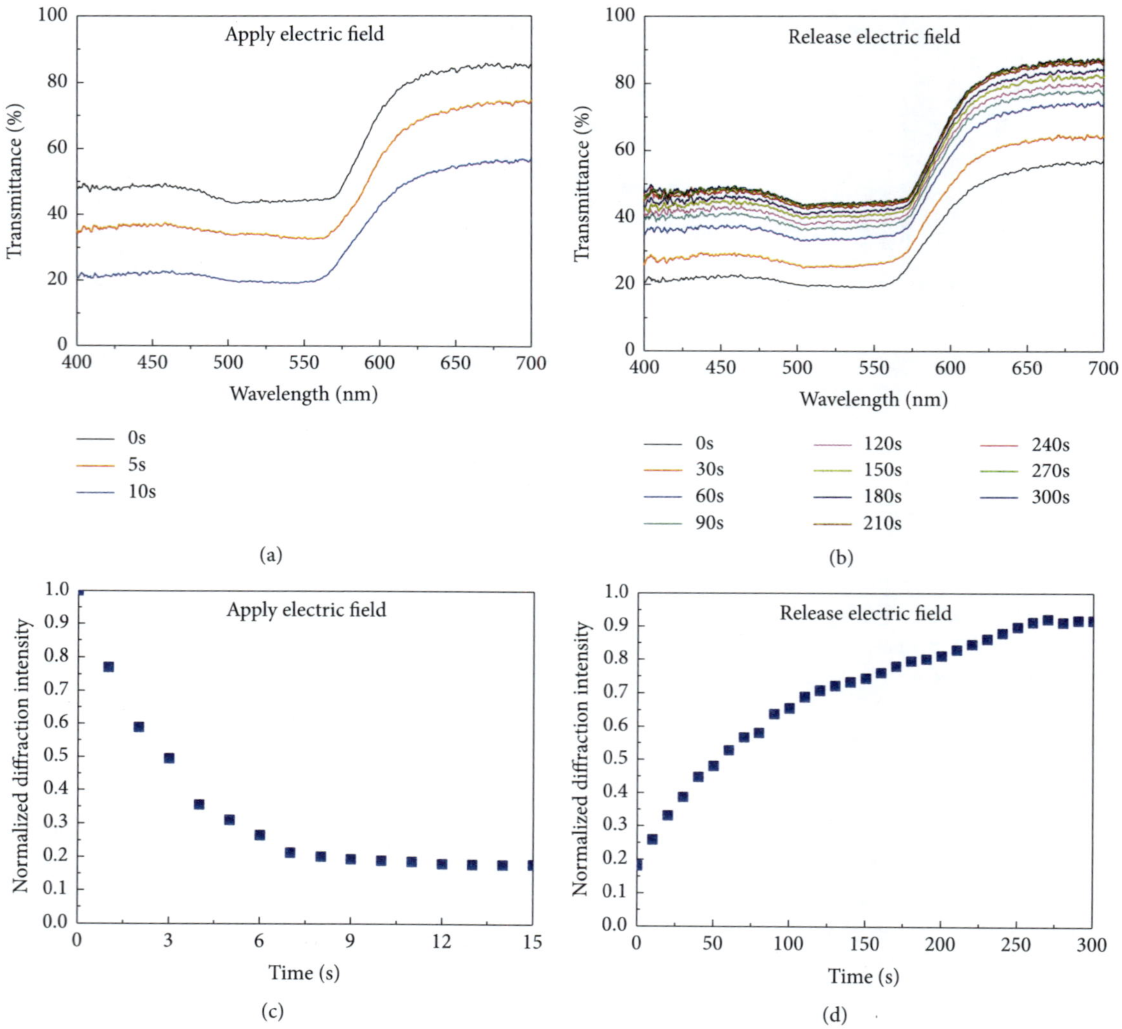

FIGURE 6: Dynamical variations of (a)/(b) the transmittance and (c)/(d) diffraction intensity of the refilled CLC template grating when (a, c) applying and (b, d) releasing the electric field.

focal conic state [21]. The refilled CLC in the nontemplate region will return from focal conic state to planer state when the electric field is released. The CLC molecules in the nontemplate region of the sample enter into a focal conic state when an electric field with 30 V is applied. The strong scattering of the focal conic state gradually reduces the transmittance, as shown in Figure 6(a). When the electric field is removed, the transmittance gradually rises back to its initial state, as shown in Figure 6(b). The electrically controlled scattering and transmittance results in the variation of the diffraction intensity with electric field [22]. Therefore, the diffraction intensity of the refilled CLC template grating can be controlled to decreases and increase with the electric field, as shown in Figures 6(c) and 6(d), respectively. In this work, the distance between the detector and the sample is 30 cm. The contrast ratio of the electrically switched diffraction can be higher if less scattered light is received by the detector, which can be achieved by increasing the distance between the detector and the sample.

4. Conclusions

In this paper, holographic interference is used to fabricate diffraction devices based on CLC polymer template gratings with circular polarization and wavelength selectivity. The refilled CLC template gratings can be operated in reflection and transmission modes. The diffractions of the reflection and transmission modes exhibit different circular polarization. By replacing the refilled CLC, the device can diffract light with specific circular polarization and wavelength. Finally, the electrical controllability of the device is demonstrated by switching the CLC texture between planer to focal conic states with an electric field.

Conflicts of Interest

The authors declare that there are no conflicts of interest regarding the publication of this paper.

Authors' Contributions

Hsien-Kuo Chin and Hui-Ying Kuo contributed equally to this paper.

Acknowledgments

This research is financially supported by clinical research grants from Kaohsiung Armed Forces General Hospital, Taiwan [no. 105-14], and the Ministry of Science and Technology of Taiwan (Contract no. MOST 103-2112-M-006-012-MY3).

References

[1] P. G. de Gennes and J. Prost, *The Physics of Liquid Crystals*, Oxford University Press, 1993.

[2] M. Mitov and N. Dessaud, "Going beyond the reflectance limit of cholesteric liquid crystals," *Nature Materials*, vol. 5, no. 5, pp. 361–364, 2006.

[3] S. P. Palto, L. M. Blinov, M. I. Barnik, V. V. Lazarev, B. A. Umanskii, and N. M. Shtykov, "Photonics of liquid-crystal structures: A review," *Crystallography Reports*, vol. 56, no. 4, pp. 622–649, 2011.

[4] H. Coles and S. Morris, "Liquid-crystal lasers," *Nature Photonics*, vol. 4, no. 10, pp. 676–685, 2010.

[5] P. Popov, L. W. Honaker, M. Mirheydari, E. K. Mann, and A. Jákli, "Chiral nematic liquid crystal microlenses," *Scientific Reports*, vol. 7, no. 1, 2017.

[6] W. Zhang, L. Zhang, X. Liang et al., "Unconventional High-Performance Laser Protection System Based on Dichroic Dye-Doped Cholesteric Liquid Crystals," *Scientific Reports*, vol. 7, no. 1, 2017.

[7] H.-C. Jau, Y. Li, C.-C. Li et al., "Light-driven wide-range nonmechanical beam steering and spectrum scanning based on a self-organized liquid crystal grating enabled by a chiral molecular switch," *Advanced Optical Materials*, vol. 3, no. 2, pp. 166–170, 2015.

[8] I. Gvozdovskyy, O. Yaroshchuk, M. Serbina, and R. Yamaguchi, "Photoinduced helical inversion in cholesteric liquid crystal cells with homeotropic anchoring," *Optics Express*, vol. 20, no. 4, pp. 3499–3508, 2012.

[9] H.-C. Jau, T.-H. Lin, R.-X. Fung, S.-Y. Huang, J.-H. Liu, and A. Y.-G. Fuh, "Optically-tunable beam steering grating based n azobenzene doped cholesteric liquid crystal," *Optics Express*, vol. 18, no. 16, pp. 17498–17503, 2010.

[10] L.-L. Ma, W. Duan, M.-J. Tang et al., "Light-driven rotation and pitch tuning of self-organized cholesteric gratings formed in a semi-free film," *Polymer*, vol. 9, no. 7, p. 295, 2017.

[11] A. Ryabchun, A. Bobrovsky, Y. Gritsai, O. Sakhno, V. Shibaev, and J. Stumpe, "Stable selective gratings in LC polymer by photoinduced helix pitch modulation," *ACS Applied Materials & Interfaces*, vol. 7, no. 4, pp. 2554–2560, 2015.

[12] P.-Z. Sun, Z. Liu, W. Wang et al., "Light-reconfigured waveband-selective diffraction device enabled by micro-patterning of a photoresponsive self-organized helical superstructure," *Journal of Materials Chemistry C*, vol. 4, no. 39, pp. 9325–9330, 2016.

[13] D. Liu, C. W. M. Bastiaansen, J. M. J. Den Toonder, and D. J. Broer, "Light-induced formation of dynamic and permanent surface topologies in chiral-nematic polymer networks," *Macromolecules* , vol. 45, no. 19, pp. 8005–8012, 2012.

[14] L. De Sio, S. Serak, N. Tabiryan, S. Ferjani, A. Veltri, and C. Umeton, "Composite holographic gratings containing light-responsive liquid crystals for visible bichromatic switching," *Advanced Materials*, vol. 22, no. 21, pp. 2316–2319, 2010.

[15] H. Yeh, G. Chen, C. Lee, and T. Mo, "Optically switchable biphotonic gratings based on dye-doped cholesteric liquid crystal films," *Applied Physics Letters*, vol. 90, no. 26, p. 261103, 2007.

[16] J. Guo, H. Cao, J. Wei et al., "Polymer stabilized liquid crystal films reflecting both right- and left-circularly polarized light," *Applied Physics Letters*, vol. 93, no. 20, p. 201901, 2008.

[17] J.-D. Lin, C.-L. Chu, H.-Y. Lin et al., "Wide-band tunable photonic bandgaps based on nematic-refilling cholesteric liquid crystal polymer template samples," *Optical Materials Express* , vol. 5, no. 6, pp. 1419–1430, 2015.

[18] J.-D. Lin, H.-L. Lin, H.-Y. Lin et al., "Widely tunable photonic bandgap and lasing emission in enantiomorphic cholesteric liquid crystal templates," *Journal of Materials Chemistry C*, vol. 5, no. 13, pp. 3222–3228, 2017.

[19] A. Bobrovsky, K. Mochalov, V. Oleinikov et al., "Optically and electrically controlled circularly polarized emission from cholesteric liquid crystal materials doped with semiconductor quantum dots," *Advanced Materials*, vol. 24, no. 46, pp. 6216–6222, 2012.

[20] S. A. Benton and V. M. Bove, *Holographic Imaging*, John Wiley & Sons, Inc., Hoboken, NJ, USA, 2008.

[21] C. Wang and T. Lin, "Bistable reflective polarizer-free optical switch based on dye-doped cholesteric liquid crystal [Invited]," *Optical Materials Express* , vol. 1, no. 8, p. 1457, 2011.

[22] Y.-C. Zheng, *ircular-polarization and wavelength selective gratings based on polymer templates of holographic cholesteric liquid crystal [Master, thesis]*, National Cheng Kung University, 2017.

13

Experimental Study on Ferromagnetic Shunt Effects on the Critical Current of BSCCO Tape in Stacked Conductors

J. Sun,[1] H. Ohara,[2] and S. Yamaguchi[1,2]

[1] *Center of Applied Superconductivity and Sustainable Energy Research (CASER), Chubu University, 1200 Matsumoto-cho, Kasugai, Aichi 487-8501, Japan*
[2] *Department of Electrical Engineering, Chubu University, 1200 Matsumoto-cho, Kasugai, Aichi 487-8501, Japan*

Correspondence should be addressed to J. Sun; j_sun@isc.chubu.ac.jp

Academic Editor: Sergei Sergeenkov

In the 200 m high temperature superconducting (HTS) cable test facility at Chubu University constructed in 2010, a three-layer structure of the tapes in the cable is employed for obtaining the high current capacity up to 2 kA. Previous study shows that the critical current is affected by the layout of the tapes such as gaps and the current feeding mode. In the stacked tape conductors, the critical current of BSCCO tapes shows strong dependence on the current feeding directions between the tapes. The critical current is improved when the opposite-direction current is applied to them and degraded for the same-direction current feeding mode because of strong magnetic field interaction between them from the transport current in each tape. This paper presents the measurements of the critical currents of a BSCCO tape in the stacked conductors with the ferromagnetic materials. By using the ferromagnetic materials surrounding the tapes, self-field of HTS tapes is affected and their critical currents are improved by more than 10%. Ferromagnetic shunt effects on the critical current of HTS tape are reported through the magnetic field analysis.

1. Introduction

Since the first DC high temperature superconducting (HTS) power cable test facility was constructed at Chubu University [1], studies and tests on the DC HTS power cable have been continued and a 200 m DC HTS cable system, called CASER-II, was set up successfully in 2010 [2]. Figure 1(a) shows a photo of the CASER-II's cable. A three-layer coaxial structure of the cable is used as in Figure 1(b). In order to improve the cable property by the tape arrangements, the performance of the HTS tapes for the DC power cable applications was investigated through the critical current measurements [3].

To obtain a high current density of a cable, a multilayer structure is needed and extremely stacked tape conductors are one of the choices to make this by using multilayer conductors as in Figure 1. The stacked structure makes the conductor compact but increases the magnetic field interaction between them. The latter would significantly affect the performance of HTS tapes in the stacked conductor. In a previous study, the critical current of BSCCO tape in the three-layer stacked conductors is dependent on current feeding mode [4]. With the opposite-direction current applied to each tape in the three-layer stacked conductors, the critical current of middle tape increases sharply by 50% due to the effects on self-field from the currents in the neighboring tapes compared to that of single one. However, the critical current degrades for the same-direction current feeding because of effects on self-field from the neighboring current in the adjacent tapes.

FMs have been used to improve the performance of BSCCO tape by covering its edges [5] and HTS cable by covering the gap between the tapes as a diverter [6]. However, the increase in the critical current is less than 7% in [5] by covering thin Ni coating layer (24 μm) on edges of a BSCCO tape. In this work, we investigated the critical current of BSSCO tape in the stacked conductors covered with and without Ferromagnetic materials (FMs) by varying current feeding mode and the critical current can be improved by more than 10% experimentally. The insulated tapes are arranged in a tape-on-tape structure and different currents

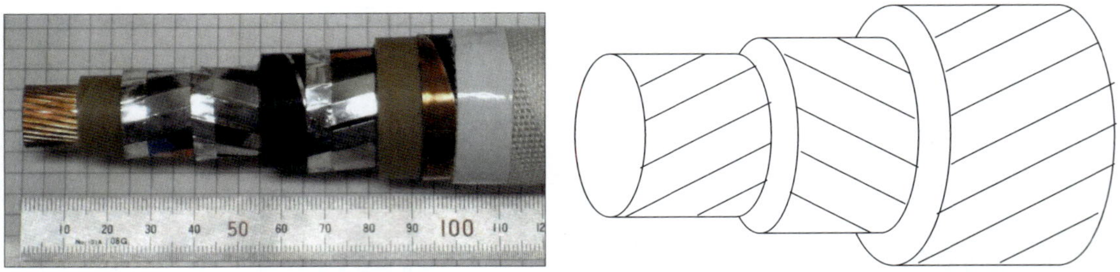

(a) (b)

FIGURE 1: (a) A photo of the CASER-II power cable. (b) A schematic layout of the coaxial three-layer HTS conductor for the CASER-II cable.

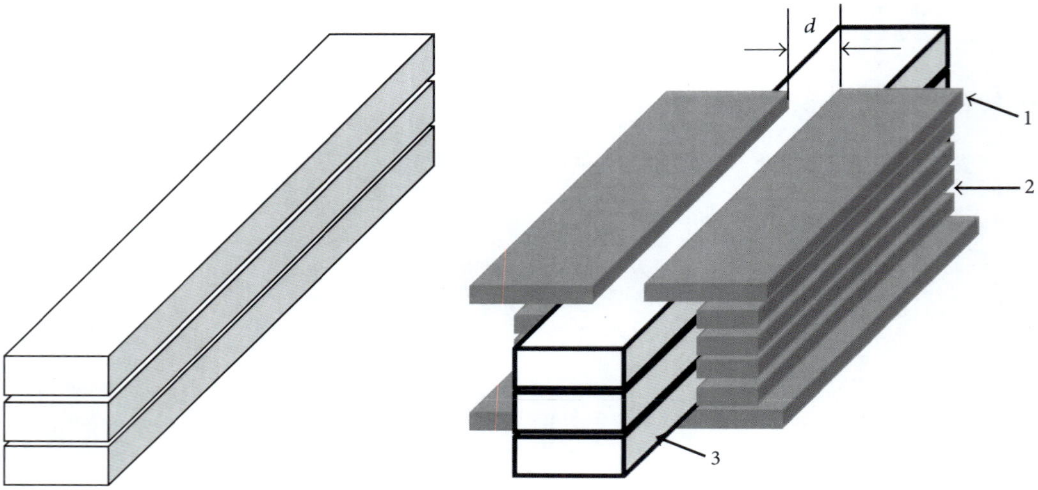

(a) (b)

FIGURE 2: A schematic layout of three-layer stacked conductor without (a) and with (b) ferromagnetic shunt materials. 1: four pieces of 5 mm width ferromagnetic sheets; 2: eight pieces of 3 mm width ferromagnetic sheets; 3: three HTS tapes; and d: a space gap between 5 mm width ferromagnetic sheets above tape wide surface.

are applied to each tape. The critical current of the target sample is measured by the four-probe method by immersing the samples into a liquid nitrogen bath. We will present the experimental results and analysis on the magnetic field interactions between tapes to improve the superconductivity for DC power cable applications.

2. Experiments

The experimental setup of the critical current measurement is similar to previous report [3]. The HTS tapes are BSCCO tapes (Bi-2223) from Sumitomo Electric Industries. Three insulated BSCCO tapes (Type HT-CA) are used in the experiment with a cross section of 4.5 mm × 0.35 mm and a length of ~270 mm [7]. The critical current of BSCCO tape is 200 A in self-field at 77 K. The tapes are arranged as a stacked structure as shown in Figure 2(a). Three voltage taps are soldered on each tape with the distance of 8 and 10 cm. The ferromagnetic sheets have a thickness of 0.3 mm. Eight 3 mm width FM sheets and four 5 mm width FM sheets are covered at edges with a space gap of d as in Figure 2(b). The ferromagnetic materials are provided from JFE steel [8].

Two power supplies (PR10-300) with output current/voltage 300 A/10 V are used to control the current feeding mode in the three-layer stacked conductor as in Figures 3(a) and 3(b). The first power supply (I_1) is used for the current feeding to the middle BSCCO tape 2 and the second power supply (I_2) for bottom and upper BSCCO tapes (1 and 3) in series. The transport current is measured by a current shunt resistor 0.25 mΩ. All voltage signals are measured by a Keithley 2700 multimeter with sensitivity of 1 μV. The I-V characteristic curves are measured through the standard four-probe method at liquid nitrogen temperature of 77 K.

3. Results and Discussion

Figures 4(a) and 4(b) present comparisons of E-I curves of a BSCCO tape for single- and three-layer stacked tape conductor for different neighboring current of ±160 A without and with FMs, respectively. The electric fields (E) are obtained by normalizing the voltages to the distance between the voltage taps. The significant difference between the three-layer stacked conductors and single one is shown when the currents in the adjacent tapes are changed. In the case of no

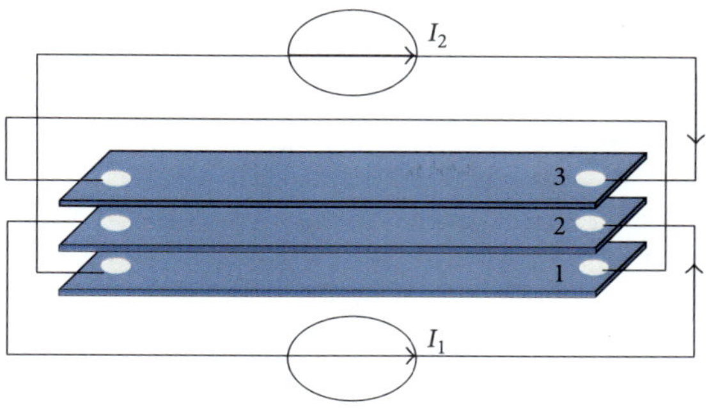

FIGURE 3: Schemes of (a) same-direction (↑↑↑) and (b) opposite-direction (↓↑↓) current feeding modes. The transport current applied to tape 1 and tape 3 is controlled by power supply (I_2) and the transport current for tape 2 is controlled by I_1. The critical current of tape 2 is measured by four-probe method.

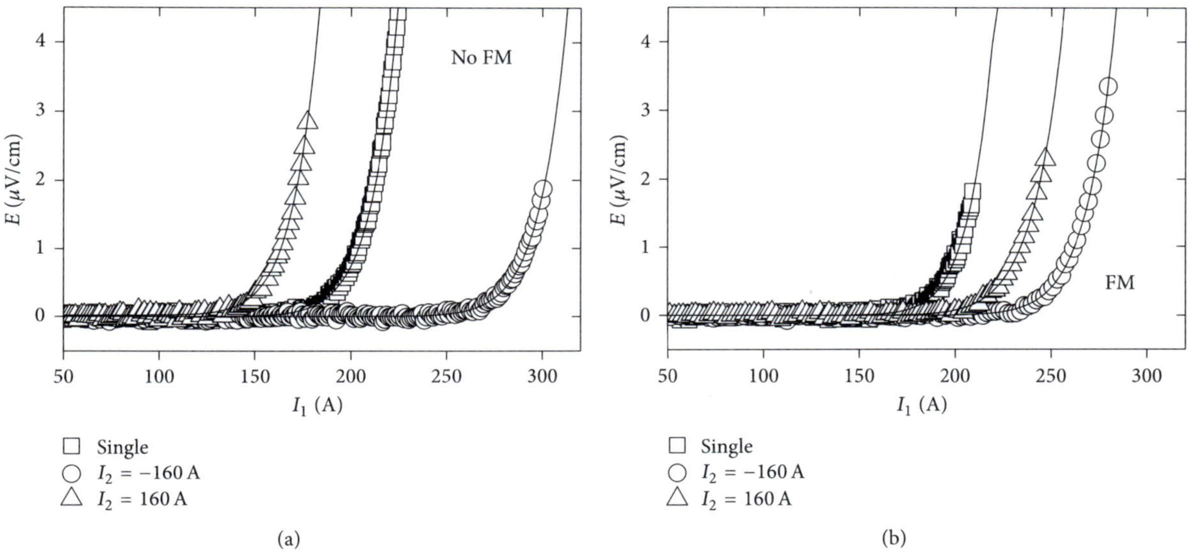

FIGURE 4: *E-I* curves of BSCCO tape for single (□) and stacked conductors (○ and △) with different currents of ±160 A applied to the adjacent BSCCO tapes without (a) and with (b) FM.

FM, the *I-V* curves shift up for opposite current of −160 A and down for the same current of 160 A, respectively. When FMs are used to cover at the edges of the stacked conductor, improvement of transport properties can be seen from *E-I* curves in Figure 4(b). To obtain the critical current, we fit the *E-I* curves by the following equation:

$$E = E_0 \times \left(\frac{I}{I_c}\right)^n, \quad (1)$$

where E is electric field, E_0 is electric field criterion of 1 μV/cm, I is transport current, I_c is critical current, and n is n-value, respectively.

Figure 5 shows the measured critical currents of BSCCO tape with respect to neighboring current without FMs. The horizontal axis is the neighboring current I_2 in the adjacent tapes 1 and 3 as in Figure 3. The positive current is the same-direction current (↑↑↑) as in Figure 3(a) and the negative current is the opposite-direction current (↓↑↓) as in Figure 3(b). The critical current of single tape (I_{c0}) is measured to be 201 A. When the positive currents are applied to the stacked conductor, the critical current of the middle tape is improved significantly. When the neighboring current becomes more than 80 A, the critical current degrades sharply by 20% as in Figure 5(b).

Figure 6 shows the measured n-value of BSCCO tapes in the stacked conductor without FMs in comparison with a single tape. The n-value shows similar trend with respect to the neighboring current as in Figure 5.

Figures 7 and 8 present the measured critical current and n-value of BSCCO tape in the stacked conductor with FMs as in Figure 2(b). The critical current increases with increasing the neighboring current I_2 for the opposite-direction current and decrease with the neighboring current I_2 for the same-direction current. The n-value shows similar behavior as the critical current with respect to the neighboring current.

Dependence of the critical current on the neighboring current is different for the stacked conductor covered with or

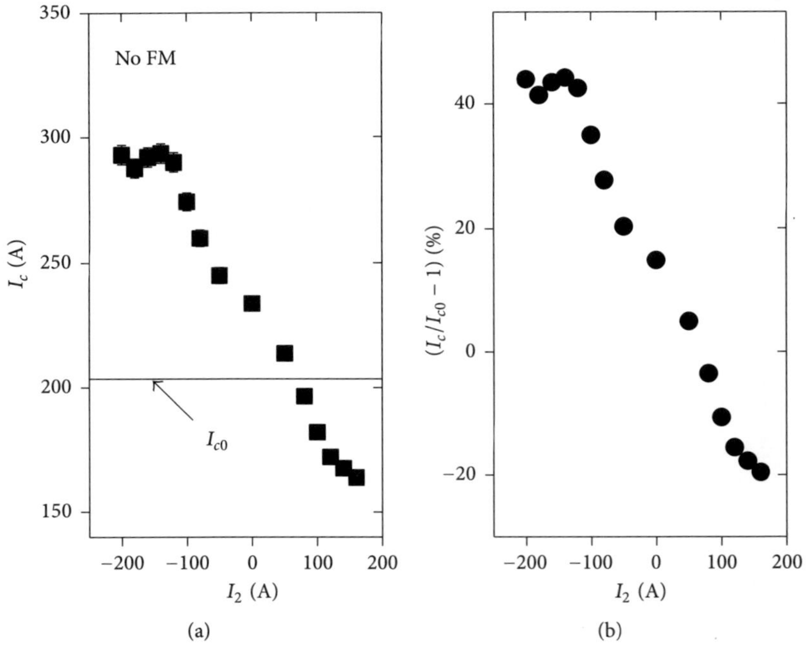

FIGURE 5: (a) I_c versus the applied current to the neighboring tapes and (b) increase of I_c with respect to the neighboring current for three-layer stacked conductor without FM.

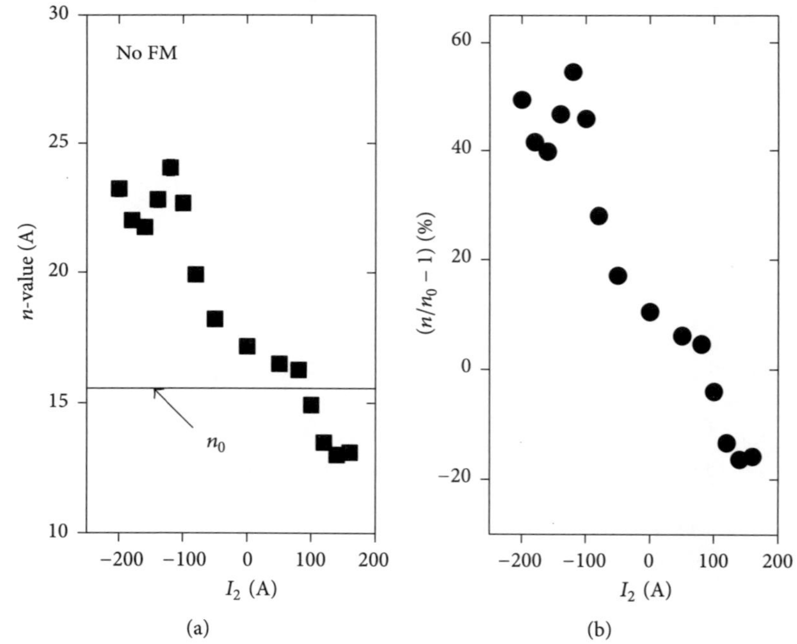

FIGURE 6: (a) n-value versus the applied current to the neighboring tapes and (b) increase of n with respect to the neighboring current for three-layer stacked conductor without FM.

without FMs. When the current of the same direction (↑↑↑) is applied to each tape, the critical current declines without FMs, which is similar with the previous report [4]. However, the critical current is improved by covering the edges of the stacked conductor by FMs as in Figures 7 and 8.

The magnetic field simulation is performed by the finite element method (FEM) ANSYS code [9] to analyze the ferromagnetic shunt effects on the self-field. Figure 9 presents the magnetic flux distribution for three-layer stacked conductor with and without FMs under the same-direction

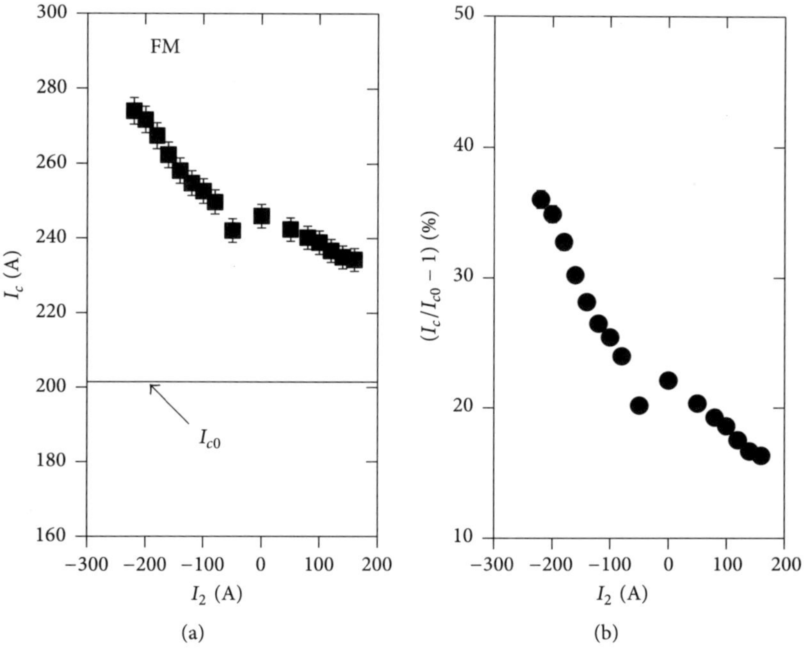

FIGURE 7: (a) I_c versus the applied current to the neighboring tapes and (b) increase of I_c with respect to the neighboring current for three-layer stacked conductor with FMs.

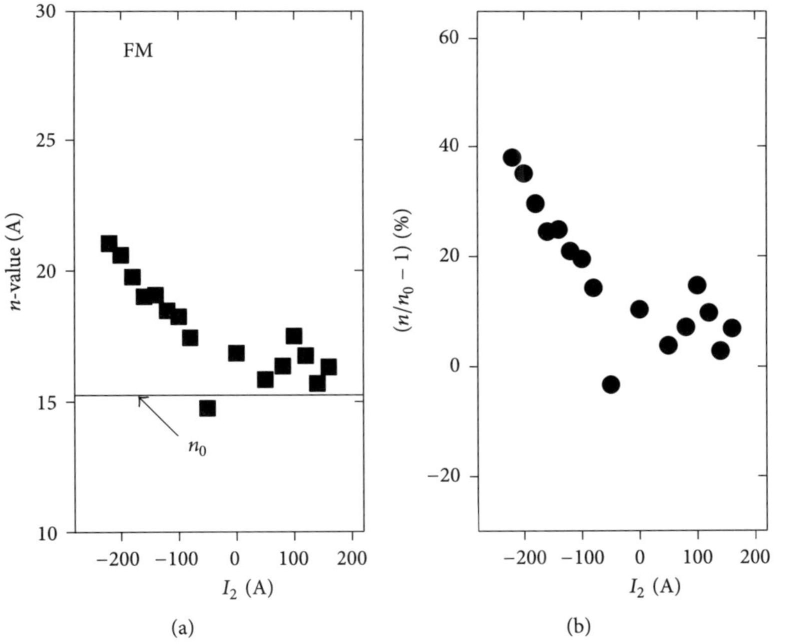

FIGURE 8: (a) n-value versus the applied current to the neighboring tapes and (b) increase of n-value with respect to the neighboring current for three-layer stacked conductor with FMs.

current (↑↑↑) of 200 A. Figure 10 presents the magnetic flux distribution for opposite-direction current of 200 A (↑↓↑). The current of 200 A is assumed to be uniformly distributed in the HTS tapes. The nonlinear magnetization B-H relations of the ferromagnetic materials are used in the calculations [8].

Compared with the magnetic field for no-FM covers, the magnetic flux is attracted through the FMs and leads to reduction of the magnetic field in the middle tape for the same-direction current as in Figure 9. However, as shown in Figure 9(b), the magnetic fields around the middle tape become neutral from neighboring current with FMs covered at the edges, which results in the great improvement of the critical current as above observed in Figures 7 and 8. In the case of the opposite-direction current, the magnetic flux

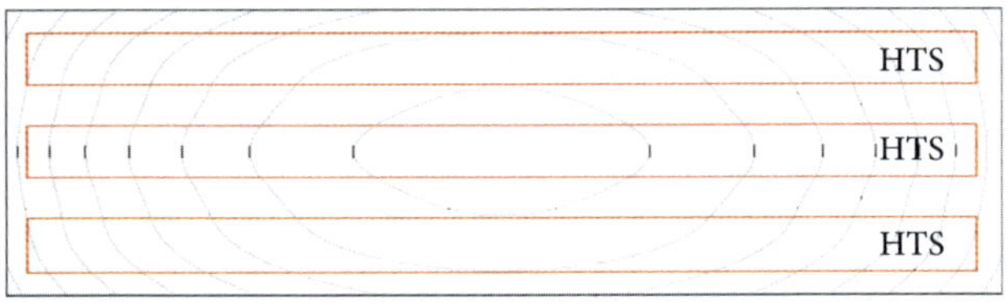

(a)

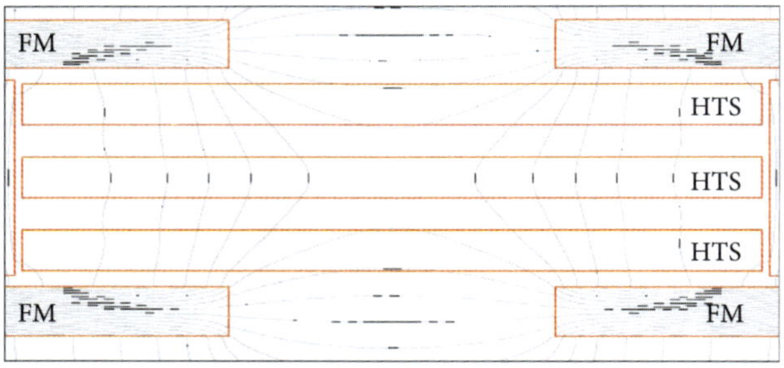

(b)

FIGURE 9: Calculated magnetic flux lines by finite element method ANSYS code [9] for three-layer stacked tape conductors (a) without FMs and (b) with FMs at the edges under the same-direction (↑↑↑) current feeding mode. The currents in each tape are $I_1 = I_2 = I_3 = 200$ A as in Figure 3(a).

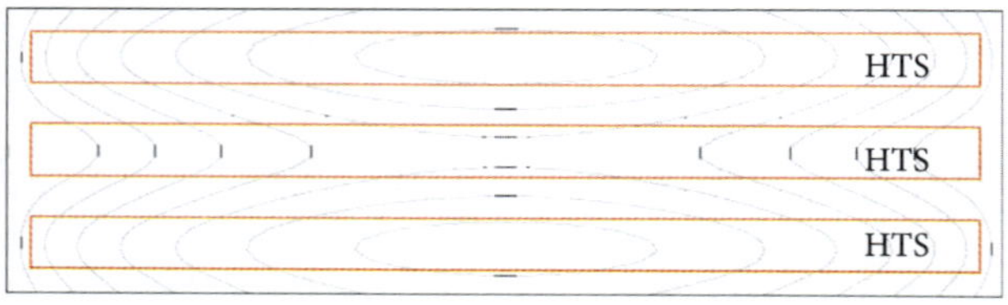

(a)

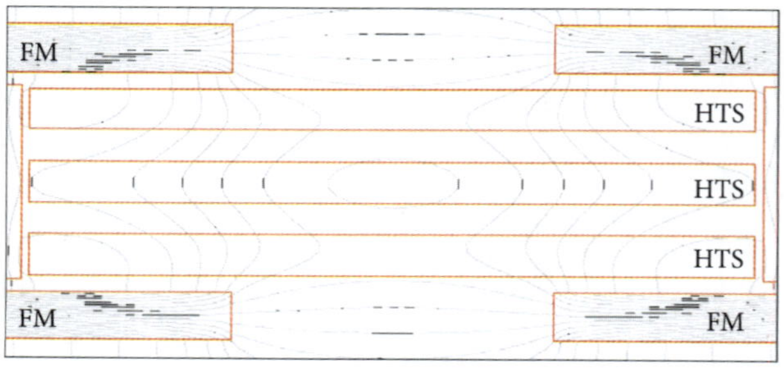

(b)

FIGURE 10: Calculated magnetic flux lines by finite element method ANSYS code [9] for three-layer stacked tape conductors (a) without FMLs and (b) with FMs at the edges under the antidirection (↑↓↑) current feeding mode. The currents in each tape are $I_1 = -I_2 = I_3 = 200$ A as in Figure 3(b).

is also absorbed by the FMs as in Figure 10(b); however, the magnetic flux distribution of stacked conductor with or without FMs is similar as in Figure 10, which leads to no improvements of the critical current or n-value as in Figures 5 and 6.

4. Conclusion

We investigated the critical current and n-value of a 200 A-level BSSCO tape in the stacked conductors covered with and without ferromagnetic materials by varying current feeding mode. With the opposite-direction (↑↓↑) current applied to each tape in the three-layer stacked tape conductors, both the critical currents and the n-values of middle tape are improved due to the effects on the self-field of the magnetic field from the neighboring current. Prominent improvement of the critical current by covering FMs at the edges around the stacked conductor comparing with uncovered one is observed when the same-direction (↑↑↑) current mode is applied due to the ferromagnetic shunt effects through the magnetic field calculations by FEM.

Competing Interests

The authors declare that they have no competing interests.

Acknowledgments

The authors acknowledge all members in CASER, Chubu University, for their useful discussion. The authors also acknowledge Professor A. Iiyoshi, the Board Chairperson of Chubu University, for his continuous encouragements through the work. The authors acknowledge Dr. Hiroshi Fujiwara, a Professor of Keio University and the President of Nano-Opt Energy Corporation, for his financial support on the construction of the CASER-II. Parts of this work were supported by a grant of Strategic Research Foundation Grant-Aided Project for Private Universities Ministry of Education, Culture, Sports, Science and Technology (MEXT).

References

[1] S. Yamaguchi, M. Hamabe, I. Yamamoto et al., "Research activities of DC superconducting power transmission line in Chubu University," *Journal of Physics: Conference Series*, vol. 97, no. 1, Article ID 012290, 2008.

[2] S. Yamaguchi, T. Kawahara, M. Hamabe et al., "Design and construction of 200meter high temperature superconducting DC power cable test facility in Chubu University," in *Proceedings of the 23rd International Cryogenic Engineering Conference and International Cryogenic Materials Conference (ICEC-ICMC '10)*, pp. 1041–1047, Wroclaw, Poland, 2011.

[3] J. Sun, S. Yamauchi, M. Sugino et al., "Critical current measurements for design of superconducting DC transmission power cable," *Physica C: Superconductivity and its Applications*, vol. 471, no. 21, pp. 1313–1316, 2011.

[4] J. Sun, H. Watanabe, M. Hamabe, N. Yamamoto, T. Kawahara, and S. Yamaguchi, "Critical current behavior of a BSCCO tape in the stacked conductors under different current feeding mode," *Physica C: Superconductivity*, vol. 494, pp. 297–301, 2013.

[5] S. Safran, M. Vojenciak, A. Gencer, and F. Gömöry, "Critical current and AC loss of DI-BSCCO tape modified by the deposition of ferromagnetic layer on edges," *IEEE Transactions on Applied Superconductivity*, vol. 20, no. 5, pp. 2294–2300, 2010.

[6] M. Vojenčiak, J. Šouc, and F. Gömöry, "Critical current and AC loss analysis of a superconducting power transmission

cable with ferromagnetic diverters," *Superconductor Science and Technology*, vol. 24, no. 7, Article ID 075001, 2011.

[7] N. Ayai, K. Yamazaki, M. Kikuchi et al., "Electrical and mechanical properties of DI-BSCCO Type HT reinforced with metallic sheathes," *IEEE Transactions on Applied Superconductivity*, vol. 19, no. 3, pp. 3014–3017, 2009.

[8] S. E. Zirka, Y. I. Moroz, P. Marketos, A. J. Moses, D. C. Jiles, and T. Matsuo, "Generalization of the classical method for calculating dynamic hysteresis loops in grain-oriented electrical steels," *IEEE Transactions on Magnetics*, vol. 44, no. 9, pp. 2113–2126, 2008.

[9] ANSYS®, Academic Teaching Advanced, Release 12.1, ANSYS Inc., http://www.ansys.com.

14

Wideband Negative Permittivity and Double Negative Fishnet-Mushroom-Like Metamaterial in X-Band Waveguide

Ahmed Mahmood,[1] Golge Ogucu Yetkin,[1] and Cumali Sabah[2,3]

[1]*Department of Electrical and Electronics Engineering, University of Gaziantep, 27310 Gaziantep, Turkey*
[2]*Department of Electrical and Electronics Engineering, Middle East Technical University-Northern Cyprus Campus (METU-NCC), Kalkanli, Guzelyurt, 99738 TRNC/Mersin 10, Turkey*
[3]*Kalkanli Technology Valley, Middle East Technical University-Northern Cyprus Campus (METU-NCC), Kalkanli, Guzelyurt, 99738 TRNC/Mersin 10, Turkey*

Correspondence should be addressed to Cumali Sabah; sabah@metu.edu.tr

Academic Editor: Charles Rosenblatt

A Fishnet-Mushroom-like metamaterial electromagnetic behaviour is represented in *S*-parameters numerically and experimentally for X-band frequencies arena. The design has introduced a dielectric substrate as a host with metallic parts. The proposed design is predicted to provide the electromagnetic band gap characterization with desired reiterative characteristic parameters, negative permittivity, and negative permeability exhibiting a double negative left-handed region, which is identified with the X-band regime with good agreement between the simulated and the measured results.

1. Introduction

Over the past decade, interest in applications metamaterials (MTMs) has grown explosively; metamaterials are a kind of a material with electromagnetic properties, which are not readily available in nature. Under certain conditions, these materials provide a simultaneous negative permittivity ($-\varepsilon$) and permeability ($-\mu$) in certain frequency range, and they exhibit reversal of Snell's Law (negative refractive index) and Doppler Effect and Cerenkov Effect around the transmission peak. Depending on the negative value of the permittivity (real part), permeability (real part), or both, metamaterials can be categorized as epsilon negative (ENG), mu negative (MNG), and/or double negative (DNG), which presented the left-handed metamaterials (LHM) they were first proposed by Veselago [1]. Conventional materials have permittivity and also permeability larger than zero in all frequency ranges and they are called double positive (DPS). They become very desirable for designing and fabricating MTMs that can achieve these properties, according to their possible applications in diverse engineering and scientific fields. The metamaterials are designed to use dielectric as host and metal containment, which have been fabricated and then tested intensively in the form of different periodical arrangement [2, 3]. Recently, the fishnet structure has attracted a great potential approved in the studies where the negative refraction index for first time was displayed in a 3D optical MTM [4, 5]. Basically, the strategy to acquire the negative permeability is to create exciting circular currents that can generate a strong magnetic resonance, while being expected to produce the negative permittivity by virtue of the electric plasma response with added continuous wires. When negative refraction index is inconsistent with the negative permeability, most of fishnet structure researches concentrated on characterizing and modeling their magnetic resonance [6–9].

The dispersion diagram of a 2D-periodic structure is produced by calculating the resonant frequencies of Eigen modes, which present differential equation solutions that simulate the fields in a single MTM cell. However, a different band gap acquired by a restricted amount of surface waves along the contour of the irreducible Brillouin zone for an infinite array of the proposed MTM structure will be

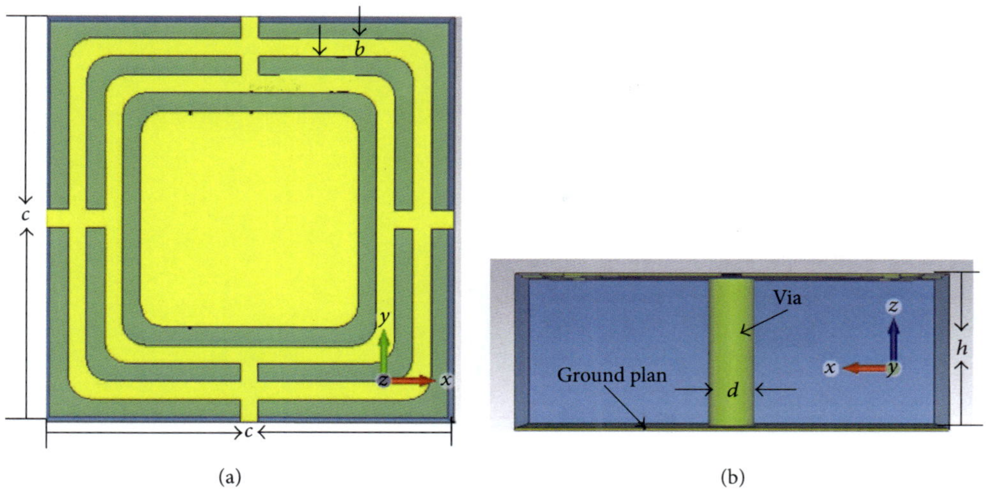

FIGURE 1: The EBG unit cell schematic. (a) Top view. (b) Side view.

produced depending on how many metamaterials arrays are presented. Therefore, only the dispersion relation can obtain the exact position of passband and band gap in the frequency spectrum.

The study presents two cases of MTMs. The first one is a wideband ($-\epsilon$) negative permittivity (ENG), whereas the second one is double negative (DNG). Both cases are the electromagnetic behaviour represented in S-parameters that are numerically and experimentally studied. The simulated and the experimental results exhibit very good agreement in the (8–12) GHz X-band regime whereas the permittivity and permeability and refraction index characteristics are extracted to explain the behaviour of the MTM cell. The proposed MTM (Fishnet-Mushroom-like) is fabricated and measured. The obtained results are compared with simulated results, which are carried out by CSTMWs' simulation program [10]. The given notes, as well as the Nicolson–Ross–Weir (NRW) procedure, are considered to obtain the properties of the designed MTM structure and the effective material parameters [11, 12].

2. Features of Electromagnetic Band Gap (EBG) Structure

2.1. Unit Cell Design. The proposed metamaterial EBG unit cell consisted of a square substrate with 4.5 mm side length and 1.6 mm thickness. The FR-4 epoxy with a permittivity $\varepsilon_r = 4.3$ and loss tangent $\delta = (0.025)$ is chosen to represent the substrate material. The metal shape is printed on the substrate front side which is composed of a squared patch surrounded by two square rings connected from their center sides like a net. A cylindrical via with a diameter of 0.5 mm is also located in the unit cell center as shown in Figure 1(b), by which the square patch and the ground plan in the bottom side are connected. Copper of electrical conductivity $\sigma = 5.8 \times 10^7$ S/m with coating thickness 0.017 mm is chosen for printing as shown in Figure 1(a). the unit cell dimension parameters are mentioned in Table 1.

TABLE 1: Structure dimensions for the EBG unit cell.

Parameter	mm
a	2.4
b	0.2
c	4.5
d	0.5
h	1.6

2.2. Dispersion Diagram. The concept of this paper states that the wave propagation direction is assumed in the same direction of the unit cells periodicity. The space harmonics share the same group velocity even if they have different phase velocities for they cannot occur separately. Each single harmonic does not satisfy the boundary conditions of the periodic structure, but their summation does so. Thus, summation is considered to be the same mode. Hence, by simulating a single proposed EBG unit cell structure, an infinite periodic structure model is carried out by CSTMWs eigenmode simulator.

The band gap characterization for EBG is shown in Figure 2. We expound the idea that the field for the first mode is TM dominant, started at zero frequency, and the eigenfrequency increases with the wave number, reaching the maximum frequency 8.45 GHz. It also decreases following the light line down to a certain frequency. The second mode starts at a higher frequency and continues increasing with a slope under the vacuum's speed of light, which is set by a straight line. The band gap is near 4 GHz for this particular structure (8.45–13.15 GHz), comparing to previous structure designs that achieved less band gap in the same arena [13, 14].

2.3. Array Structure Design and Simulations Settings. The simulations included the proposed unit cell and are ordered by varying different arrays arrangements [15], like 1 × 8-cell array and 3 × 8-, 4 × 8-, and 6 × 8-cell arrays. These

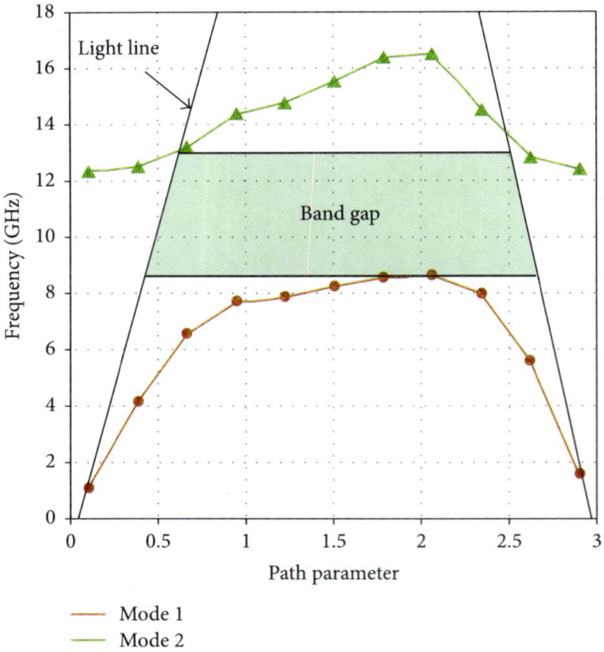

FIGURE 2: Dispersion diagram.

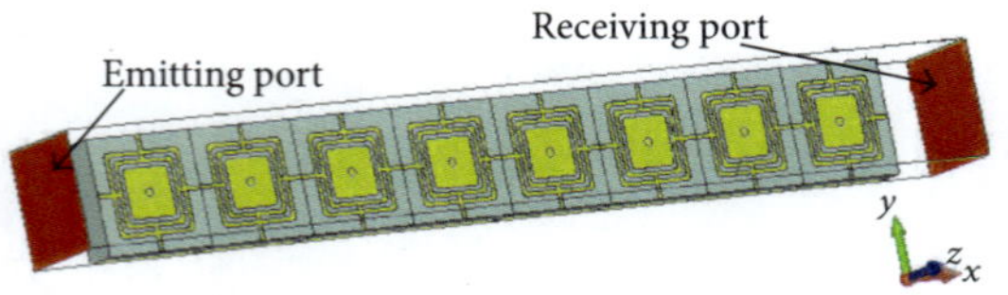

(a)

(b)

FIGURE 3: EBG array simulation settings achieved the band gap measurements. (a) EBG characterization principle; (b) waveguide boundary conditions.

arrangements are set between two waveguide ports in the x-directions in order to simulate S-parameters and obtain the reiterative characteristic parameters, and the band gap effect of varying the amount of metamaterial is so.

The EBG is analyzed and simulated by applying the sample to a surface wave propagation measurement through the setting procedures that presented in (Figures 3(a) and 3(b)).

Furthermore, microwave ports are set on the x-axis sides, on which one port is located as a transmitter generating electromagnetic wave ($E^{\rightarrow}$, $H^{\rightarrow}$) and the second port as a receiver. These boundary conditions are chosen in order to define a waveguide structure (Figure 3(b)). Moreover, a perfect magnetic boundary condition (PMC) is set in the y-directions in which infinite periodic repeating in the same directions for the eight EBG and a perfect electric boundary condition (PEC) are set in the z-directions, considering a ground plane placed in the below side of the EBG structure and with the open boundaries in the x-directions where the ports are set.

3. Results and Discussion

For an experimental reason to suit with the WR-90 dimensions and to decrease the misalignment experiment duration such as position of the fabricated sample plate, the array of 4 × 8 arrangement is chosen and fabricated as shown in Figure 4(a), whereas Figure 4(b) showed the schematic experiment.

The reflection coefficient S_{11}, transmission coefficient S_{21}, and the reiterative characteristic parameters have a negative index, whereas the permittivity and the permeability are in the X-band frequency range. The simulation results exhibit that the structure has been designed successfully; that is, the simulation and experimental data agree with each other as possible.

The experiment steps are as follows: First, the WR-90 waveguide is connected with the input and the output ports with the network analyzer by the two coaxial probes, and the network analyzer is calibrated. Second, the fabricated sample is fixed at the center of a WR-90 waveguide; all the metallic part that is used is unconnected to the waveguide inner walls. The dielectric substrate thickness has chosen being as thin as possible to minimize the dielectric loss. The selected structure dimensions which severed the optimization goal had proven the double negative performance at around X-band arena [11, 12].

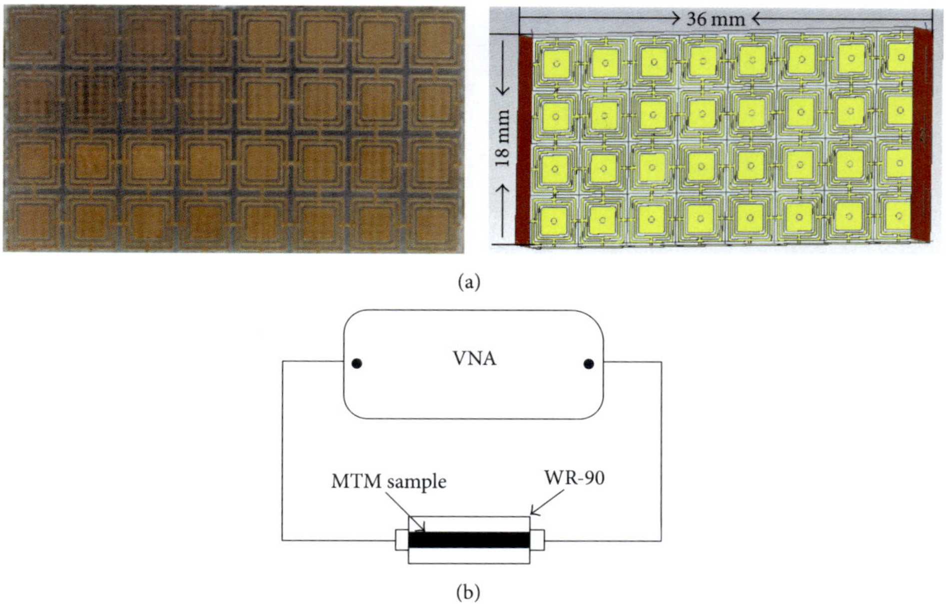

FIGURE 4: (a) The fabricated 4 × 8 arrangement sample; (b) experimental set-up schematic representation. VAN: vector network analyzer.

Referring to the method described in Section 2.2, surface waves propagating and travelling through the circuit board are detected and the amplitude and phase for the reflection and transmission data measured. Figure 5(a) shows both simulated and measured S-parameter data for the investigated structure in good agreement between them.

The experimental and the simulated results show a small difference in minor deviations between the measured and the simulated data that are pointed to manufacture allowance, which is related to PCB fabrication and the gathering process and the quality of the dielectric FR4 dispersion used as a substrate. Likewise, the misalignment experiment duration such as position of the fabricated sample plate, which cannot be neglected, port receptacle locations, and loss characteristics of the materials are regarded as another root of error. On the other hand, the extent of measurement accuracy can be elucidated through the convergence between the results of experimental and simulations outcome.

The effective parameters for the obtained investigated structure material are based on the effectively homogeneous medium occurring when the periodicity of the structure is much less than the propagation wavelength in the propagation direction. The experimented metamaterial is treated as a LHM in the region between 10.95 and 11 GHz, and the transmission peak is indicated at 10.975 GHz. It is clearly stated that the boundary conditions selected to consider the designed structure as an infinite homogeneous material in a certain region led to a frequency observation [14], where both the dielectric permittivity and magnetic permeability are simultaneously negative, which resulted in the retrieval method application.

The NRW method is used in the material parameters calculation extracted and presented in Figures 5(a) and 5(b). As the figure shows, the metamaterial structure exhibits a negative permittivity in all X-band simultaneously with negative permeability in a band between 10.950 and 11 GHz, which a high band transmission peak covered. In spite of the negative refraction index, wideband included all the X-band (see Figure 5(b)), and that type of the negative refraction index is not supporting the left-hand (LH) behaviour. As it was previously mentioned, the representation of the single negative (SNG) occurred only with negative real part of the permittivity. Consequently, both the single negative (SNG) (between 8 and 10.95 GHz and between 11 and 12 GHz) and the double negative (DNG) (10.95–11 GHz) LH region characteristics have been supported by the real part of the negative refractive index [16, 17]. Due to the electric field and the copper parts conjunction and because of strips as in the continuous-wires case, the permittivity exhibits a Drude-like behaviour. Also, the permeability shows Lorentz-like behaviour, through a strong resonant response to the magnetic field during the magnetic resonators condition.

Figures 6(a) and 6(b) show the simulated propagation of surface waves and currents due to the presence of EBG 4 × 8 structure. It is exhibited from these simulations, when surface currents are generated from (port 1) side in x-direction which then travels through the EBG structure opposite side (port 2). In Figure 6(a), the EBG structure with an operating frequency band —the single negative (SNG) (between 8 and 10.95 GHz and between 11 and 12 GHz)—was applied. As expected, the band gap effect shown in Figure 2 does not support any wave propagation at a frequency included within its band gap and suffering reduction. On the other hand, Figure 6(b) demonstrates the LHM in the region between 10.95 and 11 GHz and as a consequence of LHM behaviour the EBG structure surface presence results in a symmetric distribution of the surface current from port 1 to port 2.

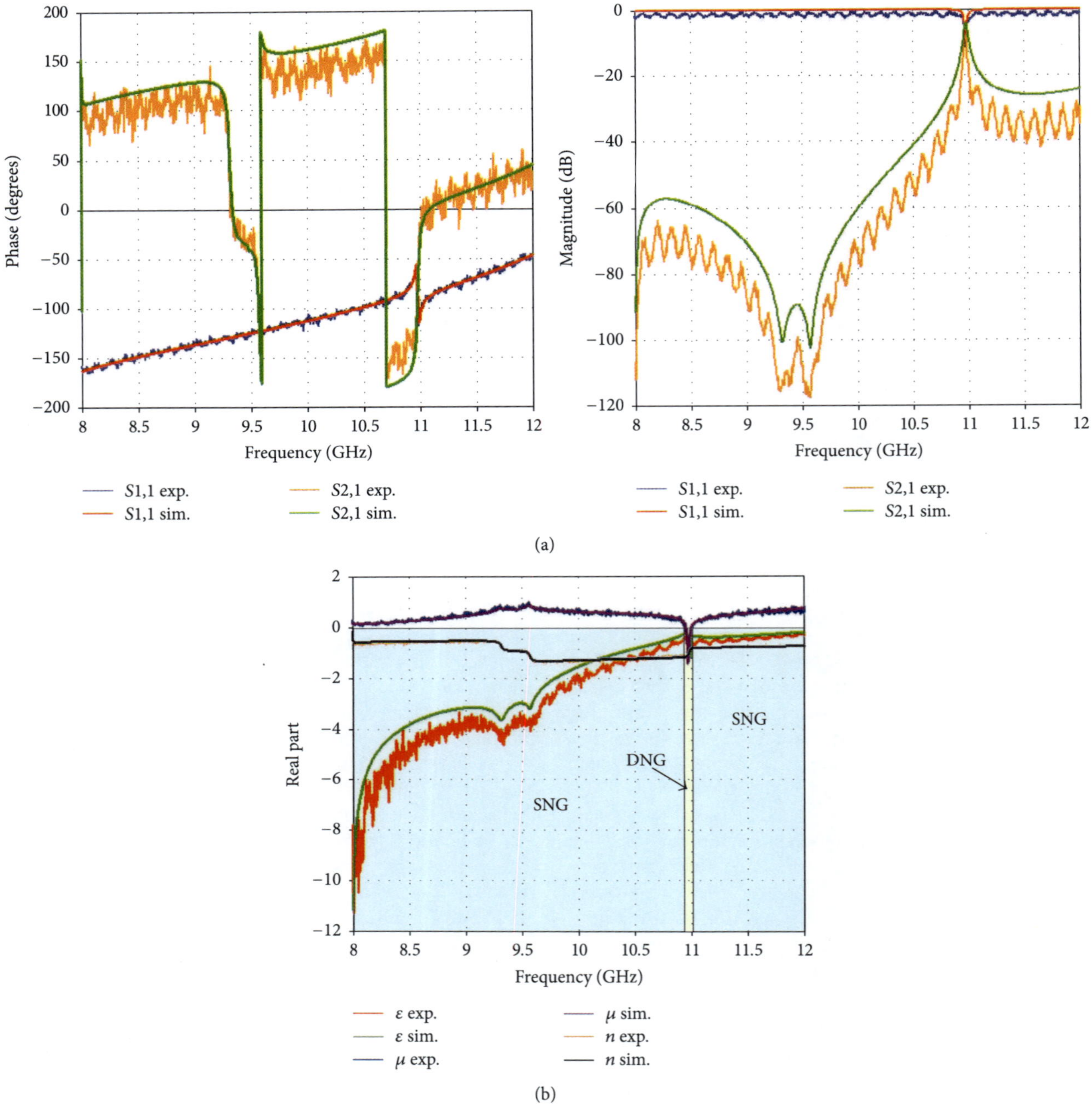

FIGURE 5: Comparison between simulation and measured results. (a) The Fishnet-Mushroom-like MTMs' S-parameters. (b) Retrieved characteristic parameters.

4. Conclusion

In this study, the proposed Fishnet-Mushroom-like MTM is fabricated and measured, and the obtained results are compared with simulated results. There are two cases of MTMs, the first one of which is ($-\epsilon$) a wideband negative permittivity (between 8 and 10.95 GHz and between 11 and 12 GHz) and the second one is double negative (DNG) (between 10.95 and 11 GHz, LH region; and 10.975 GHz is indicated as transmission peak). The electromagnetic behaviour for the said cases in question is represented in S-parameters numerically and experimentally for the X-band arena. The value of this study brings forth the fact that this structure can be used in wideband (ENG) applications such as efficient electrically small antenna enhancement applications and also for double-negativity antenna and radar cross section (RCS) property enhancement applications [18–21].

Competing Interests

The authors declare that there is no conflict of interests regarding the publication of this paper.

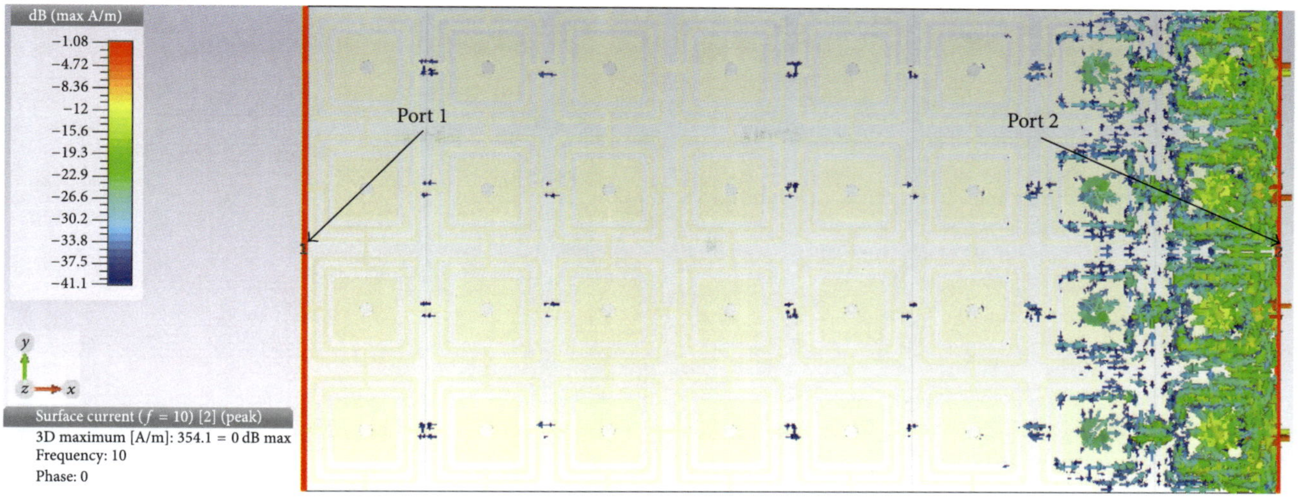

(a)

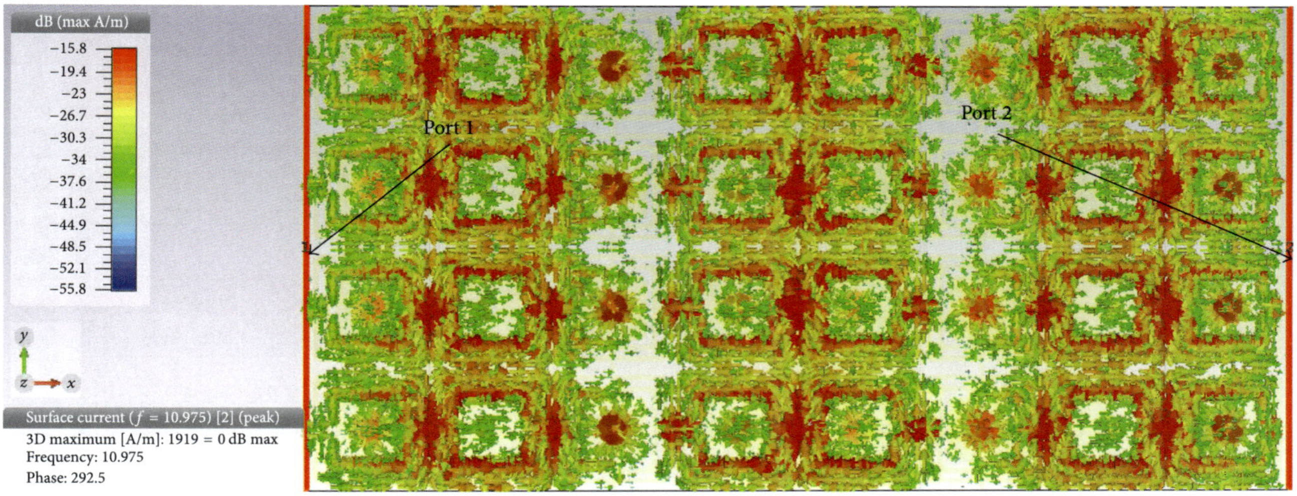

(b)

FIGURE 6: (a) Showing the simulated propagation of surface currents for 10 GHz in the SNG region. (b) Showing the simulated propagation of surface currents for 9.975 GHz in the DNG (LHM) region.

References

[1] V. G. Veselago, "The electrodynamics of substances with simultaneously negative values of ε, and μ," *Soviet Physics Uspekhi*, vol. 10, no. 4, pp. 509–514, 1968.

[2] A. I. Mackenzie, "Comparison of two AMC's on a high-permittivity substrate," in *Proceedings of the 31st International Review of Progress in Applied Computational Electromagnetics (ACES '15)*, March 2015.

[3] A. I. Mackenzie, "Microwave band gaps produced by varying numbers of mushroom metamaterial cells," in *Proceedings of the IEEE Antennas and Propagation Society International Symposium (APS '15)*, pp. 1102–1103, IEEE, Vancouver, BC, Canada, July 2015.

[4] J. Valentine, S. Zhang, T. Zentgraf et al., "Three-dimensional optical metamaterial with a negative refractive index," *Nature*, vol. 455, pp. 376–379, 2008.

[5] M. Kafesaki, I. Tsiapa, N. Katsarakis, Th. Koschny, C. M. Soukoulis, and E. N. Economou, "Left-handed metamaterials: the fishnet structure and its variations," *Physical Review B*, vol. 75, no. 23, Article ID 235114, 2007.

[6] V. D. Lam, J. B. Kim, S. J. Lee, and Y. P. Lee, "Left-handed behavior of combined and fishnet structures," *Journal of Applied Physics*, vol. 103, no. 3, Article ID 033107, 2008.

[7] A. Koray, L. Zhaofeng, S. Levent, and O. Ekmel, "Negative phase advance in polarization independent, multi-layer negative-index metamaterials," *Optics Express*, vol. 16, no. 12, pp. 8835–8844, 2008.

[8] N. T. Tung, V. D. Lam, M. H. Cho, J. W. Park, W. H. Jang, and Y. P. Lee, "Influence of the dielectric-spacer thickness on the left-handed behavior of fishnet metamaterial structure," *Photonics and Nanostructures—Fundamentals and Applications*, vol. 7, no. 4, pp. 206–211, 2009.

[9] CST Microwave Studios 2016, http://www.cst.com/.

[10] R. Ahmadian, S. Sharma, M. Rahimi, and F. B. Zarrabi, "Investigation of EBG array performance on decreasing the mutual coupling," in *Proceedings of the 4th International Conference on*

Advanced Computing and Communication Technologies (ACCT '14), pp. 359–362, February 2014.

[11] S. Sahandabadi, F. H. Kashani, and M. Fallah, "Sierpinski triangular antenna on a mushroom-like EBG metamaterial ground plane," *Journal of Telecommunication*, vol. 23, no. 2, 2014.

[12] V. S. Aravind and S. Gupta, "Compact EBG ground plane microstrip antenna for high gain applications," in *Proceedings of the 2nd International Conference on Emerging Technology Trends in Electronics, Communication and Networking (ET2ECN '14)*, pp. 1–3, IEEE, Surat, India, December 2014.

[13] S. S. Islam, M. R. I. Faruque, and M. T. Islam, "The design and analysis of a novel split-H-shaped metamaterial for multi-band microwave applications," *Materials*, vol. 7, no. 7, pp. 4994–5011, 2014.

[14] C. Sabah, A. O. Cakmak, E. Ozbay, and S. Uckun, "Transmission measurements of a new metamaterial sample with negative refraction index," *Physica B: Condensed Matter*, vol. 405, no. 14, pp. 2955–2958, 2010.

[15] C. Sabah, "Realization of polarization-angle-independent fishnet-based waveguide metamaterial comprised of octagon shaped resonators with sensor and absorber applications," *Journal of Materials Science: Materials in Electronics*, vol. 275, pp. 4777–4787, 2016.

[16] N. T. Tung, V. D. Lam, J. W. Park et al., "Single- and double-negative refractive indices of combined metamaterial structure," *Journal of Applied Physics*, vol. 106, Article ID 053109, 2009.

[17] C. Sabah and H. G. Roskos, "Numerical and experimental investigation of fishnet-based metamaterial in a X-band waveguide," *Journal of Physics D: Applied Physics*, vol. 44, no. 25, Article ID 255101, 2011.

[18] Q.-R. Zheng, Y.-M. Yan, X.-Y. Cao, and N.-C. Yuan, "High impedance ground plane (HIGP) incorporated with resistance for radar cross section (RCS) reduction of antenna," *Progress in Electromagnetics Research*, vol. 84, pp. 307–319, 2008.

[19] G. Singh, R. ni, and A. Marwaha, "A review of metamaterials and its applications," *International Journal of Engineering Trends and Technology*, vol. 19, no. 6, pp. 305–310, 2015.

[20] Q. Tang, B. Chen, W. Luo, and P. Tang, "Design of a substrate integrated waveguide aperture antenna with electromagnetic band gap structure," in *Proceedings of the IEEE International Conference on Communication Problem-Solving (ICCP '15)*, Guilin, China, October 2015.

[21] Y. Rahmat-Samii and H. Rajagopalan, "From a PEC ground plane to an EBG surface: understanding the underlying physics," in *Proceedings of the International Symposium on Antennas and Propagation (ISAP '09)*, Bangkok, Thailand, October 2009.

15

Highly Efficient Broadband Light Absorber based on Nonuniform Hyperbolic Metamaterial Film

Nina A. Zharova,[1] Alexander A. Zharov,[2] and Alexander A. Zharov Jr.[2]

[1] *Institute of Applied Physics, Russian Academy of Sciences, Nizhny Novgorod 603950, Russia*
[2] *Institute for Physics of Microstructures, Russian Academy of Sciences, Nizhny Novgorod 603950, Russia*

Correspondence should be addressed to Alexander A. Zharov; zharov@ipmras.ru

Academic Editor: Charles Rosenblatt

We develop a concept of highly efficient broadband light absorber based on nonuniform hyperbolic metamaterial. We suggest a gradual bending of the anisotropy axis inside the metamaterial that results in spatial shift of the area of resonant absorption depending on the incidence angle. In this resonant region the wavevector of light is parallel to the generatrix of resonant cone and the radiation losses are maximal because of extremely high value of refraction index. Changing the radiation frequency also shifts the spatial position of the resonant region so that high level of absorption may be achieved in wide frequency range. Using the model of nanowire medium (silver wires in silica host) we predict that 200 nm film of this hyperbolic metamaterial allows reaching almost total absorption of radiation throughout the visible band.

1. Introduction

Exotic and interesting electromagnetic properties of the so-called hyperbolic metamaterials (HMs) cause their extensive study [1–7]. HM is an anisotropic resonance medium with hyperboloid-shaped isofrequency surface. In optics HM can be realized, for example, as a metal-dielectric planar nanostructure or a lattice of metallic nanowires embedded in a dielectric matrix.

The ability to support high-index volume electromagnetic modes is one of the main advantages of HM, and it results from the subwavelength light localization, which is also a characteristic feature of surface plasmon polaritons. The similarity comes from the fact that the volume electromagnetic mode in HM is actually the superposition of interacting plasmon modes localized at the nearby metal-dielectric interfaces. The collective character of the modes results in a hybridized broadband response [8, 9].

Unique properties of HM can lead to a wide variety of applications; among these are negative refraction [10, 11], subwavelength imaging [12–17], spontaneous and thermal emission engineering [18–21], and broadband light trapping [22, 23]. Very high density of photonic states in HM [24–26] ensures effective radiative heat transfer [4–6] and allows for a *nanosized* heating/cooling devices.

Hyperbolic character of the metamaterial (MM) can be achieved only for rather high volume fraction of metal that causes high losses. The radiation absorption is especially strong in the case of high refractive index, when a wave beam propagates with its phase velocity directed along the resonance cone generatrix. Unavoidable ohmic loss is a great drawback for applications in high-resolution imaging and lithography; however this effect can be used to engineer absorbing coatings.

The functionality of the absorber (let it be an absorbing film) is defined by its absorptivity, its spatial extent (preferably, smallest), and its broadband and omnidirectional operation. The absorptivity, in turn, depends on the thickness of the film, on the radiation wavelength, and on the incidence angle. In the case of anisotropic material of the film one can also consider the orientation of the anisotropy axis. Some of the requirements are contradictory. Indeed, thin films often demonstrate resonant behaviour that leads to strong angular and wavelength dependence of the absorption coefficient; severely absorbing material with large imaginary part of the permittivity (able to absorb the radiation in small volume)

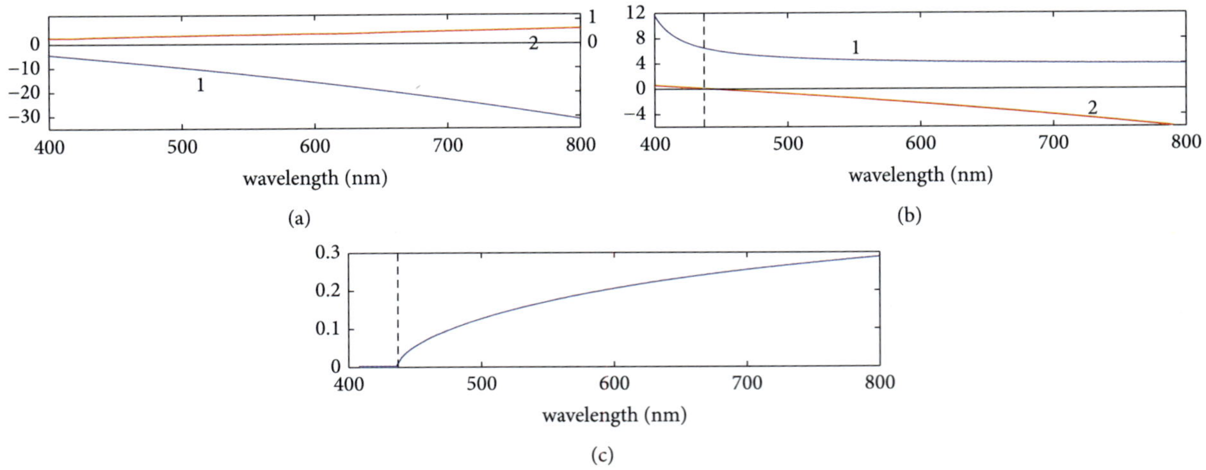

FIGURE 1: (Color online) (a) the wavelength dependence of real (curve 1) and imaginary (curve 2) parts of the permittivity of silver, ε_M (notice different scales for $\mathrm{Re}(\varepsilon_M)$ and $\mathrm{Im}(\varepsilon_M)$); (b) the dispersion of dielectric tensor components (real part) $\varepsilon_\parallel$ (curve 1) and $\varepsilon_\perp$ (curve 2) of effective wire medium (silver wires with volume fraction $\rho_M = 0.25$ in silica host); (c) the normalized critical angle, ϕ^*/π as a function of the wavelength. Dashed line marks the transition from common elliptical medium to the hyperbolic one with $\varepsilon_\parallel \varepsilon_\perp < 0$.

strongly reflects the incident wave. These contradictions make it difficult to design an absorber.

The engineering problem is twofold: first, it is necessary to avoid strong reflection when launching light from free space, and second, it is desirable to transform the incident light into small-scale radiation which is strongly damped and can be almost totally absorbed in a quite *thin* MM layer.

Usual way to improve the coupling between free space radiation and the medium is a modification of the air-medium interface, for example, using a diffraction grating. If the grating period is less than the wavelength in free space, then the scattered small-scale radiation can be effectively absorbed in the metamaterial. An interesting approach for increasing the absorption in hyperbolic media was proposed in [27] where the authors experimentally observed strongly reduced reflection from roughened metamaterial surface.

Another approach for increasing the absorption is to turn the optical axis of HM in such a way that the direction of the resonance cone generatrix would coincide with the normal to the HM interface [5–7]. In this case normally incident light directly transforms into small-scale radiation; however, rather high reflection takes place (see Figure 4) which has a detrimental effect on the absorptivity of the HM layer. In both cases, the spatial spectrum of radiation is changed mainly at the interfaces. On the other hand, the absorption can be increased due to the medium inhomogeneity.

2. Statement of the Problem

In this paper, we study the absorbing properties of the HM layer when the material structure changes inside the slab. We use the effective medium approximation and consider the propagation of the extraordinary wave in the wire medium: dielectric host with metal wires. In this case the components of effective permittivity tensor $\hat{\varepsilon}$ along the wires $\varepsilon_\parallel$ and perpendicular to them $\varepsilon_\perp$ can be defined as [28]

$$\begin{aligned} \varepsilon_\parallel &= \varepsilon_M \rho_M + \varepsilon_D (1 - \rho_M) \\ \varepsilon_\perp &= \varepsilon_D \frac{\varepsilon_M (1 + \rho_M) + \varepsilon_D (1 - \rho_M)}{\varepsilon_D (1 + \rho_M) + \varepsilon_M (1 - \rho_M)}, \end{aligned} \quad (1)$$

where ε_D is the permittivity of dielectric host, ε_M is the permittivity of metal, and ρ_M is the volume fraction of metal. Below we limit ourselves to the TM polarization of the light and use for computations the parameters of silica ($\varepsilon_D = 2.1590$) and silver with the volume fraction $\rho_M = 0.25$. The dispersion dependence of silver $\varepsilon_M(\lambda)$ is taken from [29] and depicted in Figure 1(a): notice different scales for real and imaginary parts of permittivity.

Figure 1 shows also the dependence of $\varepsilon_\parallel$, $\varepsilon_\perp$ (see Figure 1(b)), and the normalized critical angle, ϕ^*/π (see Figure 1(c)), on the radiation wavelength; the latter can be found from the condition $\tan(\phi^*) = \mathrm{Re}(\sqrt{-\varepsilon_\parallel/\varepsilon_\perp})$. ϕ^* is the angle of the resonance cone opening: the hyperbolic medium is transparent only for the waves with the wave vector $\mathbf{k}$ inside the resonance cone, i.e., when the angle between the optical axis $\mathbf{s}$ of the MM and $\mathbf{k}$ is less than ϕ^*, $\angle(\mathbf{k}, \mathbf{s}) \le \phi^*$. The refraction index for normally incident light tends to infinity when the longitudinal (along the wires) optical axis of the metamaterial makes an angle ϕ^* with the normal to the interface (the resonance cone generatrix is along the normal). Dashed line marks the critical wavelength ($\lambda^* \approx 437$ nm) at which the topological transition takes place: the effective medium transforms from common elliptical to hyperbolic state.

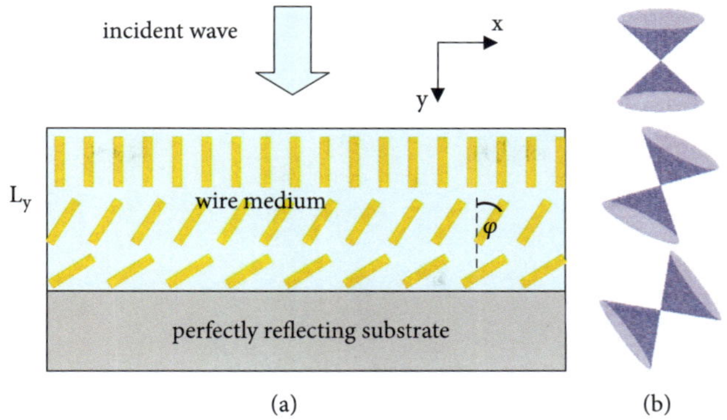

FIGURE 2: (Color online) (a) the geometry of the problem: the wire-medium slab with the thickness L_y is placed on the perfectly reflecting substrate. The components ε_{xx}, ε_{yy}, and ε_{xy} of the effective permittivity tensor depend on the orientation of the principal axes of the metamaterial, i.e., on the angle ϕ between the wires and the normal to the air-MM interface. In the slab of homogeneous metamaterial ϕ is fixed; however, we also consider the case of spatially changing orientation of the medium optical axis (see Figures 5–7). (b) Sketch of resonance cones orientation in inhomogeneous hyperbolic medium. The wave vector $\mathbf{k}$ of propagating radiation is located inside the resonance cone.

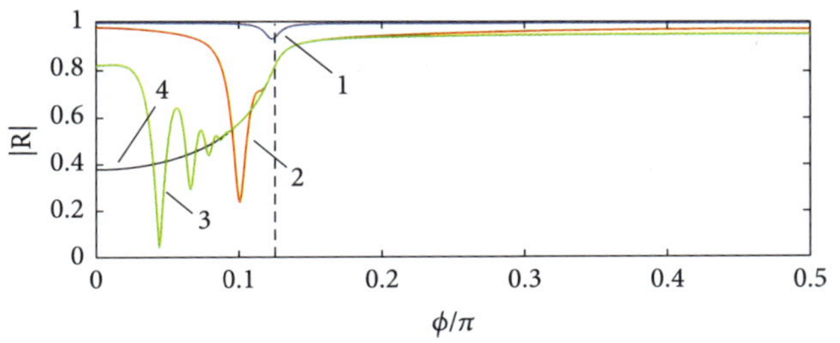

FIGURE 3: (Color online) the reflection coefficient at normal incidence as a function of the angle ϕ (see text). The dashed line indicates the specific angle $\phi = \phi^*$, at which the resonance cone generatrix is parallel to $\mathbf{n}$. Lines 1, 2, and 3 correspond to the reflection from the uniform MM film (placed on the perfectly reflecting substrate) with the thickness $L_y^{(1)} = 10$ nm, $L_y^{(2)} = 100$ nm, and $L_y^{(3)} = 1000$ nm, and the line 4 corresponds to the reflection from the HM half-space. The wavelength of the radiation is 500 nm.

3. Reflection from Homogeneous HM Slab

In the following, we consider the reflection of incident radiation from the MM slab with the thickness L_y placed on the perfectly reflecting substrate. The geometry is shown in Figure 2.

Hyperbolic medium is highly anisotropic, and the wave propagation and, in particular, the reflection coefficient depend on the mutual arrangement of the incident light wave (vector $\mathbf{k}$) and the metamaterial's normal $\mathbf{n}$ and symmetry axis $\mathbf{s}$. In this paper, we consider reflection from a uniform slab with fixed orientation of the optical axis, $\angle(\mathbf{n}, \mathbf{s}) \equiv \phi =$ const, and reflection from the MM slab in which ϕ varies linearly with the coordinate y.

Figure 3 shows the dependence of the reflection coefficient on the angle ϕ in the case of the uniform slab. The results are given for three values of the film thickness, $L_y^{(1,2,3)} = 10$, 100, and 1000 nm, and also for the reflection from the HM half-space, $L_y^{(4)} = \infty$. One can see that for the angles $\phi \approx \phi^*$ the curves 2, 3, and 4 nearly coincide. This means that rather thin MM films behave as (infinitely) thick ones when the relation $\phi = \phi^*$ is satisfied. This statement is verified for most of the visible domain, as one can see in Figures 4(a) and 4(b) (curves 3).

Figure 4 shows the spectral dependence of absorption of normally incident light. Solid lines in the panel (a) show the wavelength dependence of the absorption coefficient $D = 1 - |R|^2$ (R is the reflection coefficient, i.e., the ratio of the complex amplitudes of reflected and incident waves) in the case of metamaterial layer of finite thickness $L_y = 200$ nm, while the dotted lines correspond to the absorption coefficient $D_\infty = 1 - |R_\infty|^2$ of the MM half-space. The orientation angle is $\phi = 0$ (curve 1), $\phi = 0.1\pi$ (curve 2), and $\phi = \phi^*$ (curve 3; note that ϕ^* also depends on λ, see Figure 1(c)). Panel (b) demonstrates the wavelength dependence of the ratio D/D_∞ for these three cases. One can see that a thin layer with $\phi = \phi^*$ absorbs as an infinite one (the curve 3 actually coincides with the corresponding dotted line in panel (a) and the ratio $D/D_\infty = 1$ in the whole visible band), but considerable part of radiation is reflected. The absorption of the finite layer with fixed orientation angle (curves 1, 2) strongly depends on the wavelength and degrades for longer waves.

4. Nonuniform Absorber

Weak wavelength dependence of absorption is attractive for broadband operation, but it is achieved only for specific choice of the angle $\phi = \phi^*(\lambda)$ (curve 3 in Figures 4(a) and 4(b)). Certainly, such an absorber with the optical axis turning in accordance with the wavelength is unrealistic. However, the absorbing properties can be improved for the case of *nonuniform* film. For simplicity we consider the metamaterial with constant ρ_M and suppose that the orientation angle ϕ increases linearly (from 0 to $\pi/2$) with the propagation length. These situation cannot be achieved in common wire medium since changing the orientation of the wires changes also the density of wires and thus the value of ρ_M. On the other hand, within the effective medium

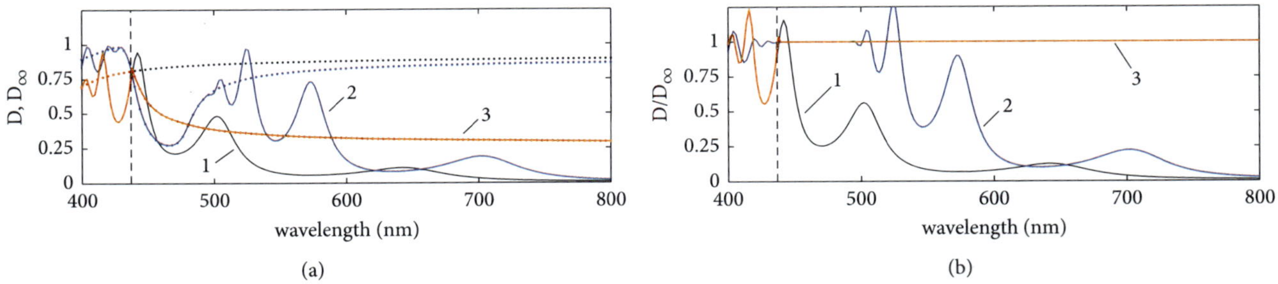

FIGURE 4: (Color online) (a) the absorption coefficient $D = 1 - |R|^2$ (solid lines) of the uniform MM layer of finite thickness $L_y = 200$ nm at normal incidence as a function of the wavelength. Dotted lines show the absorption coefficient $D_\infty = 1 - |R_\infty|^2$ for the wave incident onto the metamaterial half-space. The orientation angle is $\phi = 0$ (curve 1), $\phi = 0.1\pi$ (curve 2), and $\phi = \phi^*(\lambda)$ (curve 3). (b) The ratio D/D_∞ for these three cases.

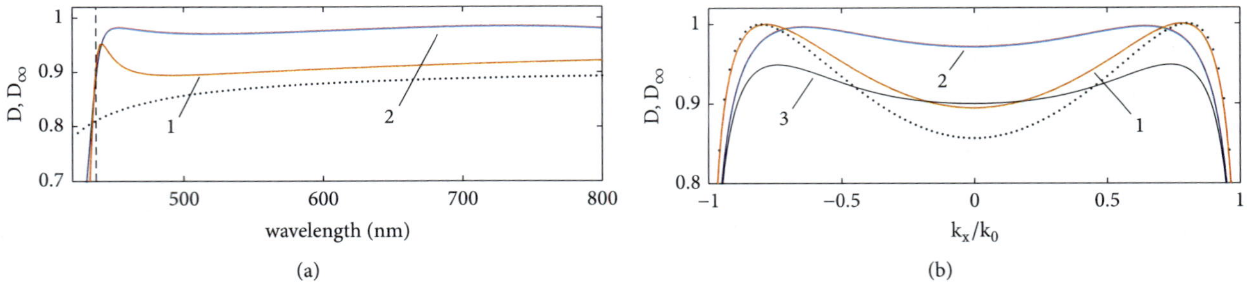

FIGURE 5: (Color online) the absorption coefficient D of the nonuniform metamaterial film as a function of the (a) wavelength (at normal incidence) and (b) tangential wavevector component of the incident light ($\lambda = 500$ nm); the thickness of the film is $L_y = 200$ nm (curve 1), $L_y = 100$ nm (curve 2), and $L_y = 50$ nm (curve 3). Dotted line is the absorption coefficient D_∞ of the metamaterial half-space.

approximation it is not necessary to use continuous wires. Elongated silver particles (meta-atoms) can serve as a metallic component of the MM. Such meta-atoms suspended in a viscous liquid can be easily aligned by means of constant electric field in order to create the proper orientational pattern [30, 31].

Bearing in mind this possible realization of such an inhomogeneous metamaterial, further we demonstrate considerable enhancement of the radiation absorption in the film with spatially varying orientation of the optical axes. As before, we use in analysis the effective medium approximation.

In our simulation we use transfer-matrix method. The whole MM layer is formally split into the set of uniform sublayers with the components of inverse permittivity tensor $\hat{\sigma} \equiv \hat{\varepsilon}^{-1}$ locally depending on the tilt angle ϕ as

$$\begin{aligned}\sigma_{xx} &= \frac{\cos^2\phi}{\varepsilon_{xx}} + \frac{\sin^2\phi}{\varepsilon_{yy}},\\ \sigma_{xy} &= \cos\phi\sin\phi\left(\frac{1}{\varepsilon_{xx}} - \frac{1}{\varepsilon_{yy}}\right),\\ \sigma_{yy} &= \frac{\sin^2\phi}{\varepsilon_{xx}} + \frac{\cos^2\phi}{\varepsilon_{yy}},\end{aligned} \tag{2}$$

where ε_{xx} and ε_{yy} are the principal-axes components of the permittivity tensor. The magnetic field $\mathbf{H} \equiv \mathbf{z}_0 H$ is locally governed by the equation

$$\begin{aligned}&\frac{\partial}{\partial x}\left(\sigma_{yy}\frac{\partial H}{\partial x}\right) - \frac{\partial}{\partial x}\left(\sigma_{xy}\frac{\partial H}{\partial y}\right) - \frac{\partial}{\partial y}\left(\sigma_{xy}\frac{\partial H}{\partial x}\right)\\ &\quad + \frac{\partial}{\partial y}\left(\sigma_{xx}\frac{\partial H}{\partial y}\right) + k_0^2 H = 0,\end{aligned} \tag{3}$$

and in each sublayer it can be represented as a sum of up- and downpropagating (along the plus- and minus-y direction) waves. The relation between the amplitudes of these waves in the neighboring sublayers is found from the continuity of tangential components of electric end magnetic fields. The thickness dy of these sublayers is the step of numerical discretization, and as such, the resulting reflection and absorption coefficients do not depend on dy (i.e., the calculations with $dy/2$ and the same total thickness of the film give only small corrections to R and D).

The transfer matrix is obtained as a result of successive application of this procedure. Using the boundary conditions at the first and last sublayers we find the required reflection coefficient.

The results shown below are obtained assuming that the orientational angle increases as $\phi = (\pi/2)y/L_y$ $(0 \le y \le L_y)$. Figure 5(a) demonstrates the wavelength dependence of the absorption coefficient $D(\lambda)$ for the nonuniform MM films with the thickness $L_y = 200$ and 100 nm. The results for thicker film are close to the case of the MM half-space, $D \approx D_\infty$. The absorption coefficient in the case of $L_y = 100$ nm is even more than D_∞ due to the interference of the incident wave and the wave which is reflected from the substrate and

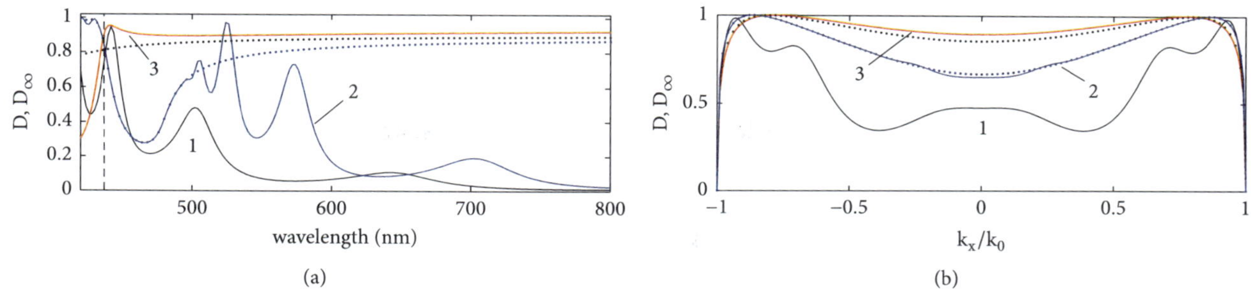

FIGURE 6: (Color online) the absorption coefficient D of the nonuniform metamaterial film (curve 3) compared with MM layer with the fixed optical axis orientation (curves 1 and 2 correspond to $\phi = 0$ and $\phi = 0.1\pi$): (a) as a function of the wavelength (at normal incidence) and (b) as a function of the tangential wavevector component of the incident light ($\lambda = 500$ nm); the thickness of the film is $L_y = 200$ nm. Dotted lines show the absorption coefficient D_∞ of the metamaterial half-space $\phi = 0$ and $\phi = 0.1\pi$.

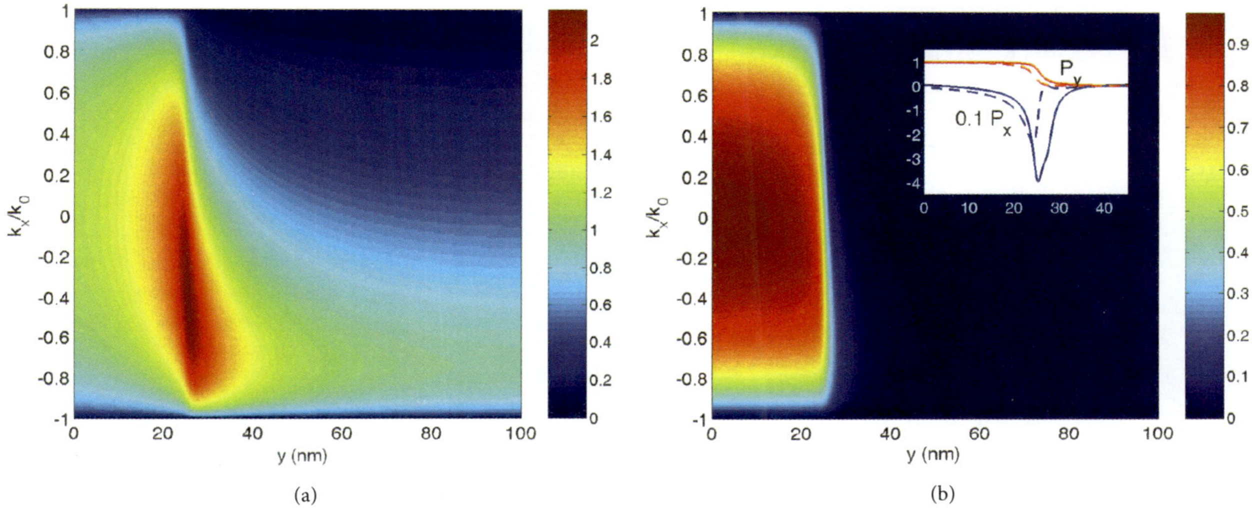

FIGURE 7: (Color online) (a) magnetic field pattern $|H_z|(y, k_x/k_0)$ for the nonuniform metamaterial film of finite thickness $L_y = 100$ nm; (b) the y component of Poynting vector, $P_y(y, k_x/k_0)$. x and y components of Poynting vector; $P_x(y)$ and $P_y(y)$ are plotted in the inset for $k_x/k_0 = -0.4$ (solid lines) and $k_x/k_0 = 0.4$ (dashed lines). Near the resonance point ($y \approx 25$ nm) P_x is tens of times higher than P_y, and the energy flows mainly in the opposite x direction. The radiation wavelength is 500 nm.

not completely damped. This interference effect results in better absorbance of the film, and the broadband character of the absorption is preserved.

The dependence of reflection and absorption on the incident angle is also an important characteristic of proposed nonuniform film, and Figure 5(b) demonstrates this function for specific wavelength $\lambda = 500$ nm and three values of the film thickness $L_y = 100$ nm (curve 1), $L_y = 200$ nm (curve 2), and $L_y = 50$ nm (curve 3, see comments below). When the component k_x of the wavevector of the incident light along the air-MM interface tends to its maximum $k_0 = 2\pi/\lambda$ the reflection sharply increases, but this is a common feature, e.g., for the reflection from dielectric half-space.

The improved absorption and broadband functionality of the proposed device are clearly seen in Figure 6, which shows the wavelength and incident angle dependence of the absorption coefficient calculated for the MM slab with fixed orientation of the optical axis (curves 1 and 2 correspond to $\phi = 0$ and $\phi = 0.1\pi$) and inhomogeneous hyperbolic metamaterial with ϕ linearly growing from 0 to $\pi/2$ (curve 3).

Figure 7 demonstrates magnetic field H_z (Figure 7(a)) and Poynting vector P_y (Figure 7(b)) pattern as a function on k_x and y. One can see that $|H_z|$ strongly depends on the incident angle, and this dependence is not symmetric with respect to $k_x = 0$. The asymmetry of the energy flow is apparent from the direction of the x component of Poynting vector P_x which is plotted in the inset of Figure 7(b), and the energy flow is mainly antialigned with the x-axis. It should be noted that, in spite of this asymmetry, the absorption coefficient $D(k_x)$ is symmetric as a function of k_x (see Figure 5(b)), and the both positive and negative k_x harmonics can be effectively absorbed.

Another important feature of the Poynting-vector pattern is its strong spatial localization. It is clear from Figure 7 that the main processes (reflection and absorption) take place in the first one-third of the film, and it seems that similar results could be achieved with half the film thickness. However, if we place the perfectly reflecting substrate at $y = 50$ nm and keep the MM structure unchanged (such that the orientation angle varies from 0 to $\pi/4$), then the absorption of the incident wave is noticeably decreased, and the absorption coefficient takes values close to $D_\infty(k_x)$ (see curve 3 in Figure 5(b)).

5. Conclusion

In conclusion, we have proposed the design of an absorbing film made of nonuniform hyperbolic metamaterial. In the film with smoothly changing orientation of the optical axis from parallel to normal to the interface, the resonance absorption takes place for each wavelength with only the position of the resonance shifted. This results in broadband operation of the absorber and also its weaker sensitivity to the incident angle. The analysis based on effective medium approximation shows that a 200-nanometer-thick film operates as a metamaterial absorbing half-space. The realization of the proposed idea in optics seems to be complicated. On the other hand, the development of nanotechnology has already led to the emergence of complicated metamaterial structures, such as flexible hyperbolic metamaterial the deformation of which evidently causes *inhomogeneous orientation of the optical axis* [32]. A method of growing tilted nanowires array [33] should be also noted.

Conflicts of Interest

The authors declare that there are no conflicts of interest regarding the publication of this paper.

Acknowledgments

This research is supported in part by Russian Foundation for Basic Research Grants no. 17-02-00281 and no. 16-02-00556.

References

[1] A. Poddubny, I. Iorsh, P. Belov, and Y. Kivshar, "Hyperbolic metamaterials," *Nature Photonics*, vol. 7, no. 12, pp. 948–957, 2013.

[2] D. Lu and Z. Liu, "Hyperlenses and metalenses for far-field super-resolution imaging," *Nature Communications*, vol. 3, article no. 1205, 2012.

[3] I. V. Iorsh, I. S. Mukhin, I. V. Shadrivov, P. A. Belov, and Y. S. Kivshar, "Hyperbolic metamaterials based on multilayer graphene structures," *Physical Review B: Condensed Matter and Materials Physics*, vol. 87, no. 7, Article ID 075416, 2013.

[4] Y. Guo, C. L. Cortes, S. Molesky, and Z. Jacob, "Broadband super-Planckian thermal emission from hyperbolic metamaterials," *Applied Physics Letters*, vol. 101, no. 13, Article ID 131106, 2012.

[5] S. M. Hashemi and I. S. Nefedov, "Wideband perfect absorption in arrays of tilted carbon nanotubes," *Physical Review B: Condensed Matter and Materials Physics*, vol. 86, no. 19, Article ID 195411, 2012.

[6] I. S. Nefedov, C. A. Valagiannopoulos, S. M. Hashemi, and E. I. Nefedov, "Total absorption in asymmetric hyperbolic media," *Scientific Reports*, vol. 3, article no. 2662, 2013.

[7] I. S. Nefedov, C. A. Valagiannopoulos, and L. A. Melnikov, "Perfect absorption in graphene multilayers," *Journal of Optics*, vol. 15, no. 11, Article ID 114003, 2013.

[8] S. V. Zhukovsky, O. Kidwai, and J. E. Sipe, "Physical nature of volume plasmon polaritons in hyperbolic metamaterials," *Optics Express*, vol. 21, no. 12, pp. 14982–14987, 2013.

[9] L. Ferrari, D. Lu, D. Lepage, and Z. Liu, "Enhanced spontaneous emission inside hyperbolic metamaterials," *Optics Express*, vol. 22, no. 4, pp. 4301–4306, 2014.

[10] J. Yao, Z. Liu, Y. Liu et al., "Optical negative refraction in bulk metamaterials of nanowires," *Science*, vol. 321, no. 5891, p. 930, 2008.

[11] A. J. Hoffman, L. Alekseyev, S. S. Howard et al., "Negative refraction in semiconductor metamaterials," *Nature Materials*, vol. 6, no. 12, pp. 946–950, 2007.

[12] Z. Jacob, L. V. Alekseyev, and E. Narimanov, "Optical hyperlens: far-field imaging beyond the diffraction limit," *Optics Express*, vol. 14, no. 18, pp. 8247–8256, 2006.

[13] J. Sun, M. I. Shalaev, and N. M. Litchinitser, "Experimental demonstration of a non-resonant hyperlens in the visible spectral range," *Nature Communications*, vol. 6, article no. 7201, 2015.

[14] T. U. Tumkur, L. Gu, J. K. Kitur, E. E. Narimanov, and M. A. Noginov, "Control of absorption with hyperbolic metamaterials," *Applied Physics Letters*, vol. 100, no. 16, Article ID 161103, 2012.

[15] T. U. Tumkur, J. K. Kitur, B. Chu et al., "Control of reflectance and transmittance in scattering and curvilinear hyperbolic metamaterials," *Applied Physics Letters*, vol. 101, no. 9, Article ID 091105, 2012.

[16] Z. Liu, H. Lee, Y. Xiong, C. Sun, and X. Zhang, "Far-field optical hyperlens magnifying sub-diffraction-limited objects," *Science*, vol. 315, no. 5819, p. 1686, 2007.

[17] K. Mantel, D. Bachstein, and U. Peschel, "Perfect imaging of hypersurfaces via transformation optics," *Optics Expresss*, vol. 36, no. 2, pp. 199–201, 2011.

[18] H. N. Krishnamoorthy, Z. Jacob, E. Narimanov, I. Kretzschmar, and V. M. Menon, "Topological transitions in metamaterials," *American Association for the Advancement of Science: Science*, vol. 336, no. 6078, pp. 205–209, 2012.

[19] Y. Guo, C. L. Cortes, S. Molesky, and Z. Jacob, "Broadband super-Planckian thermal emission from hyperbolic metamaterials," *Applied Physics Letters*, vol. 101, no. 13, Article ID 131106, 2012.

[20] S.-A. Biehs, M. Tschikin, and P. Ben-Abdallah, "Hyperbolic metamaterials as an analog of a blackbody in the near field," *Physical Review Letters*, vol. 109, no. 10, Article ID 104301, 2012.

[21] I. S. Nefedov and C. R. Simovski, "Giant radiation heat transfer through micron gaps," *Physical Review B: Condensed Matter and Materials Physics*, vol. 84, no. 19, Article ID 195459, 2011.

[22] Y. Cui, J. Xu, K. Fung et al., "Ultrabroadband light absorption by a sawtooth anisotropic metamaterial slab," *Nano Letters*, vol. 12, no. 3, pp. 1443–1447, 2012.

[23] H. Hu, D. Ji, X. Zeng, K. Liu, and Q. Gan, "Rainbow Trapping in Hyperbolic Metamaterial Waveguide," *Scientific Reports*, vol. 3, article 1249, 2013.

[24] Z. Jacob, J. Kim, G. V. Naik, A. Boltasseva, E. E. Narimanov, and V. M. Shalaev, "Engineering photonic density of states using metamaterials," *Applied Physics B: Lasers and Optics*, vol. 100, no. 1, pp. 215–218, 2010.

[25] P. Shekhar, J. Atkinson, and Z. Jacob, "Hyperbolic metamaterials: fundamentals and applications," *Nano Convergence*, vol. 1, no. 1, 2014.

[26] N. A. Zharova and A. A. Zharov, "Electromagnetic surface waves in liquid metacrystals," *Journal of the Optical Society of America B: Optical Physics*, vol. 33, no. 4, pp. 594–602, 2016.

[27] E. E. Narimanov, H. Li, Y. A. Barnakov, T. U. Tumkur, and M. A. Noginov, "Reduced reflection from roughened hyperbolic metamaterial," *Optics Express*, vol. 21, no. 12, pp. 14956–14961, 2013.

[28] L. Ferrari, C. Wu, D. Lepage, X. Zhang, and Z. Liu, "Hyperbolic metamaterials and their applications," *Progress in Quantum Electronics*, vol. 40, pp. 1–40, 2015.

[29] https://www.refractiveindex.info.

[30] A. A. Zharov and N. A. Zharova, "Liquid metacrystals," *Journal of the Optical Society of America B: Optical Physics*, vol. 31, no. 3, pp. 559–564, 2014.

[31] A. A. Zharov, A. A. Zharov, and N. A. Zharova, "Spontaneous reorientations of meta-atoms and electromagnetic spatial solitons in a liquid metacrystal," *Physical Review E: Statistical, Nonlinear, and Soft Matter Physics*, vol. 90, no. 2, Article ID 023207, 2014.

[32] T. Tumkur, G. Zhu, P. Black, Y. A. Barnakov, C. E. Bonner, and M. A. Noginov, "Control of spontaneous emission in a volume of functionalized hyperbolic metamaterial," *Applied Physics Letters*, vol. 99, no. 15, Article ID 151115, 2011.

[33] J. Zúñiga-Pérez, A. Rahm, C. Czekalla, J. Lenzner, M. Lorenz, and M. Grundmann, "Ordered growth of tilted ZnO nanowires: Morphological, structural and optical characterization," *Nanotechnology*, vol. 18, no. 19, Article ID 195303, 2007.

16

Characteristics of Local Modulation Beam Propagating through Spatial Filter System

Kewei You, [1,2,3] **Yanli Zhang**, [1,2] **Xuejie Zhang**, [1,2] **and Jianqiang Zhu** [1,2]

[1] *Shanghai Institute of Optics and Fine Mechanics, Key Laboratory of High Power Laser and Physics, Chinese Academy of Sciences, Shanghai 201800, China*
[2] *National Laboratory on High Power Laser and Physics, Shanghai 201800, China*
[3] *University of Chinese Academy of Sciences, Beijing 100049, China*

Correspondence should be addressed to Jianqiang Zhu; jqzhu@mail.shcnc.ac.cn

Academic Editor: Jorg Fink

As local defects may significantly harm beam quality and affect safe operation, a systematic analysis of the ability of a spatial filter to alleviate these adverse effects is required. Thus, the evolutional characteristics of a beam modulated by a local defect propagating through a spatial filter system at an image reply plane and a downstream plane are analyzed in detail. Modulation stripes appear at the image reply plane; these are caused by the pinhole cutoff effect. The modulation degree increases with increasing defect size. The maximum intensification factor can reach 3.2 under certain conditions. Thus, the defect size should be restricted to a reasonable size for safe operation with a specified pinhole size. Moreover, a maximal value appears at the downstream plane, and the intensity enhances with increasing defect size. To ensure beam quality, the maximum allowable defect size and angle of the spatial filter should meet special constraints. The maximum allowable defect size is calculated based on practical configuration parameters.

1. Introduction

In a high-power laser system, the laser beam passes through numerous optical components. The amplitude and phase modulations caused by damage defects and contamination particles can result in downstream intensification, which can damage certain optical components in the systems [1–5]. As one of the crucial factors restricting the output power of high-power laser systems, the research of defects in optical components has aroused great interest for a long time. To reduce the detrimental effects of optical component defects, several effective measures have been put into force, such as developing new optical component processing technology and strengthening environmental cleanliness [6, 7]. On the other hand, the spatial filter serves as an essential part of a high-power laser system [8–11]. An appropriate filter pinhole has been selected to filter out high-frequency modulation, sequentially sharply minimizing nonlinear self-focusing effect [12, 13]. Although the function of a spatial filter to control beam quality has been investigated extensively, previous studies have concentrated mainly filtering effect with whole beam phase or amplitude distortion, such as large-scale phase aberrations related to intensity distribution [10], wave front defects involving spatial periods [14], and flat-topped multi-Gaussian beams with apertures [15]. There are still some researches about controlling features of spatial filter on nonlinear growth with small soft Gaussian scatterers perturbation [13], but little work is done about local circular defects which is commonly found from the near-field measurement in practical systems.

It is well known that local diffraction modulation may cause a great threat to expensive downstream optical components, so much attention should be paid to how the modulated beam is suppressed by transmitting a spatial filter. Since the local modulation beam may contain a low-frequency component and has a wider spectrum than an unmodulated beam, the characteristics of local modulation beam propagation through a spatial filter deserve to be discussed in detail. In this paper, a systematic analysis of the propagating characteristics for an image and subsequent

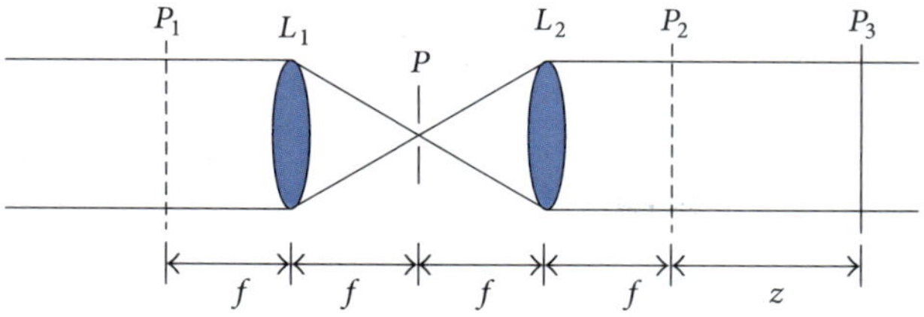

FIGURE 1: Spatial imaging configuration.

plane is conducted. Based on this analysis, some specific cleanliness rules about amplifier and optical components are put forward, thereby providing useful guidance for the inspection of optical components.

2. Theoretical Model

Figure 1 depicts a typical spatial imaging configuration that uses the imaging properties of confocal lens pairs. According to a practical spatial filter system in a multipass main amplifier system, the focus lengths of lens L_1 and L_2 are both f. The distance from input plane P_1 to L_1 and the distance from image reply plane P_2 to L_2 are also f. P represents the pinhole plane and the Fourier spectrum plane of the input field. P_3 is the observation plane, and the distance from P_2 is z.

According to the basic theory of Fourier Optics, the input and output light fields of an ideal spatial filter have the following relation:

$$E_2(x_2, y_2) = \mathfrak{F}^{-1}\left[T\left(\sqrt{x_p^2 + y_p^2}\right)\mathfrak{F}(E_1(x_1, y_1))\right] \quad (1)$$

$$T\left(\sqrt{x_p^2 + y_p^2}\right) = \begin{cases} 1 & \sqrt{x_p^2 + y_p^2} \le \dfrac{d}{2} \\ 0 & \sqrt{x_p^2 + y_p^2} > \dfrac{d}{2}, \end{cases} \quad (2)$$

where E_1 and E_2 are light fields at the input plane and image relay plane; T is the transmission function of the filter pinhole; d is the diameter of the filter pinhole; x_1, y_1, x_2, y_2, x_p, and y_p are the coordinates at the input plane, image relay plane, and filter pinhole plane, respectively, and operators $\mathfrak{F}$ and $\mathfrak{F}^{-1}$ represent the Fourier transform and inverse Fourier transform, respectively. Suppose the input beam is modulated by a local circular defect, and the defect is located at the center of the input beam. Then, the distribution can be expressed as

$$\begin{aligned} E_1 &= U_1(x_1, y_1) \\ &\quad + (-U_1(x_1, y_1) + U_1(x_1, y_1) \times \exp(-i\phi)) \\ &\quad \times \operatorname{circ}\left(\sqrt{x_1^2 + y_1^2}\right) \end{aligned} \quad (3)$$

$$\operatorname{circ}\left(\sqrt{x_1^2 + y_1^2}\right) = \begin{cases} 1 & \sqrt{x_1^2 + y_1^2} \le R \\ 0 & \sqrt{x_1^2 + y_1^2} > R. \end{cases} \quad (4)$$

As shown in (3), the input beam can be represented as the superposition of the background beam and a local supplement beam. The background beam can be treated as a large-aperture uniform light field ($U_1(x_1, y_1)$). The supplement beam is caused by a defect with radius R, and ϕ denotes the phase modulation. On the basis of the Babinet theorem, the analysis of a beam modulated by a small defect is related to the circular aperture diffraction. Thus, the analysis of the circular aperture diffraction is made first. The distribution at the filter pinhole plane is given by

$$\begin{aligned} E'(x_p, y_p) &= \mathfrak{F}\left(E'_1(x_1, y_1)\right) \\ &= \frac{\pi R^2 A'_1}{j\lambda f}\left[2\frac{J_1(kr_pR/f)}{kr_pR/f}\right], \end{aligned} \quad (5)$$

where $E'_1(x_1, y_1) = A'_1(\sqrt{x_1^2 + y_1^2} \le R)$, the radius of circular aperture is also R, r_p is the radial coordinate at the filter pinhole plane, and only intensity is considered. According to (1), the circular aperture diffraction light field at the image relay plane can be written as

$$\begin{aligned} E'_2(x_2, y_2) &= \frac{1}{j\lambda f}\iint E'(x_p, y_p) \\ &\quad \times \exp\left(-\frac{jkx_px_2}{f} - \frac{jky_py_2}{f}\right)dx_pdy_p. \end{aligned} \quad (6)$$

According to (3), the light field at the image relay plane can be given by

$$\begin{aligned} &E_2(x_2, y_2) \\ &\quad = U_2(x_2, y_2) \\ &\quad\quad + \left(-E'_2(x_2, y_2) + E'_2(x_2, y_2) \times \exp(-i\phi)\right), \end{aligned} \quad (7)$$

where $U_2(x_2, y_2)$ is the imaging light field of the background beam and the other represents the imaging light field of the local supplement beam.

3. Numerical Simulations and Result Analysis

Spatial filtering can effectively remove the fastest-growing high-frequency components of the instabilities by using the cutoff frequency provided by the filter pinhole. The diameter of a pinhole in a spatial filter system can be expressed as [17]

$$d = 2f\theta, \quad (8)$$

where θ is the angular radius of the focal spot relative to the center of the lens. The cutoff frequency of a spatial filter is $K_c = \pi d/\lambda f$ [13]. Thus, the frequency that is higher than K_c will be filtered out, making the propagating beam smooth. Combining with (8), $K_c = 2\pi\theta/\lambda$. Thus, the frequency that can propagate through the filter pinhole really depends on θ.

Numerical calculations are carried out given that the simulated tenth-order super-Gaussian beam at 1053 nm has a radius of 185 mm at 1% of the maximum intensity. The focal length of the lens is 11350 mm, and defects with pure phase modulation π are in the range of $R = 0.5$–3.5 mm. θ is 200 μrad, and the diameter of the filter pinhole is 4.6 mm. The intensification factor in a beam section at z is defined by the intensity ratio I/I_0. I is the maximum intensity at the observation plane, and I_0 represents the input intensity on the axis.

Based on the expression of the Airy disk width at the focal plane of the lens $D = 1.22\lambda f/w$, w is the radius of the transmission beam. We can see that the width is inversely proportional to the radius. Then, the large-aperture beam propagates through the filter pinhole with little distortion and possesses a narrow spectrum. However, spectrum broadening resulting from a small local defect will be cut off by the filter pinhole, causing intensity fluctuations. We performed numerical simulations on beam characteristics at the image relay plane according to (7). Affected by local defects, a nonuniform intensity distribution appears at the image relay plane, which is mainly concentrated in the region of the imaging light field of the local supplement beam. Then, we define this region as a modulation region. Figure 2 presents the intensification factor varying with increasing defect radius. It is found that when the defect radius is 3.2 mm, the Airy disk width of the local supplement beam is approximately 4.6 mm, and the intensification factor grows to a maximum of 3.2. It is clearly seen that the sizes of the Airy disk and pinhole ($\theta = 200\,\mu$rad) are basically equal. Therefore, when the cutoff of the pinhole is just at the first notch of the Airy diffraction pattern of the local supplement beam, the imaging light fields of the background beam and the local supplement beam may produce a strong interaction with each other, resulting in the enhanced growth of intensity fluctuations. However, when $R < 3.2$ mm, the cutoff position of the pinhole is located within the Airy disk of the local supplement beam. The smaller the defect size is, the more area of the Airy disk is cut off, and the more energy loss occurs. Thus, the modulation degree of the beam decreases as the defect size decreases.

As depicted in Figure 2, there exist turning points ($R =$ 1.4 and $R = 2.1$ mm), and the intensification factors are both 1.4. At the same time, the defect sizes satisfy the expression $R = 2/3 \times R_{\max}$. $R_{\max}$ denotes the defect size that has the most serious impact on the beam. As shown in Figure 3, the turning point indicates the boundary of two different intensity distributions. We will give a simple explanation for this phenomenon. With increasing defect size, the cutoff position gets closer to the first notch of the Airy diffraction pattern, generating an increasing intensity modulation in the modulation region. Moreover, according to (7), the intensity on the axis can be simplified to $I_2 = |A_2 - 2A_2'|^2$, where A_2 and A_2' are the amplitudes of U_2 and E_2', respectively. With increasing A_2', I_2 drops and then increases. The intensity on the edge of the modulation region changes little. Thus, there must exist a defect ($R = 2.1$ mm) that can make the intensity at the center and edge of the modulation region nearly equal (Figure 3(b)). When $R < 2.1$ mm, the spatial filter plays a role in removing the modulation region. Thus, the center of the light field is displayed as a dark region. Meanwhile,

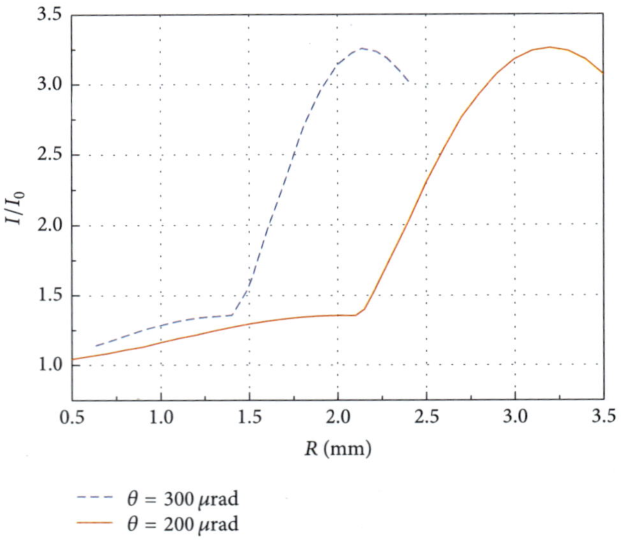

FIGURE 2: Intensification factor varying with increasing defect size at image relay plane.

the background beam is also perturbed. When $R > 2.1$ mm, there is a sharp intensity modulation in the beam center that is surrounded by a circle of dark stripes. We conducted a preliminary experiment to verify the above conclusions. By changing the sizes of the defects, the distributions of the stripes can be captured by Charge-Coupled Device (CCD) (Figure 4). Compared with the simulation calculations, both results are approximately in agreement, demonstrating the credibility of the theory to a certain extent under existing experimental conditions. More accurate experiments will be carried out in the future.

The quality of the downstream beam when using the filter pinhole is shown in Figure 5. It is found that a maximal value exists along the propagation distance. The value is larger than the one at the image relay plane. Figures 6(a) and 6(b) demonstrate the maximal value and corresponding location that evolve with increasing defect size. In Figure 6(a), we can see that the bigger the defect size, the greater the intensification factor when $R \le 3.2$ mm. The trend is close to the linear change. Meanwhile, the corresponding location becomes gradually closer to the image relay plane, which may induce damage to optical components just placed at the image relay plane, such as the reflector in SG-II system. In addition, Figure 6(b) shows that as the size of the defect increases, restraining of characteristic high-frequency modulation by the spatial filter has been greatly weakened. Thus, the location of the maximum value gradually approaches the position of the modulation beam propagating in free space when $R >$ 3.2 mm. Thus, $R = 3.2$ mm is a turning point in the process of changes in the maximal value location.

To ensure the quality of the propagation beam and the safety of optical components, the intensification factor is limited to 1.4 [18]. By analyzing the beam quality at the image relay plane and downstream plane, which is affected by different sizes of local defects, a rule identifying the maximum defect size for a given spatial filter that bounds the safe

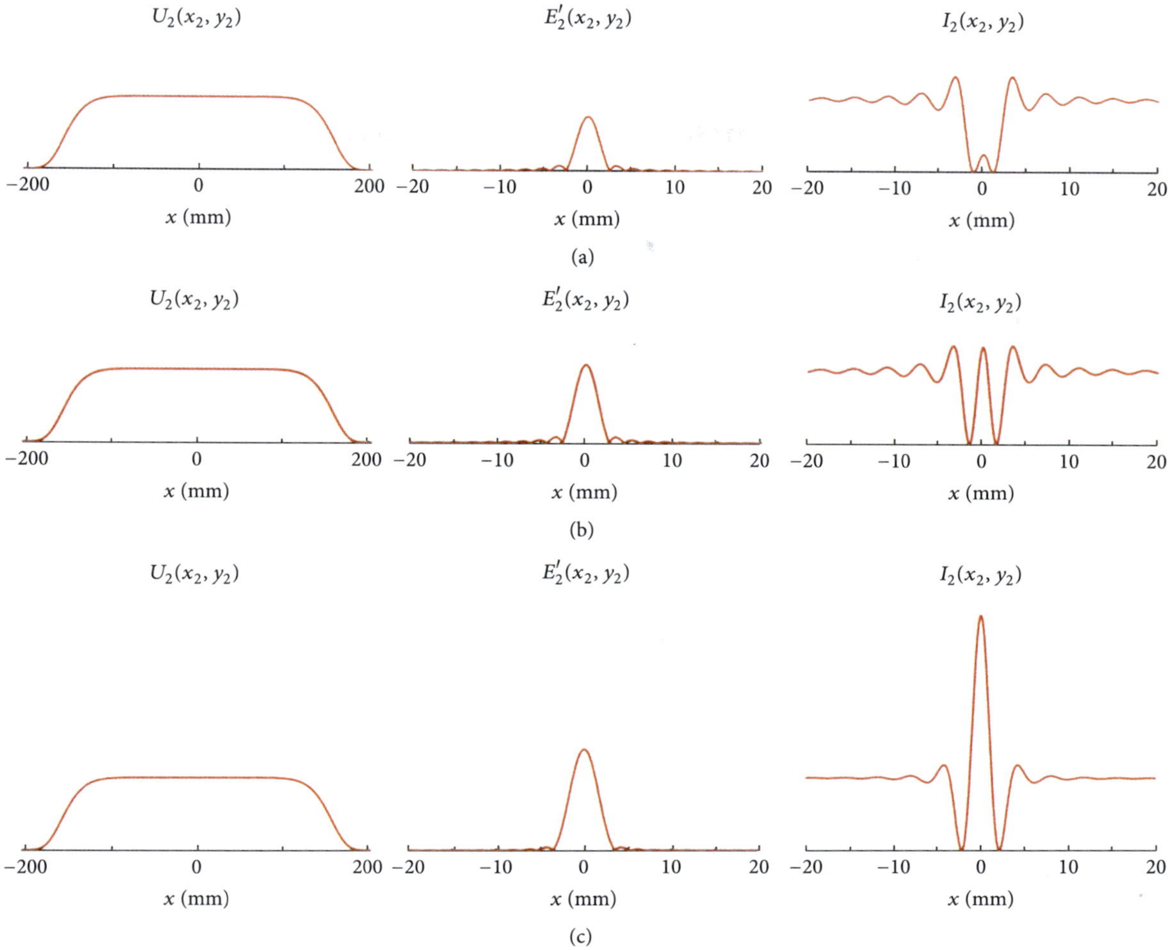

FIGURE 3: Intensity profiles at image relay plane. (a) $R = 1.5$ mm, (b) $R = 2.1$ mm, and (c) $R = 3.2$ mm.

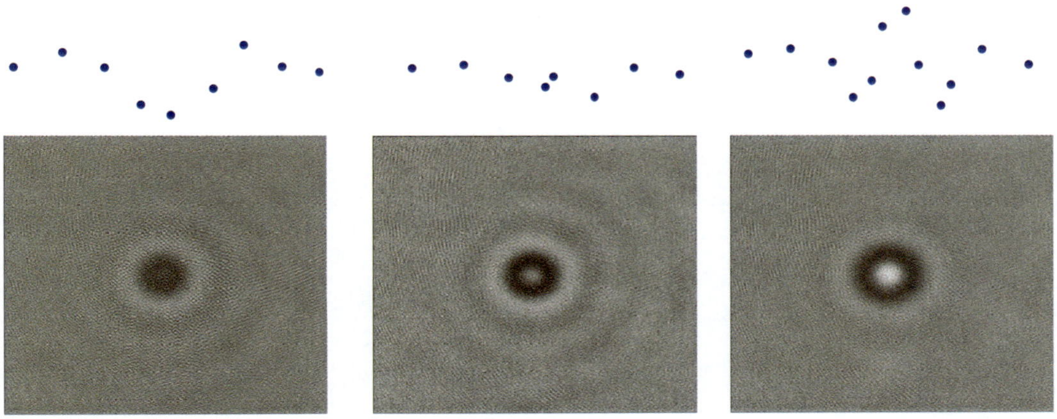

FIGURE 4: Experimental results. (a) One-dimensional intensity distributions. Blue dotted lines represent experimental results, and red solid lines represent calculation results. (b) Light intensity distributions acquired by CCD.

operating range is given by the maximum allowable defect times half-angle of spatial filter ≤240 mm-μrad. Namely, if there is a defect larger than 1.2 mm, the optical component located at the imaging reply plane or downstream plane must be paid more attention, and some effective processes should be taken if necessary. In addition, the maximum allowable defect radiuses in different high-power laser systems are listed in Table 1, including the results in the SG-II high-power laser facility. We can observe from the table that the larger the filter pinhole is, the weaker constraints it has on the modulation

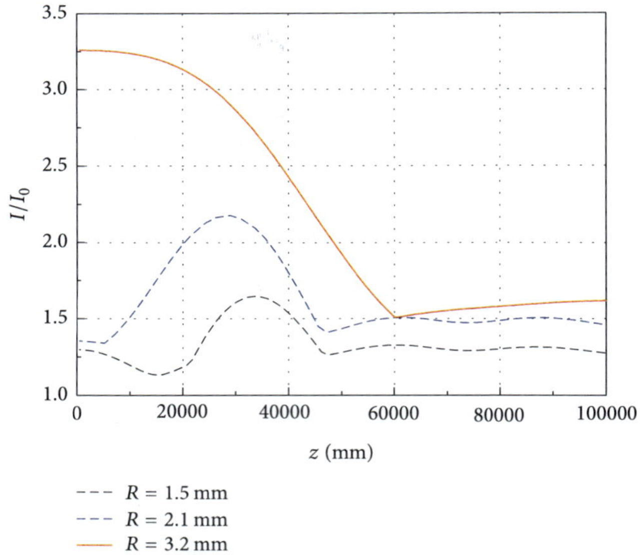

FIGURE 5: Intensification factor varying with distance from image relay plane.

TABLE 1: Maximum allowable defect radiuses in different high-power laser systems.

Filter pinhole (μrad)	Defect radius (mm)
80	3.0
109	2.2
128 [16]	1.8
141 [16]	1.7
150 [6]	1.6
200 [6]	1.2

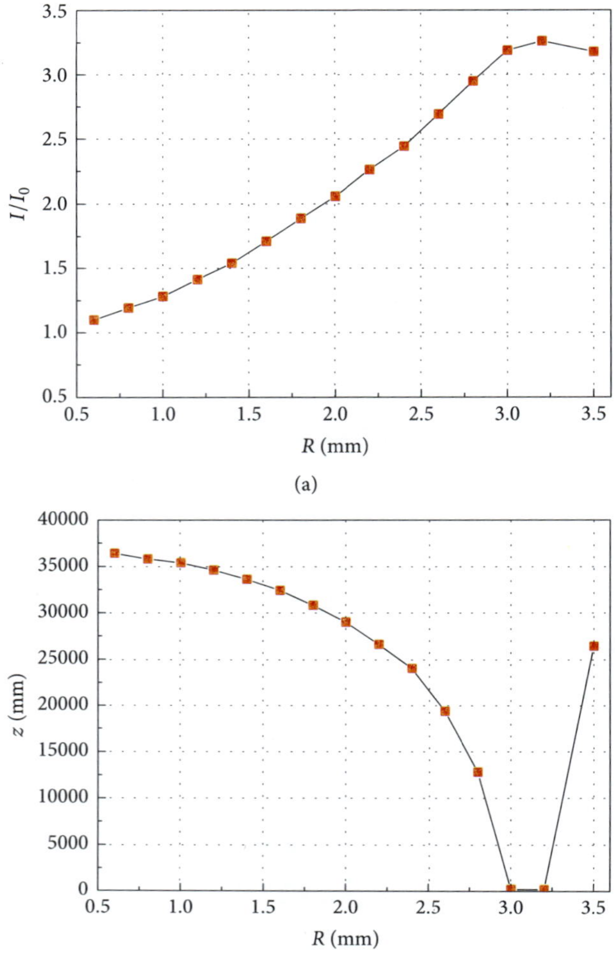

FIGURE 6: Maximum intensification factor and corresponding location varying with defect size along downstream light field: (a) maximum intensification factor and (b) corresponding location.

beam, and the smaller the defect size allowed. Therefore, the constraints of defect size and the selection of filter pinhole parameters should be put forward with regard to the quality of the propagation beam.

4. Conclusion

The imaging property of a spatial filter system is used to minimize the intensity fluctuations. However, the spatial filter does not work effectively because of the broadening spectrum caused by local defects. The relation between the sizes of filter pinhole and local defects is presented. Based on the above, the beam quality at the image relay plane and downstream plane deteriorates significantly as the defect size increases. Ultimately, to control low-frequency beam modulation, the fatal defect size is calculated according to spatial filter parameters. We believe that these results will help in giving specifications for components and determining which defects should be eliminated during operating high-power lasers.

Conflicts of Interest

The authors declare that there are no conflicts of interest regarding the publication of this paper.

References

[1] S. Mainguy, B. Le Garrec, and M. Josse, "Downstream impact of flaws on the LIL/LMJ laser lines," in *Proceedings of the Laser-Induced Damage in Optical Materials*, Boulder, Colo, USA, September 2005.

[2] M. L. Spaeth, K. R. Manes, and D. H. Kalantar, "Description of the NIF Laser," *Fusion Science and Technology*, vol. 69, pp. 25–145, 2016.

[3] B. Y. Wang, J. Y. Zhang, S. Shi, K. W. You, and J. Q. Zhu, "Effects of side subsurface defects induced by CNC machine on the gain spatial distribution in neodymium phosphate glass," *High Power Laser Science Engineering*, vol. 4, no. 1, 2016.

[4] H. Bercegol, G. J. Exarhos, A. H. Guenther et al., "The impact of laser damage on the lifetime of optical components in fusion lasers," in *Proceedings of the XXXV Annual Symposium*

on Optical Materials for High Power Lasers: Boulder Damage Symposium, vol. 5273, Boulder, Colo, USA, 2004.

[5] D. F. Zhao, L. Wang, Z. Q. Lin, P. Shao, and J. Q. Zhu, "SG-II-Up prototype final optics assembly: optical damage and clean-gas control," *High Power Laser Science and Engineering*, vol. 3, 2015.

[6] M. L. Spaeth, K. R. Manes, and J. Honig, "Cleanliness for the NIF 1ω laser amplifiers," *Fusion Science and Technology*, vol. 69, no. 1, pp. 250–264, 2016.

[7] J. Honig, "Cleanliness improvements of National Ignition Facility amplifiers as compared to previous large-scale lasers," *Optical Engineering*, vol. 43, no. 12, pp. 2904–2911, 2004.

[8] P. M. Celliers, K. G. Estabrook, R. J. Wallace et al., "Spatial filter pinhole for high-energy pulsed lasers," *Applied Optics*, vol. 37, no. 12, pp. 2371–2378, 1998.

[9] J. E. Murray, D. Milam, C. D. Boley, K. G. Estabrook, and J. A. Caird, "Spatial filter pinhole development for the national ignition facility," *Applied Optics*, vol. 39, no. 9, pp. 1389–1405, 2000.

[10] J. T. Hunt, P. A. Renard, and W. W. Simmons, "Improved performance of fusion lasers using the imaging properties of multiple spatial filters," *Applied Optics*, vol. 16, no. 4, pp. 779–782, 1977.

[11] R. J. Korniski, L. R. Gardner, K. P. Thompson, English Jr., and J. L. Miller, "Spatial filter lens design for the main laser of the National ignition facility," in *Proceedings of the International Optical Design Conference*, vol. 3482, Kona, Hawaii, USA, 1998.

[12] P. P. Sun, D. Liu, Y. L. Zhang, X. Y. Li, Y. Zhang, and J. Q. Zhu, "Evolution of low-frequency noise passing through a spatial filter in a high power laser system," *Science China Physics, Mechanics & Astronomy*, vol. 54, no. 3, pp. 411–415, 2011.

[13] J. T. Hunt, J. A. Glaze, W. W. Simmons, and P. A. Renard, "Suppression of self-focusing through low-pass spatial filtering and relay imaging," *Applied Optics*, vol. 17, no. 13, pp. 2053–2057, 1978.

[14] M. Bray, A. Liard, and G. Chabassier, "Laser MegaJoule optics (I): new methods of optical specification," in *Proceedings of the 1999 Optical Fabrication and Testing*, vol. 3739, pp. 449–460, May 1999.

[15] Y. Gao, B. Zhu, D. Liu, and Z. Lin, "Propagation of flat-topped multi-Gaussian beams through a double-lens system with apertures," *Optics Express*, vol. 17, no. 15, pp. 12753–12766, 2009.

[16] H. J. Liu, F. Jing, Q. Li, Z. T. Peng, and D. X. Hu, "The effect of spatial filter pinhole on output beams quality in high power laser," *Journal of Yunnan University*, vol. 184, no. 27, 2009.

[17] M. Born and E. Wolf, *Principles of optics: electromagnetic theory of propagation, interference and diffraction of light*, Cambridge University Press, Cambridge, UK, 7th edition, 1999.

[18] G. Feng and S. Zhou, "Discussion of comprehensive evaluation on laser beam quality," *Zhongguo Jiguang*, vol. 36, no. 7, pp. 1643–1653, 2009.

17

Critical Temperature for Ordered-Disordered Phase Transformation in Cu_3Au under Pressure

Pham Dinh Tam,[1] Bui Duc Tinh,[2,3] Nguyen Quang Hoc,[3] and Pham Duy Tan[4]

[1]*Le Quy Don University of Technology, 100 Hoang Quoc Viet, Cau Giay District, Hanoi, Vietnam*
[2]*Institute of Research and Development, Duy Tan University, 03 Quang Trung, Da Nang, Vietnam*
[3]*Hanoi National University of Education, 136 Xuan Thuy Street, Cau Giay District, Hanoi, Vietnam*
[4]*College of Armor, Tam Dao, Vinh Phuc, Vietnam*

Correspondence should be addressed to Bui Duc Tinh; tinhbd@hnue.edu.vn

Academic Editor: Jörg Fink

We use the statistical moment method to study the dependence of the critical temperature T_c for Cu_3Au on pressure in the interval from 0 to 30 kbar. The calculated mean speed of changing critical temperature to pressure is 1.8 K/kbar. This result is in a good agreement with the experimental data.

1. Introduction

The ordered-disordered phase transformation in Cu_3Au under pressure has been studied by experimental methods such as electrical-resistance measurements made while the sample is at high temperature and under pressure [1] and the X-ray diffraction and resistance measurement. The transitional process in the Cu_3Au alloy from the disordered to the ordered state and the relaxation process in the phase change in Cu_3Au have been investigated by measuring the time dependence of the X-ray superstructure line width and the electrical resistivity [2]. The ordering kinetics of the order-disorder phase transition in Cu_3Au has been investigated by X-ray diffraction [3].

A coarse-grained model for a Cu_3Au system undergoing an order-disorder transition is constructed. The model is characterized by a Ginzburg-Landau Hamiltonian with a three-component order parameter and the symmetry of the Cu_3Au system. The ordering dynamics of this model subjected to a temperature quench are then studied by use of Langevin dynamics. The model is analyzed with a generalization of the recently developed first-principles theory of unstable thermodynamic systems [4]. The ordered-disordered phase transition in alloy Cu_3Au also is investigated theoretically by applying statistical methods for ordered phenomena such as the Kirkwood method, the pseudopotential method, and the pseudochemical method [5, 6]. However, these works only considered the dependence of ordered parameter on temperature and considered the critical temperature at zero pressure.

In this paper, the dependence of critical temperature on pressure in alloy Cu_3Au is studied by using the model of effective metals and the statistical moment method (SMM). We obtained a rather simple equation describing this dependence. Our numerical calculations are in a good agreement with the experimental data.

2. Hemholtz Free Energy of Cu_3Au Alloy by Using the SMM

Using the thermodynamic theory of alloy in [5, 7], we analyze the ordered alloy Cu_3Au into a combination of four effective metals Cu^*1, Cu^*2, Au^*1, and Au^*2. Then, the Helmholtz free energy of alloy Cu_3Au can be calculated through the Helmholtz free energy of these effective metals and has the form

$$\psi_{Cu_3Au} = \frac{1}{4}\left(P_{Cu}^{(1)}\psi_{Cu^*1} + 3P_{Cu}^{(2)}\psi_{Cu^*2} + P_{Au}^{(1)}\psi_{Au^*1} + 3P_{Au}^{(2)}\psi_{Au^*2}\right) - TS_c, \quad (1)$$

where $P_{\alpha}^{(\beta)}$ (α = Cu, Au; β = 1, 2) is the probability so that the atom α occupies the lattice site of β-type and these probabilities are determined in [8], $\psi_{Cu^*1}, \psi_{Cu^*2}, \psi_{Au^*1}$, and ψ_{Au^*2} are the Helmholtz free energy of effective metals Cu^*1, Cu^*2, Au^*1, and Au^*2, respectively, and S_c is the configurational entropy.

The Helmholtz free energy of effective metals $\alpha * \beta$ (α = Cu, Au; β = 1, 2) is calculated by the SMM analogously as for pure metals [9] and is equal to

$$\psi_{\alpha*\beta} = 3R\left\{\frac{u_{\alpha*\beta}}{6k_B} + T\left\{X_{\alpha*\beta} + \ln\left[1 - \exp\left(-2X_{\alpha*\beta}\right)\right]\right\}\right\},$$
$$u_{\alpha*\beta} = u_\alpha + \frac{P_{\alpha\alpha'}}{C_\alpha}\left(\Delta_{\alpha\alpha'}^{(0)} - 2\omega\right), \quad (2)$$
$$X_{\alpha*\beta}\sqrt{\frac{k_{\alpha*\beta}}{m_{\alpha*\beta}}}, \quad k_{\alpha*\beta} = k_\alpha + 3\frac{P_{\alpha\alpha'}}{C_\alpha}\Delta_{\alpha\alpha'}^{(2)},$$

where u_α, k_α are parameters of the pure metal α [9], $P_{\alpha\alpha'}$ is the probability so that the atom of α-type and the atom of α'-type (α, α' = Cu, Au; $\alpha \neq \alpha'$) are side by side, and ω is the ordered energy and is determined by [6]

$$\omega = \frac{1}{2}\left(\varphi_{\text{Cu-Cu}} + \varphi_{\text{Au-Au}}\right) - \varphi_{\text{Cu-Au}}, \quad (3)$$

where $\varphi_{\text{Cu-Cu}}$, $\varphi_{\text{Au-Au}}$, and $\varphi_{\text{Cu-Au}}$ are the interaction potential between atoms Cu-Cu, Au-Au, and Cu-Au, respectively, on same distance and $\Delta_{\alpha\alpha'}^{(0)}$ and $\Delta_{\alpha\alpha'}^{(2)}$ are the difference of interaction potentials and the difference of derivatives of second degree for interaction potential to displacement of atom pairs $\alpha' - \alpha'$ and $\alpha - \alpha$, respectively, on same distance a.

Substituting (2) into (1), we obtain the expression of the Helmholtz free energy for alloy Cu_3Au as follows:

$$\psi_{Cu_3Au} = \frac{1}{4}\left(3\psi_{Cu} + \psi_{Au}\right) + 6R\left[3T\left(\frac{X_{Cu}}{k_{Cu}} - \frac{X_{Au}}{k_{Au}}\right)\Delta_{\text{Cu-Au}}^{(2)} - \frac{\omega}{k_B}\right]P_{\text{Cu-Au}} - TS_c, \quad (4)$$

where $X_\alpha = x_\alpha \coth x_\alpha$, $x_\alpha = (\hbar/2k_BT)\sqrt{k_\alpha/m_\alpha}$ (α = Cu, Au), m_α is the mass of atom α, ψ_{Cu} and ψ_{Au} are the Helmholtz free energies of pure metals Cu and Au, respectively, and the configurational entropy of alloy has the form [6]

$$S_c = -\frac{R}{4}\left(P_{Cu}^{(1)}\ln P_{Cu}^{(1)} + 3P_{Cu}^{(2)}\ln P_{Cu}^{(2)} + P_{Au}^{(1)}\ln P_{Au}^{(1)} + 3P_{Au}^{(2)}\ln P_{Au}^{(2)}\right). \quad (5)$$

3. Critical Temperature for Cu_3Au Alloy under Pressure

The ordered-disordered phase transition in alloy Cu_3Au is the phase transition of first type [8], where the following relations are satisfy simultaneously:

$$\left(\frac{\partial\psi_{Cu_3Au}}{\partial\eta}\right)_{\eta=\eta_0} = 0, \quad (6)$$

$$\left(\psi_{Cu_3Au}\right)_{\eta=\eta_0} = \left(\psi_{Cu_3Au}\right)_{\eta=0}, \quad (7)$$

where η is the equilibrium long-range ordered parameter at the temperature T and pressure p and is determined from condition (6) and η_0 is the equilibrium long-range ordered parameter at the critical temperature T_c.

The probabilities $P_{\alpha}^{(\beta)}$ and $P_{\alpha\alpha'}$ are represented through the ordered parameter η by the following relations [6, 8]:

$$P_{Au}^{(1)} = \frac{1}{4} + \frac{3}{4}\eta,$$
$$P_{Au}^{(2)} = \frac{1}{4} - \frac{1}{4}\eta,$$
$$P_{Cu}^{(1)} = \frac{3}{4} - \frac{3}{4}\eta, \quad (8)$$
$$P_{Cu}^{(2)} = \frac{3}{4} + \frac{1}{4}\eta,$$
$$P_{AuCu} = \frac{3}{16} + \frac{1}{16}\eta^2 + \varepsilon_{AuCu},$$

where ε_{AuCu} is the correlational parameter. This parameter has small value and is ignored.

Substituting (4) into (6) and (7), paying attention to (8), and carrying out some calculations, we obtain two equations in order to determine η_0 and T_c as follows:

$$\left[3\left(\frac{X_{Cu}}{k_{Cu}} - \frac{X_{Au}}{k_{Au}}\right)\Delta_{\text{Cu-Au}}^{(2)} - \frac{\omega}{k_BT}\right]\eta = -\frac{1}{4} \cdot \ln\frac{(1+3\eta)(3+\eta)}{(1-\eta)(3-3\eta)}, \quad (9)$$

$$\left[3\left(\frac{X_{Cu}}{k_{Cu}} - \frac{X_{Au}}{k_{Au}}\right)\Delta_{\text{Cu-Au}}^{(2)} - \frac{\omega}{k_BT}\right]\eta_0^2 = \frac{2}{3}\left[3\ln 3 - 4\ln 4 - \left(\frac{3}{4} - \frac{3}{4}\eta_0\right)\ln\left(\frac{3}{4} - \frac{3}{4}\eta_0\right) - 3\left(\frac{3}{4} + \frac{1}{4}\eta_0\right)\ln\left(\frac{3}{4} + \frac{1}{4}\eta_0\right) - \left(\frac{1}{4} + \frac{3}{4}\eta_0\right)\ln\left(\frac{1}{4} + \frac{3}{4}\eta_0\right) - 3\left(\frac{1}{4} - \frac{1}{4}\eta_0\right)\ln\left(\frac{1}{4} - \frac{1}{4}\eta_0\right)\right] - \Delta(a, T_c), \quad (10)$$

where $\Delta(a, T_c) = (2/RT)[\psi_{Cu}(a) - \psi_{Cu}(a')] + (2/3RT)[\psi_{Au}(a) - \psi_{Au}(a')]$ and a and a' are the lattice parameters

TABLE 1: Solutions of (13) at different pressures ($\omega/k_B = 910.6$K).

p (kbar)	0	5	10	15	20	25	30
a (10^{-10} m)	2.7618	2.7591	2.7563	2.7536	2.7509	2.7480	2.7453
T_c (K)	665	676	686	695	704	711	718

of alloy Cu_3Au at the critical temperature T_c in the ordered zone and the disordered zone, respectively.

From (9) we find the dependence of η on temperature and pressure as follows:

$$\frac{\omega}{k_B T} = \frac{1}{4\eta} \ln \frac{(1+3\eta)(3+\eta)}{(1-\eta)(3-3\eta)} + 3\left[\left(\frac{X_{Cu}}{k_{Cu}} - \frac{X_{Au}}{k_{Au}}\right)\Delta^{(2)}_{Cu\text{-}Au}\right]_{T,p}. \quad (11)$$

Second term in right side of (10) depends on temperature and pressure. At phase transition point in (9), $T = T_c$ and $\eta = \eta_0$. Therefore, from (9) and (10) we find the equation in order to determine η_0 as follows:

$$-\eta_0 \ln \frac{(1+3\eta_0)(3+\eta_0)}{3(1-\eta_0)^2} = -4\Delta(a, T_c) + \frac{8}{3}\Big[3\ln 3 - 4\ln 4 - \left(\frac{3}{4} - \frac{3}{4}\eta_0\right)\ln\left(\frac{3}{4} - \frac{3}{4}\eta_0\right) - 3\left(\frac{3}{4} + \frac{1}{4}\eta_0\right)\ln\left(\frac{3}{4} + \frac{1}{4}\eta_0\right) - \left(\frac{1}{4} + \frac{3}{4}\eta_0\right)\ln\left(\frac{1}{4} + \frac{3}{4}\eta_0\right) - 3\left(\frac{1}{4} - \frac{1}{4}\eta_0\right)\ln\left(\frac{1}{4} - \frac{1}{4}\eta_0\right)\Big]. \quad (12)$$

Because the parameters a and a' are somewhat different, $\Delta(a, T_c)$ has very small contribution to (12). Therefore, $\Delta(a, T_c)$ approximately does not depend on temperature and pressure and is determined at the critical point and zero pressure.

Using the expressions of ψ_α and a in [9, 10] at the temperature $T = T_c = 665$ K and pressure $p = 0$, we obtain $\Delta(a, T_c) = 0.6526\eta_0^2$.

Substituting this value of $\Delta(a, T_c)$ into (12), we find the ordered parameter $\eta_0 = 0.37$. Substituting this value of η_0 into (11), the dependence of critical temperature T_c on pressure has the form

$$k_B T_c = \left\{1.207 + 3\left[\left(\frac{X_{Cu}}{k_{Cu}} - \frac{X_{Au}}{k_{Au}}\right)\Delta^{(2)}_{Cu\text{-}Au}\right]_{T_c,p}\right\}^{-1}\omega. \quad (13)$$

4. The Results and Discussion

At the critical temperature T_c (~100 K), X_{Cu}, X_{Au} are very near unit and we can take $X_{Cu} = X_{Au} = 1$. On the other hand,

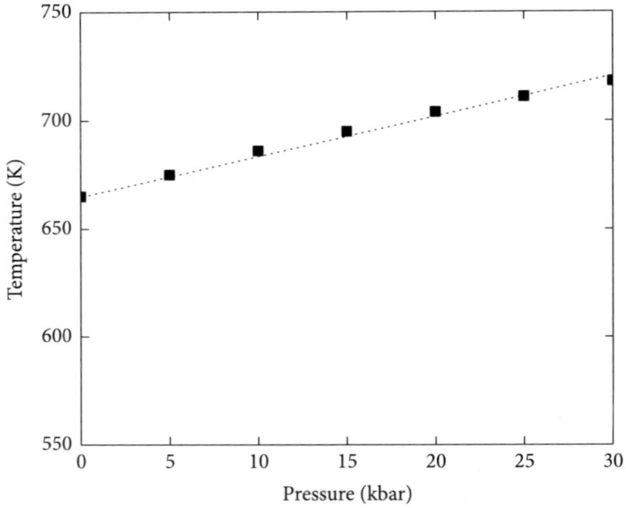

FIGURE 1: The dependence of the critical temperature T_c for alloy Cu_3Au on pressure.

from [11] we find $\Delta^{(2)}_{Cu\text{-}Au} = (1/6)(k_{Au} - k_{Cu})$. So, (13) has the following simple form:

$$\frac{k_B T_c}{\omega} = \left[1.207 + \frac{1}{2}\frac{(k_{Au} - k_{Cu})^2}{k_{Au}k_{Cu}}\right]^{-1}. \quad (14)$$

Applying the potential Lennard-Jones $(n - m)$ [12] to interactions Cu-Cu and Au-Au and the expression of k_α in [11], we have

$$\frac{(k_{Au} - k_{Cu})^2}{k_{Au}k_{Cu}} = Aa^{2.5}X(a) + \frac{1}{Aa^{2.5}X(a)} - 2, \quad (15)$$

where $A = 0.052$, $X(a) = (1 - 0.02a^{3.5})/(1 - 0.002a^6)$, a is measured in unit of 10^{-10} m.

From (14), (15) and the equation of parameter a for alloy Cu_3Au in [10], we find the dependence of the critical temperature T_c on pressure. Our numerical calculations of the dependence of $T_c(p)$ with the values of pressure from 0 to 30 kbar are given in Table 1 and represented in Figure 1.

From Figure 1 we see that, in the interval of pressure from 7 to 21 kbar, the critical temperature T_c depends near linearly on pressure with the mean speed of changing $\Delta T/\Delta p \approx 1.8$ K/kbar. This result agrees with experiments [1], where the rate of change of the critical temperature with pressure is 2.1 K/kbar from 7 to 21 kbar. The kinetics of the order transformation below T_c are adequately described by the homogeneous reaction rate equation and an activation volume of 6.8 cm^3/mole of atoms. The magnitude of this activation volume indicates that the formation of vacancies on the gold sublattice is the rate-limiting step in the homogeneous ordering process.

If ignoring the second term in right side of (11) (this term depends on pressure and temperature), we obtain the expression of ordered parameter η calculated by other statistical methods [8].

In conclusion, the obtained dependence of critical temperature on pressure (see (14)) in alloy Cu_3Au has simple analytic form and numerical result in a good agreement with the experimental data.

Conflicts of Interest

The authors declare that they have no conflicts of interest.

Acknowledgments

This work was carried out by the financial support from HNUE, the Le Quy Don University of Technology.

References

[1] M. C. Franzblau and R. B. Gordon, "The Order-Disorder Transformation in Cu," *Journal of Applied Physics*, vol. 38, no. 1, pp. 103–110, 1967.

[2] T. Hashimoto, K. Nishimura, Y. Takeuchi, and J. Phys, "Dynamics on transitional ordering process in Cu3Au alloy from disordered state to ordered state," *Journal of The Physical Society of Japan*, vol. 45, 9 pages, 1978.

[3] K. Torli, T. Tamaki, N. Wakabayashi, and J. Phys, "Order-disorder kinetics of Cu3Au studied by X-Ray diffraction," *Jounal of The Physical Society of Japan*, vol. 59, 9 pages, 1990.

[4] Z. Lai, "Theory of ordering dynamics for Cu3Au," *Physical Review B: Covering Condensed Matter and Materials Physics*, vol. 41, no. 13, 1990.

[5] P. D. Tam and N. Q. Hoc, "Effective Metal's Model for Alloy," in *Proceedings of the 6th National Congress on Physics*, vol. 1, pp. 125–128, Hanoi, 2005.

[6] L. V. Panina, K. Mohri, K. Bushida, and M. Noda, "Giant magneto-impedance and magneto-inductive effects in amorphous alloys (invited)," *Journal of Applied Physics*, vol. 76, no. 10, pp. 6198–6203, 1994.

[7] K. Masuda-Jindo, V. V. Hung, and P. D. Tam, "Application of statistical moment method to thermodynamic quantities of metals and alloys," *Calphad*, vol. 26, no. 1, pp. 15–32, 2002.

[8] A. A. Smirnov, *Molecular Dynamic Theory of Metals*, Nauka, Moscow, Russia, 1966.

[9] K. Masuda-Jindo, V. V. Hung, and P. D. Tam, "Thermodynamic quantities of metals investigated by an analytic statistical moment method," *Physical Review B: Condensed Matter and Materials Physics*, vol. 67, article 094301, no. 9, 2003.

[10] P. D. Tam, "The lattice spacings for binary alloys AB," *Communication in Physics*, vol. 2, pp. 78–83, 1998.

[11] P. D. Tam, "The melting temperature for binary alloys AB at various pressures," *VNU Journal of Science: Natural Sciences and Technology*, vol. 2, no. 35, 1999.

[12] S. Zhen and G. J. Davies, "Calculation of the Lennard-Jones $n-m$ potential energy parameters for metals," *Physica Status Solidi*, vol. 78, no. 2, pp. 595–605, 1983.

18

Passively Q-Switched Erbium-Doped Fiber Laser via Evanescent Field Interaction with Few-Layer Molybdenum Ditelluride

Man Wu and Ying Li

SZU-NUS Collaborative Innovation Center for Optoelectronic Science & Technology, International Collaborative Laboratory of 2D Materials for Optoelectronics Science and Technology of Ministry of Education, College of Optoelectronic Engineering, Shenzhen University, Shenzhen 518060, China

Correspondence should be addressed to Man Wu; manmwu@szu.edu.cn

Guest Editor: Xinxing Zhou

We demonstrate an all-fiberized passively Q-switched erbium-doped fiber laser (EDFL) via evanescent field interaction between molybdenum ditelluride saturable absorber (SA) and guided mode of the D-shaped fiber. By integrating the few-layer molybdenum ditelluride prepared by CVD method onto the side-polished fiber, the SA can be realized by the strong interaction between the evanescent field of the waveguide and the nonlinear optical material. The proposed passively Q-switched EDFL could deliver output pulses at 1566 nm wavelength with pulse width of 5.03 μs, a repetition rate of 13.9 kHz, a pulse energy of 150.6 nJ, and an output power of 2.1 mW when pumped by a 980 nm laser diode of 180 mW.

1. Introduction

Passively Q-switched fiber lasers have been extensively used in optical communication, optical sensing, industrial processing, etc., for the high beam quality, high stability, high energy, and the compact setups [1–5]. The saturable absorber (SA) is indispensable for the passively Q-switched fiber laser for the quality factor Q of the optical cavity can be modulated by the intensity-dependent nonlinearity of SAs [1, 2]. Inspired by the requirements of the compact, cost-effective, stable Q-switched fiber laser, the researchers from optics, optical material, and some other research community have paid much attention on the fields. As we know, the semiconductor saturable absorber mirrors (SESAMs) have been commercialized for the precise control of the absorption wavelength, saturation threshold, modulation depth, and relaxation time. However, SESAMs suffer from the limited bandwidth and complicated preparation procedures [5]. With the emergence of the low-dimensional materials [6–15], especially the two-dimensional (2D) nanomaterials [3, 4], their unique characteristics, such as the ultrafast recovery time, high damage threshold, and broadband nonlinear optical response, have provided an ideal platform to improve the laser performance. With the remarkable progress of graphene, other 2D materials, such as transition metal dichalcogenides (TMDs) [8–11], topological insulators [12, 13], and black phosphorus [16, 17], have also attracted growing research attention resulting from their excellent physical/chemical properties. However, graphene possesses a relatively low modulation depth, and topological insulators are limited to complex preparation processes, and black phosphorus is prone to oxidation and humidity. In contrast to them, TMDs have outstanding layer-dependent semiconducting and nonlinear optical properties.

TMDs, characterized by the chemical formula MX_2, where M is a transition metal (commonly Mo, W) and X is a group VI element (S, Se, and Te), are a family of highly anisotropic layered semiconductor materials. For the TMD monolayer, it is composed of two hexagonal planes of X atoms and an intermediate hexagonal plane of M atoms coordinated by covalent bonds with the X atoms in a trigonal prismatic arrangement [8, 18]. With the technical evolution, the nonlinear optical performance of TMDs, such as MoS_2, WS_2, $MoSe_2$, and WSe_2, has been investigated. Compared to other TMDs, telluride material has not attracted adequate attentions. In addition, it is found that $MoTe_2$, in contrast to MoS_2 and $MoSe_2$, has smaller bandgap, which implied it can exhibit broadband absorption. Most recently,

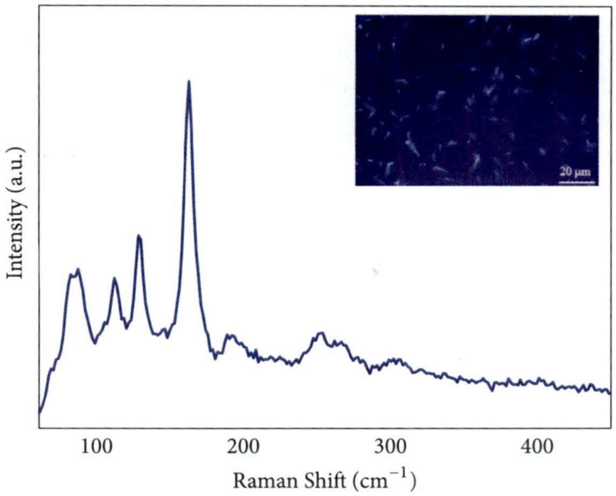

FIGURE 1: Raman spectra of the few-layer $MoTe_2$, and the inset shows its SEM image.

Mao et al. reported the nonlinear optical properties of few-layer $MoTe_2$ fabricated by a liquid exfoliation method and achieved soliton mode-locking operations in an EDFL [18]. Wang et. al. realized the high energy soliton pulse output with magnetron-sputtering-deposition-grown $MoTe_2$ saturable absorber [19]. Liu et. al. have implemented a $MoTe_2$-based passively Q-switched EDFL via magnetron sputtering technique [20]. Yan et. al. have also investigated the nonlinear optical performance of $MoTe_2$ around 3 μm in a solid-state laser system [21]. However, the material quality and laser energy based on $MoTe_2$ still need improvement.

In this contribution, we have prepared the few-layer $MoTe_2$ via the chemical vapor deposition method and fabricated an all-fiberized SA by transferring the $MoTe_2$ onto the D-shaped fiber. With the help of the strong interaction between evanescent field and the nonlinear optical material, the all-fiberized Q-switched fiber laser can be realized, and the high energy pulsed fiber laser can be delivered.

2. Materials Preparation and Characterizations

The $MoTe_2$ has been prepared via the chemical vapor deposition method. The quartz boat containing 5 mg Te powders is placed upstream of the tube with a distance of 14 cm from the heating center, while the second boat containing 0.5 mg MoO_3 powders is placed on the heating center of the tube. The substrate is placed above the second boat and faced up. The mixed gas of H_2/Ar (95% Ar) is flowing with 100 sccm for 30 min as the cleaning gas before heating and with 30-100 sccm as the carrier gas during the heating process. The pressure is in atmospheric environment. The furnace is heated to 720°C in 20 min and then maintained at given temperature for 30 min before it is naturally cooled. Then the few-layer $MoTe_2$ can be prepared.

The Raman spectra of prepared $MoTe_2$ show the few-layer nature of the material, which agree well with the earlier finding [22]. As shown in the inset of Figure 1, the scanning electron microscope (SEM) images suggest the 2D nature of the $MoTe_2$. With the grown $MoTe_2$, we have characterized its nonlinear absorption performance via the Z-scan technique with a 1560 nm mode-locked fiber laser whose repetition rate is ~20 MHz and pulse width is ~1.5 ps. Figure 2(a) shows the typical open-aperture curve of the $MoTe_2$ sample, which manifests that the material can exhibit nonlinear saturable absorption characteristics. In addition, we have given the relationship between incident intensity and the nonlinear transmittance, as shown in Figure 2(b), by fitting the experimental results with the following equation [11]:

$$\alpha(I_n) = \frac{\alpha_s}{1 + I_n/I_s} + \alpha_{ns} \quad (1)$$

where $\alpha(I_n)$ is the absorption coefficient, α_s is the saturable loss, α_{ns} is nonsaturable absorbance, and I and I_s are input and the saturation intensities, respectively. The fitted modulation depth and saturable intensity are 3% and 6 MW/cm^2, respectively.

3. Experimental Results and Discussions

To validate the nonlinear optical response of the $MoTe_2$, we have fabricated a fiberized SA by integrating the $MoTe_2$ onto the D-shaped fiber. Moreover, we followed the methods of Du et al. [11] to design and build an EDFL with a ring cavity configuration modulated by the $MoTe_2$ SA, as shown in Figure 3. In the experimental setup, the overall cavity length is about 8.6 m, which consists of a piece of high concentration erbium doped fiber (LIEKKI Er80-8/125) with length about 0.8 m, which is pumped by a commercial 980 nm laser diode (LD), a polarization independent isolator (PI-ISO) to guarantee the unidirectional laser operation, an intracavity polarization controller (PC) to tune the cavity birefringence, and some other single mode fiber (SMF-28) optical components. We adopted a 980/1550 WDM coupler to introduce the 980 nm pump laser into the fiber ring cavity. Different from the work with transmission-type SA [11], we introduced a $MoTe_2$ SA via evanescent field interaction between guided wave and the $MoTe_2$, which can make the whole laser setup compact in an all-fiberized way. In addition, the evanescent field interaction can improve the interaction

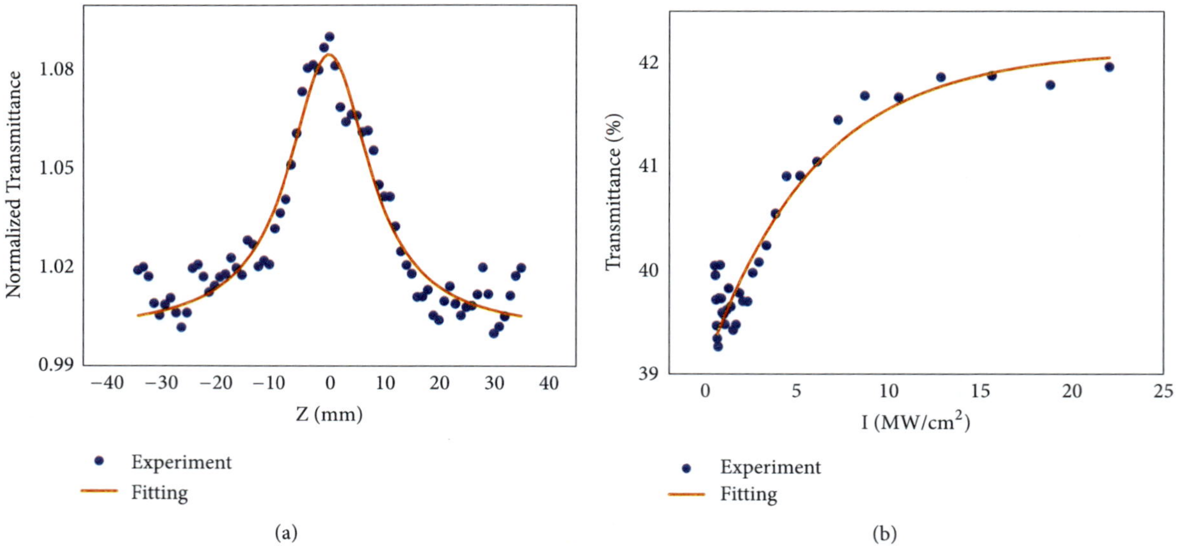

FIGURE 2: The experimental data of the nonlinear transmittance of the $MoTe_2$.

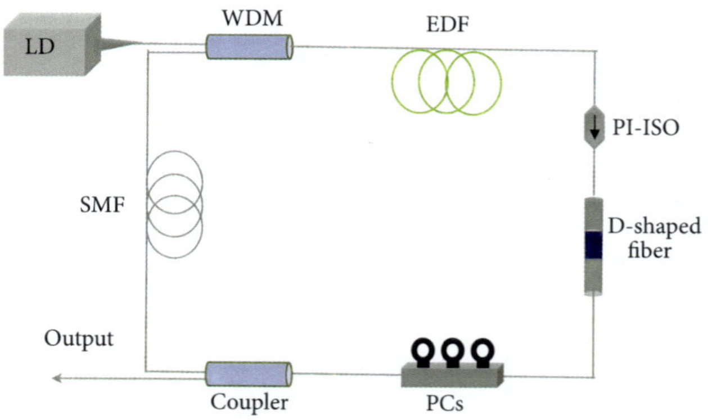

FIGURE 3: The experimental setup of the Q-switched fiber laser based on $MoTe_2$.

length and alleviate the material damage. To monitor the laser performance, an optical spectrum analyser (Ando AQ-6317B) and a real-time oscilloscope with a bandwidth of 4 GHz (Agilent DSO9404A) together with a photodetector (MC PD-12D) have been employed to simultaneously monitor the optical spectrum and the temporal evolution of the output pulse train. In addition, the output power of the Q-switched fiber laser was measured with a power meter (Ando, AQ2140).

In this experiment, the self-started Q-switching operation can be obtained by increasing the pump power up to around 60 mW. Figure 4 summarizes the typical operation characteristics of the Q-switching state at a pump power of 180 mW. Figure 4(a) shows the optical spectrum centered at 1566 nm delivered from the Q-switched fiber laser. Figure 4(b) shows the typical Q-switched pulse train with a repetition rate of 13.9 kHz. It can be seen that no peak intensity modulation had been observed on the pulse train, illustrating the high stability of Q-switching operation. The inset of Figure 4(b) shows the profile of the Q-switched pulse with a symmetric Gaussian-like shape with a full width at half maximum (FWHM) of 5.03 μs.

We have also investigated the power-dependent pulse duration, repetition rate, output power, and the pulse energy of the Q-switched fiber laser. Figure 5(a) presents the evolution curves of the repetition rate and pulse duration with the pump power. It shows typical features of the Q-switching operation. By increasing the pump power, the repetition rate will linearly increase from 4.9 kHz to 13.9 kHz, while the pulse duration varied in the range of 32.0 μs to 5.03 μs. Figure 5(b) shows the relationship between the average output power and the pulse energy with the pump power. The average output power increased almost linearly with the pump power. Moreover, at the maximum pump power of 180 mW, the maximum obtained pulse energy is up to 150.6 nJ, which is higher than or comparable to that of the Q-switch pulses obtained in the Er-doped fiber lasers with other SAs, such as CNTs, graphene, and TI, and higher than that of the reported results based on $MoTe_2$ SA [17–19]. The improved pulse energy can result from the effective interaction between the evanescent field of the fiber and the few-layer $MoTe_2$, which can provide a good platform to alleviate the damage threshold and tune the linear or nonlinear interaction with an all-fiberized configuration. Although the microfiber can

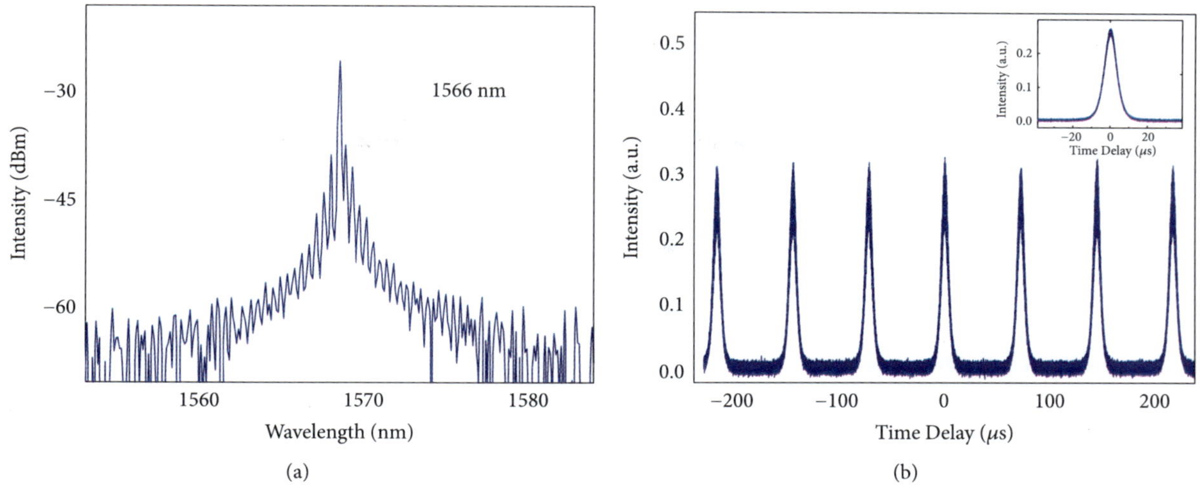

FIGURE 4: (a) The spectra of the Q-switched fiber laser based on $MoTe_2$. (b) The pulse train and the profile of the Q-switched pulse shown inset.

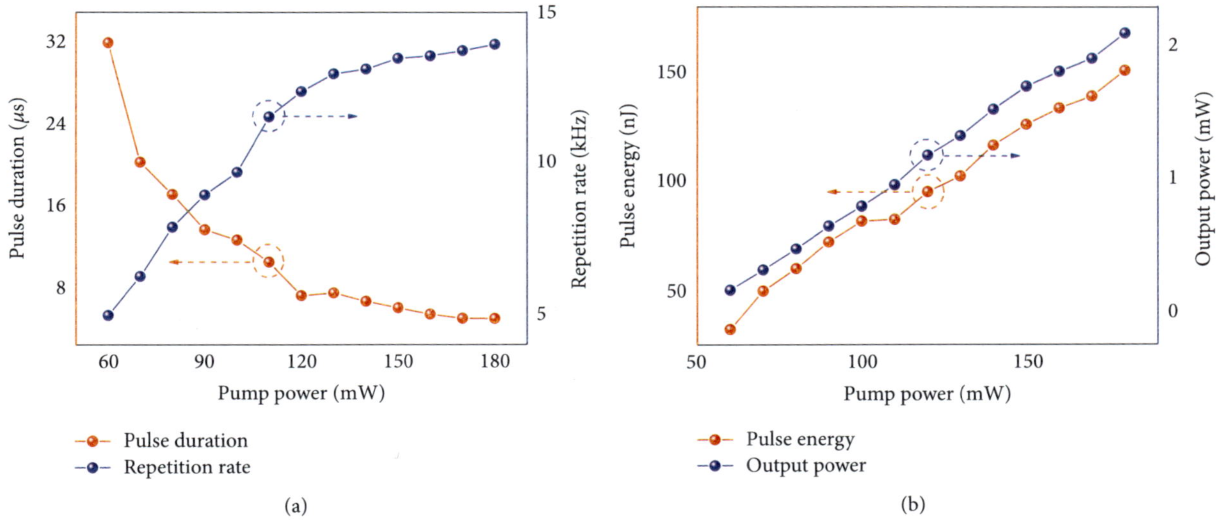

FIGURE 5: (a) The relationship between pulse duration and repetition rate with pump power. (b) The trend curves of output power and pulse energy with the pump power.

have strong evanescent field to interact with the material [19, 20], the waveguide will be easy to introduce large loss and strong nonlinearity for the small core size. However, the D-shaped fiber can provide a robust platform for the evanescent field interactions between light field propagating along the D-shaped fiber and the few-layer $MoTe_2$.

4. Conclusions

We have successfully demonstrated an all-fiberized passively Q-switched EDFL by using an evanescent field interaction based $MoTe_2$. The $MoTe_2$ has been prepared via chemical vapor deposition method and then integrated with D-shaped fiber to act as the nonlinear optical modulators. Maximum output pulse energy of 150.6 nJ has been obtained at a repetition rate of 13.9 kHz from the laser cavity under the input pump power of 180 mW at 980 nm. The experimental results can not only broaden the nonlinear application of $MoTe_2$, but provide an inroad for other TMDs nonlinear materials and their hybrid systems.

Conflicts of Interest

The authors declare that there are no conflicts of interests regarding the publication of this paper.

Acknowledgments

The authors would like to acknowledge Dr. Donglin Lu from Xiangtan University for providing the $MoTe_2$ sample for the experiment.

References

[1] F. Bonaccorso, Z. Sun, T. Hasan, and A. C. Ferrari, "Graphene photonics and optoelectronics," *Nature Photonics*, vol. 4, no. 9, pp. 611–622, 2010.

[2] M. E. Fermann and I. Hartl, "Ultrafast fibre lasers," *Nature Photonics*, vol. 7, no. 11, pp. 868–874, 2013.

[3] Q. Bao, H. Zhang, Y. Wang et al., "Atomic-layer craphene as a saturable absorber for ultrafast pulsed lasers," *Advanced Functional Materials*, vol. 19, no. 19, pp. 3077–3083, 2009.

[4] Z. Sun, T. Hasan, F. Torrisi et al., "Graphene mode-locked ultrafast laser," *ACS Nano*, vol. 4, no. 2, pp. 803–810, 2010.

[5] J. A. Alvarez-Chavez, H. L. Offerhaus, J. Nilsson, P. W. Turner, W. A. Clarkson, and D. J. Richardson, "High-energy, high-power ytterbium-doped Q-switched fiber laser," *Optics Expresss*, vol. 25, no. 1, pp. 37–39, 2000.

[6] S. Yamashita, Y. Inoue, S. Maruyama et al., "Saturable absorbers incorporating carbon nanotubes directly synthesized onto substrates and fibers and their application to mode-locked fiber lasers," *Optics Expresss*, vol. 29, no. 14, pp. 1581–1583, 2004.

[7] F. Wang, A. G. Rozhin, V. Scardaci et al., "Wideband-tuneable, nanotube mode-locked, fibre laser," *Nature Nanotechnology*, vol. 3, no. 12, pp. 738–742, 2008.

[8] Q. H. Wang, K. Kalantar-Zadeh, A. Kis, J. N. Coleman, and M. S. Strano, "Electronics and optoelectronics of two-dimensional transition metal dichalcogenides," *Nature Nanotechnology*, vol. 7, no. 11, pp. 699–712, 2012.

[9] S. Wang, H. Yu, H. Zhang et al., "Broadband few-layer MoS_2 saturable absorbers," *Advanced Materials*, vol. 26, no. 21, pp. 3538–3544, 2014.

[10] H. Zhang, S. B. Lu, J. Zheng et al., "Molybdenum disulfide (MoS_2) as a broadband saturable absorber for ultra-fast photonics," *Optics Express*, vol. 22, no. 6, pp. 7249–7260, 2014.

[11] L. Du, G. Jiang, L. Miao et al., "Few-layer rhenium diselenide: an ambient-stable nonlinear optical modulator," *Optical Materials Express* , vol. 8, no. 4, p. 926, 2018.

[12] C. Zhao, H. Zhang, X. Qi et al., "Ultra-short pulse generation by a topological insulator based saturable absorber," *Applied Physics Letters*, vol. 101, p. 118, 2012.

[13] C. Zhao, Y. Zou, Y. Chen et al., "Wavelength-tunable picosecond soliton fiber laser with Topological Insulator: Bi_2Se_3 as a mode locker," *Optics Express*, vol. 20, no. 25, pp. 27888–27895, 2012.

[14] S. Chen, Q. Wang, C. Zhao, and Y. Li, "Stable single-longitudinal-mode fiber ring laser using topological iInsulator-based saturable absorber," *Journal of Lightwave Technology*, vol. 32, pp. 4438–4444, 2014.

[15] G. Jiang, Y. Jin, L. Miao et al., "Tunable Gold Nanorods Q-Switcher for Pulsed Er-Doped Fiber Laser," *IEEE Photonics Journal*, vol. 9, no. 5, pp. 1–9, 2017.

[16] Z. Qin, G. Xie, H. Zhang et al., "Black phosphorus as saturable absorber for the Q-switched Er: ZBLAN fiber laser at 2.8 μm," *Optics Express*, vol. 23, no. 19, pp. 24713–24718, 2015.

[17] Y. Chen, G. Jiang, S. Chen et al., "Mechanically exfoliated black phosphorus as a new saturable absorber for both Q-switching and mode-locking laser operation," *Optics Express*, vol. 23, no. 10, pp. 12823–12833, 2015.

[18] D. Mao, B. Du, D. Yang et al., "Nonlinear Saturable Absorption of Liquid-Exfoliated Molybdenum/Tungsten Ditelluride Nanosheets," *Small*, vol. 12, no. 11, pp. 1489–1497, 2016.

[19] J. Wang, Z. Jiang, H. Chen et al., "High energy soliton pulse generation by a magnetron-sputtering-deposition-grown Moe_2," *Photonics Research*, vol. 6, no. 6, pp. 535–541, 2018.

[20] M. Liu, W. Liu, P. Yan, S. Fang, H. Teng, and Z. Wei, "High-power $MoTe_2$-based passively Q-switched erbium-doped fiber laser," *Chinese Optics Letters*, vol. 16, no. 2, Article ID 020007, 2018.

[21] Z. Yan, T. Li, S. Zhao et al., "$MoTe_2$ saturable absorber for passively Q-switched Ho,Pr:$LiLuF_4$ laser at 3 μm," *Optics & Laser Technology*, vol. 100, pp. 261–264, 2018.

[22] M. Grzeszczyk, K. Gołasa, M. Zinkiewicz et al., "Raman scattering of few-layers $MoTe_2$," *2D Materials*, vol. 3, no. 2, Article ID 025010, 2016.

19

Quantum Simulations of Charge Separation at a Model Donor-Acceptor Interface: Role of Delocalization and Local Packing

Allen Kelley, Kush Patel, and Eric R. Bittner

Department of Chemistry, University of Houston, Houston, TX 77204, USA

Correspondence should be addressed to Eric R. Bittner; ebittner@central.uh.edu

Academic Editor: Zhixin Hu

Organic Polymer-based photovoltaic systems offer a viable alternative to more standard solid-state devices for light-harvesting applications. In this study, we investigate the electronic dynamics of a model organic photovoltaic (OPV) heterojunction consisting of polyphenylene vinylene (PPV) oligomers and a [6, 6]-phenyl C61-butyric acid methyl ester (PCBM) blend. Our approach treats the classical molecular dynamics of the atoms within an Ehrenfest mean-field treatment of the π-π^* singly excited states spanning a subset of donor and acceptor molecules near the phase boundary of the blend. Our results indicate that interfacial electronic states are modulated by C=C bond stretching motions and that such motions induce avoided crossings between nearby excited states thereby facilitating transitions from localized excitonic configurations to delocalized charge-separated configurations on an ultrafast time-scale. The lowest few excited states of the model interface rapidly mix allowing low frequency C-C out-of-plane torsions to modulate the potential energy surface such that the system can sample both intermolecular charge-transfer and charge-separated electronic configurations on sub-100 fs time scales. Our simulations support an emerging picture of carrier generation in OPV systems in which interfacial electronic states can rapidly decay into charge-separated and current producing states via coupling to vibronic degrees of freedom.

1. Introduction

Advances in both materials and device fabrication have led to the development of highly efficient organic polymer-based photovoltaic cell (OPV) in which the power conversion efficiency is in excess of 10-11% under standard solar illumination and efficiencies as high as 12% in multijunction OPVs [1]. This increase in power conversion efficiency indicates that mobile charge carriers can be efficiently generated and collected in well-optimized devices; however, the underlying photophysical mechanism for converting highly bound molecular (Frenkel) excitons into mobile and asymptotically free photocarriers remains elusive in spite of vigorous, multidisciplinary research activity [2–18].

It is generally recognized that the local microstructure of the interface between the donor and acceptor materials composing a bulk heterojunction cell plays a central role in determining the overall efficiency of an OPV device. Generally speaking, charge separation in OPV materials occurs when excitons near the donor/acceptor (DA) interface dissociate into either charge-transfer or charge-separated states.

$$D^* : A \longrightarrow D^+ : A^- \text{ or } D^+ \cdots A^- \quad (1)$$

These two states are distinguished by whether or not the positively charged (hole) on an electron donor molecule is directly adjacent to the acceptor anion, termed a charge-transfer state (CT), or if the two charges are separated by one or more molecular units, termed a charge-separated state (CS). In the former (CT) case, the two charges are pinned to the D : A interface by Coulombic forces which are only weakly screened by the low dielectric nature of conjugated organic systems. Curiously, recent ultrafast spectroscopic measurements on optimized polymer : fullerene-based OPV systems indicate that charge carriers begin to appear ≤ 100-fs following photoexcitation [2, 12, 15, 19–22]. Moreover

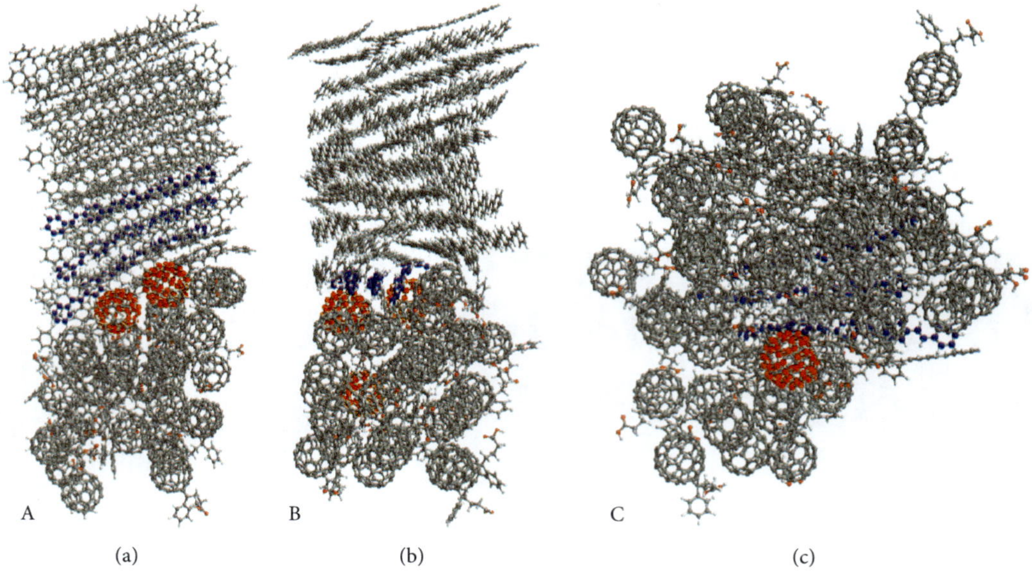

FIGURE 1: Snapshot of the three MD simulation cells: A, B, and C. The red and blue highlighted molecules denote the π-active units in our system.

experiments by Gélinas et al., in which Stark-effect signatures in transient absorption spectra were analysed to probe the local electric field as charge separation proceeds, indicate that electrons and holes separate by as much as 40 Å over the first 100 fs and evolve further on a picosecond time scales to produce unbound and freely mobile charge pairs [4].

It is generally recognized the electronic states at DA interfaces are delocalized over multiple molecular units and may well extend deeply into the bulk. As a result, the spatial extent of these states presents a profound challenge to even state-of-the-art ab initio quantum chemical methods. Moreover, one can not ignore the role of structural fluctuations and intermolecular vibrations in modulating the electronic energies. Recently, Bittner and Silva proposed that energy level fluctuations due to thermal motions can lead to strong quantum mechanical coupling between excitonic and charge-separated states [23]. Central to this theory is the fact that the golden-rule transition rates between states can be determined by examining energy gap fluctuations. This model unifies a number of seeming disparate observations but lacks a desired atomistic basis.

Here we employ a hybrid molecular mechanics/quantum mechanics (MM/QM) approach to sample the electronic excited states of a model bulk heterojunction system. Our approach provides an explicit, semiempirical description of the excited π-electronic states of select molecular units near the interface between phase segregated donor and acceptor molecular units. Our simulations indicate that interfacial electronic states are modulated by C=C bond stretching motions and that such motions induce avoided crossings between nearby excited states thereby facilitating transitions from localized excitonic configurations to delocalized charge-separated configurations on an ultrafast time-scale.

2. Results

Our results build upon a model of charge separation in donor : acceptor blends that we recently reported upon [24]. In this study, we considered a model blend of poly(p-phenylene vinylene) (PPV) donor oligomers and phenyl-C61-butyric acid methyl ester (PCBM) and defined the "π-active" region by selecting (by hand) a single PCBM and three nearby PPV chains to assess the penetration of a localized electronic state into the bulk region. In this paper we model a phase separated interface between PCBM and PPV oligomers. While the semiempirical treatment does facilitate including up to a few hundred π electrons, we do need to be judicious in choosing the "π-active" region.

Figure 1 shows representative molecular configurations of the three cases studied in this paper. Each corresponds to a periodic simulation cell in the xy plane following equilibration at 100 K and 1 atm pressure and they are representative of typical interfacial configurations. In each figure, the red and blue colored spheres represent atoms included in the quantum chemical description. In Case A, we selected 2 interfacial PCBMs and 3 nearby π-active PPV oligomers that penetrate into the bulk polymer region, including a total of 230 C atoms. In Case B, we selected 3 PCBM and 3 nearby PPV oligomers expanding the π-active molecules that form the interfacial heterojunction, including a total of 288 C atoms. Cases A and B differ solely in the initial selection of the PPV oligomers included in the quantum mechanical subspace. For comparison, in Case C we selected 1 PCBM and 3 PPV oligomers that penetrate into the bulk. In all cases, we employ periodic boundary conditions in xyz as in our previous study [24]. This simulation was set up to be very similar to case A, only adding a single PCBM molecule to the interfacial region.

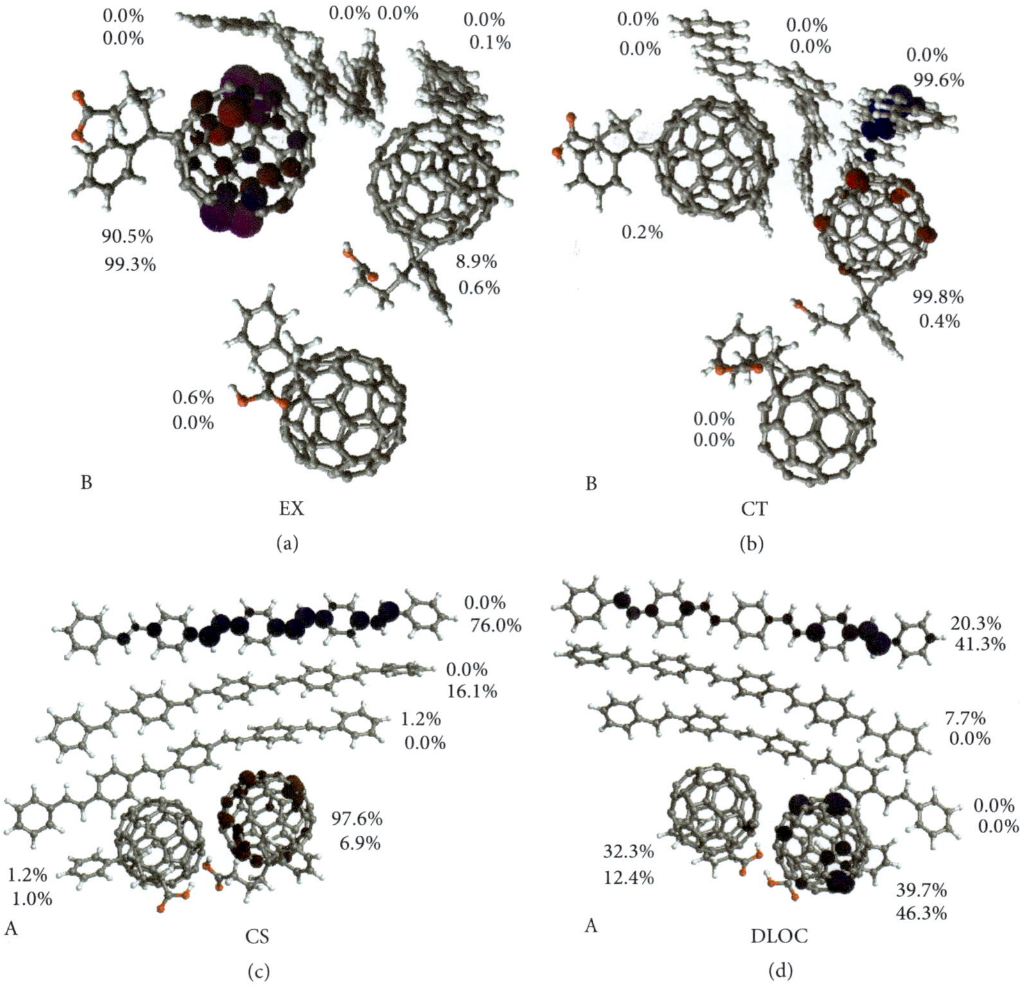

FIGURE 2: The dominant adiabatic states from simulations A (c, d) and B (a, b) as shown in Figure 1 The red and blue numbers denote the electron/hole density as a percent on the indicated molecule. The states correspond to the x-axis in Figure 3. The four snapshots shown represent typical states for our systems. (a, b) from left to right present an exciton located on the PCBM molecule and a charge-transfer state. (c, d) from left to right show a charge-separated state and a partly delocalized state.

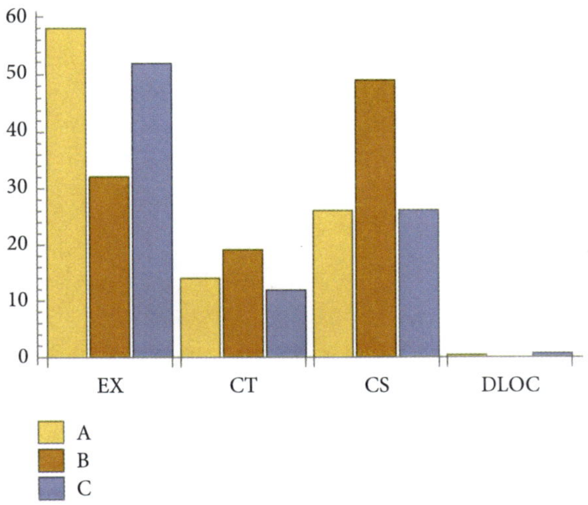

FIGURE 3: Histogram plot of the populations of adiabatic states in systems A, B, and C taken every fs. Each cluster represents a different classification of the adiabatic states as shown in Figure 2.

These three cases are certainly not an exhaustive study of all possible ways that an excitation can localize at an interface and in general one should include all the C atoms as being part of the π system as to include the possibility for both localized and delocalized excitations. However, such study is not computationally feasible and we select these to test the extent that low-lying excitations can penetrate into the bulk region, how this penetration may facilitate charge separation, and the extent that this plays into the mixing between charge-transfer, excitonic, and charge-separated configurations.

Over the course of eight 10 ps simulations, the lowest lying excitation samples a variety of electronic states ranging from localized excitons to charge-separated, charge-transfer, and delocalized states. Figure 2 shows the charge density for four typical eigenstates encountered over the course of a given simulation. We characterize these as follows:

(i) Excitonic (EX) states characterized as having >50% of the electron/hole density on a single molecule.

The exciton state populates ≈ 58% of the states for simulation A, ≈32% for simulation B, and ≈52% for simulation C.

(ii) Charge-Transfer states (CT) characterized as having >50% of the electron/hole density occupying adjacent molecules. The charge-transfer state populates ≈ 14% of the states for simulation A, ≈19% for simulation B, and ≈12% for simulation C.

(iii) Charge-Separated (CS) states characterized as having >50% of the electron/hole density occupying a PCBM and a PPV separated by a single molecule. The charge-separated state are populated ≈ 26% for simulation A, ≈49% for simulation B, and ≈26% for simulation C.

(iv) Delocalized states (DLOC), characterized as having <50% of the electron/hole density on a single PPV or PCBM molecule.

For the most part, the lowest eigenstates in simulations A and B do fall into the categories above. In simulation C, however, the lowest electronic excited state spends about 10% of its time as an eigenstate that cannot be neatly binned into the above categories and corresponds to a fairly equal mixing between excitonic and charge-transfer states and excitonic and charge-separated states.

The striking difference between the blend (C) and the phase separated cases (A and B) is that the stability of localized excitons is very much determined by the proximity between donor and acceptor molecules. In both A and C, the π electrons spend over half their time in excitonic configurations and less than 30% of their time in charge-separated configurations. In both A and C, the 3 PPV chains included in the π system are stacked into the bulk region away from the PCBMs. In contrast in B, charge-separated configurations, defined as a buildup of negative charge on the more distant PCBM, dominate. Only rarely in all cases was the excited state considered as "delocalized."

The energy of the lowest excited state following excitation at $t = 0$ fs for simulations A and B is shown in Figure 4(a). After excitation there is very little energetic relaxation in all of the systems simulated. The simulations appear to cycle through many adiabatic states in a short period of time leaving the impression of a weak electron-phonon coupling. This can be rationalized as the electron/hole density often delocalize over multiple molecules and many conjugated C-C bonds. Another striking effect of the systems is the large number of avoided crossings that occur between the lowest lying states. There is also a 20 fs oscillation in the CI energies, driving the systems excited states into many regions of strong coupling. The 20 fs oscillation also appears in the autocorrelation and the Fourier transform of the gap energies, contributing the C=C bond stretching modes around ≈1600 cm^{-1}. The oscillation is contributed to small thermally activated fluctuations within the simulation, showing that even at 100 K the thermal fluctuations possess sufficient energy to bring these states into regions of strong electronic coupling.

In Figure 4(b) we show a histogram of the 5 lowest excitation energies accumulated over 40 ps of simulation time following promotion to the lowest excited singlet state. The energies take a Gaussian distribution around their respective means and the distributions overlap. On average there is an energy *difference* between 50 and 70 meV between the first and second excited state energies. This suggests that, on average, there is a dense nearly continuous spectrum of charge-separated excited states that can easily be brought into strong electronic coupling by small fluctuations the system brought about by nuclear motion. *In other words, even though the lowest excited state generally can be considered to be an interfacial charge-transfer state, small nuclear motions rapidly mix electronic configurations allowing such states to become delocalized charge-separated states.*

We next consider the origins of the energy fluctuations evidenced in Figure 4(a). While we only show two 200 fs segments of eight 10 ps simulations over this period, one can see that the excitation energies are modulated and cover a small range. The autocorrelation plots of the band gap energies, shown in Figure 5, show that the correlation times for the three simulations are very short ≈ 8 fs meaning that the system changes rapidly enough that the oscillations observed are independent of one another. By taking the Fourier transform of the gap the IR active modes that contribute to the modulation of the excitation energies inside of the systems are found as shown in Figure 5. The modulation of the excitation energy appears to be heavily dependent upon the torsion, C=C, and C-H stretching modes. In each of the plots three distinct regions can be seen, the low frequency torsional modes occur between 200 and 500 cm^{-1}, the C=C stretching modes occur between 1300 and 1800 cm^{-1}, and the C-H stretching modes occur between 2800 and 3300 cm^{-1}. We conclude that small-scale vibronic fluctuations in the molecular structures and orientations produce significant energetic overlap between different adiabatic states to drive the system from purely excitonic to purely charge-transfer on a rapid time-scale. This is evidenced in the progression of the excitation energies, as small fluctuations in these modes can easily bring the excited states into strong coupling regimes.

According to the model outlined below the mean (Δ_0) and variance ($\overline{V}$) of the energy gap distributions shown in Figure 4 can be used as input to estimate the state to state transition rate for a two-level system. We take $T_d^{-1} \approx \overline{V}/\hbar$ as an estimate of the decoherence time and we introduce τ as the natural lifetimes of each state. The results are shown in Table 1. The estimated transition rates are consistent with the observations that the systems rapidly sample a wide number of possible configurations over the course of the molecular dynamics simulation. On average, the state to state couplings of 56 meV for simulation A and 48 meV for simulation B are comparable to the average energy gaps between the lowest excited states. The strong electronic coupling allows for rapid transitions; however, larger couplings also imply shorter electronic decoherence times, effectively quenching the ability of charges to separate by tunneling.

3. Discussion

We present here the results of hybrid QM/MM simulations of the excited states of model PPV/PCBM heterojunction interfaces. Our results indicate that varying the blend ratio and

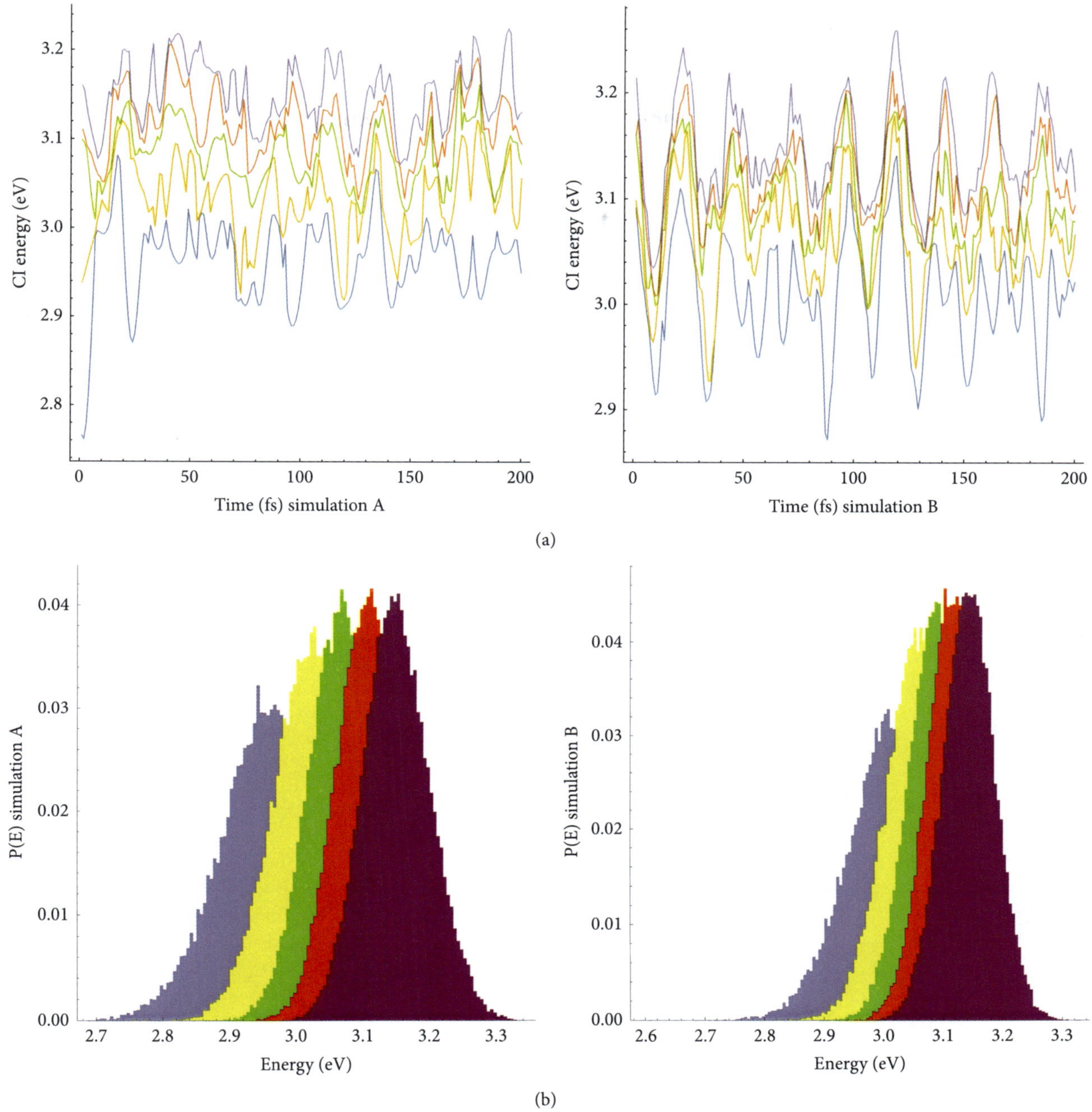

FIGURE 4: (a) Excitation energies for simulations A and B for the first 200 fs following excitation to the lowest SCI state. Each colored line corresponds to the excitation energy from the ground electronic state to one of the lowest 5 singlet excited states. (b) Histogram distributions of the 5 lowest excitation energy levels over a 200 ps simulation. Throughout the simulation, colors of each histogram correspond to the time-histories plotted in (a).

placement of the molecules comprising the heterojunction greatly affect the distribution of states yet have little effect upon the rate constants of the system. We also propose that thermal noise can rapidly change the character of the lowest lying excited state from purely excitonic to charge-separated on a time-scale of sub-100 fs.

Simulations A and C have a very similar placement of molecules, only differing in that simulation A adds a PCBM molecule to the heterojunction. The addition of the PCBM only slightly changes the distribution of states as seen in Figure 4. The exciton states continue to be the most favoured state inside the system, even slightly increasing in probability, while the delocalized states slightly decrease in probability. Simulation B completely changes the heterojunction, placing three PPV and two PCBM molecules at the interface as seen in Figure 1. The states generated by simulation B are radically different from those seen in simulation A and C as seen in Figure 4. The probability of finding the system in the exciton

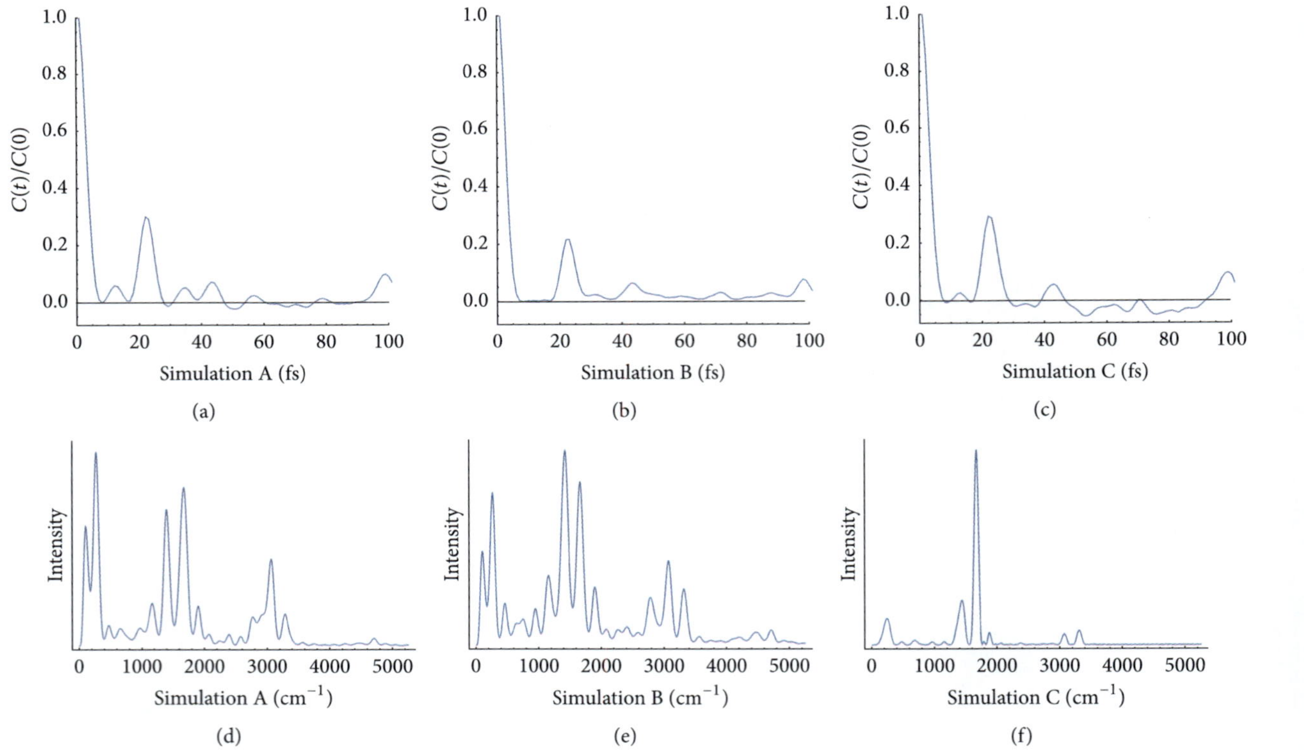

FIGURE 5: (a–c) Autocorrelation of the energy gap between first and second excited states for simulations A, B, and C. The correlation times for all three simulations is 8 fs. (d–f) These plots show the active IR active modes that contribute to the modulation in the CI energies of the systems. Three distinct regions are visible in each plot. The low frequency torsional modes occur between 200 and 500 cm^{-1}, the C=C stretching modes occur between 1300 and 1800 cm^{-1}, and the C-H stretching modes occur between 2800 and 3300 cm^{-1}.

TABLE 1: Estimated excited state ($i \rightarrow f$) transition rates and vibronic couplings for simulations A, B, and C.

Transition	Δ_o (eV)	$\langle V \rangle$ (eV)	T_d (fs)	k^{-1} (fs)
A				
1→2	0.070	0.050	13.16	19.5
1→3	0.12	0.056	11.75	32.9
1→4	0.16	0.063	10.45	38.9
B				
1→2	0.050	0.042	15.67	18.9
1→3	0.09	0.049	13.43	29.3
1→4	0.12	0.052	12.66	40.0
C				
1→2	0.11	0.057	11.41	21.8
1→3	0.18	0.076	8.66	28.6
1→4	0.26	0.076	8.70	50.0

state is dramatically reduced while the charge-separated state becomes predominant. This result is quite interesting as it highlights that the complexity of simulating heterojunctions resides not only in the size of the system but also on how the donor/accepter interface is chosen.

All three systems start with an exciton localized on the PCBM and dissociating into a charge-transfer state with the hole (or electron) delocalized over multiple polymer units before localizing to form charge-separated states. There is a wide range of electronic states tightly clustered within a small energy band, allowing small changes in local bond lengths to have a dramatic role in modulating the electronic couplings between excited states. We speculate that the dramatic shift in population seen in simulation B can be caused by disorder in the PPV molecules reducing the band gap by 20 meV. The PPV molecules comprising the interface region undergo large distortions in the C-C torsion angles allowing the molecules to cycle through a larger range of configurations inside of a short time interval. The presence of more π active PPV molecules at the interface also appears to lead to more avoided crossing regions and the ability of the system to more efficiently dissociate excitons into charge-transfer and charge-separated states to a distance to where their Coulombic attraction is comparable to the thermal energy. While the finite size of our system prevents further dissociation of the charges, the results are suggestive that such interstate crossing events driven by bond-fluctuations can efficiently separate the charges. The results presented here corroborate recent ultrafast experimental evidence suggesting that free polarons can form on an ultrafast time scales (sub-100 fs) and that thermally activated low frequency torsional modes are key in effective electron hole separation in PPV/PCBM heterojunctions.

4. Methods

4.1. Simulation Methods. Our simulations employ a modified version of the TINKER molecular dynamics (MD) package [25] in which the MM3 [26] intramolecular bonding parameters are allowed to vary with the local π-electronic density as described by a Parisier-Parr-Pople (PPP) semiempirical Hamiltonian [27, 28]. Similar approaches have been described by Lobaugh and Rossky [29] and Muller and Warshel [30] to include electronic dynamics into an otherwise classical force field description.

We assume that the nuclear dynamics of all atoms in the system evolve according to a classical description

$$m_i \ddot{r}_i(t) = F_i(\mathbf{r}(t), P(t)), \tag{2}$$

where $F_i(\mathbf{r}(t), P(t))$ is the force acting on atom i at time t and $P(t)$ is the instantaneous electronic density for the π-electrons. The Fock operator for the π electrons is given by

$$\begin{aligned} \mathcal{F}_{\mu\mu} &= W_\mu + \frac{1}{2}P^\circ_{\mu\mu}(\mu\mu \mid \mu\mu) \\ &\quad + \sum_{\rho\neq\mu}\left(P_{\rho\rho} - Z_\rho\right)V\left(r_{\mu\rho}\right), \\ \mathcal{F}_{\mu\nu} &= \beta_{\mu\nu}\left(r_{\mu\nu}\right) - \frac{1}{2}P^\circ_{\mu\nu}(\mu\mu \mid \nu\nu), \end{aligned} \tag{3}$$

where W_μ is the valence state potential for a given atom type and we make the zero-differential overlap (ZDO) assumption to write the electronic repulsion integrals as

$$(\mu\nu \mid \rho\sigma) = (\mu\mu \mid \rho\rho)\,\delta_{\mu\nu}\delta_{\rho\sigma}. \tag{4}$$

The Greek indices correspond to atom-centred basis orbitals. Note that we specifically distinguish P° as the ground state (Hartree-Fock) electronic density. Both the hopping integrals $\beta_{\mu\nu}$ and Coulomb repulsion integrals are determined using empirical formulas and depend upon the intermolecular bonding between atoms, accounting for nonplanarity. The molecular orbitals are eigenstates of $\mathcal{F}$

$$\mathcal{F}\psi_n = \epsilon_n \psi_n, \tag{5}$$

where ψ_n is given as a linear combination of the atom-centred orbitals

$$\psi_n = \sum_\mu c_{n\mu}\chi_\mu. \tag{6}$$

From the $\{c_{n\mu}\}$ coefficients, we obtain the ground state density matrix

$$P^\circ_{\mu\nu} = 2\sum_{n_{occ}} c_{n\mu}c_{n\nu}. \tag{7}$$

The electronic energy is given by

$$E_{HF} = \mathrm{Tr}\left[\mathcal{F}P^\circ\right] \tag{8}$$

and is minimized by imposing the condition

$$\langle[\mathcal{F}, P^\circ]\rangle = 0. \tag{9}$$

Electronic $\pi \rightarrow \pi^*$ excitations are subsequently introduced by using configuration interaction (singles) (CI-S) theory where we write the excited state wave function as a linear combination of elementary excitations between occupied and unoccupied molecular orbitals

$$\begin{aligned} |\Psi_k\rangle &= \sum_{h\in occ}\sum_{e\in unocc} C^{eh}_k |\psi_h\psi_e\rangle \\ &= \sum_{\mu\nu}\left(\sum_{eh} C^{eh}_k c_{e\mu}c_{h\nu}\right)|\chi_\nu\chi_\mu\rangle \\ &= \sum_{\mu\nu} C^{\mu\nu}_k |\chi_\nu\chi_\mu\rangle, \end{aligned} \tag{10}$$

where the $\{C^{\mu\nu}_k\}$ coefficients give the electron/hole amplitude in the atom-centred basis. These are eigenstates of the CI-S Hamiltonian and we used a total of 10 occupied and 10 unoccupied Hartree-Fock molecular orbitals to construct the electron/hole configurations for the CI calculations. The excited state bond charge density matrix is constructed by assuming that electron densities are added to the unoccupied orbitals and hole densities are subtracted from the Hartree-Fock ground state; namely,

$$P_{\mu\nu} = P^\circ_{\mu\nu} + P^e_{\mu\nu} - P^h_{\mu\nu}, \tag{11}$$

where

$$\begin{aligned} P^e_{\mu\nu} &= \sum_\sigma C^{\mu\sigma}_k C^{\nu\sigma}_k \\ P^h_{\mu\nu} &= \sum_\sigma C^{\sigma\mu}_k C^{\sigma\nu}_k. \end{aligned} \tag{12}$$

In the former, we sum over the occupied orbitals (holes) and in the latter we sum over unoccupied orbitals. $P_{\mu\nu}$ is interpreted as the "bond charge" matrix whereby the diagonal elements give the net π-electron count on site μ and the off-diagonal $P_{\mu\nu}$ terms give the π-bond order between sites μ and ν. Consequently, excitations systematically move electron density from atom to atom and change the local bond orders.

Within the force field, bond force field parameters are modified by assuming a linear interpolation between single and double bonds. For example, we write the bond force constant between C atoms i and j as

$$k = k_2 - \delta k\left(1 - P_{ij}\right), \tag{13}$$

where k_2 is the force constant for a C=C double bond when $P_{ij} = 1$, k_1 is the bond force constant for a C-C single bond when $P_{ij} = 0$, and $\delta k = (k_2 - k_1)$. In a similar definition, we interpolate the equilibrium bond lengths between single- and double bonds as

$$l = l_2 + \delta l\left(1 - P_{ij}\right). \tag{14}$$

This treatment is not limited to C-C bonds but is applied to all heavy atoms contributing to the π system. While the MM3 parameter set was originally intended to account for the electronic ground state, we assume that the parameters are transferable to the lower lying excited states. Modifying the equilibrium bond orders and bond lengths allows the MM calculation to respond to the migration of charge.

During the equilibration steps we assume the system to be in its electronic ground state, after which we excite the system to the first excited state and allow the system to respond to the change in the electronic density within the adiabatic Born-Oppenheimer approximation. It is important to note that the excited state we prepare is not the state which carries the most oscillator strength to the ground nor do we account for nonadiabatic surface hopping-type transitions in our approach [31–33]. The dynamics simulations shown reflect the longer-time fate of the lowest lying excited state populations and sample possible configurations that can be accessed by the system. The combination of a classical MD force field with a semiempirical description of a selected few molecules within the system seems to be a suitable compromise between a fully ab initio approach which would be limited to only a few molecules and short simulation times and a fully classical MD description which would neglect any transient changes in the local electronic density [22].

4.2. Estimating State to State Rates. We can estimate the state → state rates using the model by Bittner et al. [23, 24]. Consider a two-state system with coupling λ in which the energy gap $\Delta(t)$ fluctuates in time around its average $\overline{\Delta}$. In a two-state basis the Hamiltonian can be written as

$$H = \frac{\Delta(t)}{2}\widehat{\sigma_z} + \lambda\widehat{\sigma_x}, \tag{15}$$

where $\widehat{\sigma_k}$ are Pauli matrices. Note that (15) can be transformed such that fluctuations are in the off-diagonal coupling, becoming

$$H = \frac{\Delta_0}{2}\widehat{\sigma_z} + \delta V(t)\widehat{\sigma_x}, \tag{16}$$

where $\Delta_0 = \overline{\Delta} + \lambda$ and $\delta\overline{V}(t) = 0$. The fluctuations in the electronic energy levels are attributed to thermal and bond-vibrational motions of polymer chains which can be related to the spectral density, $S(\omega)$, via

$$\overline{V}^2 = \delta\overline{V}^2(t) = \int_{-\infty}^{+\infty}\frac{d\omega}{2\pi}S(\omega). \tag{17}$$

Averaging over the environmental noise, we can write the average energy gap as $\hbar\overline{\Omega} = \sqrt{\Delta_0^2 + \overline{V}^2}$ with eigenstates

$$\begin{aligned} |\Psi+\rangle &= \cos\theta\,|0\rangle + \sin\theta\,|1\rangle \\ |\Psi-\rangle &= -\sin\theta\,|0\rangle + \cos\theta\,|1\rangle, \end{aligned} \tag{18}$$

where $\tan 2\theta = |\overline{V}|/\Delta_0$ defines the mixing angle between original kets. Consequently, by analysing energy gap fluctuations, we can obtain an estimate of both the coupling between states as well as transition rates. To estimate the average transition rate between states the equations of motion for the reduced density matrix for a two-level system coupled to a dissipative environment are used.

$$\begin{aligned} \dot{\rho}_{11} &= -\frac{i}{\hbar}V(\rho_{21}-\rho_{12}) - \frac{1}{\tau_1}\rho_{11} \\ \dot{\rho}_{22} &= \frac{i}{\hbar}V(\rho_{21}-\rho_{12}) - \frac{1}{\tau_2}\rho_{22} \\ \dot{\rho}_{12} &= -\frac{i}{\hbar}V(\rho_{22}-\rho_{11}) - \frac{1}{T_d}\rho_{12} - \frac{\Delta_0}{i\hbar}\rho_{12} \\ \dot{\rho}_{21} &= \frac{i}{\hbar}V(\rho_{22}-\rho_{11}) - \frac{1}{T_d}\rho_{21} + \frac{\Delta_0}{i\hbar}\rho_{21}. \end{aligned} \tag{19}$$

τ_1 and τ_2 have been introduced as the lifetimes of each state and T_d is the decoherence time for the quantum superposition. The decoherence time can be related to the spectral density via $T_d^{-1} = \overline{V}/\hbar$. Taking T_d to be short compared to the lifetimes of each state, we can write the population of the initial states as

$$\rho_{11}(t) = \exp\left[-\left(\frac{1}{\tau_1} - k\right)\right], \tag{20}$$

where k is the average state to state transition rate. If we integrate over all time we obtain an equation of the form

$$\int_0^{\infty}\rho_{11}(t)\,dt = \left(\frac{1}{\tau_1} + k\right)^{-1} \tag{21}$$

suggesting a form for the exact solution of (6). Taking the Laplace transform of the equations of motion (6) and assuming that our initial population is in state 1 ($\rho_{11}(0) = 1$) the equations of motion become a series of algebraic equations

$$\begin{aligned} -1 &= -\frac{i}{\hbar}V(\rho_{21}-\rho_{12}) - \frac{1}{\tau_1}\rho_{11} \\ 0 &= \frac{i}{\hbar}V(\rho_{21}-\rho_{12}) - \frac{1}{\tau_2}\rho_{22} \\ 0 &= -\frac{i}{\hbar}V(\rho_{22}-\rho_{11}) - \frac{1}{T_d}\rho_{12} - \frac{\Delta_0}{i\hbar}\rho_{12} \\ 0 &= \frac{i}{\hbar}V(\rho_{22}-\rho_{11}) - \frac{1}{T_d}\rho_{21} + \frac{\Delta_0}{i\hbar}\rho_{21} \end{aligned} \tag{22}$$

which produces a rate constant of the form

$$k = 2\frac{\overline{V}^2}{\hbar^2}\frac{T_d}{(T_d\Delta_0/\hbar)^2 + 1}. \tag{23}$$

Conflicts of Interest

The authors declare no financial conflicts of interest.

Authors' Contributions

Eric R. Bittner conceived the project. Allen Kelley performed the simulations and Kush Patel assisted with analysis. All authors contributed to the Results and Discussion and Kush Patel assisted with the drafting of the Methods section.

Acknowledgments

The work at the University of Houston was funded in part by the National Science Foundation (CHE-1362006, CHE-1664971, and MRI-1531814) and the Robert A. Welch Foundation (E-1337).

References

[1] Z. He, C. Zhong, S. Su, M. Xu, H. Wu, and Y. Cao, "Enhanced power-conversion efficiency in polymer solar cells using an inverted device structure," *Nature Photonics*, vol. 6, no. 9, pp. 591–595, 2012.

[2] M. Tong, N. E. Coates, D. Moses, A. J. Heeger, S. Beaupré, and M. Leclerc, "Charge carrier photogeneration and decay dynamics in the poly(2,7-carbazole) copolymer PCDTBT and in bulk heterojunction composites with PC70 BM," *Physical Review B: Condensed Matter and Materials Physics*, vol. 81, no. 12, Article ID 125210, 2010.

[3] L. G. Kaake, D. Moses, and A. J. Heeger, "Coherence and uncertainty in nanostructured organic photovoltaics," *The Journal of Physical Chemistry Letters*, vol. 4, no. 14, pp. 2264–2268, 2013.

[4] S. Gélinas, A. Rao, A. Kumar et al., "Ultrafast long-range charge separation in organic semiconductor photovoltaic diodes," *Science*, vol. 343, no. 6170, pp. 512–516, 2014.

[5] E. Collini and G. D. Scholes, "Coherent intrachain energy migration in a conjugated polymer at room temperature," *Science*, vol. 323, no. 5912, pp. 369–373, 2009.

[6] S. Mukamel, "Comment on "coherence and uncertainty in nanostructured organic photovoltaics"," *The Journal of Physical Chemistry A*, vol. 117, no. 40, pp. 10563-10564, 2013.

[7] N. R. Monahan, K. W. Williams, B. Kumar, C. Nuckolls, and X. Zhu, "Direct observation of entropy-driven electron-hole pair separation at an organic semiconductor interface," *Physical Review Letters*, vol. 114, no. 24, 2015.

[8] F. Provencher, N. Bérubé, A. W. Parker et al., "Direct observation of ultrafast long-range charge separation at polymer-fullerene heterojunctions," *Nature Communications*, vol. 5, article no. 4288, 2014.

[9] G. S. Engel, T. R. Calhoun, E. L. Read et al., "Evidence for wavelike energy transfer through quantum coherence in photosynthetic systems," *Nature*, vol. 446, no. 7137, pp. 782–786, 2007.

[10] D. Beljonne, E. Hennebicq, C. Daniel et al., "Excitation migration along oligophenylenevinylene-based chiral stacks: delocalization effects on transport dynamics," *The Journal of Physical Chemistry B*, vol. 109, no. 21, pp. 10594–10604, 2005.

[11] H. Tamura, E. R. Bittner, and I. Burghardt, "Exciton dissociation at donor-acceptor polymer heterojunctions: quantum nonadiabatic dynamics and effective-mode analysis," *The Journal of Chemical Physics*, vol. 126, no. 2, p. 021103, 2007.

[12] G. Grancini, M. Maiuri, D. Fazzi et al., "Hot exciton dissociation in polymer solar cells," *Nature Materials*, vol. 12, no. 1, pp. 29–33, 2013.

[13] A. Troisi, "How quasi-free holes and electrons are generated in organic photovoltaic interfaces," *Faraday Discussions*, vol. 163, pp. 377–392, 2013.

[14] H. Tamura, J. G. Ramon, E. R. Bittner, and I. Burghardt, "Phonon-driven ultrafast exciton dissociation at donor-acceptor polymer heterojunctions," *Physical Review Letters*, vol. 100, no. 10, 2008.

[15] C.-X. Sheng, T. Basel, B. Pandit, and Z. V. Vardeny, "Photoexcitation dynamics in polythiophene/fullerene blends for photovoltaic applications," *Organic Electronics*, vol. 13, no. 6, pp. 1031–1037, 2012.

[16] X. Yang, T. E. Dykstra, and G. D. Scholes, "Photon-echo studies of collective absorption and dynamic localization of excitation in conjugated polymers and oligomers," *Physical Review B: Condensed Matter and Materials Physics*, vol. 71, no. 4, 2005.

[17] H. Tamura and I. Burghardt, "Ultrafast charge separation in organic photovoltaics enhanced by charge delocalization and vibronically hot exciton dissociation," *Journal of the American Chemical Society*, vol. 135, no. 44, pp. 16364–16367, 2013.

[18] Y. Gao, T. P. Martin, E. T. Niles, A. J. Wise, A. K. Thomas, and J. K. Grey, "Understanding morphology-dependent polymer aggregation properties and photocurrent generation in polythiophene/fullerene solar cells of variable compositions," *The Journal of Physical Chemistry C*, vol. 114, no. 35, pp. 15121–15128, 2010.

[19] N. S. Sariciftci and A. J. Hebger, "Reversible, metastable, ultrafast photoinduced electron transfer from semiconducting polymers to buckminsterfullerene and in the corresponding donor/acceptor heterojunctions," *International Journal of Modern Physics B*, vol. 08, no. 03, pp. 237–274, 1994.

[20] N. Banerji, S. Cowan, M. Leclerc, E. Vauthey, and A. J. Heeger, "Exciton formation, relaxation, and decay in PCDTBT," *Journal of the American Chemical Society*, vol. 132, no. 49, pp. 17459–17470, 2010.

[21] N. Banerji, "Sub-picosecond delocalization in the excited state of conjugated homopolymers and donor-acceptor copolymers," *Journal of Materials Chemistry C*, vol. 1, no. 18, pp. 3052–3066, 2013.

[22] A. E. Jailaubekov, A. P. Willard, J. R. Tritsch et al., "Hot charge-transfer excitons set the time limit for charge separation at donor/acceptor interfaces in organic photovoltaics," *Nature Materials*, vol. 12, no. 1, pp. 66–73, 2013.

[23] E. R. Bittner and C. Silva, "Noise-induced quantum coherence drives photo-carrier generation dynamics at polymeric semiconductor heterojunctions," *Nature Communications*, vol. 5, article no. 3119, 2014.

[24] E. R. Bittner and A. Kelley, "The role of structural fluctuations and environmental noise in the electron/hole separation kinetics at organic polymer bulk-heterojunction interfaces," *Physical Chemistry Chemical Physics*, vol. 17, no. 43, pp. 28853–28859, 2015.

[25] J. W. Ponder, "TINKER: Software Tools for Molecular Design," http://dasher.wustl.edu/tinker/, 2004.

[26] N. L. Allinger, F. Li, L. Yan, and J. C. Tai, "Molecular mechanics (MM3) calculations on conjugated hydrocarbons," *Journal of Computational Chemistry*, vol. 11, no. 7, pp. 868–895, 1990.

[27] J. A. Pople, "Electron interaction in unsaturated hydrocarbons," *Transactions of the Faraday Society*, vol. 49, pp. 1375–1385, 1953.

[28] R. Pariser and R. G. Parr, "A semi-empirical theory of the electronic spectra and electronic structure of complex unsaturated molecules. II," *The Journal of Chemical Physics*, vol. 21, no. 5, pp. 767–776, 1953.

[29] J. Lobaugh and P. J. Rossky, "Solvent and intramolecular effects on the absorption spectrum of betaine-30," *The Journal of Physical Chemistry A*, vol. 104, no. 5, pp. 899–907, 2000.

[30] R. P. Muller and A. Warshel, "Ab initio calculations of free energy barriers for chemical reactions in solution," *The Journal of Physical Chemistry C*, vol. 99, no. 49, pp. 17516–17524, 1995.

[31] J. C. Tully, "Molecular dynamics with electronic transitions," *The Journal of Chemical Physics*, vol. 93, no. 2, pp. 1061–1071, 1990.

[32] J. C. Tully and R. K. Pkeston, "Trajectory surface hopping approach to nonadiabatic molecular collisions: the reaction of H^+ with D_2," *The Journal of Chemical Physics*, vol. 55, no. 2, pp. 562–572, 1971.

[33] G. Granucci and M. Persico, "Critical appraisal of the fewest switches algorithm for surface hopping," *The Journal of Chemical Physics*, vol. 126, no. 13, Article ID 134114, 2007.

20

PMMA Sandwiched Bi_2Te_3 Layer as a Saturable Absorber in Mode-Locked Fiber Laser

Guobao Jiang,[1,2] Yuan Zhou,[1] Lulu Wang,[1] and Ying Chen[1,2]

[1]*School of Electronic Information and Electrical Engineering, Changsha University, Changsha 410003, China*
[2]*Key Laboratory for Micro-/Nano-Optoelectronic Devices of Ministry of Education, School of Physics and Electronics, Hunan University, Changsha 410082, China*

Correspondence should be addressed to Ying Chen; yingchen@ccsu.edu.cn

Guest Editor: Xinxing Zhou

In this paper, we fabricated a PMMA sandwiched Bi_2Te_3 self-assembly layer as a saturable absorber device, which was used as a passive mode locker for ultrafast pulse generation at the telecommunication band. Nanosheets of Bi_2Te_3 as a bulk topological insulator were successfully synthesized through a solvothermal treatment and self-assemble method to form a thin film at a water-air interface. In order to transfer the Bi_2Te_3 self-assembly layer to the optical fiber end, we design a construction of two PMMA layers sandwiched self-assembly layer. By incorporating this saturable absorber into an erbium-doped fiber laser, femtosecond mode-locking operation was experimentally demonstrated. The output pulse width is about 505 fs. Our results indicate that PMMA sandwiched topological insulator layer structure could be an improvement technology in traditional PMMA transfer method and could be used as a long-term stable saturable absorber for the passively mode locking lasers.

1. Introduction

Topological insulator is a state of quantum matter with insulating bulk states like an ordinary insulator and highly conducting and massless spin-helical surface states on its surface [1–7]. These unique properties, which are derived from the combination of spin-orbit interactions and time-reversal symmetry, enable topological insulators as efficient saturable absorber materials for the generation of ultrafast pulse [8, 9]. Compared with traditional saturable absorbers, such as semiconductor saturable absorber mirrors (SESAMs) [10, 11], single-walled carbon nanotubes (SWCNTs) [12], graphene [13], transitional metal dichalcogenides [14, 15], black phosphorus [16, 17], topological insulator saturable absorber possesses many good qualities, such as low saturable absorption threshold, large modulation depth [18, 19], short recovery time [20, 21], and wavelength independent saturable absorption [22]. To fabricate saturable absorbers with advanced performance, it is necessary to search for a facile and simple transfer technique of topological insulator nanosheets onto target substrates like optical fiber-ferrules or optical quartz substrate.

Several methods have been proposed for the transfer of saturable absorbers onto the optical fiber-ferrules, such as optical deposition [23], ink jet print [24], drop cast [25], and polymer composite [26]. Drop cast is favored for its simple and facility. However, its coffee ring effect is difficult to effectively restrain while drying the solute. Ink jet print method is tedious and prone to error and requires for precision instruments and skillful professional operators. Optical deposition was used to deposit carbon nanotubes and graphene onto cores of optical fiber ends. Though the method is simple, it requires precise control of the optical power and does not possess controllability of the deposition layer properties.

In this paper, we present a simple approach to improve the Bi_2Te_3 (one kind of topological insulator) transfer method. The Bi_2Te_3 nanosheets were deposited on the PMMA film, and then another PMMA film was placed onto the Bi_2Te_3 nanosheets to isolate air and water. By this process, the stability of Bi_2Te_3 based saturable absorber was greatly increased. The sandwiched structure of Bi_2Te_3 film was placed onto fiber end facet and introduced into fiber ring laser. Ultrafast pulse was obtained with sandwiched structure of Bi_2Te_3 film

FIGURE 1: SEM image of Bi_2Te_3 nanosheets.

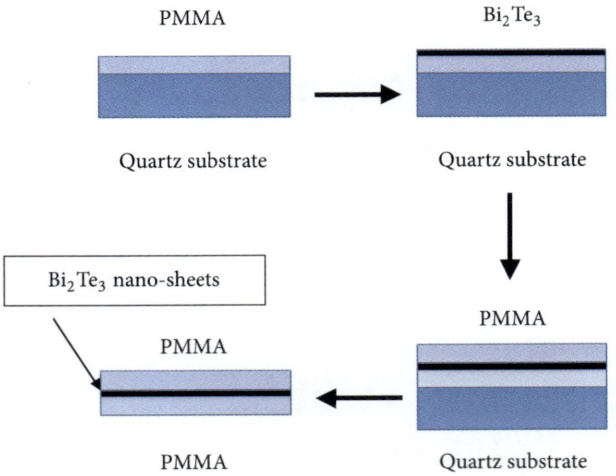

FIGURE 2: Fabrication of PMMA sandwiched Bi_2Te_3 nanosheets saturable absorber.

saturable absorber. It has a pulse width of about 505 fs and a repetition rate of 13.14 MHz. The stable mode-locking operation lasted at least 6 hours. All these results show that this PMMA sandwiched Bi_2Te_3 structure is an effective and stable saturable absorber, which could have potential applications in ultrafast pulse generation.

2. Saturable Absorber Preparation

2.1. Synthesis of Bi_2Te_3 Nanosheets. Bi_2Te_3 nanosheets were synthesized following a solvothermal method of Wang et al. [27]. Using a typical synthesis, a stoichiometric ratio of bismuth chloride ($BiCl_3$) and sodium selenide (Na_2TeO_3) was dissolved in ethylene glycol with vigorous stirring. Then the mixture was transferred into the Teflon-lined stainless-steel autoclave and heated to 200°C. The autoclave was maintained at the reaction temperature for 36 h and then cooled to room temperature naturally. The black powders were collected by filtering, washed with distilled water and ethanol, and finally dried at 60°C in vacuum overnight. The as-grown and washed powders were dispersed in ethanol solution.

2.2. Characterization of Bi_2Te_3 Nanosheets. Scanning electronic microscope (SEM) measurements were carried out to characterize the morphology of Bi_2Te_3 nanosheets. The SEM image shows that the Bi_2Te_3 sample has a uniform size and shape in Figure 1. The obtained products are predominantly hexagonal-based plates, which matches well with the regular hexagonal lattice structure of Bi_2Te_3.

2.3. Fabrication of Bi_2Te_3 Based Saturable Absorber. Figure 2 shows the fabrication of Bi_2Te_3 based saturable absorber. The PMMA solution was spin coated onto quartz substrate and then the Bi_2Te_3 dispersion was also spin coated onto the PMMA film. Thus the Bi_2Te_3 based saturable absorber was fabricated. To enhance the mechanical strength and stability of the saturable absorber, another PMMA film was placed onto the Bi_2Te_3 nanosheets to isolate air and water. The sandwiched structure was peeled down from the quartz substrate. The Bi_2Te_3 based saturable absorber was completely made.

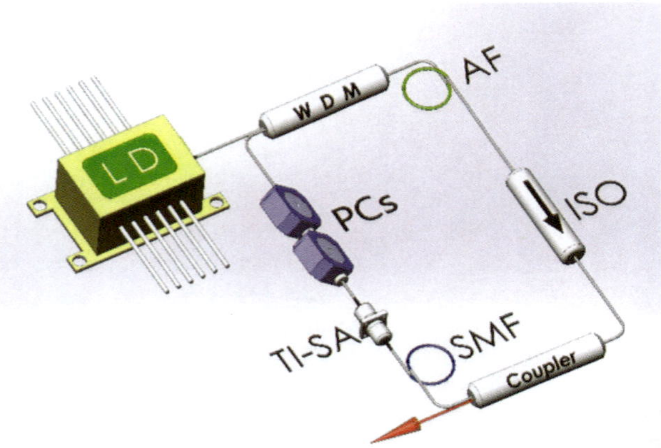

FIGURE 3: The schematic of mode-locked fiber laser.

3. Bi_2Te_3 Saturable Absorber Based Fiber Laser

3.1. Design of Mode-Locked Fiber Laser. To construct a passively mode-locked ring fiber laser, the experimental configuration as in Figure 3 is employed, including the standard fiber-optic components such as wavelength division multiplexer (WDM), polarization controller (PC), coupler, optical isolator, and active fiber (AF). The fiber laser has a ring cavity configuration with a total cavity length of 15.22 m, which comprises a piece of 1 m erbium-doped fiber (EDF, LIEKKI Er 80-8/125) as AF with group velocity dispersion (GVD) of -20 ps^2/km and the pigtails of fiber-optic components. All these pigtails are standard single mode fiber (SMF-28) with GVD of -23 ps^2/km at 1550 nm. The pump, which was generated from a 975 nm laser diode (LD) source, is coupled into the cavity through a 980/1550 wavelength-division multiplexer (WDM), and a 10% fiber coupler is employed to output the pulsed laser. A polarization independent isolator (PI-ISO) is used to force the unidirectional operation of the ring cavity, and a polarization controller (PC) is used to fine adjust the polarization state of circulating light and the in-cavity

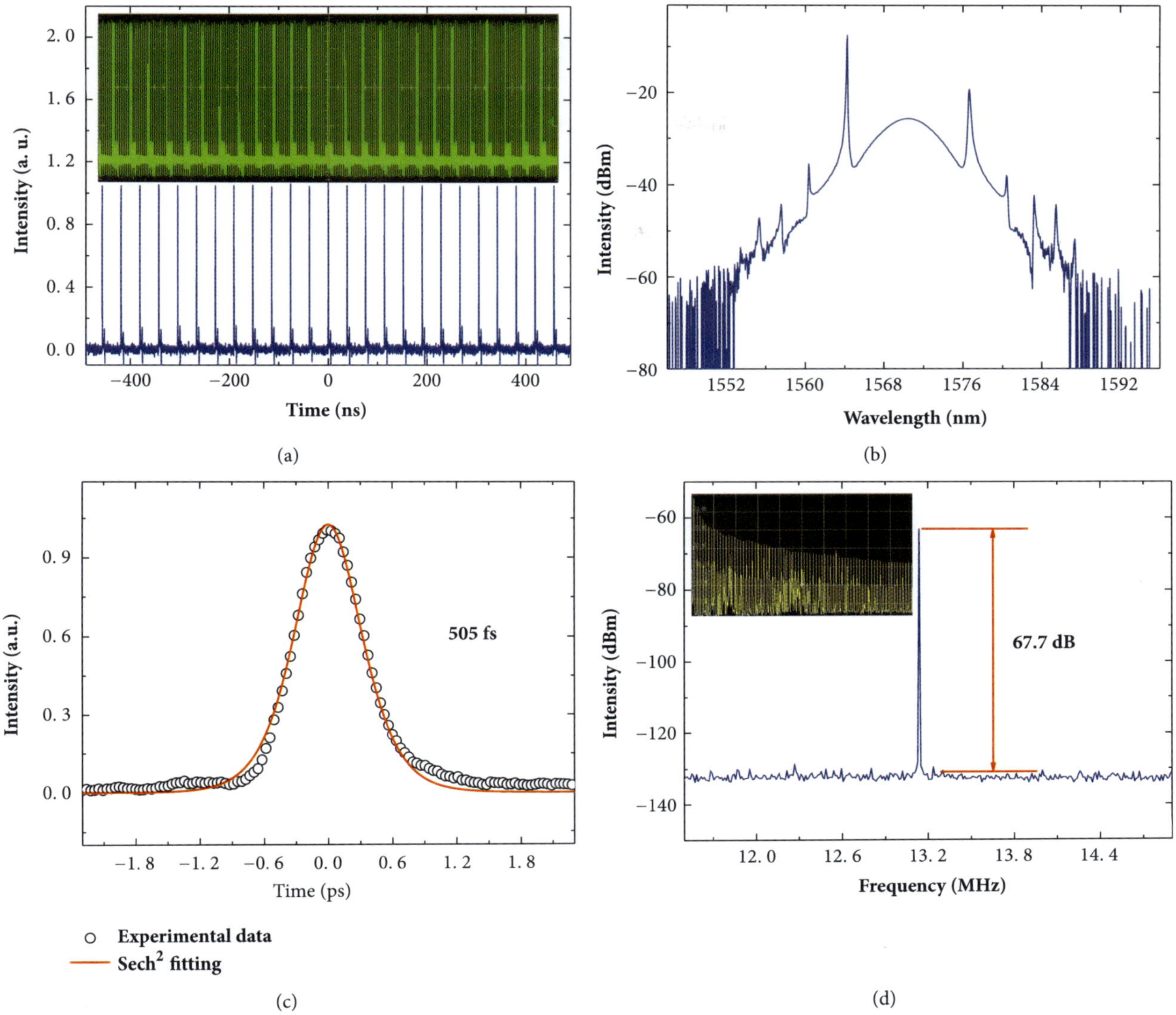

FIGURE 4: Single pulse mode-locking results of fiber laser.

birefringence. An optical spectrum analyzer (Ando AQ-6317B) and an oscilloscope (Tektronix TDS3054B) combined with a 5 GHz photo-detector (Thorlabs SIR5) are employed to simultaneously monitor the optical spectra and temporal profile of the output pulse. The pulse duration is measured with a commercial second harmonic generation autocorrelator.

3.2. Mode-Locking Results and Discussions. The mode-locking operation could be self-started with a mode-locking threshold of 63.7 mW. Figure 4 summarizes the characteristics of single soliton pulse of our fiber laser at pump power of 85.4 mW. Figure 4(a) shows that the single pulse train with a repetition rate of 13.14 MHz, which matches well with the cavity length, indicates that the laser operates in the mode-locking state. The corresponding optical spectrum, as shown in Figure 4(b), has obvious Kelly sidebands spectral indicating that the fiber laser operates in soliton regime. It has a central wavelength of 1570.45 nm and a 3 dB bandwidth of 5.35 nm. Correspondingly, as shown in Figure 4(d), the measured autocorrelation (AC) trace can be well fitted by hyperbolic secant function with a full width at half maximum (FWHM) of 756 fs, showing that the real pulse width is about 505 fs. The time-bandwidth product is 0.328, indicating that the obtained soliton pulse is almost transform limited. Figure 4(d) is corresponding measured RF spectrum with resolution bandwidth (RBW) of 10 Hz; the signal-to-noise ratio (SNR) of our fiber laser is up to 67.7 dB indicating that our fiber laser operates in a relatively stable regime. Moreover, as can be seen in long scale RF spectrum (the insert), there is not any excrescent frequency components except the fundamental and harmonic frequency components, further confirming the stability of our fiber laser and single soliton operation.

To investigate the long-term stability of the single soliton operation, we recorded the output spectra every hour over 6 h with fixed experimental setup, such as pump power of 90 mW, as shown in Figure 5. Neither the central wavelength drift

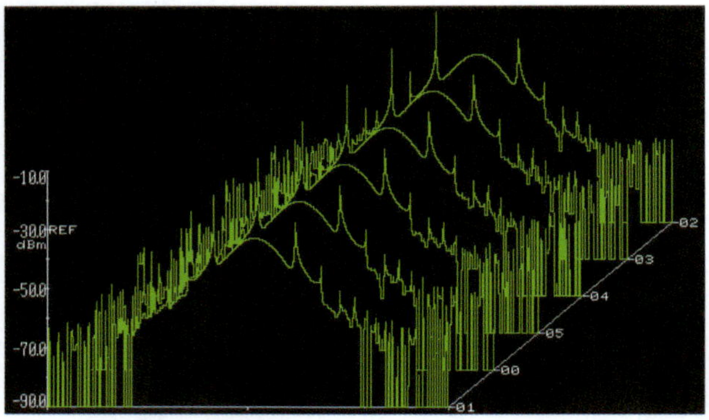

FIGURE 5: Long-term stability of optical spectra.

nor new wavelength component was observed during our measurement and always exhibits the same profiles, showing excellent repeatability and superimposability. All these results confirm the fiber laser possesses a reasonably good stability that is suitable for practical applications.

During the entire measurement, neither the central wavelength drifting nor new wavelength components were observed, revealing that the mode-locked fiber laser shows long-term stability, which benefits from the PMMA sandwiched Bi_2Te_3 nanosheets saturable absorber. There are many other advantages to this novel PMMA-TI-PMMA structure. First, the spin coated layer spontaneously generated at the interface between the solution and air can remain very uniform. Second, the spin coated layer can be effectively protected by double PMMA layers, and the original morphology will be automatically kept while being transferred and installed. Third, the saturable absorber device can be effectively protected from oxidation, since it is isolated from air and water due to the double PMMA sandwich structure. Also, this mode-locked result shows no distinct different with previous reports [28–30], which means that this transfer process has no adverse effect to the Bi_2Te_3 based saturable absorber. All of these results lead to a conclusion: the PMMA sandwiched structure could protect well the Bi_2Te_3 based saturable absorber and ensure no recede of the performance of saturable absorber.

4. Conclusions

Here, we fabricated a PMMA sandwiched Bi_2Te_3 nanosheets structure process. Based on this PMMA sandwiched Bi_2Te_3 nanosheets saturable absorber, we have demonstrated stable mode-locked fiber laser. Ultrafast pulse with a pulse width about 505 fs was obtained from the mode-locked fiber laser. These results suggest that the PMMA sandwiched Bi_2Te_3 nanosheets structure is a good saturable absorber candidate for ultrafast pulse generation. Also this technique could be developed to fabricate other stable two-dimensional materials based devices.

Conflicts of Interest

The authors declare that they have no conflicts of interest.

Acknowledgments

This work was partially supported by the Natural Science Foundation of Hunan Province, China (Grant no. 2018JJ2455).

References

[1] J. E. Moore and L. Balents, "Topological invariants of time-reversal-invariant band structures," *Physical Review B: Condensed Matter and Materials Physics*, vol. 75, no. 12, 2007.

[2] L. Fu, C. L. Kane, and E. J. Mele, "Topological insulators in three dimensions," *Physical Review Letters*, vol. 98, no. 10, 2007.

[3] L. Fu and C. L. Kane, "Topological insulators with inversion symmetry," *Physical Review B: Condensed Matter and Materials Physics*, vol. 76, no. 4, 2007.

[4] J. C. Y. Teo, L. Fu, and C. L. Kane, "Surface states and topological invariants in three-dimensional topological insulators: Application to Bil-x Sbx," *Physical Review B: Condensed Matter and Materials Physics*, vol. 78, no. 4, 2008.

[5] D. Hsieh, D. Qian, L. Wray et al., "A topological Dirac insulator in a quantum spin Hall phase," *Nature*, vol. 452, no. 7190, pp. 970–974, 2008.

[6] H. Zhang, C.-X. Liu, X.-L. Qi, X. Dai, Z. Fang, and S.-C. Zhang, "Topological insulators in Bi_2 Se_3, Bi_2 Te_3 and Sb_2 Te_3 with a single Dirac cone on the surface," *Nature Physics*, vol. 5, no. 6, pp. 438–442, 2009.

[7] Y. L. Chen, J. G. Analytis, J.-H. Chu et al., "Experimental realization of a three-dimensional topological insulator, Bi_2Te_3," *Science*, vol. 325, no. 5937, pp. 178–181, 2009.

[8] C. Zhao, H. Zhang, X. Qi et al., "Ultra-short pulse generation by a topological insulator based saturable absorber," *Applied Physics Letters*, vol. 101, no. 21, 2012.

[9] C. Zhao, Y. Zou, Y. Chen et al., "Wavelength-tunable picosecond soliton fiber laser with Topological Insulator: Bi_2Se_3 as a mode locker," *Optics Express*, vol. 20, no. 25, pp. 27888–27895, 2012.

[10] U. Keller, "Recent developments in compact ultrafast lasers," *Nature*, vol. 424, no. 6950, pp. 831–838, 2003.

[11] G. Steinmeyer, D. H. Sutter, L. Gallmann, N. Matuschek, and U. Keller, "Frontiers in ultrashort pulse generation: Pushing the limits in linear and nonlinear optics," *Science*, vol. 286, no. 5444, pp. 1507–1512, 1999.

[12] F. Wang, A. G. Rozhin, V. Scardaci et al., "Wideband-tuneable, nanotube mode-locked, fiber laser," *Nature Nanotechnology*, vol. 3, pp. 738–742, 2008.

[13] F. Bonaccorso, Z. Sun, T. Hasan, and A. C. Ferrari, "Graphene photonics and optoelectronics," *Nature Photonics*, vol. 4, no. 9, pp. 611–622, 2010.

[14] S. Wang, H. Yu, H. Zhang et al., "Broadband few-layer MoS_2 saturable absorbers," *Advanced Materials*, vol. 26, no. 21, pp. 3538–3544, 2014.

[15] J. Du, Q. Wang, G. Jiang et al., "Ytterbium-doped fiber laser passively mode locked by few-layer Molybdenum Disulfide (MoS2) saturable absorber functioned with evanescent field interaction," *Scientific Reports*, vol. 4, no. 1, 2015.

[16] F. Xia, H. Wang, and Y. Jia, "Rediscovering black phosphorus as an anisotropic layered material for optoelectronics and electronics," *Nature Communications*, vol. 5, p. 4458, 2014.

[17] Y. Chen, G. Jiang, S. Chen et al., "Mechanically exfoliated black phosphorus as a new saturable absorber for both Q-switching and mode-locking laser operation," *Optics Express*, vol. 23, no. 10, pp. 12823–12833, 2015.

[18] C. Zhao, H. Zhang, X. Qi et al., "Ultra-short pulse generation by a topological insulator based saturable absorber," *Applied Physics Letters*, vol. 101, Article ID 211106, 2012.

[19] Y. Chen, C. Zhao, S. Chen et al., "Large energy, wavelength widely tunable, topological insulator Q-switched erbium-doped fiber laser," *IEEE Journal of Selected Topics in Quantum Electronics*, vol. 20, no. 5, pp. 315–322, 2014.

[20] S. Chen, C. Zhao, Y. Li et al., "Broadband optical and microwave nonlinear response in topological insulator," *Optical Materials Express* , vol. 4, no. 4, pp. 587–596, 2014.

[21] G. Jiang, J. Yi, L. Miao et al., "Bismuth Telluride nanocrystal: broadband nonlinear response and its application in ultrafast photonics," *Scientific Reports*, vol. 8, no. 1, 2018.

[22] S. Chen, Y. Chen, M. Wu, Y. Li, C. Zhao, and S. Wen, "Stable Q-Switched Erbium-Doped Fiber Laser Based on Topological Insulator Covered Microfber," *IEEE Photonics Technology Letters*, vol. 26, pp. 987–990, 2014.

[23] M. Wu, Y. Chen, H. Zhang, and S. Wen, "Nanosecond Q-switched erbium-doped fiber laser with wide pulse-repetition-rate range based on topological insulator," *IEEE Journal of Quantum Electronics*, vol. 50, no. 6, pp. 393–396, 2014.

[24] B. S.-Y. Ung, B. Weng, R. Shepherd, D. Abbott, and C. Fumeaux, "Inkjet printed conductive polymer-based beam-splitters for terahertz applications," *Optical Materials Express* , vol. 3, no. 9, pp. 1242–1249, 2013.

[25] Y. Chen, C. Zhao, H. Huang et al., "Self-assembled topological insulator: Bi_2Se_3 membrane as a passive Q-switcher in an erbium-doped fiber laser," *Journal of Lightwave Technology*, vol. 31, no. 17, pp. 2857–2863, 2013.

[26] H. Zhang, Q. Bao, D. Tang, L. Zhao, and K. Loh, "Large energy soliton erbium-doped fiber laser with a graphene-polymer composite mode locker," *Applied Physics Letters*, vol. 95, no. 14, Article ID 141103, 2009.

[27] Q. Wang, Y. Chen, L. Miao et al., "Wide spectral and wavelength-tunable dissipative soliton fiber laser with topological insulator nano-sheets self-assembly films sandwiched by PMMA polymer," *Optics Express*, vol. 23, no. 6, pp. 7681–7693, 2015.

[28] Q. Wang, Y. Chen, G. Jiang et al., "Drop-casted self-assembled topological insulator membrane as an effective saturable absorber for ultrafast laser photonics," *IEEE Photonics Journal*, vol. 7, no. 2, 2015.

[29] J. Liu, M. Wu, B. Huang et al., "Widely Wavelength-Tunable Mid-Infrared Fluoride Fiber Lasers," *IEEE Journal of Selected Topics in Quantum Electronics*, vol. 24, no. 3, pp. 1–7, 2018.

[30] B. Huang, P. Tang, J. Yi et al., "Resonantly pumped Er:YAG laser Q-switched by topological insulator nanosheets at 1617 nm," *Optical Materials*, vol. 71, 2016.

Magnetoresistance Effect in NiFe/BP/NiFe Vertical Spin Valve Devices

Leilei Xu,[1,2] Jiafeng Feng,[3] Kangkang Zhao,[2] Weiming Lv,[4] Xiufeng Han,[3] Zhongyuan Liu,[4] Xiaohong Xu,[1] He Huang,[2] and Zhongming Zeng[2]

[1]*School of Chemistry and Materials Science, Key Laboratory of Magnetic Molecules and Magnetic Information Materials, Ministry of Education, Shanxi Normal University, Linfen 41004, China*
[2]*Key Laboratory of Nanodevices and Applications, Suzhou Institute of Nano-Tech and Nano-Bionics, Chinese Academy of Sciences, Ruoshui Road 398, Suzhou 215123, China*
[3]*Beijing National Laboratory for Condensed Matter Physics, Institute of Physics, Chinese Academy of Sciences, Beijing 100190, China*
[4]*State Key Laboratory of Metastable Materials Science and Technology, Yanshan University, Qinhuangdao 066004, China*

Correspondence should be addressed to Xiaohong Xu; xuxiaohong_ly@163.com and Zhongming Zeng; zmzeng2012@sinano.ac.cn

Academic Editor: Veer P. S. Awana

Two-dimensional (2D) layered materials such as graphene and transition metal dichalcogenides are emerging candidates for spintronic applications. Here, we report magnetoresistance (MR) properties of a black phosphorus (BP) spin valve devices consisting of thin BP flakes contacted by NiFe ferromagnetic (FM) electrodes. The spin valve effect has been observed from room temperature to 4 K, with MR magnitudes of 0.57% at 4 K and 0.23% at 300 K. In addition, the spin valve resistance is found to decrease monotonically as temperature is decreased, indicating that the BP thin film works as a conductive interlayer between the NiFe electrodes.

1. Introduction

Two-dimensional (2D) nanomaterials such as single-layer graphene and transition metal dichalcogenides (TMDs) have attracted great attention as building blocks for future (opto)electronic technologies due to their specific layered structures and novel physical properties [1–3]. Recently, the 2D nanomaterials have also been demonstrated to have potential for application in the field of spintronics [4–11]. The 2D materials have been largely researched as nonmagnetic interlayer of spin valve, which is similar to traditional magnetic tunneling junctions consisting of two ferromagnetic (FM) layers separated by a nonmagnetic insulating spacer, usually Al_2O_3 and MgO, and the resistance depends on the magnetization orientation of two ferromagnetic electrodes [12, 13]. The first experimental work for realization of spin-transport phenomenon is reported by Tombros et al. in graphene-based planar spin valve structure [4]. Later on, magnetoresistance (MR) was measured at room temperature in graphene vertical spin valve [7, 8]. Subsequent works reported spin-dependent transport in h-BN and transition metal dichalcogenides (TMDs), such as MoS_2 and WS_2 [9–11]. These studies suggest that the 2D nanomaterials may be promising for spintronic applications. Recently, a few-layer black phosphorus (BP), a newly identified 2D nanomaterial, has been demonstrated to be an appealing candidate material owing to its exotic physical properties such as thickness-dependent tunable band gap and high carrier mobility [14–18]. Interestingly, theoretical study predicted a relatively large MR ratio in the BP-based spin valve structure [19–21]. However, so far, there have been no reports of MR effect in BP-based spin valve.

In this work, we report on the fabrication and spin valve effect in the BP-based device. 2D BP is sandwiched by two permalloy electrodes (Py and $Ni_{81}Fe_{19}$) and the BP layer serves as nonmagnetic spacer layer as shown in Figure 1(a). The devices show spin valve effect from room temperature to low temperature with a MR of 0.57% at 4 K. The temperature dependence of the device resistance reveals that the BP layer works as a metallic layer between two FM electrodes.

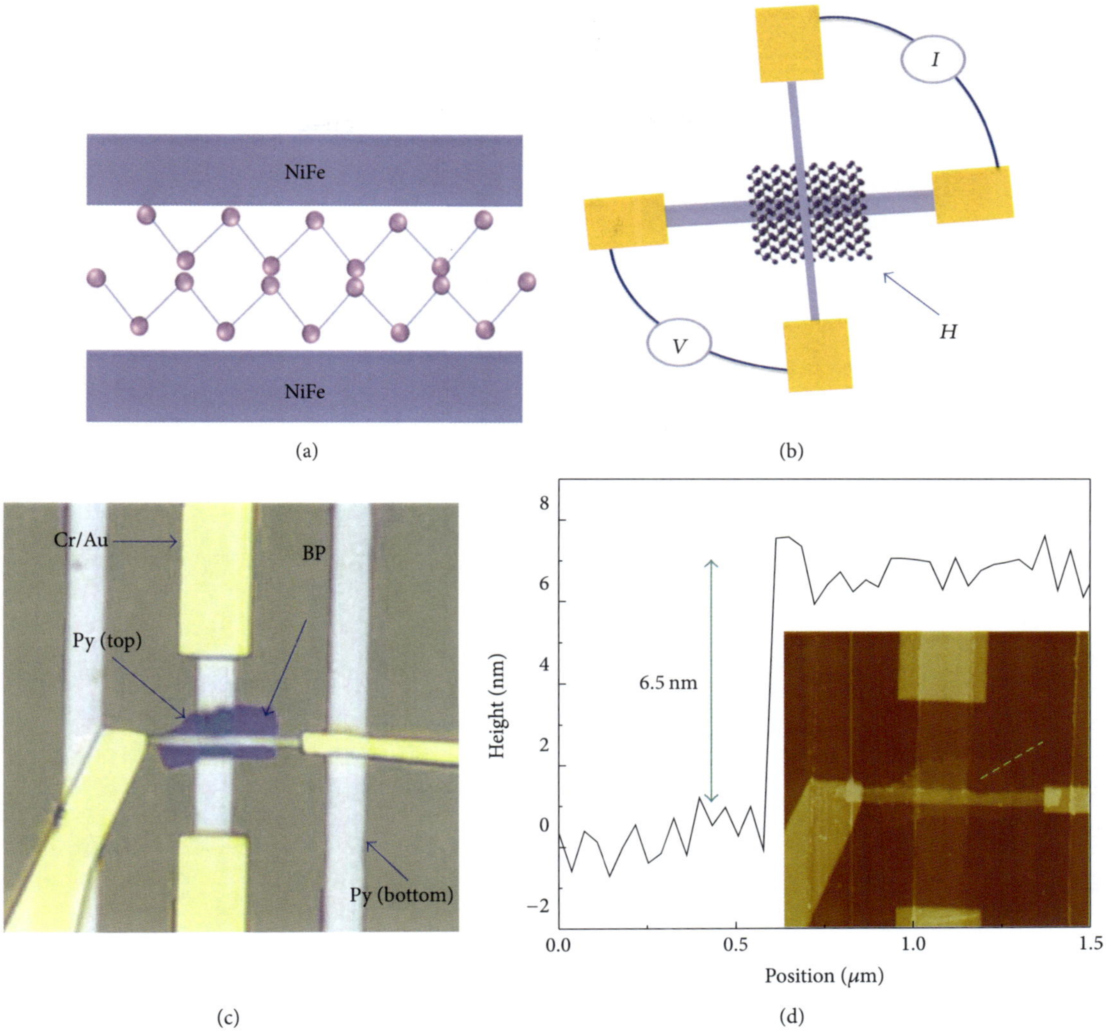

FIGURE 1: ((a) and (b)) Structure and measurement structure of BP-based vertical spin valve device, consisting of bottom NiFe electrode, 2D-BP spacer, and top NiFe electrode. (c) Optical macrograph of NiFe/BP/NiFe spin valve device. (d) AFM graph of device.

2. Methods

2.1. Material and Device Fabrication. The BP crystals were synthesized from red phosphorus under high temperature of 1000°C and high pressure of 2 GPa. Thin BP flake was obtained by mechanically exfoliating BP crystal using adhesive tape (scotch tape), and then the flake was transferred onto the prepatterned Py (bottom) electrodes on SiO_2/Si substrate. The bottom electrodes were fabricated by e-beam lithography (EBL) and a lift-off procedure after e-beam evaporating Py with thickness of ~30 nm. In the subsequent process, top Py electrodes with thickness of ~50 nm were fabricated by another run of EBL, metal deposition and lift-off process. Finally, the bottom and top FM electrodes were connected with large electrodes by EBL and Cr (5 nm)/Au (60 nm) deposition.

2.2. Device Characterization and Measurement Setup. The devices were measured with a four-terminal setup, where the bias currents flow perpendicular to the device plane of the spin valve. The magnetic field was applied in-plane at 45° to the direction of the Py ferromagnetic electrodes as shown in Figure 1(b). The BP flake was initially identified by optical microscopy in Figure 1(c) and then further confirmed by atomic force microscopy (AFM). Figure 1(d) shows an atomic force microscope (AFM) image of one device, revealing that the thickness of BP for the device is ~6.5 nm. The transport measurements for the BP-based spin valve devices were performed using Physical Properties Measurements System (PPMS) made by Quantum Design.

3. Results and Discussion

3.1. Current-Voltage Characteristics and Spin Valve Effect. Figure 2(a) displays the current-voltage (I-V) curves of one typical device for various temperatures. The linear I-V curves indicate the Ohmic contact characteristics of the BP flake and FM electrodes. At room temperature, the resistance-area

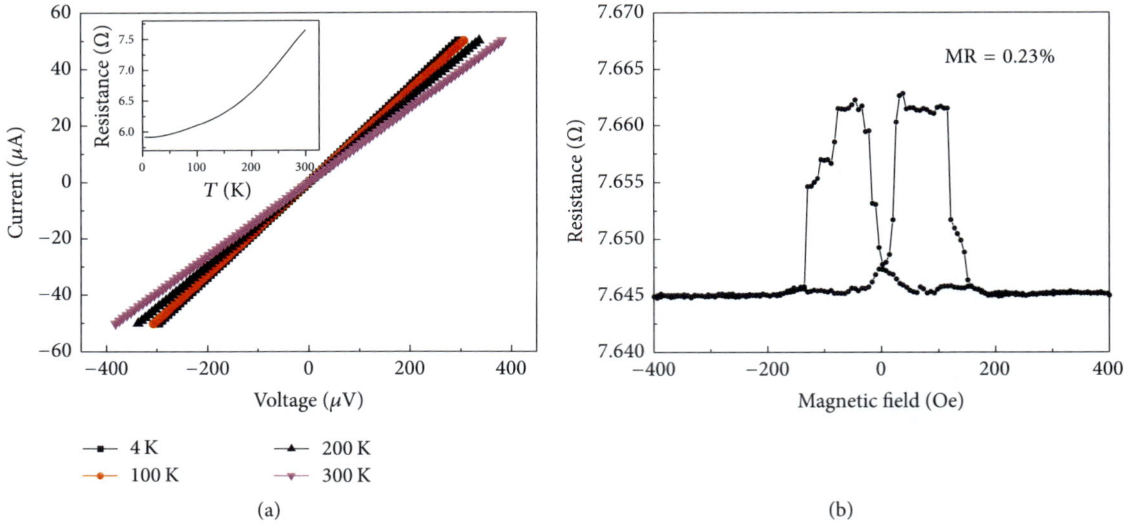

FIGURE 2: Characterizations of NiFe/BP/NiFe spin valve device. (a) The current-voltage curves of device at different temperature from 4 K to 300 K and resistance curve as a function of temperature at zero magnetic field in inset figure (a). (b) Resistance curves of device when external magnetic field is applied at 300 K and the value of MR is 0.23% for the structure of Py/BP/Py spin valve.

(RA) product of the device is on the order of ~10^{-11} Ω·m^2, which is smaller than that in the monolayer MoS_2 device (~10^{-10} Ω·m^2 [10]). This difference may be related to the different band gaps (~0.5 eV in our device while ~1.87 eV in [10]). Note that the resistance decreases with reducing temperature, indicating that the BP behaves as a metal in the spin valve structure as shown in inset of Figure 2(a). The results suggest that the thin BP flake behaves as a conducting thin film rather than a tunnel barrier between the two FM electrodes, which is consistent with the previous works in MoS_2 and WS_2 based spin valve [10, 11].

Spin valve effect is characterized by measuring the resistance as a function of magnetic field. By sweeping the magnetic field, the resistance can be tuned into the high-resistance (R_{AP}) state and low-resistance (R_P) state since it depends on the orientation of the magnetization of the FM electrodes. The magnetoresistance is defined as MR = 100 × $(R_{AP} - R_P)/R_P$, where R_{AP} and R_P are the resistances when the magnetization vectors of two Py electrodes are antiparallel and parallel to each other, respectively. Note that the widths of the top and bottom electrodes were designed to be 500 nm and 2 μm as shown in Figure 1(c), respectively. This yields a large difference in coercivity between two FM electrodes. Thus the bottom electrode is easier to magnetize than the top one under the application of magnetic field owing to weaker shape anisotropy. The resistance as a function of magnetic field for a representative Py/BP/Py spin valve at room temperature is shown in Figure 2(b). As the magnetic field scan from −400 Oe to 400 Oe, the top and bottom electrodes switch in sequence, resulting in the observation of a resistance plateau. The MR value of the device is determined to be 0.23% at RT.

3.2. Temperature Dependence of the Spin Valve Effect. Figure 3(a) shows a series of MR curves for a representative Py/BP/Py spin valve at various temperatures ranging from 4.2 K to 300 K. The maximum MR value is 0.57% at 4 K. A simple relation between the MR and the polarization of the FM electrodes for a junction can be approximated as MR = $2P_1P_2/(1 - P_1P_2)$, where P_1 and P_2 are the electron spin polarization of the two FM metals, respectively [10]. Assuming that two FM electrodes have the same composition, then the polarization ($P_1 \approx P_2 = P$) of the two Py electrode is estimated to be ~5%, which is comparable to the other values reported previously [10], but smaller than that ($P \sim 0.3$) in the Py/Al_2O_3 interface. A possible reason for this reduced polarization is the exposure to air of the Py surface prior to the application of the BP layer. Some contaminants are inevitably adsorbed on the surface and air exposure of Py may produce antiferromagnetic NiO, significantly decreasing the spin polarization [22]. Future well-controlled fabrication process in situ without air exposure of the interfaces may improve the interface quality, maximizing the MR effect.

The magnitude of the MR monotonically decreases as the temperature is increased, as shown in Figure 3(b). The decrease in MR amplitude at higher temperature may be attributed to the inelastic scattering with phonons, magnetic impurity scattering, surface states, and thermal smearing of electron energy distribution in the FM metals [23]. The data are found to follow Bloch's law, where the spin polarization is described by $P(T) = P(0)(1 - \alpha T^{3/2})$. By fitting the data with MR relation by considering the temperature dependence of spin polarization, the material-dependent constant α can be estimated to be $5.9 \times 10^{-5}\ K^{-3/2}$. This value is comparable to that of 3~5 × $10^{-5}\ K^{-3/2}$ reported in the literature [24]. Temperature-dependent relation of parallel and antiparallel resistance are showed in Figure 3(c), which further indicated BP not as insulting barrier layer in the MTJ, but as metallic layer in vertical spin valve.

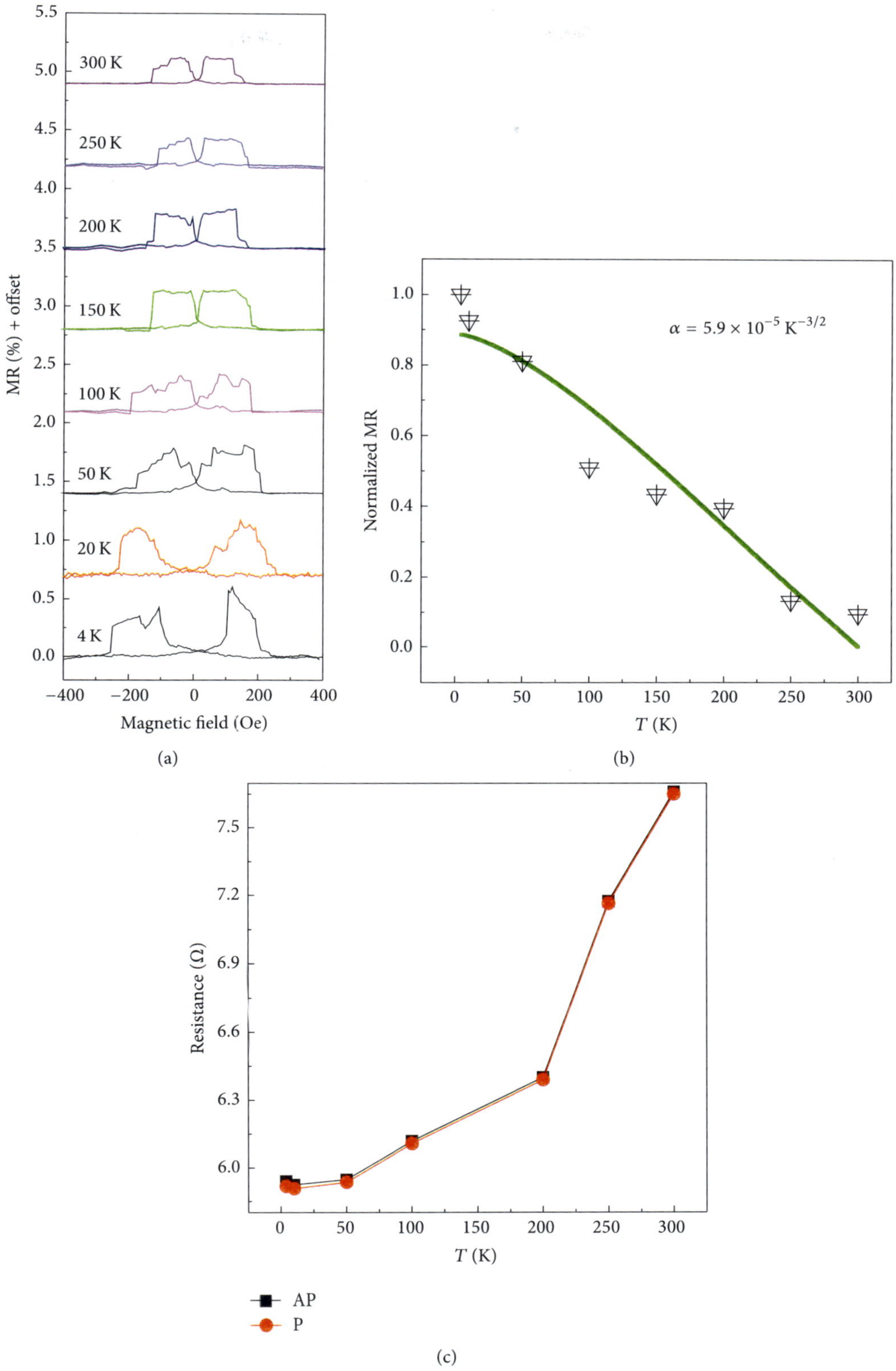

FIGURE 3: Temperature dependence of spin valve effect in NiFe/BP/NiFe device. (b) Normalized MR ratio as a function of temperature and the solid line is the fitting to Bloch's law. (c) Resistance of device, corresponding to low resistance (R_P) and high resistance (R_{AP}).

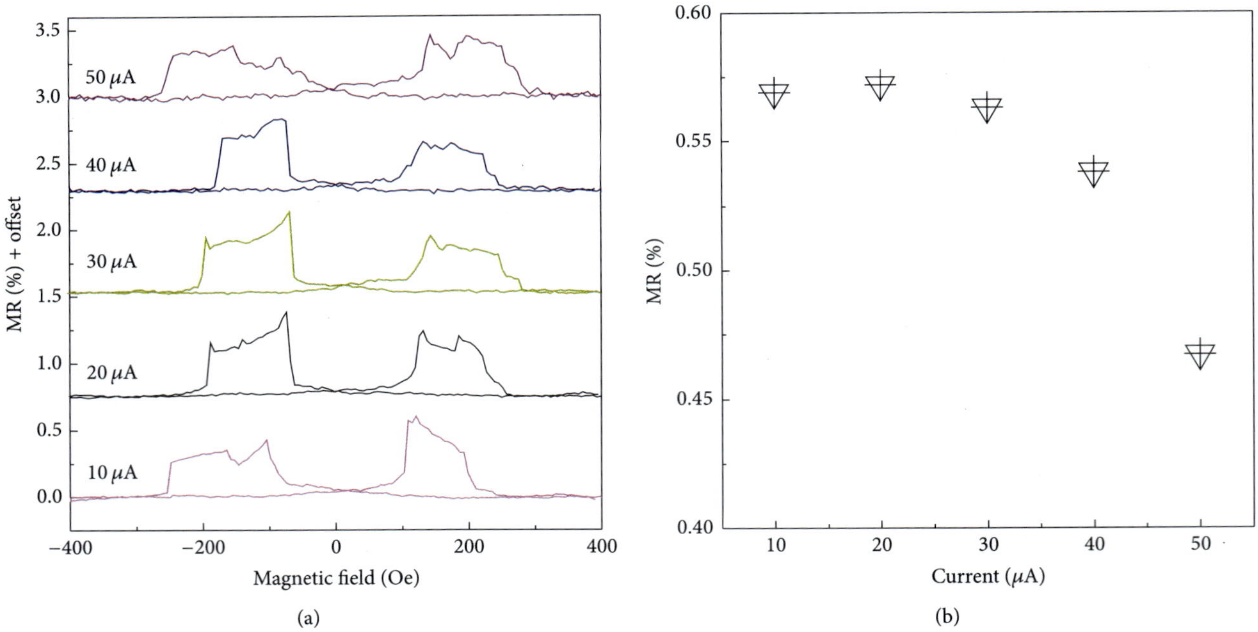

FIGURE 4: (a) Magnetoresistance curve at various bias currents. (b) MR ratio as a function of bias current.

3.3. Bias Current Dependence of the Spin Valve Effect. Finally, we investigated the bias current dependence of the spin valve effect. Figure 4(a) shows the resistance as a function of magnetic field at various bias currents from 10 μA to 50 μA at 4 K. The amplitude of MR value is found to be decreased as the bias current increases as shown in Figure 4(b). We attribute the decrease in MR value at larger bias current to the spin excitations localized at the interfaces between the FM electrodes and the BP interlayer [25] as well as the localized trap states in the BP interlayer [26] at the BP interlayer played a metallic property role in vertical spin valve. The results provide a possible approach to use the emerging BP nanomaterials for future spintronics applications such as magnetic memory and logic devices.

Competing Interests

The authors declare that they have no competing interests.

Acknowledgments

The work was supported by the National Natural Science Foundation of China (Grants nos. 11274343 and 91222110), National Science Fund for Distinguished Young Scholars (Grant no. 51025103), and Program for New Century Excellent Talents in University (NCET-13-0993).

References

[1] M. Osada and T. Sasaki, "Two-dimensional dielectric nanosheets: novel nanoelectronics from nanocrystal building blocks," *Advanced Materials*, vol. 24, no. 2, pp. 210–228, 2012.

[2] K. S. Novoselov, A. K. Geim, S. V. Morozov et al., "Electric field in atomically thin carbon films," *Science*, vol. 306, no. 5696, pp. 666–669, 2004.

[3] B. Radisavljevic, A. Radenovic, J. Brivio, V. Giacometti, and A. Kis, "Single-layer MoS_2 transistors," *Nature Nanotechnology*, vol. 6, no. 3, pp. 147–150, 2011.

[4] N. Tombros, C. Jozsa, M. Popinciuc, H. T. Jonkman, and B. J. Van Wees, "Electronic spin transport and spin precession in single graphene layers at room temperature," *Nature*, vol. 448, no. 7153, pp. 571–574, 2007.

[5] W. Han, W. H. Wang, K. Pi et al., "Electron-hole asymmetry of spin injection and transport in single-layer graphene," *Physical Review Letters*, vol. 102, no. 13, Article ID 137205, 2009.

[6] W. Han, K. Pi, W. Bao et al., "Electrical detection of spin precession in single layer graphene spin valves with transparent contacts," *Applied Physics Letters*, vol. 94, no. 22, Article ID 222109, 2009.

[7] E. Cobas, A. L. Friedman, O. M. J. Van'T Erve, J. T. Robinson, and B. T. Jonker, "Graphene as a tunnel barrier: graphene-based magnetic tunnel junctions," *Nano Letters*, vol. 12, no. 6, pp. 3000–3004, 2012.

[8] M. Z. Iqbal, M. W. Iqbal, J. H. Lee, Y. S. Kim, S.-H. Chun, and J. Eom, "Spin valve effect of NiFe/graphene/NiFe junctions," *Nano Research*, vol. 6, no. 5, pp. 373–380, 2013.

[9] A. Dankert, M. Venkata Kamalakar, A. Wajid, R. S. Patel, and S. P. Dash, "Tunnel magnetoresistance with atomically thin two-dimensional hexagonal boron nitride barriers," *Nano Research*, vol. 8, no. 4, pp. 1357–1364, 2015.

[10] W. Wang, A. Narayan, L. Tang et al., "Spin-valve effect in NiFe/MoS_2/NiFe junctions," *Nano Letters*, vol. 15, no. 8, pp. 5261–5267, 2015.

[11] M. Z. Iqbal, M. W. Iqbal, S. Siddique, M. F. Khan, and S. M. Ramay, "Room temperature spin valve effect in NiFe/WS_2/Co junctions," *Scientific Reports*, vol. 6, Article ID 21038, 2016.

[12] H. X. Wei, Q. H. Qin, M. Ma, R. Sharif, and X. F. Han, "80% tunneling magnetoresistance at room temperature for thin Al-O barrier magnetic tunnel junction with CoFeB as free and reference layers," *Journal of Applied Physics*, vol. 101, no. 9, Article ID 09B501, 2007.

[13] S. Yuasa, T. Nagahama, A. Fukushima, Y. Suzuki, and K. Ando, "Giant room-temperature magnetoresistance in single-crystal Fe/MgO/Fe magnetic tunnel junctions," *Nature Materials*, vol. 3, no. 12, pp. 868–871, 2004.

[14] L. Li, Y. Yu, G. J. Ye et al., "Black phosphorus field-effect transistors," *Nature Nanotechnology*, vol. 9, no. 5, pp. 372–377, 2014.

[15] F. Xia, H. Wang, and Y. Jia, "Rediscovering black phosphorus as an anisotropic layered material for optoelectronics and electronics," *Nature Communications*, vol. 5, article 4458, 2014.

[16] A. Castellanos-Gomez, L. Vicarelli, E. Prada et al., "Isolation and characterization of few-layer black phosphorus," *2D Materials*, vol. 1, no. 2, Article ID 025001, 2014.

[17] A. Castellanos-Gomez, M. Buscema, R. Molenaar et al., "Deterministic transfer of two-dimensional materials by all-dry viscoelastic stamping," *2D Materials*, vol. 1, no. 1, Article ID 011002, 2014.

[18] B. Wan, B. Yang, Y. Wang et al., "Enhanced stability of black phosphorus field-effect transistors with SiO_2 passivation," *Nanotechnology*, vol. 26, no. 43, Article ID 435702, 2015.

[19] M. Chen, Z. Yu, Y. Wang, Y. Xie, J. Wang, and H. Guo, "Nonequilibrium spin injection in monolayer black phosphorus," *Physical Chemistry Chemical Physics*, vol. 18, no. 3, pp. 1601–1606, 2016.

[20] Y. Anugrah, M. C. Robbins, P. A. Crowell, and S. J. Koester, "Determination of the Schottky barrier height of ferromagnetic contacts to few-layer phosphorene," *Applied Physics Letters*, vol. 106, no. 10, Article ID 103108, 2015.

[21] M. Venkata Kamalakar, B. N. Madhushankar, A. Dankert, and S. P. Dash, "Low schottky barrier black phosphorus field-effect devices with ferromagnetic tunnel contacts," *Small*, vol. 11, no. 18, pp. 2209–2216, 2015.

[22] M. R. Fitzsimmons, T. J. Silva, and T. M. Crawford, "Surface oxidation of Permalloy thin films," *Physical Review B*, vol. 73, no. 1, Article ID 014420, 2006.

[23] J. J. Åkerman, I. V. Roshchin, J. M. Slaughter, R. W. Dave, and I. K. Schuller, "Origin of temperature dependence in tunneling magnetoresistance," *Europhysics Letters*, vol. 63, no. 1, pp. 104–110, 2003.

[24] C. H. Shang, J. Nowak, R. Jansen, and J. S. Moodera, "Temperature dependence of magnetoresistance and surface magnetization in ferromagnetic tunnel junctions," *Physical Review B*, vol. 58, no. 6, pp. R2917–R2920, 1998.

[25] S. Zhang, P. M. Levy, A. C. Marley, and S. S. P. Parkin, "Quenching of magnetoresistance by hot electrons in magnetic tunnel junctions," *Physical Review Letters*, vol. 79, no. 19, pp. 3744–3747, 1997.

[26] E. Y. Tsymbal, O. N. Mryasov, and P. R. LeClair, "Spin-dependent tunnelling in magnetic tunnel junctions," *Journal of Physics Condensed Matter*, vol. 15, no. 4, pp. R109–R142, 2003.

Speed of the Goldstone Sound Mode of an Atomic Fermi Gas Loaded on a Square Optical Lattice with a Non-Abelian Gauge Field in the Presence of a Zeeman Field

Zlatko Koinov[1] **and Israel Chávez Villalpando**[2]

[1] *Department of Physics and Astronomy, University of Texas at San Antonio, San Antonio, TX 78249, USA*
[2] *Instituto de Investigaciones en Materiales, Universidad Nacional Autónoma de México, Apartado Postal 20-364, 01000 Mexico Distrito Federal, Mexico*

Correspondence should be addressed to Zlatko Koinov; zlatko.koinov@utsa.edu

Academic Editor: Oleg Derzhko

The speed of the Goldstone sound mode of a spin-orbit-coupled atomic Fermi gas loaded in a square optical lattice with a non-Abelian gauge field in the presence of a Zeeman field is calculated within the Gaussian approximation and from the Bethe-Salpeter equation in the generalized random phase approximation. It is found that (i) there is no sharp change of the slope of the Goldstone sound mode across the topological quantum phase transition point and (ii) the Gaussian approximation significantly overestimates the speed of sound of the Goldstone mode compared to the value provided by the BS formalism.

1. Introduction

It is well known that topological superfluids are new states of matter that can be observed in two-dimensional atomic Fermi gases with strong Rashba spin-orbit coupling (SOC), conventional s-wave pairing, and out-of-plane magnetic field (Zeeman field) which breaks time-reversal symmetry. In what follows, we study the speed of sound of a two-component pseudospin-$\uparrow,\downarrow$ fermionic gas loaded on a square optical lattice with a non-Abelian gauge field $\mathbf{A} = (\alpha\sigma_y, -\alpha\sigma_x)$ in the presence of a Zeeman field, where α is independently tunable parameter, and $\sigma_i, i = x, y, z$ are the Pauli matrices. The external non-Abelian gauge field effectively creates a SOC in the gas. The pseudospin of atoms can couple with the Zeeman field and the orbital degrees of freedom of atoms, which gives rise to a topological quantum phase transition (TQPT) between gapped (topological trivial) and gapless (topological nontrivial) superfluid states. In other words, the spectrum of the Hamiltonian becomes gapless at some set of points in the momentum space. Due to absence of symmetry breaking across the TQPTs, it remains experimentally challenging to detect these phase transitions. It is experimentally possible to detect TQPTs by measuring the dynamic response of the bulk under an external force [1], by observing the edge modes [2, 3], or the momentum distributions of atoms across the phase transition boundary using a noise-correlation imaging [4] and momentum-resolved spectroscopy [5, 6].

Recent numerical calculations of the sound speed of atomic Fermi gases with s-wave attraction, synthetic Rashba SOC, and out-of-plane Zeeman field in two- and three-dimensional *free space* [7–9] show sharp changes of the slope of the Goldstone sound mode across the phase transition points, and this can be used to detect TQPTs. Our goal is to study whether a similar behavior of the sound speed occurs in the lattice case if the SOC is created by the non-Abelian gauge field. We found that the sharp change in the slope does not exist in the presence of lattice geometry.

It is widely accepted among the cold-atom community that the system of a two-component Fermi gas with s-wave attraction in square optical lattice with an external non-Abelian gauge field (or a synthetic Rashba SOC) and out-of-plane Zeeman field can be described by a negative-U Hubbard model [10–26]. We restrict the discussion to the case of atoms confined to the lowest-energy band (single-band model). We assume that there are $M = N_\uparrow + N_\downarrow$ atoms distributed along N sites, and the filling factor $f = f_\uparrow + f_\downarrow = M/N$ is smaller than unity. The corresponding polarization is $P = (f_\uparrow - f_\downarrow)/f$. The Hamiltonian for an uniform system is $\widehat{H} = \widehat{H}_0 + \widehat{H}_U$, where

$$\widehat{H}_0 = -\sum_{\langle i,j\rangle,\sigma,\sigma'} \psi^\dagger_{i,\sigma} t^{\sigma\sigma'}_{i,j} \psi_{j,\sigma'} - \mu \sum_{i,\sigma} \widehat{n}_{i,\sigma} + \widehat{H}_Z. \quad (1)$$

The out-of-plane Zeeman field is described by the term $\widehat{H}_Z$:

$$\widehat{H}_Z = h \sum_i \left(\psi^\dagger_{i,\uparrow}, \psi^\dagger_{i,\downarrow}\right) \sigma_z \begin{pmatrix} \psi_{i,\uparrow} \\ \psi_{i,\downarrow} \end{pmatrix}. \quad (2)$$

The Hubbard part of the Hamiltonian is

$$\widehat{H}_U = -U \sum_i \widehat{n}_{i,\uparrow} \widehat{n}_{i,\downarrow}. \quad (3)$$

Here, μ is the chemical potential, and $\widehat{n}_{i,\sigma} = \psi^\dagger_{i,\sigma}\psi_{i,\sigma}$ is the density operator on site i. The Fermi operator $\psi^\dagger_{i,\sigma} = (1/N)\sum_{\mathbf{k}} \exp(-\imath \mathbf{k}.\mathbf{r}_i)\psi^\dagger_{\mathbf{k},\sigma}$ ($\psi_{i,\sigma} = (1/N)\sum_{\mathbf{k}} \exp(\imath \mathbf{k}.\mathbf{r}_i)\psi_{\mathbf{k},\sigma}$) creates (destroys) a fermion on the lattice site $\mathbf{r}_i$ with pseudospin projection σ. The sum $\sum_{\langle ij\rangle}$ means sum over the nearest-neighbor sites of the 2D lattice. $t^{\sigma\sigma'}_{i,j} = t_{i,j}\exp[-\imath\theta^{\sigma\sigma'}_{i,j}]$ describes the hoping of the atoms between site j and site i, and $\theta^{\sigma\sigma'}_{i,j} = \int_{\mathbf{r}_j}^{\mathbf{r}_i} \mathbf{A}.d\mathbf{r}$ is accumulated via the Peierls substitution phase factor. The strength of the on-site Hubbard interaction is $U > 0$, which corresponds to attractive interaction.

Hamiltonian (1) in a momentum representation assumes the form $\widehat{H}_0 = \sum_{\mathbf{k}} \psi^\dagger_{\mathbf{k}}[\xi(\mathbf{k})\sigma_0 - h\sigma_z + J_x(\mathbf{k})\sigma_x + J_y(\mathbf{k})\sigma_y]\psi_{\mathbf{k}}$. Here, $\psi^\dagger_{\mathbf{k}} = (\psi^\dagger_{\mathbf{k},\uparrow}, \psi^\dagger_{\mathbf{k},\downarrow})$, $\xi(\mathbf{k}) = \epsilon(\mathbf{k}) - \mu$, $\epsilon(\mathbf{k})$ is the tight-binding energy, σ_0 is the identity matrix, and $J_{x,y}(\mathbf{k})$ characterize the SOC.

In the case of a square lattice the vectors $\mathbf{k} = (k_x, k_y)$ (in units a^{-1}) are restricted to be within the Brillouin zone, i.e., a square defined by the following four points $(\pm\pi/a, \pm\pi/a)$. The tight-binding energy and $J_{x,y}$ are given by

$$\begin{aligned} \epsilon(\mathbf{k}) &= -2t\cos(\alpha)\left[\cos(k_x) + \cos(k_y)\right], \\ J_x(\mathbf{k}) &= -2t\sin(\alpha)\sin(k_y), \\ J_y(\mathbf{k}) &= 2t\sin(\alpha)\sin(k_x), \end{aligned} \quad (4)$$

where t is the strength of the nearest-neighbor hopping (in what follows, the lattice spacing a and spacing t are set to unity).

The opening or closing of the gap can be achieved by varying the tunable parameter α along with the Zeeman field. We first assume that the opening or closing of the gap is achieved by varying the parameter α, while by changing the out-of-plane Zeeman field the polarization is kept fixed. We also discuss the second possibility, when one can close or open the gap by changing the Zeeman field while keeping the parameter α fixed. In this case the polarization changes during the transition from gapless to gapped superfluid states.

The single-particle and collective excitations of the system under consideration manifest themselves as poles of the single-particle and two-particle Green's functions, but the corresponding expressions for the Green's functions cannot be evaluated exactly because the interaction part of the Hubbard Hamiltonian is quartic in the fermion fields. The simplest way to solve this problem is to apply the so-called mean field decoupling of the quartic interaction. According to the standard mean field theory on the Hubbard Hamiltonian, the on-site interaction is decoupled (up to Hartree-Fock correction terms) as

$$-U\widehat{n}_{i,\uparrow}\widehat{n}_{i,\downarrow} \approx \Delta\left(\psi_{i,\downarrow}\psi_{i,\uparrow} + \psi^\dagger_{i,\uparrow}\psi^\dagger_{i,\downarrow} + \frac{\Delta}{U}\right), \quad (5)$$

where the order parameter $\Delta = -U\langle\psi_{i,\downarrow}\psi_{i,\uparrow}\rangle$ is a real number.

Instead of applying the mean field decoupling, we shall transform the quartic terms to a quadratic form by introducing a four-component boson field which mediates the interaction of fermions in the same manner as in quantum electrodynamics, where the photons mediate the interaction of electric charges. Green's functions are thermodynamic averages of the $\widehat{T}_u$-ordered products of field operators. The standard procedure for calculating Green's functions is to apply Wick's theorem. This enables us to evaluate the $\widehat{T}_u$-ordered products of field operators as perturbation expansions involving only wholly contracted field operators. These expansions can be summed formally to yield different equations of Green's functions. The main disadvantage of this procedure is that the validity of the equations must be verified diagram by diagram. For this reason, we shall use the method of Legendre transforms [27] of the generating functional for connected Green's functions to derive the Schwinger-Dyson [28, 29] (SD) equation $G^{-1} = G^{(0)-1} - \Sigma$ for the poles of the single-particle Green's function, G, and the Bethe-Salpeter [30, 31] (BS) equation $[K^{(0)-1} - I]\Psi = 0$ for the poles of the poles of the two-particle Green's function, K. Here, $G^{(0)}$ is the free single-particle propagator, Σ is the fermion self-energy, I is the BS kernel, and the two-particle free propagator $K^{(0)} = GG$ is a product of two fully dressed single-particle Green's functions. The kernel of the BS equation is defined as a sum of the direct interaction, $I_d = \delta\Sigma^F/\delta G$, and the exchange interaction $I_{exc} = \delta\Sigma^H/\delta G$, where Σ^F and Σ^H are the Fock and the Hartree parts of the fermion self-energy Σ. Since the fermion self-energy depends on the two-particle Green's function, the positions of both poles must to be obtained by solving the SD and BS equations self-consistently. Instead of solving self-consistently the SD and BS equations, we shall employ the generalized random phase approximation (GRPA). In this approximation, the single-particle excitations are obtained in the mean field approximation, while the collective modes are obtained by solving the BS equation in which single-particle Green's functions are calculated in Hartree-Fock approximation, and the BS kernel is obtained by summing ladder and bubble diagrams.

2. Functional-Integral Formalism

2.1. Field-Theoretical Approach to the Hubbard Model. The functional-integral formulation of the Hubbard model requires the representation of the Hubbard interaction $-U\sum_i \widehat{n}_{i,\uparrow}\widehat{n}_{i,\downarrow}$ in terms of squares of fermion operators. This can be done by employing a certain Hubbard-Stratonovich transformation. This field-theoretical approach to the Hubbard Hamiltonian has already been used to describe the collective modes of ultracold ^6Li-^{40}K mixture in a square optical lattice (see [32]). Hamiltonian (1) in the present study has two more interactions. It is possible to include both the SOC and the Zeeman field into initial definition of the free fermion Green's function, and therefore most of the general equations in [32] remain unchanged. In particular, the general expressions for the SD and BS equations are the same as those without the SOC and the Zeeman field, but with more complicated fermion self-energy and BS kernel.

As it is well known, Green's functions in the functional-integral approach are defined by means of the so-called generating functional with sources for the boson and fermion fields, but the corresponding functional integrals cannot be evaluated exactly because the interaction part of the Hubbard Hamiltonian is quartic in the Grassmann fermion fields. However, we can transform the quartic terms to a quadratic form by introducing a model system which consists of a four-component boson field $A_\alpha(z)$ interacting with fermion fields $\widehat{\overline{\psi}}(y) = \widehat{\Psi}^\dagger(y)/\sqrt{2}$ and $\widehat{\psi}(x) = \widehat{\Psi}(x)/\sqrt{2}$, where

$$\widehat{\Psi}(x) = \begin{pmatrix} \psi_\uparrow(x) \\ \psi_\downarrow(x) \\ \psi_\uparrow^\dagger(x) \\ \psi_\downarrow^\dagger(x) \end{pmatrix}, \qquad \widehat{\Psi}^\dagger(y) = \left(\psi_\uparrow^\dagger(y)\,\psi_\downarrow^\dagger(y)\,\psi_\uparrow(y)\,\psi_\downarrow(y)\right). \tag{6}$$

Here $\alpha = 1,2,3,4$, $x = (\mathbf{r}_i, u)$, $y = (\mathbf{r}_j, u')$, and $z = (\mathbf{r}_k, v)$ are composite variables, where $\mathbf{r}_i$, $\mathbf{r}_j$, and $\mathbf{r}_k$ are the lattice site vectors. According to imaginary-time (Matsubara) formalism the variables $0 \le v, u, u' \le \beta = (k_B T)^{-1}$; k_B is the Boltzmann constant ($\hbar = k_B = 1$).

The field operators (6) allow us to define generalized single-particle Green's function by using a tensor product of these two matrices. The corresponding Green function, represented by a 4×4 matrix, includes all possible thermodynamic averages:

$$\widehat{G}(x_1; y_2) = -\left\langle \widehat{T}_u\left(\widehat{\Psi}(x_1) \otimes \widehat{\overline{\Psi}}(y_2)\right)\right\rangle. \tag{7}$$

The action of the above-mentioned model system is assumed to be of the following form $S = S_0^{(F)} + S_0^{(B)} + S^{(F-B)}$, where $S_0^{(F)} = \widehat{\overline{\psi}}(y)\widehat{G}^{(0)-1}(y;x)\widehat{\psi}(x)$, $S_0^{(B)} = (1/2)A_\alpha(z)D_{\alpha\beta}^{(0)-1}(z,z')A_\beta(z')$, $S^{(F-B)} = \widehat{\overline{\psi}}(y)\widehat{\Gamma}_\alpha^{(0)}(y,x \mid z)\widehat{\psi}(x)A_\alpha(z)$. The fermion action $S_0^{(F)}$ corresponds to the Hamiltonian $\widehat{H}_0$. The corresponding inverse Green function of free fermions $\widehat{G}^{(0)-1}(y;x)$ is given by the following 4×4 matrix:

$$\widehat{G}^{(0)-1}(y;x) = \sum_{\mathbf{k},\omega_m} \exp\left[\imath\mathbf{k}.(\mathbf{r}_i - \mathbf{r}_{i'}) - \omega_m(u-u')\right] \cdot \widehat{G}^{(0)-1}(\mathbf{k}, \imath\omega_m). \tag{8}$$

The symbol $\sum_{\omega_m}$ is used to denote $\beta^{-1}\sum_m$ (for fermion fields $\omega_m = (2\pi/\beta)(m+1/2)$; $m = 0, \pm1, \pm2, \ldots$).

The Fourier transform of noninteracting Green's function is

$$\widehat{G}^{(0)-1}(\mathbf{k}, \imath\omega_m) = \begin{pmatrix} \imath\omega_m - \xi(\mathbf{k}) - h & -\left(J_x(\mathbf{k}) + \imath J_y(\mathbf{k})\right) & 0 & 0 \\ -\left(J_x(\mathbf{k}) - \imath J_y(\mathbf{k})\right) & \imath\omega_m - \xi(\mathbf{k}) + h & 0 & 0 \\ 0 & 0 & \imath\omega_m + \xi(\mathbf{k}) + h & -\left(J_x(\mathbf{k}) - \imath J_y(\mathbf{k})\right) \\ 0 & 0 & -\left(J_x(\mathbf{k}) + \imath J_y(\mathbf{k})\right) & \imath\omega_m + \xi(\mathbf{k}) - h \end{pmatrix}. \tag{9}$$

After including the Zeeman field and the SOC interaction into free fermion Green's function (9), one can perform the same steps as in [32] to come up with the following single-particle Green function:

$$\widehat{G}^{-1}(1;2) = \begin{pmatrix} G_{11}^{(0)-1} & G_{12}^{(0)-1} & 0 & -\Delta_{i_1,i_2} \\ G_{21}^{(0)-1} & G_{22}^{(0)-1} & \Delta_{i_1,i_2} & 0 \\ 0 & \Delta_{i_1,i_2} & G_{33}^{(0)-1} & G_{31}^{(0)-1} \\ -\Delta_{i_1,i_2} & 0 & G_{41}^{(0)-1} & G_{44}^{(0)-1} \end{pmatrix}, \tag{10}$$

where $G_{ij}^{(0)-1} = G_{ij}^{(0)-1}(\mathbf{r}_{i_1} - \mathbf{r}_{i_2}; u_1 - u_2)$, and $\Delta_{i_1,i_2} \equiv \Delta\delta(\mathbf{r}_{i_1} - \mathbf{r}_{i_2})$, where Δ is a constant in space. Physically, it describes a superfluid state of Cooper pairs with zero momentum when the pairing is only between atoms with equal and opposite momenta.

2.2. Mean Field Approximation. The Fourier transform of zero-temperature single-particle Green's function (10) in the mean field approximation is

$$\widehat{G}_{MF}(\mathbf{k}, \imath\omega_m) = \begin{pmatrix} \imath\omega_m - \xi(\mathbf{k}) - h & -(J_x(\mathbf{k}) + \imath J_y(\mathbf{k})) & 0 & \Delta \\ -(J_x(\mathbf{k}) - \imath J_y(\mathbf{k})) & \imath\omega_m - \xi(\mathbf{k}) + h & -\Delta & 0 \\ 0 & -\Delta & \imath\omega_m + \xi(\mathbf{k}) + h & -(J_x(\mathbf{k}) - \imath J_y(\mathbf{k})) \\ \Delta & 0 & -(J_x(\mathbf{k}) + \imath J_y(\mathbf{k})) & \imath\omega_m + \xi(\mathbf{k}) - h \end{pmatrix}^{-1}. \quad (11)$$

The matrix elements $\widehat{G}^{MF}_{n_1 n_2}(\mathbf{k}, \imath\omega_m)$ $(n_1, n_2 = 1, 2, 3, 4)$ of $\widehat{G}_{MF}(\mathbf{k}, \omega)$ are given by

$$G^{MF}_{n_1 n_2}(\mathbf{k}, \omega) = \frac{A_{n_1 n_2}(\mathbf{k})}{\omega - \omega(\mathbf{k}) + \imath 0^+} + \frac{B_{n_1 n_2}(\mathbf{k})}{\omega + \omega(\mathbf{k}) - \imath 0^+} + \frac{C_{n_1 n_2}(\mathbf{k})}{\omega - \Omega(\mathbf{k}) + \imath 0^+} + \frac{D_{n_1 n_2}(\mathbf{k})}{\omega + \Omega(\mathbf{k}) - \imath 0^+}, \quad (12)$$

where the corresponding expressions for $A_{n_1 n_2}(\mathbf{k})$, $B_{n_1 n_2}(\mathbf{k})$, $C_{n_1 n_2}(\mathbf{k})$, and $D_{n_1 n_2}(\mathbf{k})$ can be obtained by inverting matrix (11) and setting $T \longrightarrow 0$.

For a fixed filing factor $f = f_\uparrow + f_\downarrow$ and at a given Zeeman field h, the chemical potential μ and the gap Δ have to be determined by solving the mean field number and gap equations:

$$f = 1 - \sum_{\mathbf{k}} \left[\frac{1}{2} - F(\Omega(\mathbf{k}))\right] \frac{\xi(\mathbf{k})}{\Omega(\mathbf{k})} \cdot \left(1 + \frac{S(\mathbf{k}) + h^2}{\sqrt{(S(\mathbf{k}) + h^2)\xi^2(\mathbf{k}) + h^2\Delta^2}}\right) - \sum_{\mathbf{k}} \left[\frac{1}{2} - F(\omega(\mathbf{k}))\right] \frac{\xi(\mathbf{k})}{\omega(\mathbf{k})} \cdot \left(1 - \frac{S(\mathbf{k}) + h^2}{\sqrt{(S(\mathbf{k}) + h^2)\xi^2(\mathbf{k}) + h^2\Delta^2}}\right), \quad (13)$$

$$\frac{1}{U} = \sum_{\mathbf{k}} \left[\frac{1}{2} - F(\Omega(\mathbf{k}))\right] \frac{1}{2\Omega(\mathbf{k})} \cdot \left(1 + \frac{h^2}{\sqrt{(S(\mathbf{k}) + h^2)\xi^2(\mathbf{k}) + h^2\Delta^2}}\right) + \sum_{\mathbf{k}} \left[\frac{1}{2} - F(\omega(\mathbf{k}))\right] \frac{1}{2\omega(\mathbf{k})} \cdot \left(1 - \frac{h^2}{\sqrt{(S(\mathbf{k}) + h^2)\xi^2(\mathbf{k}) + h^2\Delta^2}}\right). \quad (14)$$

Here, $S(\mathbf{k}) = J_x^2(\mathbf{k}) + J_y^2(\mathbf{k})$, $F(\omega)$ is the Fermi-Dirac distribution function, and the following notations have been introduced:

$$\Omega(\mathbf{k}) = \sqrt{S(\mathbf{k}) + \xi^2(\mathbf{k}) + \Delta^2 + h^2 + 2\sqrt{(S(\mathbf{k}) + h^2)\xi^2(\mathbf{k}) + h^2\Delta^2}}, \quad (15)$$

$$\omega(\mathbf{k}) = \sqrt{S(\mathbf{k}) + \xi^2(\mathbf{k}) + \Delta^2 + h^2 - 2\sqrt{(S(\mathbf{k}) + h^2)\xi^2(\mathbf{k}) + h^2\Delta^2}}. \quad (16)$$

The number equation (13) follows from the definitions of the filing factors $f_{\uparrow,(\downarrow)} = \sum_{\mathbf{k}} \sum_{\imath\omega_m} G^{MF}_{11,(22)}(\mathbf{k}, \imath\omega_m)$ through corresponding mean field Green's functions. It is known that the gap equation can be obtained by minimizing the mean field Helmholtz free energy with respect to Δ.

As we mentioned before, in order to keep the polarization fixed, one has to adjust the corresponding Zeeman field. This means that μ, Δ, and h have to be calculated by solving the number and the gap equations, along with the following equation:

$$f_\downarrow - f_\uparrow = \sum_{\mathbf{k}} \left[\frac{1}{2} - F(\Omega(\mathbf{k}))\right] \cdot \frac{h}{\Omega(\mathbf{k})} \left(1 + \frac{\xi^2(\mathbf{k}) + \Delta^2}{\sqrt{(S(\mathbf{k}) + h^2)\xi^2(\mathbf{k}) + h^2\Delta^2}}\right) + \sum_{\mathbf{k}} \left[\frac{1}{2} - F(\omega(\mathbf{k}))\right] \cdot \frac{h}{\omega(\mathbf{k})} \left(1 - \frac{\xi^2(\mathbf{k}) + \Delta^2}{\sqrt{(S(\mathbf{k}) + h^2)\xi^2(\mathbf{k}) + h^2\Delta^2}}\right). \quad (17)$$

Another interesting feature is that because of the SOC term in Hamiltonian (1) the pairing field contains both singlet and triplet component. The singlet $\Phi_{\downarrow\uparrow}(\mathbf{k}) = -\Phi_{\uparrow\downarrow}(\mathbf{k}) = \langle \psi_{\mathbf{k}\downarrow}\psi_{-\mathbf{k}\uparrow} \rangle$ and triplet $\Phi_{\uparrow\uparrow}(\mathbf{k}) = \langle \psi_{\mathbf{k}\uparrow}\psi_{-\mathbf{k}\uparrow} \rangle$, $\Phi_{\downarrow\downarrow}(\mathbf{k}) = \langle \psi_{\mathbf{k}\downarrow}\psi_{-\mathbf{k}\downarrow} \rangle$ amplitudes, obtained by means of Green's function elements G^{MF}_{23}, G^{MF}_{13}, and G^{MF}_{24}, are

$$\Phi_{\downarrow\uparrow}(\mathbf{k}) = \frac{\Delta\left[1/2 - f_F(\Omega(\mathbf{k}))\right]}{2\Omega(\mathbf{k})} \left[1 + \frac{h^2}{\sqrt{S(\mathbf{k})\xi^2(\mathbf{k}) + h^2\left[\Delta^2 + \xi^2(\mathbf{k})\right]}}\right]$$

$$+\frac{\Delta\left[1/2-f_F(\omega(\mathbf{k}))\right]}{2\omega(\mathbf{k})}\left[1-\frac{h^2}{\sqrt{S(\mathbf{k})\,\xi^2(\mathbf{k})+h^2\left[\Delta^2+\xi^2(\mathbf{k})\right]}}\right];$$

$$\Phi_{\uparrow\uparrow}(\mathbf{k})=\frac{\Delta J(\mathbf{k})\left[h-\xi(\mathbf{k})\right]}{2\sqrt{S(\mathbf{k})\,\xi^2(\mathbf{k})+h^2\left[\Delta^2+\xi^2(\mathbf{k})\right]}}\left[\frac{1/2-f_F(\Omega(\mathbf{k}))}{\Omega(\mathbf{k})}-\frac{1/2-f_F(\omega(\mathbf{k}))}{\omega(\mathbf{k})}\right];$$

$$\Phi_{\downarrow\downarrow}(\mathbf{k})=\frac{\Delta J^*(\mathbf{k})\left[h+\xi(\mathbf{k})\right]}{2\sqrt{S(\mathbf{k})\,\xi^2(\mathbf{k})+h^2\left[\Delta^2+\xi^2(\mathbf{k})\right]}}\left[\frac{1/2-f_F(\Omega(\mathbf{k}))}{\Omega(\mathbf{k})}-\frac{1/2-f_F(\omega(\mathbf{k}))}{\omega(\mathbf{k})}\right]. \tag{18}$$

With the help of the pairing amplitudes, one can calculate the total condensate fraction $f_c = f_s + f_{tr}$, where f_s and f_{tr} are the singlet and the triplet contributions, correspondingly. At zero temperature, we obtain

$$f_s=\frac{2}{N}\sum_{\mathbf{k}}\left|\Phi_{\downarrow\uparrow}(\mathbf{k})\right|^2=\frac{\Delta^2}{8N}\sum_{\mathbf{k}}\left[\frac{1}{\Omega(\mathbf{k})}\left(1+\frac{h^2}{\sqrt{S(\mathbf{k})\,\xi^2(\mathbf{k})+h^2\left[\Delta^2+\xi^2(\mathbf{k})\right]}}\right)+\frac{1}{\omega(\mathbf{k})}\left(1-\frac{h^2}{\sqrt{S(\mathbf{k})\,\xi^2(\mathbf{k})+h^2\left[\Delta^2+\xi^2(\mathbf{k})\right]}}\right)\right]^2, \tag{19}$$

$$f_{tr}=\frac{1}{N}\sum_{\mathbf{k}}\left(\left|\Phi_{\uparrow\uparrow}(\mathbf{k})\right|^2+\left|\Phi_{\downarrow\downarrow}(\mathbf{k})\right|^2\right)=\frac{\Delta^2}{8N}\cdot\sum_{\mathbf{k}}\frac{S(\mathbf{k})\left[h^2+\xi^2(\mathbf{k})\right]}{S(\mathbf{k})\,\xi^2(\mathbf{k})+h^2\left[\Delta^2+\xi^2(\mathbf{k})\right]}\left[\frac{1}{\Omega(\mathbf{k})}-\frac{1}{\omega(\mathbf{k})}\right]^2. \tag{20}$$

2.3. The Bethe-Salpeter Equation for the Collective Excitations in the Generalized Random Phase Approximation. The spectrum of the collective modes will be obtained by solving the BS equation in the GRPA. As we have already mentioned, the kernel of the BS equation is a sum of the direct $I_d = \delta\Sigma^F/\delta G$ and exchange $I_{exc} = \delta\Sigma^H/\delta G$ interactions, written as derivatives of the Fock and the Hartree parts of the self-energy. Thus, in the GRPA the corresponding equation for the BS amplitude $\Psi^{\mathbf{Q}}_{n_2,n_1}$ is given by (see [32])

$$\Psi^{\mathbf{Q}}_{n_2 n_1}=K^{(0)}\left(\begin{matrix}n_1 & n_3\\ n_2 & n_4\end{matrix}\mid\omega(\mathbf{Q})\right)\left[I_d\begin{pmatrix}n_3 & n_5\\ n_4 & n_6\end{pmatrix}+I_{exc}\begin{pmatrix}n_3 & n_5\\ n_4 & n_6\end{pmatrix}\right]\Psi^{\mathbf{Q}}_{n_6,n_5}, \tag{21}$$

where $I_d\left(\begin{smallmatrix}n_1 & n_3\\ n_2 & n_4\end{smallmatrix}\right) = -\Gamma^{(0)}_\alpha(n_1,n_3)D^{(0)}_{\alpha\beta}\Gamma^{(0)}_\beta(n_4,n_2)$ and $I_{exc}\left(\begin{smallmatrix}n_1 & n_3\\ n_2 & n_4\end{smallmatrix}\right) = (1/2)\Gamma^{(0)}_\alpha(n_1,n_2)D^{(0)}_{\alpha\beta}\Gamma^{(0)}_\beta(n_4,n_3)$ are the direct and the exchange interactions, correspondingly. The two-particle propagator $K^{(0)}$ in the GRPA is defined as follows:

$$\begin{aligned}K^{(0)}\left(\begin{matrix}n_1 & n_3\\ n_2 & n_4\end{matrix}\mid\omega,\mathbf{Q}\right)&\equiv K_{n_1n_3n_4n_2}(\omega,\mathbf{Q})=\int\frac{d\Omega}{2\pi}\cdot\int\frac{d^2\mathbf{k}}{(2\pi)^2}G^{MF}_{n_1n_3}(\mathbf{k}+\mathbf{Q},\Omega+\omega(\mathbf{Q}))\,G^{MF}_{n_4n_2}(\mathbf{k},\Omega)\\
&=\int\frac{d^2\mathbf{k}}{(2\pi)^2}\left[\frac{A_{n_1n_3}(\mathbf{k}+\mathbf{Q})\,B_{n_4n_2}(\mathbf{k})}{\omega-(\omega(\mathbf{k}+\mathbf{Q})+\omega(\mathbf{k}))}-\frac{B_{n_1n_3}(\mathbf{k}+\mathbf{Q})\,A_{n_4n_2}(\mathbf{k})}{\omega+(\omega(\mathbf{k}+\mathbf{Q})+\omega(\mathbf{k}))}\right.\\
&\quad+\frac{C_{n_1n_3}(\mathbf{k}+\mathbf{Q})\,D_{n_4n_2}(\mathbf{k})}{\omega-(\Omega(\mathbf{k}+\mathbf{Q})+\Omega(\mathbf{k}))}-\frac{D_{n_1n_3}(\mathbf{k}+\mathbf{Q})\,C_{n_4n_2}(\mathbf{k})}{\omega+(\Omega(\mathbf{k}+\mathbf{Q})+\Omega(\mathbf{k}))}\\
&\quad+\frac{A_{n_1n_3}(\mathbf{k}+\mathbf{Q})\,D_{n_4n_2}(\mathbf{k})}{\omega-(\omega(\mathbf{k}+\mathbf{Q})+\Omega(\mathbf{k}))}-\frac{B_{n_1n_3}(\mathbf{k}+\mathbf{Q})\,C_{n_4n_2}(\mathbf{k})}{\omega+(\omega(\mathbf{k}+\mathbf{Q})+\Omega(\mathbf{k}))}\\
&\quad\left.+\frac{C_{n_1n_3}(\mathbf{k}+\mathbf{Q})\,B_{n_4n_2}(\mathbf{k})}{\omega-(\Omega(\mathbf{k}+\mathbf{Q})+\omega(\mathbf{k}))}-\frac{D_{n_1n_3}(\mathbf{k}+\mathbf{Q})\,A_{n_4n_2}(\mathbf{k})}{\omega+(\Omega(\mathbf{k}+\mathbf{Q})+\omega(\mathbf{k}))}\right].\end{aligned} \tag{22}$$

The BS equation (21) can be written in the matrix form as $(\widehat{I}+U\widehat{Z})\widehat{\Psi}=0$, where $\widehat{I}$ is the unit matrix, the matrix $\widehat{Z}$ is a 16×16 matrix, and the transposed matrix of $\widehat{\Psi}$ is given by

$$\widehat{\Psi}^T=\left(\Psi^{\mathbf{Q}}_{1,1}\;\Psi^{\mathbf{Q}}_{1,2}\;\Psi^{\mathbf{Q}}_{1,3}\;\Psi^{\mathbf{Q}}_{1,4}\;\Psi^{\mathbf{Q}}_{2,1}\;\Psi^{\mathbf{Q}}_{2,2}\;\Psi^{\mathbf{Q}}_{2,3}\;\Psi^{\mathbf{Q}}_{2,4}\;\Psi^{\mathbf{Q}}_{3,1}\;\Psi^{\mathbf{Q}}_{3,2}\;\Psi^{\mathbf{Q}}_{3,3}\;\Psi^{\mathbf{Q}}_{3,4}\;\Psi^{\mathbf{Q}}_{4,1}\;\Psi^{\mathbf{Q}}_{4,2}\;\Psi^{\mathbf{Q}}_{4,3}\;\Psi^{\mathbf{Q}}_{4,4}\right). \tag{23}$$

At a given vector $\mathbf{Q}$ from the Brillouin zone, the collective excitation spectrum $\omega(\mathbf{Q})$ is obtained by the condition $\det|\widehat{I}+U\widehat{M}(\omega(\mathbf{Q}),\mathbf{Q}))| = 0$ to have a nontrivial solution of the BS equation. By applying simple matrix algebra, the above secular 16×16 determinant can be simplified to a 10×10 BS determinant $\det|BS(\omega(\mathbf{Q}),\mathbf{Q})| = 0$, where the BS matrix is given by

$$BS(\omega,\mathbf{Q}) = \begin{pmatrix} A_{2\times 2}(\omega,\mathbf{Q}) & B_{2\times 8}(\omega,\mathbf{Q}) \\ C_{8\times 2}(\omega,\mathbf{Q}) & D_{8\times 8}(\omega,\mathbf{Q}) \end{pmatrix}. \quad (24)$$

The elements $A_{i,j}$, $B_{i,j}$, $C_{i,j}$, and $D_{i,j}$ of the blocks $A_{2\times 2}(\omega,\mathbf{Q})$, $B_{2\times 8}(\omega,\mathbf{Q})$, $C_{8\times 2}(\omega,\mathbf{Q})$, and $D_{8\times 8}(\omega,\mathbf{Q})$ are given in the Appendix in terms of the propagators $K_{n_1 n_3 n_4 n_2}(\omega,\mathbf{Q})$.

In contrast to our functional-integral formalism, one can use the Hubbard-Stratonovich transformation to introduce the energy gap as an order parameter field. In this approach, one can integrate out the fermion fields and to arrive at an effective action. The next steps are to consider the state which corresponds to the saddle point of the effective action and to write the effective action as a series in powers of the fluctuations and their derivatives. The exact result can be obtained by explicitly calculating the terms up to second order in the fluctuations and their derivatives. This approximation is called the Gaussian approximation. Within this approximation, the collective excitation spectrum $\omega(\mathbf{Q})$ is defined by the 2×2 Gaussian secular determinant

$$\det\left|A_{2\times 2}(\omega(\mathbf{Q}),\mathbf{Q})\right| = 0, \quad (25)$$

while within the BS formalism the secular determinant is more complicated ($(\omega(\mathbf{Q}),\mathbf{Q})$ dependence is understood):

$$\det\left|A_{2\times 2} - B_{2\times 8}.D^{-1}_{8\times 8}.C_{8\times 2}\right| = 0 \quad (26)$$

3. Numerical Results

There are various mean field quantities of physical interest, such as the chemical potential, the pairing gap, the singlet and triplet pairing amplitudes, and the singlet and triplet condensate fractions. We focus on the zero-temperature case assuming two different filling factors of $f = 0.4$ and $f = 0.6$. The strength of the attractive interaction is $U/t = 4$, and for both filling factors the polarization is fixed to be $P = 0.48$.

In Figure 1, we plot the chemical potential μ, the gap Δ, and the corresponding Zeeman field h, as functions of the SOC parameter α, obtained by solving the mean field equations (13)-(17). In Figure 2 we have shown the singlet and the triplet condensate fractions, as functions of the SOC parameter α, obtained by solving the mean field equations along with definitions (19) and (20). It can be seen that at the value $\alpha = \pi/2$, where $\xi(\mathbf{k}) = -\mu$ and $S(\mathbf{k}) = 4t^2[\sin^2(k_x) + \sin^2(k_y)]$, the chemical potential has a minimum, while the gap and the Zeeman field reach their maximum values. As expected, the singlet and triplet condensate fraction-graphs have maximum at the same α as the gap.

Next, we have calculated the speed of sound as a function of the SOC parameter α. The sound speed c is proportional to the slope u of the Goldstone sound mode $c = uat/\hbar$. The slope u has been numerically obtained using the collective-mode dispersion $\omega(\mathbf{Q})$ for small wave vectors in Q_x-direction.

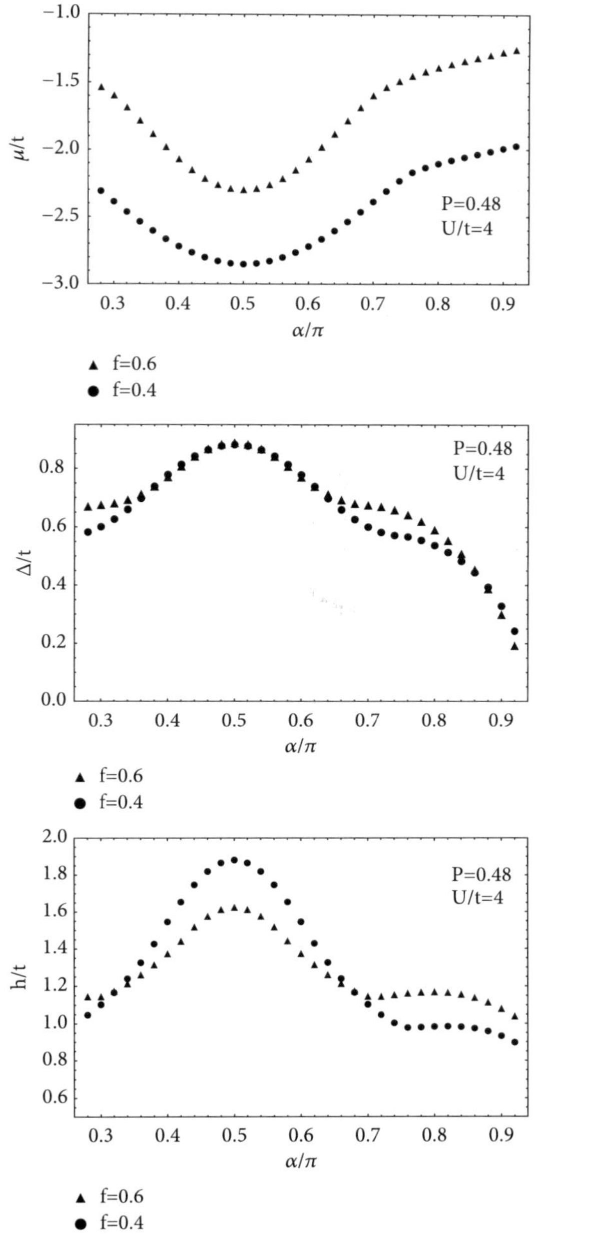

FIGURE 1: Chemical potential μ, pairing gap Δ, and the Zeeman field h of a Fermi gas in a square optical lattice subject to a non-Abelian gauge field $\mathbf{A} = (\alpha\sigma_y, -\alpha\sigma_x)$. The graphs are obtained by solving the mean field equations (13)-(17) using two different filling factors: $f = 0.4$ and $f = 0.6$. The on-site attractive strength is $U = 4t$, and the polarization is $P = 0.48$.

It is clear that the single-particle dispersion, given by (16), becomes precisely zero at some special points $\mathbf{k}_0$ in momentum space satisfying the condition $S(\mathbf{k}_0) = 0$ when $h = \sqrt{(\epsilon(\mathbf{k}_0) - \mu)^2 + \Delta^2}$. The condition $S(\mathbf{k}_0) = 0$ is met at the following points: $(0,0)$, $(0,\pm\pi)$, $(\pm\pi,0)$, and $(\pm\pi,\pm\pi)$. For $U/t = 4$, $P = 0.48$, and a filling factor $f = 0.4$, we have two points at which the single-particle dispersion

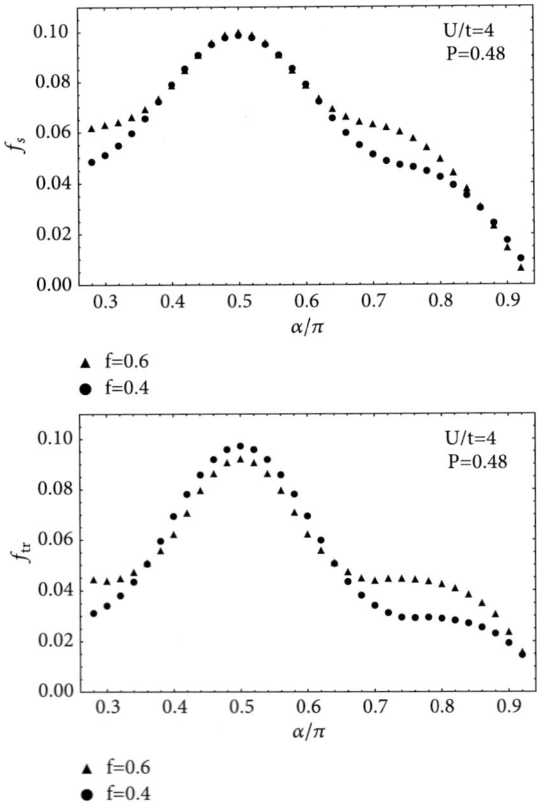

FIGURE 2: The singlet f_s and the triplet f_{tr} contributions to the total condensed fraction $f = f_s + f_{tr}$ of a Fermi gas in a square optical lattice subject to a non-Abelian gauge field $\mathbf{A} = (\alpha\sigma_y, -\alpha\sigma_x)$. The graphs are obtained by solving the mean field equations along with definitions (19) and (20). The system parameters are the same as in Figure 1.

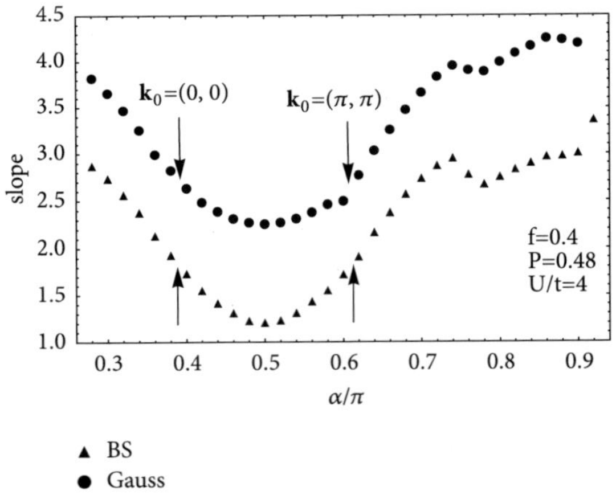

FIGURE 3: The slope of the Goldstone sound mode of a Fermi gas in a square optical lattice subject to a non-Abelian gauge field $\mathbf{A} = (\alpha\sigma_y, -\alpha\sigma_x)$. The graphs are obtained by applying the Gaussian and the BS secular determinants. The system parameters are the same as in Figure 1. The arrows show the values of α for which the opening and closing of a gap take place.

$\omega(\mathbf{k}_{0,1}) = 0$. The first one is at the center of the Brillouin zone $\mathbf{k}_0 = (0,0)$, and $\omega(\mathbf{k}_0) = 0$ for α/π between 0.39 and 0.40. At the second point, $\mathbf{k}_1 = (\pi,\pi)$, we found also the existence of opening and closing of a gap in momentum space when $0.61 < \alpha/\pi < 0.62$. Therefore, at both points one should have a TQPT between superfluid states with different topologies. During the TQPT the different phases of matter are not characterized by an order parameter, but rather an integer number, the Chern number, describing the system as a whole. This is because the different phases across the TQPT are characterized by different topologies of Fermi surface instead of being classified by different symmetries.

In Figure 3, we have shown the slope calculated by using the Gaussian and the BS secular determinants (25) and (26), correspondingly. It is worth mentioning the significant difference between the slope obtained within the Gaussian approximation and the slope obtained by the BS secular determinant. The arrows show the points where the opening and closing of a gap take place. As can be seen, the sharp change in the slope does not exist in the presence of a non-Abelian gauge field.

The boundaries for the TQPT in [7–9] were accessed by varying the Zeeman filed at a constant strength of the synthetic Rashba SOC. It was found that on a topological trivial side, where $h < h_c = \sqrt{(\epsilon(\mathbf{k}_0) - \mu)^2 + \Delta^2}$, the slope is suppressed by the Zeeman field, while in the nontrivial side, where $h > h_c = \sqrt{(\epsilon(\mathbf{k}_0) - \mu)^2 + \Delta^2}$, the slope is enhanced by the Zeeman field. As mentioned in [7, 8], this is quite unusual since we generally expect that the superfluidity should be suppressed by the Zeeman field.

We have fixed the non-Abelian parameter $\alpha = 0.38\pi$, and we have accessed the opening and the closing of a gap in momentum space at the point $\mathbf{k}_0 = (0,0)$ by varying the Zeeman filed. To do this, first we used (13)-(14) to calculate numerically the chemical potential and the gap as functions of the Zeeman field. The results are shown in Figure 4. The TQPT takes place for $h_c/t = 1.39$ and $h_c/t = 1.49$. The insert shows the behavior of the slope close to $h_c/t = 1.49$. We found that there is no sharp change in the slope across the two TQPT points.

4. Discussion

In this paper, we are concerned with the question which naturally arises here as to whether there is a possibility of detecting TQPTs by monitoring the behavior of slope of the sound mode if a non-Abelian gauge field is used instead of a synthetic Rashba SOC. The answer is not trivial because the non-Abelian gauge field is taken into account via the Peierls substitution. As a result, the tight-binding energy $\epsilon(\mathbf{k}) = -2t\cos(\alpha)[\cos(k_x) + \cos(k_y)]$ does depend on the parameter α. The second difference is that the nondiagonal part of $\theta_{i,j}^{\sigma\sigma'}$, i.e., $\sigma \neq \sigma'$, describes the spin-flipped hopping between site i and site j. In the case of a synthetic Rashba SOC we do not have spin-flipped hopping, and the hopping to the nearest-neighbor site does not depend on the strength of Rashba SOC.

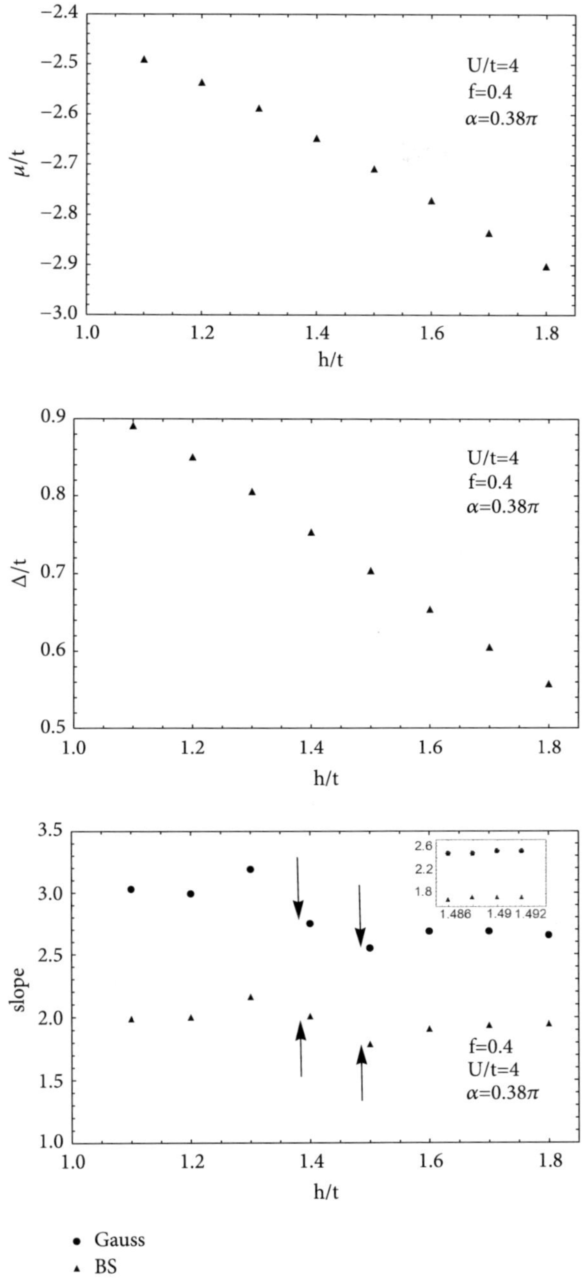

FIGURE 4: Chemical potential μ, pairing gap Δ, and the slope of the Goldstone sound mode of a Fermi gas in a square optical lattice subject to a non-Abelian gauge field with $\alpha = 0.38\pi$ as a function of the Zeeman field h. The chemical potential and the gap are obtained by solving the mean field number and gap equations (13)-(14). The arrows show the values of the Zeeman field for which the opening and closing of a gap take place. The insert shows the slope close to $h_c/t = 1.49$.

In summary, we have derived the BS equation in the GRPA for the collective excitation energy of a Fermi gas loaded on a square optical lattice with a non-Abelian gauge field in the presence of a Zeeman field. To the best of our knowledge, there is no other calculation of the speed of sound in a lattice geometry as a function of the non-Abelian parameter with which we can make comparisons. According to our numerical calculations, there is no sharp change of the slope of the Goldstone sound mode across the phase transition point. It is found that the Gaussian approximation significantly overestimates the speed of sound of the Goldstone mode compared to the value obtained within the BS formalism.

Appendix

Blocks of the BS Matrix

The two-particle propagator in BS equation (21) is defined as

$$\begin{aligned} K^{(0)}\begin{pmatrix} n_1 & n_3 \\ n_2 & n_4 \end{pmatrix} \mid \omega, \mathbf{Q}) &\equiv K_{n_1 n_3 n_4 n_2}(\omega, \mathbf{Q}) \\ &= \sum_{\mathbf{k}} \int \frac{d\Omega}{2\pi} G^{MF}_{n_1 n_3}(\mathbf{k}+\mathbf{Q}, \Omega+\omega) G^{MF}_{n_4 n_2}(\mathbf{k}, \Omega). \end{aligned} \tag{A.1}$$

The elements $A_{i,j}$, $B_{i,j}$, $C_{i,j}$, and $D_{i,j}$ of the blocks $A_{2\times2}(\omega,\mathbf{Q})$, $B_{2\times8}(\omega,\mathbf{Q})$, $C_{8\times2}(\omega,\mathbf{Q})$, and $D_{8\times8}(\omega,\mathbf{Q})$ are given in terms of the propagators $K_{n_1 n_3 n_4 n_2}(\omega,\mathbf{Q})$ as follows:

$$\begin{aligned} A_{1,1} &= 1 + \frac{U_0}{2}\left(K_{1144} + K_{2233} - K_{1234} - K_{2143}\right), \\ A_{1,2} &= \frac{U_0}{2}\left(K_{1234} + K_{2413} - K_{1414} - K_{2323}\right), \\ A_{2,1} &= \frac{U_0}{2}\left(K_{3142} + K_{4231} - K_{3232} - K_{4141}\right), \\ A_{2,2} &= 1 + \frac{U_0}{2}\left(K_{3322} + K_{4411} - K_{3412} - K_{4321}\right), \\ B_{1,1} &= \frac{U_0}{2}\left(K_{1124} - K_{2123}\right), \\ B_{1,2} &= \frac{U_0}{2}\left(K_{2323} - K_{1324}\right), \\ B_{1,3} &= \frac{U_0}{2}\left(K_{1214} - K_{2213}\right), \\ B_{1,4} &= \frac{U_0}{2}\left(K_{1434} - K_{2433}\right), \\ B_{1,5} &= -\frac{1}{2} - \frac{U_0}{2}\left(K_{1144} - K_{2143}\right), \\ B_{1,6} &= \frac{U_0}{2}\left(K_{1344} - K_{2343}\right), \\ B_{1,7} &= \frac{U_0}{2}\left(K_{1444} + K_{2223} - K_{1224} - K_{2443}\right), \\ B_{1,8} &= \frac{U_0}{2}\left(K_{1334} + K_{2113} - K_{1114} - K_{2333}\right), \\ B_{21} &= \frac{U_0}{2}\left(K_{3122} - K_{4121}\right), \end{aligned} \tag{A.2}$$

$$B_{2,2} = -\frac{1}{2} - \frac{U_0}{2}\left(K_{3322} - K_{4321}\right),$$

$$B_{2,3} = \frac{U_0}{2}\left(K_{3212} - K_{4211}\right),$$

$$B_{2,4} = \frac{U_0}{2}\left(K_{3432} - K_{4431}\right),$$

$$B_{2,5} = \frac{U_0}{2}\left(K_{4141} - K_{3142}\right),$$

$$B_{2,6} = \frac{U_0}{2}\left(K_{3342} - K_{4341}\right),$$

$$B_{2,7} = \frac{U_0}{2}\left(K_{3442} + K_{4221} - K_{3222} - K_{4441}\right),$$

$$B_{2,8} = \frac{U_0}{2}\left(K_{3332} + K_{4111} - K_{3112} - K_{4331}\right),$$

(A.3)

$$C_{1,1} = U_0\left(K_{1232} - K_{1142}\right),$$

$$C_{1,2} = U_0\left(K_{1412} - K_{1322}\right),$$

$$C_{2,1} = U_0\left(K_{3232} - K_{3142}\right),$$

$$C_{2,2} = -1 - U_0\left(K_{3322} - K_{3412}\right),$$

$$C_{3,1} = U_0\left(K_{2231} - K_{2141}\right),$$

$$C_{3,2} = U_0\left(K_{2411} - K_{2321}\right),$$

$$C_{4,1} = U_0\left(K_{4233} - K_{4143}\right),$$

$$C_{4,2} = U_0\left(K_{4413} - K_{4323}\right),$$

$$C_{5,1} = -1 - U_0\left(K_{1144} - K_{1234}\right),$$

(A.4)

$$C_{5,2} = U_0\left(K_{1414} - K_{1324}\right),$$

$$C_{6,1} = U_0\left(K_{3234} - K_{3144}\right),$$

$$C_{6,2} = U_0\left(K_{3414} - K_{3324}\right),$$

$$C_{7,1} = \frac{U_0}{2}\left(K_{1141} + K_{3233} - K_{1231} - K_{3143}\right),$$

$$C_{7,2} = \frac{U_0}{2}\left(K_{1321} + K_{3413} - K_{1411} - K_{3413}\right),$$

$$C_{8,1} = \frac{U_0}{2}\left(K_{2142} + K_{4234} - K_{2232} - K_{4144}\right),$$

$$C_{8,2} = \frac{U_0}{2}\left(K_{2322} + K_{4414} - K_{2412} - K_{4324}\right),$$

$$D_{1,1} = 1 - U_0 K_{1122},$$

$$D_{1,2} = U_0 K_{1322},$$

$$D_{1,3} = -U_0 K_{1212},$$

$$D_{1,4} = -U_0 K_{1432},$$

$$D_{1,5} = U_0 K_{1142},$$

$$D_{1,6} = -U_0 K_{1342},$$

$$D_{1,7} = U_0\left(K_{1222} - K_{1442}\right),$$

$$D_{1,8} = U_0\left(K_{1112} - K_{1332}\right),$$

$$D_{2,1} = -U_0 K_{3122},$$

$$D_{2,2} = 1 + U_0 K_{3322},$$

$$D_{2,3} = -U_0 K_{3212},$$

$$D_{2,4} = -U_0 K_{3432},$$

$$D_{2,5} = U_0 K_{3142},$$

$$D_{2,6} = -U_0 K_{3342},$$

$$D_{2,7} = -U_0\left(K_{3222} - K_{3442}\right),$$

$$D_{2,8} = U_0\left(K_{3112} - K_{3332}\right),$$

$$D_{3,1} = -U_0 K_{2121},$$

$$D_{3,2} = U_0 K_{2321},$$

$$D_{3,3} = 1 - U_0 K_{2211},$$

$$D_{3,4} = -U_0 K_{2431},$$

$$D_{3,5} = U_0 K_{2141},$$

$$D_{3,6} = -U_0 K_{2341},$$

$$D_{3,7} = U_0\left(K_{2221} - K_{2441}\right),$$

$$D_{3,8} = U_0\left(K_{2111} - K_{2331}\right),$$

$$D_{4,1} = -U_0 K_{4123},$$

$$D_{4,2} = U_0 K_{4323},$$

$$D_{4,3} = -U_0 K_{4213},$$

$$D_{4,4} = 1 - U_0 K_{4433},$$

$$D_{4,5} = U_0 K_{4143},$$

$$D_{4,6} = -U_0 K_{4343},$$

$$D_{4,7} = U_0\left(K_{4223} - K4443\right),$$

$$D_{4,8} = U_0\left(K_{4113} - K_{4333}\right),$$

$$D_{5,1} = -U_0 K_{1124},$$

$$D_{5,2} = U_0 K_{1324},$$

$$D_{5,3} = -U_0 K_{1214},$$

$$D_{5,4} = -U_0 K_{1434},$$

$$D_{5,5} = 1 + U_0 K_{1144},$$

$$D_{5,6} = -U_0 K_{1344},$$

$$D_{5,7} = U_0\left(K_{1224} - K1444\right),$$

$$D_{5,8} = U_0 \left(K_{1114} - K_{1334}\right),$$
$$D_{6,1} = -U_0 K_{3124},$$
$$D_{6,2} = U_0 K_{3324},$$
$$D_{6,3} = -U_0 K_{3214},$$
$$D_{6,4} = -U_0 K_{3434},$$
$$D_{6,5} = U_0 K_{3144},$$
$$D_{6,6} = 1 - U_0 K_{3344},$$
$$D_{6,7} = U_0 \left(K_{3224} - K3444\right),$$
$$D_{6,8} = U_0 \left(K_{3114} - K_{3334}\right),$$
$$D_{7,1} = \frac{U_0}{2} \left(K_{1121} - K_{3123}\right),$$
$$D_{7,2} = \frac{U_0}{2} \left(K_{3323} - K_{1321}\right),$$
$$D_{7,3} = \frac{U_0}{2} \left(K_{1211} - K_{3213}\right),$$
$$D_{7,4} = \frac{U_0}{2} \left(K_{1431} - K_{3433}\right),$$
$$D_{7,5} = \frac{U_0}{2} \left(K_{3143} - K_{1141}\right),$$
$$D_{7,6} = \frac{U_0}{2} \left(K_{1342} - K_{3343}\right),$$
$$D_{7,7} = -1 - \frac{U_0}{2} \left(K_{1221} + K_{3443} - K_{1441} - K_{3223}\right),$$
$$D_{7,8} = \frac{U_0}{2} \left(K_{1331} + K_{3113} - K_{1111} - K_{3333}\right),$$
$$D_{8,1} = \frac{U_0}{2} \left(K_{2122} - K_{4124}\right),$$
$$D_{8,2} = \frac{U_0}{2} \left(K_{4324} - K_{2322}\right),$$
$$D_{8,3} = \frac{U_0}{2} \left(K_{2212} - K_{4214}\right),$$
$$D_{8,4} = \frac{U_0}{2} \left(K_{2432} - K_{4434}\right),$$
$$D_{8,5} = \frac{U_0}{2} \left(K_{4144} - K_{2142}\right),$$
$$D_{8,6} = \frac{U_0}{2} \left(K_{2342} - K_{4344}\right),$$
$$D_{8,7} = \frac{U_0}{2} \left(K_{2442} + K_{4224} - K_{2222} - K_{4444}\right)$$
$$D_{8,8} = -1 - \frac{U_0}{2} \left(K_{2112} + K_{4334} - K_{2332} - K_{4114}\right). \tag{A.5}$$

Conflicts of Interest

The authors declare that they have no conflicts of interest.

Acknowledgments

Israel Chávez Villalpando gratefully acknowledges funding from CONACyT Grant no. 291001 and UNAM-DGAPA-PAPIIT IN102417.

References

[1] H. Pino, E. Alba, J. Taron, J. J. Garcia-Ripoll, and N. Barberán, "Hall response of interacting bosonic atoms in strong gauge fields: from condensed to fractional-quantum-Hall states," *Physical Review A: Atomic, Molecular and Optical Physics*, vol. 87, no. 5, Article ID 053611, 2013.

[2] N. Goldman, J. Beugnon, and F. Gerbier, "Detecting chiral edge states in the hofstadter optical lattice," *Physical Review Letters*, vol. 108, no. 25, Article ID 255303, 2012.

[3] N. Goldman, J. Beugnon, and F. Gerbier, "Identifying topological edge states in 2D optical lattices using light scattering," *The European Physical Journal Special Topics*, vol. 217, no. 1, pp. 135–152, 2013.

[4] M. Greiner, C. A. Regal, J. T. Stewart, and D. S. Jin, "Probing pair-correlated fermionic atoms through correlations in atom shot noise," *Physical Review Letters*, vol. 94, no. 11, Article ID 110401, 2005.

[5] L. Tarruell, D. Greif, T. Uehlinger, G. Jotzu, and T. Esslinger, "Creating, moving and merging Dirac points with a Fermi gas in a tunable honeycomb lattice," *Nature*, vol. 483, no. 7389, pp. 302–305, 2012.

[6] Z. Fu, L. Huang, Z. Meng et al., "Radio-frequency spectroscopy of a strongly interacting spin-orbit-coupled Fermi gas," *Physical Review A: Atomic, Molecular and Optical Physics*, vol. 87, no. 5, Article ID 053619, 2013.

[7] L. He and X.-G. Huang, "Nonperturbative effects on the ferromagnetic transition in repulsive Fermi gases," *Physical Review A: Atomic, Molecular and Optical Physics*, vol. 85, no. 4, Article ID 043618, 2012.

[8] L. He and X.-G. Huang, "Superfluidity and collective modes in Rashba spin-orbit coupled Fermi gases," *Annals of Physics*, vol. 337, pp. 163–207, 2013.

[9] Y. Xu, R.-L. Chu, and C. Zhang, "Anisotropic weyl fermions from the quasiparticle excitation spectrum of a 3D fulde-ferrell superfluid," *Physical Review Letters*, vol. 112, no. 13, Article ID 136402, 2014.

[10] I. I. Satija, D. C. Dakin, J. Y. Vaishnav, and C. W. Clark, "Physics of a two-dimensional electron gas with cold atoms in non-Abelian gauge potentials," *Physical Review A: Atomic, Molecular and Optical Physics*, vol. 77, no. 4, Article ID 043410, 2008.

[11] A. Bermudez, N. Goldman, A. Kubashiak, M. Lewenstein, and M. A. Martin-Delgado, "Non-abelian optical lattices: anomalous quantum hall effect and dirac fermions," *Physical Review Letters*, vol. 103, Article ID 035301, 2009.

[12] A. Bermudez, N. Goldman, A. Kubasiak, M. Lewenstein, and M. A. Martin-Delgado, "Topological phase transitions in the non-Abelian honeycomb lattice," *New Journal of Physics*, vol. 12, Article ID 033041, 2010.

[13] N. Goldman, A. Kubasiak, P. Gaspard, and M. Lewenstein, "Ultracold atomic gases in non-Abelian gauge potentials: the

case of constant Wilson loop," *Physical Review A: Atomic, Molecular and Optical Physics*, vol. 79, no. 2, Article ID 023624, 2009.

[14] A. Kubasiak, P. Massignan, and M. Lewenstein, "Topological superfluids on a lattice with non-Abelian gauge fields," *EPL (Europhysics Letters)*, vol. 92, no. 4, Article ID 46004, 2010.

[15] L. Wang and L. Fu, "Interaction-induced merging of Dirac points in non-Abelian optical lattices," *Physical Review A: Atomic, Molecular and Optical Physics*, vol. 87, no. 5, Article ID 053612, 2013.

[16] M. Iskin, "Superfluid phases of ultracold Fermi gases on a checkerboard superlattice," *Physical Review A: Atomic, Molecular and Optical Physics*, vol. 88, no. 5, Article ID 013631, 2013.

[17] M. Burrello, I. C. Fulga, E. Alba, L. Lepori, and A. Trombettoni, "Topological phase transitions driven by non-Abelian gauge potentials in optical square lattices," *Physical Review A: Atomic, Molecular and Optical Physics*, vol. 88, no. 5, Article ID 053619, 2013.

[18] Q. Sun, G.-B. Zhu, W.-M. Liu, and A.-C. Ji, "Spin-orbit coupling effects on the superfluidity of a Fermi gas in an optical lattice," *Physical Review A: Atomic, Molecular and Optical Physics*, vol. 88, no. 6, Article ID 063637, 2013.

[19] Y. Zhang and C. Zhang, "Bose-einstein condensates in spin-orbit-coupled optical lattices: flat bands and superfluidity," *Physical Review A*, vol. 87, Article ID 023611, 2013.

[20] F. Lin, C. Zhang, and V. W. Scarola, "Emergent kinetics and fractionalized charge in 1D spin-orbit coupled flatband optical lattices," *Physical Review Letters*, vol. 112, no. 11, Article ID 110404, 2014.

[21] D. Toniolo and J. Linder, "Superfluidity breakdown and multiple roton gaps in spin-orbit-coupled Bose-Einstein condensates in an optical lattice," *Physical Review A: Atomic, Molecular and Optical Physics*, vol. 89, no. 6, Article ID 061605(R), 2014.

[22] C. Qu, M. Gong, and C. Zhang, "Fulde-Ferrell-Larkin-Ovchinnikov or Majorana superfluids: the fate of fermionic cold atoms in spin-orbit-coupled optical lattices," *Physical Review A*, vol. 89, Article ID 053618, 2014.

[23] Y. Xu, C. Qu, M. Gong, and C. Zhang, "Competing superfluid orders in spin-orbit-coupled fermionic cold-atom optical lattices," *Physical Review A*, vol. 89, Article ID 013607, 2014.

[24] H. Zhai, "Degenerate quantum gases with spin-orbit coupling: a review," *Reports on Progress in Physics*, vol. 78, no. 2, Article ID 026001, 25 pages, 2015.

[25] H.-K. Tang, X. Yang, J. Sun, and H.-Q. Lin, "Berezinskii-Kosterlitz-Thoules phase transition of spin-orbit coupled Fermi gas in optical lattice," *Europhysics Letters*, vol. 107, Article ID 40003, 2014.

[26] M. Iskin, "Topological superfluids on a square optical lattice with non-Abelian gauge fields: effects of next-nearest-neighbor hopping in the BCS-BEC evolution," *Physical Review A: Atomic, Molecular and Optical Physics*, vol. 93, Article ID 013608, 2016.

[27] C. Itzykson and J. B. Zuber, *Quantum Field Theory*, McGraw-Hill, New York, NY, USA, 1980.

[28] J. Schwinger, "The theory of quantizied fields," *Physical Review A: Atomic, Molecular and Optical Physics*, vol. 82, p. 914, 1951.

[29] F. J. Dyson, "The *S* matrix in quantum electrodynamics," *Physical Review A: Atomic, Molecular and Optical Physics*, vol. 75, no. 11, pp. 1736–1755, 1949.

[30] H. A. Bethe and E. E. Salpeter, "A relativistic equation for bound-state problems," *Physical Review A: Atomic, Molecular and Optical Physics*, vol. 82, p. 309, 1951.

[31] H. A. Bethe and E. E. Salpeter, "A relativistic equation for bound-state problems," *Physical Review A: Atomic, Molecular and Optical Physics*, vol. 84, p. 1232, 1951.

[32] Z. G. Koinov, "Collective modes of an ultracold ^{6}Li-^{40}K mixture in an optical lattice," *Advances in Condensed Matter Physics*, vol. 2015, Article ID 952852, 12 pages, 2015.

23

Unprecedented Integral-Free Debye Temperature Formulas: Sample Applications to Heat Capacities of ZnSe and ZnTe

R. Pässler

Institut für Physik, Technische Universität Chemnitz, 09107 Chemnitz, Germany

Correspondence should be addressed to R. Pässler; passler@physik.tu-chemnitz.de

Academic Editor: Oleg Derzhko

Detailed analytical and numerical analyses are performed for combinations of several complementary sets of measured heat capacities, for ZnSe and ZnTe, from the liquid-helium region up to 600 K. The isochoric (harmonic) parts of heat capacities, $C_{Vh}(T)$, are described within the frame of a properly devised four-oscillator hybrid model. Additional anharmonicity-related terms are included for comprehensive numerical fittings of the isobaric heat capacities, $C_p(T)$. The contributions of Debye and non-Debye type due to the low-energy acoustical phonon sections are represented here for the first time by unprecedented, integral-free formulas. Indications for weak electronic contributions to the cryogenic heat capacities are found for both materials. A novel analytical framework has been constructed for high-accuracy evaluations of Debye function integrals via a couple of integral-free formulas, consisting of Debye's conventional low-temperature series expansion in combination with an unprecedented high-temperature series representation for reciprocal values of the Debye function. The zero-temperature limits of Debye temperatures have been detected from published low-temperature $C_p(T)$ data sets to be significantly lower than previously estimated, namely, 270 (±3) K for ZnSe and 220 (±2) K for ZnTe. The high-temperature limits of the "true" (harmonic lattice) Debye temperatures are found to be 317 K for ZnSe and 262 K for ZnTe.

1. Introduction

Since the invocation of the concept of apparently characteristic, material-specific temperature parameters, Θ, within Debye's classical paper [1] on specific heats of solids, one was concerned with a large variety of quotations of corresponding Θ_D values (so-called "Debye temperatures") within numerous thermophysical research papers, including various representative review articles [2–5] and books [6–10]. A brief inspection of the enormous Θ_D data material published hitherto, however, showed readily that it is obviously not possible to find unique Θ_D values for the individual materials in question, which should apply to duly broad temperature regions. In order to save, nevertheless, the apparently rather popular concept of Debye temperatures, it became thus already long ago a widely accepted custom to adopt the physically somewhat problematic notion of variable (T-dependent) Debye temperatures, $\Theta_D(T)$. Accordingly, Debye's heat capacity model function [1] used to be employed in practice in the general form [3, 4, 6, 10–15]

$$\frac{C_p(T)}{C_{Vh}(\infty)} \equiv \kappa(T) = \frac{3}{(x_D(T))^3} \int_0^{x_D(T)} dx \frac{x^4 e^x}{(e^x - 1)^2} \leq 1, \quad (1)$$

where the upper limits of integration are given by the ratios, $x_D(T) = \Theta_D(T)/T$, of adjustable Debye temperatures versus the respective lattice temperatures and $C_{Vh}(\infty)$ represents the familiar Dulong-Petit limiting value for the isochoric lattice heat capacity in harmonic approximation (i.e., in particular $C_{Vh}(\infty) = 6R$, for binary materials).

In the course of numerical fittings of measured (isobaric) heat capacities, $C_p(T)$, on the basis of (1), it has continually been found that, in contrast with Debye's original suggestion [1], proper simulations of such heat capacity curves can in fact be realized only by admitting rather strong T-dependencies of the material-specific Debye temperatures, $\Theta_D(T)$. This statement applies above all to the liquid-helium-hydrogen region, where the adjusted $\Theta_D(T)$ values are as a rule rapidly falling [3–5, 12–17] from their $T \rightarrow 0$ limiting levels, $\Theta_D(0)$, to certain minimum values, $\Theta_{D\min}$, which used to be located in the vicinities of $T \approx \Theta_D(0)/15$. Furthermore, towards higher temperatures, many $\Theta_D(T)$ curves show a more or less pronounced increase up to certain material-specific maxima, $\Theta_{D\max}$, the actual magnitudes of which are in many cases significantly higher than the respective $\Theta_{D\min}$ levels. Such typical nonmonotonic behaviours of $\Theta_D(T)$ curves in the cryogenic region have been found in particular for Si and Ge [18–21] as well as for a large variety of III–V [11, 12, 15, 22–30], II–VI [14, 31–38], and I–VII [16, 39, 40] materials.

Globally one can thus assess that Debye's original idea, according to which heat capacities of solids might be represented by Debye function integrals (of type (1)), whose upper limits of integration should be based on fixed, material-specific Θ_D values, is largely illusionary. Actually, not a single binary or ternary material could be found hitherto, within a wealth of thermophysical researches, for which it would have been possible to simulate in adequate way the temperature dependence of measured heat capacities, from the liquid-helium region up to the vicinity of room temperature, in terms of a unique Θ_D value. This means, among other things, that there is hardly a chance to come to appropriate quantitative descriptions of the temperature dependencies of heat capacities mainly via constructions of various, highly elaborate analytical models [41–44] for integrals of Debye's type and imputing, at the same time, only some fixed (T-independent) values for Debye temperatures (like $\Theta_D = 632$ K for GaN [43] or $\Theta_D = 920$ K for ZnO [44] and $\Theta_D = 946$ K for MgO [44]).

The basic cause of the occurrence of relatively deep minima, $\Theta_{D\min}$, in the cryogenic region had been indicated already many years ago by Schrödinger [2] to be obviously due to the onset of thermal activation of *short*-wavelength transverse acoustical (TA) phonons, which are quantum-theoretically manifested by pronounced peaks (singularities) in the respective material-specific phonon density of states (PDOS) spectra (see, e.g., the calculated $g_P(\varepsilon)$ spectra shown in [45, 46] for ZnSe and ZnTe). Such drastic deviations of physically realistic PDOS spectra from Debye's naïve (quadratic) PDOS model function [1], $g_D(\varepsilon) \propto \varepsilon^2$, are continually confirmed to be indeed the main cause of the obviously typical, nonmonotonic (non-Debye) behaviours of $\Theta_D(T)$ curves.

Of course, when a representative set of properly determined $\Theta_D(T)$ values is actually available for one or the other material in question, one can readily calculate the respective heat capacities, either via direct numerical evaluations of the integral occurring in (1) or by interpolations between discrete $\widehat{x}_D$ values quoted in some of the more or less comprehensive $C(\widehat{x}_D)$ or $C(\widehat{x}_D)/C_{Vh}(\infty)$ tables presented by various authors [6, 8, 47–49] (see also the list of presently obtained high-accuracy sample values given in Table 7).

Considerably more problematic and difficult, however, is a proper computational solution of the hitherto largely ignored *inverse* problem, namely, the reliable determination of effective Debye temperatures, $\Theta_D(T)$, on the basis of measured heat capacities, $C_p(T)$. In view of the longstanding absence of corresponding integral-free formulas, it was the custom in thermophysical investigations over a long time to resort to some of the tables quoted above and to perform thus approximate estimations of the respective x_D-ratios via interpolations between discrete $\widehat{x}_D$-ratios quoted for the couples of the two neighbouring $C_p(\widehat{T})$ (or $C_p(\widehat{T})/C_{Vh}(\infty)$) values. Such conventional interpolation procedures were still feasible as long as one used to be concerned only with relative small sets (limited to few dozens) of measured $C_p(T)$ data points. However, in contemporary literature, one is repeatedly concerned with rather large sets (of hundreds) of $C_p(T)$ data points (like the presently analyzed $C_p(T)$ data sets given in [46] for ZnSe and ZnTe). With respect to the latter such clumsy point-by-point interpolation procedures appear to be largely obsolete. A practicable computational alternative, however, requires the availability of a duly elaborate analytical framework for the *inverse* relationship, which should hence enable performing straightforward evaluations of the actual $x_D(T) = \Theta_D(T)/T$-ratios directly on the basis of given heat capacity values, $C_p(T)$. It is the main purpose of the present study to develop here (within Appendix B) such a representative analytical framework of integral-free formulas and to present corresponding sample results for Debye temperatures of ZnSe and ZnTe.

Concerning the successive development of this novel analytical apparatus for Debye temperatures, we would still like to note that some partial results in form of asymptotic (approximate) Debye temperature expressions had already been published in two preceding papers, namely, in [13] for the low-temperature region ($T < \Theta_D(T)/12$),

$$\Theta_D(T) \longrightarrow \Theta_D^{l_0}(T) = T \cdot \left(\frac{4\pi^4 C_{Vh}(\infty)}{5C_p(T)}\right)^{1/3}, \qquad (2)$$

and in [15] for the complementary range of higher temperatures ($T > \Theta_D(T)/12$),

$$\Theta_D(T) \approx \Theta_D^{h_0}(T)$$
$$= T \cdot \sqrt[2]{-35 + 5 \cdot \sqrt[2]{49 + 56\left(\frac{C_{Vh}(\infty)}{C_p(T)} - 1\right)}}. \qquad (3)$$

However, due to a lack of space, the somewhat unusual derivation procedure for this highly useful high-temperature formula (i.e., Eq. (26) in [15]) could not be included into the former paper. A duly detailed display of the corresponding analytical derivation procedure is now made up, among other things, in Appendix B.

In Section 2 we briefly sketch the basic arguments and equations of the analytical framework for theoretical calculations of isobaric heat capacities. In Section 3 we display a reformulated, *integral-free* version of the multioscillator hybrid model [14, 21, 50]. This is based mainly on the usage of novel, unprecedented analytical formulas (derived in Appendix A) for the contributions of the continuous, long-wavelength sections of transverse acoustical (TA) phonons to the resulting lattice heat capacities. In Section 4 we adopt (in analogy to [14]) a special 4-peak Einstein oscillator constellation of the hybrid model, and we use the respective analytical heat capacity expression for careful fittings of compatible sets of low- and high-temperature data available for ZnSe and ZnTe (up to 600 K). In Section 5 we perform, on the basis of a couple of precision formulas derived in Appendix B, transformations of the experimental heat capacity data in question (including the corresponding theoretical $C_p(T)$ and $C_{Vh}(T)$ curves), into the respective Debye temperature representations. The results are discussed in Section 6.

2. Basic Equations for Temperature Dependencies of Heat Capacities in Solids

Within the frame of the harmonic lattice regime, the temperature dependencies of the isochoric heat capacities, $C_{Vh}(T)$, are well known to be given by corresponding heat capacity shape functions, $\kappa_h(T)$, of the general form [4, 6, 45, 50–55]

$$\frac{C_{Vh}(T)}{C_{Vh}(\infty)} \equiv \kappa_h(T) = \int d\varepsilon g_P(\varepsilon)\left(\frac{\varepsilon/2k_BT}{\sinh(\varepsilon/2k_BT)}\right)^2 \leq 1, \tag{4}$$

where we have denoted by $\varepsilon = \hbar\omega$ the energies of the individual phonons and $g_P(\varepsilon)$ represents the material-specific PDOS spectral function [3, 4, 19, 56]. The latter is supposed within the present context to have been normalized to unity [6, 45, 52, 53, 55]; that is, $\int d\varepsilon g_P(\varepsilon) = 1$.

Within the frame of numerical analyses of experimental heat capacity data, however, one is continually concerned with a basic theoretical complication due to the inherent differences between the temperature dependencies of theoretical $C_{Vh}(T)$ functions, on the one hand, and those of measured (isobaric) heat capacities, $C_p(T)$, on the other hand. The latter are well known to be throughout somewhat *higher*, $C_p(T) > C_{Vh}(T)$, than their theoretical (isochoric) counterparts pertaining to the *harmonic* lattice regime [4, 50, 54]. The respective differences, $C_p(T) - C_{Vh}(T) > 0$, are usually found to be very small from 0 up to temperatures of order T_h, where the heat capacity amounts to about 50% of the Delong-Petit limiting value, $C_p(T_h) \cong C_{Vh}(T_h) = C_{Vh}(\infty)/2$. On the other hand, one is usually concerned with a relatively strong monotonic increase of these differences, $d(C_p(T) - C_{Vh}(T))/dT > 0$, at higher temperatures ($T > T_h$). These differences are generally ascribed to cumulative effects of lattice expansion and lattice anharmonicities [3, 11, 18, 21, 23, 39, 50, 54, 56, 57]. We have repeatedly found within a larger series of heat capacity studies [14, 15, 21, 50, 54] that, within regions from T_h up to temperatures of order $5T_h$ (at least), the rapid increase of these differences can be simulated in good approximation by a proportionality of type $C_p(T) - C_{Vh}(T) \propto T \times (C_{Vh}(T))^2 \propto T \times (\kappa_h(T))^2$ (note that the structure of the latter is analogous to the known Nernst-Lindemann formula [2, 3, 58], $C_p(T) - C_V(T) \propto T \times (C_p(T))^2$, for the differences between isobaric and isochoric heat capacities). Admitting, furthermore, that at sufficiently high temperatures (up to melting points, if necessary) some additional higher-order power terms of T may come into play, it appeared reasonable to simulate the T-dependencies of isobaric heat capacities by a duly general algebraic expression of the form [14, 15, 21, 50, 54, 59]

$$C_p(T) = C_{Vh}(\infty)\left[\kappa_h(T) + (\kappa_h(T))^2 \sum_{n=1}^{(\infty)} A_n T^n\right] + c_1 T, \tag{5}$$

where the empirical Taylor series expansion coefficients, A_n ($n = 1, 2, \ldots$), are quantifying the material-specific weights of contributions of the individual, anharmonicity-related T^n power terms [15, 21, 54, 57]. Furthermore we have still included into (5) an additional linear term, $C^{(E)}(T) = c_1 T$, which has to represent possible contributions of a degenerate electronic system [4, 6, 13, 23, 60–63] to the measured heat capacities. Such electronic contributions ("Sommerfeld terms" [4, 6, 60]) are known to be rather strong (tending to become even the dominant ones, at sufficiently low temperatures) in metals, due to large concentrations of conduction electrons. In contrast to this, such electronic contributions are as a rule rather weak in semiconductors and insulators due to the relatively low concentrations of free electrons associated with lattice imperfections (impurities). Accordingly, the actual strengths of such electronic contributions can vary considerably even between different samples of one and the same semiconductor material [4, 62]. We will find below (in Section 4) indications for some weak electronic contributions to the isobaric heat capacities measured at liquid-helium temperatures [46], for both ZnSe and ZnTe.

3. Multioscillator Hybrid Model and Its Integral-Free Representation

Physically realistic, material-specific phonon density of states (PDOS) functions, $g_P(\varepsilon)$ (like those shown, e.g., in Figure 3 of [45] or in Figure 4 of [46], for ZnSe and ZnTe), which are determining the T-dependencies of the respective harmonic heat capacity shape functions, $\kappa_h(T)$ (see (4)), are known to consist in general of numerous peaks (critical points) in combination with various continuous curve sections and/or gaps between them. Many fine details of such material-specific PDOS spectra, however, are practically not detectible via analyses of experimental heat capacity data [3, 64]. Their possible influence on thermal properties used to be strongly reduced by the thermal averaging process (in (4)), so that they are not discernable from the background of accidental experimental uncertainties (being usually of the order ±1%). Consequently, it is as a rule possible to realize rather good fittings of available $C_p(T)$ data on the basis of relatively simple theoretical model functions for PDOS spectra [64], provided that the latter are sufficiently general and flexible for reasonable reproductions of various *prominent* features, at least. This concerns above all the liquid-helium-hydrogen region [13, 64], where the lattice heat capacities are known to be represented in good approximation by Taylor series of *odd*-order terms [13, 16, 18, 23, 35, 39, 65]

$$C_p^{(L)}(T) \longrightarrow C_{Vh}(T) = c_3 T^3 + c_5 T^5 + \cdots . \tag{6}$$

This limiting $C_{Vh}(T)$ behaviour is the automatic consequence of the well-known fact that the low-energy (TA phonon) tails of PDOS spectra for three-dimensional crystals are generally given by Taylor series [13, 16, 18, 23, 35, 39, 64, 65] of exclusively *even*-order power terms, $g_P^{(\text{cont.})}(\varepsilon) = \gamma_2\varepsilon^2 + \gamma_4\varepsilon^4 + \cdots$, of the phonon energy. An incorporation of a corresponding *continuous* low-energy section into theoretical $g_P(\varepsilon)$ model functions is therefore indispensable for their actual applicability to the cryogenic region. At the same time it is also clear that such continuous low-energy tail sections are naturally ending (i.e., $g_P^{(\text{cont.})}(\varepsilon \geq \varepsilon_1) = 0$) at the positions of the first critical points [14, 21, 50, 64], ε_1, which are usually corresponding to the first peaks of short-wavelength TA phonons (note that, for materials with zinc blende structure like ZnSe and ZnTe, the energy positions of the first critical points are corresponding to those of the L-points on the first Brillouin zone boundaries, i.e., $\varepsilon_1 = \varepsilon_{\text{TA}}(L)$ [7, 32, 36, 45, 46, 66, 67]).

3.1. General Analytical Framework of Multioscillator Hybrid Models. Less critical than the general ansatz for the $0 < \varepsilon < \varepsilon_1$ region is the modelling of $g_P(\varepsilon)$ spectra for energies higher than the first critical points, $\varepsilon > \varepsilon_1$. Numerous analyses have shown that excellent simulations of $C_{Vh}(T) \propto \kappa_h(T)$ dependencies (4), from cryogenic to room temperatures, can be realized on the basis of combinations of several (macroscopic) Einstein oscillators [68], $\propto w_n\delta(\varepsilon - \varepsilon_n)(n = 1 \text{ to } n_E)$, the ordered energy positions of which, $\varepsilon_1 < \varepsilon_2 < \cdots < \varepsilon_{n_E}$, including their actual weights, w_n, are to be properly adjusted in the course of least-mean-square fitting processes for heat capacities. Thus one can take the normalized PDOS function in (4) to be generally representable by a multioscillator hybrid model function of the sufficiently general form [14, 21, 50]

$$g_P(\varepsilon) = \vartheta(\varepsilon_1 - \varepsilon)\sum_{k=1}^{2} w_{Ck}\frac{(2k+1)\varepsilon^{2k}}{\varepsilon_1^{2k+1}} + \sum_{n=1}^{n_E} w_n\delta(\varepsilon - \varepsilon_n), \tag{7}$$

where we have denoted by $\vartheta(\varepsilon_1 - \varepsilon)$ (= 1 for $\varepsilon < \varepsilon_1$ and 0 for $\varepsilon > \varepsilon_1$) the step function, which realizes the cut of the continuous part of the $g_P(\varepsilon)$ function just at the first singular point. By integrating (7) over the whole spectrum (from 0 beyond the highest Einstein oscillator), we see that the supposed normalization of the $g_P(\varepsilon)$ function to unity is equivalent to a normalization condition of the form [14, 21, 50]

$$1 = \int_0^{(\to\infty)} d\varepsilon g_P(\varepsilon) = \sum_{k=1}^{2} w_{Ck} + \sum_{n=1}^{n_E} w_n, \tag{8}$$

for the whole set of weighting factors. This condition is to be strictly fulfilled within any fitting process of experimental data on the basis of the $g_P(\varepsilon)$ model function (7).

Inserting, finally, the $g_P(\varepsilon)$ function (see (7)) into the integral (4) and denoting by $\Theta_n \equiv \varepsilon_n/k_B$ ($n = 1$ to n_E) the characteristic Einstein temperatures corresponding to the energy positions of the individual macroscopic oscillators we obtain for the normalized (harmonic) heat capacity shape function the general expression [14, 21, 50]

$$\kappa_h(T) = \sum_{k=1}^{2} w_{Ck}\kappa_{Ck}(x_1(T)) + \sum_{n=1}^{n_E} w_n\left(\frac{\Theta_n/2T}{\sinh(\Theta_n/2T)}\right)^2, \tag{9}$$

where the $\kappa_{Ck}(x_1(T))$ functions (being due to the two continuous components, $\propto \varepsilon^{2k}$, in (7)) are given in form of integrals of Debye and non-Debye type [14, 21]

$$\kappa_{Ck}(x_1(T)) = \frac{2k+1}{(x_1(T))^{2k+1}}\int_0^{x_1(T)} dx\frac{x^{2k+2}e^x}{(e^x-1)^2}. \tag{10}$$

Here we have denoted by

$$x_1(T) \equiv \frac{\Theta_1}{T} \tag{11}$$

the ratio of the characteristic phonon temperature associated with the first (lowest) special point, Θ_1, versus lattice temperature, T (in analogy with the notation used within Debye's theory, $x_D(T) \equiv \Theta_D/T$; cf. (1) and Appendix A).

Informative byproducts of numerical fittings of measured $C_p(T)$ data sets via (5), in combinations with (9) for the harmonic heat capacity shape function, $x_h(T)$ (including (10), for the $\kappa_{Ck}(x_1(T))$ components), are the *moments*, $\mu_P^{(m)}$, of material-specific PDOS spectral functions [3, 4, 16, 18, 19, 21, 50, 52, 56, 69–72], which are generally defined by

$$\mu_P^{(m)} \equiv \int d\varepsilon g_P(\varepsilon)\varepsilon^m, \quad \text{for } -3 < m \neq 0. \tag{12}$$

Inserting the $g_P(\varepsilon)$ hybrid model function (see (7)) into the integral for $\mu_P^{(m)}$ (see (12)), we obtain for these moments the model-specific expression [14, 21]

$$\mu_P^{(m)} = \left(\frac{3w_{C1}}{3+m} + \frac{5w_{C2}}{5+m}\right)\varepsilon_1^m + \sum_{n=1}^{n_E} w_n\varepsilon_n^m. \tag{13}$$

3.2. Integral-Free Formulas for the Contributions of Long-Wave Acoustical (TA) Phonons. On the basis of the hitherto used original (i.e., integral) expressions (see (10)) for the two low-phonon-energy contributions, $\kappa_{C1/C2}(x_1(T))$ (in (9)), the corresponding least-mean-square analysis procedures performed in several preceding studies [14, 21, 50] turned out to be somewhat heavy and time-consuming. This disadvantage of the former hybrid model versions [14, 21, 50] was one of the main motivations for performing here (in Appendix A) a comprehensive analytical study aiming at the construction of more practicable, integral-free formulas for the integrals (10) of Debye and non-Debye type.

Consider first the integral (10) for the low-energy component of Debye type, $\kappa_{Ck=1}(x_1(T))$ (see Figure 1). A detailed analyses (in Section A.2 of Appendix A) has shown that the corresponding $\kappa_{C1}(x_1)$ dependence (see (A.13)) can be represented in good approximation in terms of a proper combination (A.16) of truncated low- and high-temperature expansions, $\kappa_{C1l}^{n_D=1}(x_1 \geq x_{c1(D)})$ (see (A.14)) and $\kappa_{C1h}^{m_R=7}(x_1 \leq x_{c1(D)})$ (see (A.15)), assuming a switch between the two complementary branches just at their crossing point, $x_{c1(D)} = 6.5135$ (cf. the upper inset of Figure 1). This means for the low-temperature branch, $x_1(T) \equiv \Theta_1/T \geq x_{c1(D)} = 6.5135$,

$$\kappa_{C1}(x_1) \longrightarrow \kappa_{C1l}^{n_D=1}(x_1) = \frac{4\pi^4}{5x_1^3} - \frac{3x_1}{e^{x_1}-1} - 12e^{-x_1}\left(1 + \frac{3}{x_1} + \frac{6}{x_1^2} + \frac{6}{x_1^3}\right), \quad (14)$$

and for the high-temperature branch, $0 \leq x_1(T) \equiv \Theta_1/T \leq x_{c1(D)} = 6.5135$,

$$\kappa_{C1}(x_1) \longrightarrow \kappa_{C1h}^{m_R=7}(x_1) = \left[1 + \sum_{m=1}^{m_R=7} c_{2m}^{DR} x_1^{2m}\right]^{-1} \quad (15)$$

(the respective expansion coefficients of which, c_{2m}^{DR}, are listed in Table 6).

Concerning the integral (10) for the second low-energy component, $\kappa_{Ck=2}(x_1(T))$ (see Figure 1), we have found within the frame of a detailed study (in Section A.2, of Appendix A) that the corresponding $\kappa_{C2}(x_1)$ dependence of non-Debye type (see (A.17)) can be represented in good approximation in terms of the combination (A.24) of the truncated low- and high-temperature expansions, $\kappa_{C2l}^{n_N=1}(x_1 \geq x_{c2(N)})$ (see (A.18)) and $\kappa_{C2h}^{m_R=6}(x_1 \leq x_{c2(N)})$ (see (A.23)), assuming again a switch between the two complementary branches at the respective crossing point, $x_{c2(N)} = 7.0962$ (cf. the lower inset of Figure 1). This means for the low-temperature branch, $x_1(T) \equiv \Theta_1/T \geq x_{c2(N)} = 7.0962$, explicitly

$$\kappa_{C2}(x_1) \longrightarrow \kappa_{C2l}^{n_N=1} = \frac{80\pi^6}{21x_1^5} - \frac{5x_1}{e^{x_1}-1} - 30e^{-x_1}\left(1 + \frac{5}{x_1} + \frac{20}{x_1^2} + \frac{60}{x_1^3} + \frac{120}{x_1^4} + \frac{120}{x_1^5}\right), \quad (16)$$

and for the high-temperature branch, $0 \leq x_1(T) \equiv \Theta_1/T \leq x_{c2(N)} = 7.0962$,

$$\kappa_{C2}(x_1) \longrightarrow \kappa_{C2h}^{m_R=6}(x_1) = \left[1 + \sum_{m=1}^{m_R=6} c_{2m}^{NR} x_1^{2m}\right]^{-1} \quad (17)$$

(the respective expansion coefficients of which, c_{2m}^{NR}, are listed in Table 6).

The large scale representations of the $\kappa_{C1}(x_1)$ and $\kappa_{C2}(x_1)$ curves in Figure 1 are suggesting apparently perfect coincidences between the exact (solid) curves and the respective approximate curve sections. At the same time we see, from the insets of Figure 1, that the approaches to the exact $\kappa_{Ck}(x_1)$ functions are really excellent only in the regions $x_1 < 5$ and $x_1 > 9$, whereas one is concerned in the vicinities of the crossing points, that is, at $x_1 \approx x_{c1(D)} = 6.5135$, for the Debye-like component, and at $x_1 \approx x_{c2(N)} = 7.0962$, for the non-Debye component, with certain overestimations of the exact $\kappa_{Ck}(x_1)$ values by the approximate formulas (up to a maximum order of 0.008%, for $k = 1$, or of 0.015%, for $k = 2$). Yet, such computational inaccuracies are by 1 to 2 orders of magnitude smaller than the typical uncertainties of measured heat capacities, $C_p(T)$. Thus, these computational inaccuracies are usually negligible within the frame of practical applications.

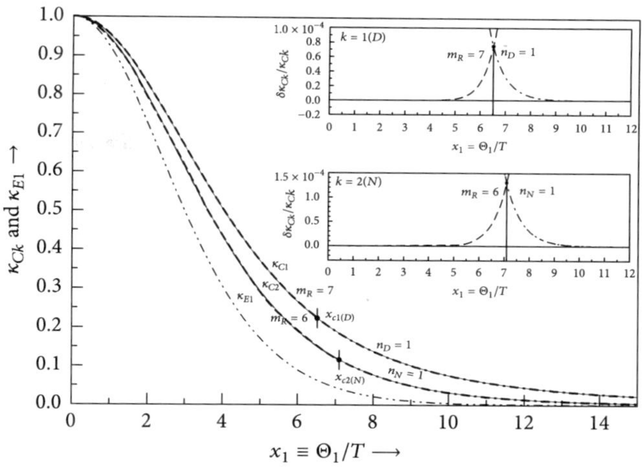

FIGURE 1: Graphical representation of the exact x_1-dependencies (—) of the normalized low-phonon-energy contributions to lattice heat capacities of Debye and non-Debye type, $\kappa_{C1}(x_1)$ and $\kappa_{C2}(x_1)$ (see (10)), in comparison with the respective low-temperature approximations (·–·–·–·), due to the functions $\kappa_{C1l}^{n_D=1}(x_1 \geq x_{c1(D)})$ (see (14)) or $\kappa_{C2l}^{n_N=1}(x_1 \geq x_{c2(D)})$ (see (16)), the associated high-temperature approximations (– – – – –), due to the functions $\kappa_{C1h}^{m_R=7}(x_1 \leq x_{c1(D)})$ (see (15)) or $\kappa_{C2h}^{m_R=6}(x_1 \leq x_{c2(D)})$ (see (17)), and the normalized heat capacity contribution of the first Einstein oscillator (–···–), $\kappa_{E1}(x_1) = ((x_1/2)/\sinh(x_1/2))^2$.

3.3. Connection with Conventional Cryogenic Heat Capacity Parameters. At sufficiently low (liquid-helium) temperatures, where $x_1(T) \equiv \Theta_1/T > 12$, the first terms ($\propto x_1^{-3}$ or $\propto x_1^{-5}$) in the respective low-temperature expressions (see (14) and (16)) are strongly dominating. Thus the Debye-like and non-Debye components, $\kappa_{Ck}(x_1(T))$ ((see (10) for $k = 1$ and 2, resp.), tend here to the known $\propto T^3$ and $\propto T^5$ asymptotes [14, 21]

$$\kappa_{C1}(x_1(T)) \longrightarrow \left(\frac{4\pi^4}{5}\right) \cdot \left(\frac{T}{\Theta_1}\right)^3, \quad \kappa_{C2}(x_1(T)) \longrightarrow \left(\frac{80\pi^6}{21}\right) \cdot \left(\frac{T}{\Theta_1}\right)^5. \quad (18)$$

The asymptote of the $k = 1$ component is of Debye type [1], whereas the asymptote of the $k = 2$ component is responsible for the onset of the typical non-Debye behaviour. The latter is frequently emerging already at very low temperatures (of order $T \approx \Theta_D(0)/50$), where the Einstein oscillator terms in (9) for the $\kappa_h(T)$ dependence are practically negligible (owing to their plateau behaviour, $(\Theta_n/(2T\sinh(\Theta_n/2T)))^2 \to (\Theta_n/T)^2\exp(-\Theta_n/T) \to 0$; cf. also the $\kappa_{E1}(x_1)$ curve in Figure 1). In this way we satisfy ourselves that the low-temperature dependence of the harmonic lattice heat capacity, $C_{Vh}(T) = C_{Vh}(\infty)\kappa_h(T)$ (see (4)), reduces, as expected, to a couple of two odd-order terms [13, 16, 18, 23, 35, 39, 65], $C_{Vh}(T) \to c_3T^3 + c_5T^5$ (see (6)). The respective expansion coefficients, c_3 and c_5, are readily seen to be given in terms of the weighting factors w_{C1} and w_{C2}, the characteristic phonon temperature associated with the first Einstein peak, $\Theta_1 \equiv \varepsilon_1/k_B$, and the classical Dulong-Petit heat capacity value, $C_{Vh}(\infty)$, by the relations

$$c_3 = w_{C1}\frac{4\pi^4}{5\Theta_1^3}C_{Vh}(\infty),$$
$$c_5 = w_{C2}\frac{80\pi^6}{27\Theta_1^5}C_{Vh}(\infty). \quad (19)$$

Comparing, finally, the analytical expression for the limiting (cubic) asymptote implied by the present formalism, $C_{Vh}(T) \to c_3T^3 = w_{C1}(4\pi^4/5)(T/\Theta_1)^3C_{Vh}(\infty)$ (due to (19)), with its conventional predecessor, $C_{Vh}(T) \to (4\pi^4/5)(T/\Theta_D(0))^3C_{Vh}(\infty)$ (due to Debye's theory [1–3, 8, 13]), we see that these two $C_{Vh}(T \to 0)$ expressions are coincident when the weighting factor for the Debye-like component, w_{C1}, is just given by the cube [14, 21] of the ratio between the characteristic phonon temperature associated with the first Einstein peak, Θ_1, and the $T \to 0$ limit value, $\Theta_D(0)$ of the conventional Debye temperature; that is,

$$w_{C1} = \left(\frac{\Theta_1}{\Theta_D(0)}\right)^3. \quad (20)$$

4. Sample Analyses of Heat Capacity Data Sets for ZnS and ZnTe

For the sake of numerical analyses of given experimental heat capacity data within the frame of a multioscillator hybrid model like that displayed above, it is still necessary to specify the material-specific constellation (in particular the total number, n_E) of Einstein oscillators to be taken into consideration. A relatively simple version in form of a *three-oscillator-based* hybrid model [21] had been used in [50] for a large variety of III–V and II–VI materials with partly different crystal structures. Within the present (more detailed) study, for ZnSe and ZnTe, we shall consider a structure-specific constellation of *four* discrete oscillators (in analogy to [14]), which offers a physically adequate association with the contributions of the 6 phonon branches in zinc blende structure materials (consisting of 2 TA branches and 1 LA branch, for the lower parts, and 2 TO branches and 1 LO branch, for the upper parts, of the respective PDOS spectra [7, 32, 36, 45, 46, 66, 67]). Let us briefly sketch the basic arguments in favour of the corresponding four-oscillator model [14].

4.1. Four-Oscillator Constellation Chosen for ZnSe and ZnTe. In the course of numerous analyses of experimental $C_p(T)$ data sets by different versions of the multioscillator hybrid model [14, 21, 50] we have repeatedly made the experience that a necessary condition for the contributions of two oscillator peaks to be actually discernible, in the course of least-mean-square fitting processes, consists in a sufficiently large distance (of more than 10%, at least) between their respective (ε_{n+1} versus ε_n) energy positions. Considering under this aspect the known positions of the dominating optical (ε_{LO} versus ε_{TO}) phonon peaks [7, 32, 36, 45, 46, 66, 67] in ZnSe and ZnTe we see that their relative distances are markedly smaller than 10%, so that their individual contributions to the measured $C_p(T)$ dependencies are practically indiscernible. Consequently, the contributions of the three optical phonon branches can be lumped together, being thus representable within the four-oscillator hybrid model [14] by a *single* oscillator with the highest energy, $\varepsilon_4 \equiv k_B\Theta_4 = k_B\Theta_{TO\&LO}$, and a fixed weight of

$$w_4 = w_{TO\&LO} = \frac{1}{2}. \quad (21)$$

Qualitatively quite different is the state of affairs for the three acoustical phonon branches. From the theoretically calculated PDOS spectra [7, 32, 36, 45, 46, 66, 67] one can see that the distances of the dominating short-wave LA peaks from the prominent TA peaks (as well as from LO/TO peaks) are larger than 30%. The contributions of the dominating LA peaks, $\varepsilon_3 \equiv k_B\Theta_3 = k_B\Theta_{LA}$ (= centres of gravity of the LA branches), are thus well discernible from the contributions of TA phonons as well as from those of the LO/TO phonons. This corresponds in reasonable approximation to a fixed weight of

$$w_3 = w_{LA} = \frac{1}{6}, \quad (22)$$

for the ε_3-oscillator. Finally, with respect to the 2 TA phonon branches, we observe that the energy differences, for example, between the special points $\varepsilon_{TA}(L)$ and $\varepsilon_{TA}(X)$, are larger than 20% [7, 32, 36, 45, 46, 66, 67], so that their individual contributions have good chances to be still discernible within the frame of duly careful numerical analysis procedures. This observation suggests representing the whole variety of short-wave TA phonons by two discrete model oscillators, $\varepsilon_n \equiv k_B\Theta_n = k_B\Theta_{TAn}$ ($n = 1$ and 2). Concerning their actual weights, w_1 and w_2, we observe that, owing to the preceding fixations of the weights for the LA and LO/TO oscillators (i.e., $w_4 = 1/2$ (21) and $w_3 = 1/6$ (22)), the global normalization condition (8) for the weights of all contributions associated with the two TA phonon branches (i.e., w_1 and w_2, in combination with w_{C1} and w_{C2}) reduces, consequently, to a normalization condition of the TA phonon-related weighting factors of the special form [14]

$$w_{C1} + w_{C2} + w_1 + w_2 = w_{TA} = \frac{1}{3}. \quad (23)$$

Accordingly, only 3 of these 4 weighting factors are actually playing the role of *in*dependent fitting parameters;

that is, the magnitude of one of them is automatically fixed by the set of the three other ones (e.g., $w_{C2} = 1/3 - w_{C1} - w_1 - w_2$). In summary, within this special 4-oscillator hybrid model [14], which turns out to be obviously well suited for binary materials with zinc blende structure, the normalized heat capacity shape function, $\kappa_h(T)$ (9), reduces to a parameterized function of just 7 independent parameters (consisting of the 4 discrete phonon temperatures Θ_n, $n = 1$ to 4, in combination with the three weighting factors w_1, w_2, and w_{C1}).

Furthermore we would still like to observe that, in many cases, one knows at least approximate values for the $T \to 0$ limiting values of Debye temperatures, $\Theta_D(0)$ (either as results of former analyses of cryogenic heat capacity data [13, 34] or from independent estimations on the basis of elastic constants [35, 73–77]). This preliminary knowledge allows us, if necessary, to substitute the weight, w_{C1}, of the Debye-like component, $\kappa_{C1}(x_1(T))$, by $w_{C1} \to (\Theta_1/\Theta_D(0))^3$ (according to (20)).

4.2. Additional Empirical Parameters Involved by the Isobaric Heat Capacities. It has already been pointed out above (in Section 2) that, for numerical simulations of measured isobaric heat capacity data, $C_p(T)$, by means of (5), it is generally necessary to envisage the inclusion of further empirical parameters (like A_n, $n = 1, 2, \ldots$, and/or c_1) into the fitting process. Such additional parameters have been introduced above for a quantification of the deviations, $C_p(T) - C_{Vh}(T) > 0$, of measured $C_p(T)$ values from the theoretical (harmonic) lattice heat capacities, $C_{Vh}(T) \propto \kappa_h(T)$ (see (4)), due to lattice anharmonicities and possible electronic contributions. The actual inclusion (or exclusion) of one or the other additional parameter depends, of course, on the concrete constellation of the whole set of experimental $C_p(T)$ data in consideration.

The crucial experimental basis of the present heat capacity analyses, for ZnSe and ZnTe, was provided by Kremer et al. [46] in form of unusually comprehensive and informative sets of $C_p^{\text{low}}(T)$ data points comprising the regions from 2 K up to 277 K, for natural ZnSe, and up to about 150 K, for natural ZnTe (see the sets of $C_p^{\text{low}}(T)$ and $C_p^{\text{low}}(T)/T^3$ data points [46] represented by empty circles in Figures 2 and 3, resp., and cf. Figure 8 in [46]). Yet, particularly from the inset of Figure 2 we see that the sequence of the respective $\rho_p(T) \equiv C_p(T)/T^3$ data points (○) for ZnSe are indicative of some *non*monotonic (local minimum) behaviour in the liquid-helium region. This is at variance in particular to earlier $C_p^{\text{low}}(T)$ data (+) given for the interval 1.7 K to 24.8 K by Birch (see Table 1 of [34]), which showed an exclusively monotonic behaviour of the sequence of the respective $\rho_p(T) \equiv C_p(T)/T^3$ data points (+) in the same region.

The more or less pronounced nonmonotonic behaviours of Kremer's $\rho_p(T)$ data points for natural ZnSe (Figure 2) and ZnTe (Figure 3), in the vicinities of 3.7 K or 2.6 K, respectively, can be ascribed to relatively weak (nevertheless not completely negligible) contributions of degenerate electronic systems to the measured $C_p^{\text{low}}(T)$ dependencies (in analogy to similar observations reported for semiconductor materials like Si [61], Ge [62], several III–V materials [23], and $FeGa_3$

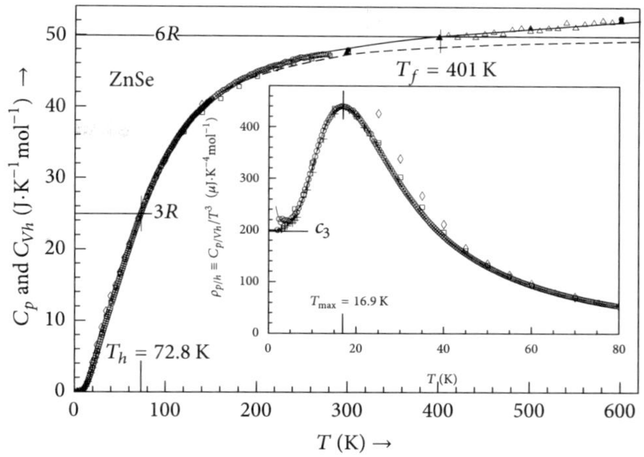

FIGURE 2: Fitting of the comprehensive set of $C_p^{\text{low}}(T)$ data (○) provided for ZnSe by Kremer et al. [46] in combination with a series of directly measured $C_p^{\text{high}}(T)$ values (△) given for 400 K ≤ T ≤ 600 K by Pashinkin and Malkova (Table 1 in [78]). For comparisons with other data available for ZnSe, we have also visualized the low-temperature data points given by Irwin and Lacombe [33] (◊), Birch [34] (+), and Sirota et al. [79] (□), including the smoothed high-temperature data points quoted by Gadzhiev et al. [80] (●) and Pashinkin and Malkova [78] (▲) [$C_p(T)$ curve (—; see (5)) and $C_{Vh}(T)$ curve (- - - - -; see (4), in combination with (9) and (14) to (17)].

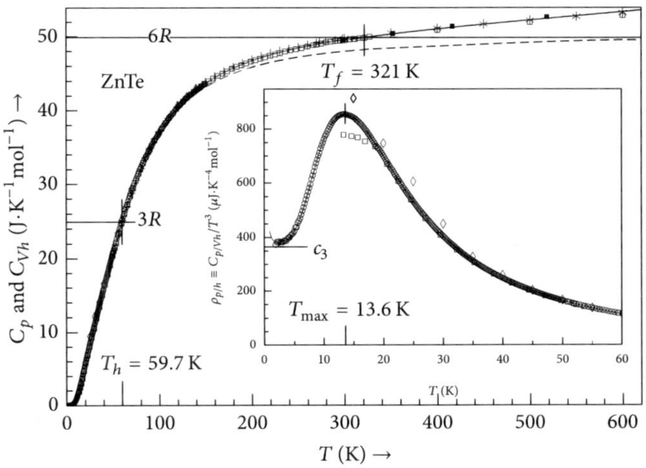

FIGURE 3: Fitting of the set of $C_p^{\text{low}}(T)$ data (○) provided for ZnTe by Kremer et al. [46] in combination with the upper section (102 K < T < 327 K) of the $C_p^{\text{low}}(T)$ data set given by Gavrichev et al. [38] (□), a set of approximate $C_p^{\text{high}}(T)$ data points due to Gadzhiev et al. [80] (■) (which we have redigitalized from their Figure 1), and a series of equidistant (smoothed) $C_p^{\text{high}}(T)$ data points (∗) available from the SGTE data review [81]. For comparisons with other data available for ZnTe we have also visualized the low-temperature data points given by Demidenko and Maltsev [31] (+) and Irwin and Lacombe [33] (◊), including the smoothed high-temperature data points quoted by Gavrichev et al. [82] (□) [$C_p(T)$ curve (—; see (5)) and $C_{Vh}(T)$ curve (- - - - -; see (4), in combination with (9) and (14) to (17)].

Table 1: Sets of the adjusted model-specific parameter values, which have been obtained (for properly fixed $\Theta_D(0)$ values) via combined least-mean-square fittings of the presently considered sets of low- and high-temperature isobaric heat capacity data, $C_p(T)$, for ZnSe (Figure 2) and for ZnTe (Figure 3), using (5) in combination with (9) and (14) to (17). In the lower part of Table 1 we have still quoted a series of characteristic, material-specific quantities (cf. the text), the actual values of which are resulting in unambiguous way from the fitted parameter values listed previously.

Material	ZnSe	ZnTe
T-ranges (K)	2 to 600	2 to 600
$\Theta_D^{(\text{fixed})}(0)$ (K)	**270**	**220**
$\Theta_{1(\text{TA})}$ (K)	78.32	61.93
$\Theta_{2(\text{TA})}$ (K)	109.4	88.01
$\Theta_{3(\text{LA})}$ (K)	213.0	167.1
$\Theta_{4(\text{TO\&LO})}$ (K)	314.4	263.4
w_1	0.0632	0.0698
w_2	0.2349	0.2331
A_1 (K^{-1})	0.539×10^{-4}	0.751×10^{-4}
A_2 (K^{-2})	0.600×10^{-7}	1.011×10^{-7}
c_1 ($J{\cdot}K^{-2}mol^{-1}$)	0.120×10^{-3}	0.045×10^{-3}
$w_{C1} = (\Theta_{1(\text{TA})}/\Theta_D(0))^3$	0.0244	0.0223
$w_{C2} = 1/3 - w_{C1} - w_1 - w_2$	0.0108	0.0081
c_3 ($J{\cdot}K^{-4}mol^{-1}$)	0.1975×10^{-3}	0.3651×10^{-3}
c_5 ($J{\cdot}K^{-6}mol^{-1}$)	0.667×10^{-6}	1.626×10^{-6}
$\mu_P^{(1)}$ (meV)	19.43	16.05
$\mu_P^{(2)}$ ($(\text{meV})^2$)	448.2	308.0
Θ_P (K)	**225.5**	**185.8**
Δ_P	0.43	0.45
$\Theta_{Dh}(\infty)$ (K)	**317.1**	**262.9**
$C_p(T_r)$ (J/K)	48.19	49.48
$S_p(T_r)$ (J/K)	72.12	82.12
$H_p(T_r) - H_p(0)$ (kJ)	10.188	10.980

[63]). Such electronic contributions are analytically represented within the present context by the respective linear (electronic) term, $C^{(E)}(T) = c_1 T$, in (5).

The subsequent least-mean-square fitting processes are thus to be performed on the basis of (5) in combination with (9), (14) to (17), and (20). For present purposes we shall limit the sets of high-temperature $C_p(T)$ data, which are to be actually included into the fitting processes, to temperatures up to 600 K (for both materials under study). Owing to these limitations we found that couples of just the first *two* anharmonicity-related expansion coefficients, A_1 and A_2 (in (5)), are sufficient for good simulations of the respective high-temperature curve sections. The present least-mean-square fitting processes are thus involving simultaneous adjustments of the (more or less strongly correlated) magnitudes of altogether 10 model-specific parameters.

4.3. Combined Fittings of Low- and High-Temperature Data and Results. Concerning the case of ZnSe (Figure 2) we have performed a simultaneous fit of the comprehensive set of $C_p^{\text{low}}(T)$ data (○) given by Kremer et al. [46] in combination with a series of directly measured $C_p^{\text{high}}(T)$ values (△) given for the region $400\,\text{K} \le T \le 600\,\text{K}$ by Pashinkin and Malkova (cf. Table 1 in [78]). The resulting total set of adjusted parameters is listed in Table 1. For comparisons of the presently analyzed data with some former ones given by various authors we have still shown in Figure 2 the fine $C_p^{\text{low}}(T < 25\,\text{K})$ data points due to Birch [34] (+), the more or less strongly deviating $C_p^{\text{low}}(T)$ data points due to Adachi et al. [10, 33] (◊) and Sirota et al. [79] (□), two $C_p^{\text{high}}(T)$ data points given (for $T = 300$ and 600 K) by Gadzhiev et al. [80] (●), and several smoothed $C_p^{\text{high}}(T)$ values estimated by Pashinkin and Malkova [78] (▲).

For the case of ZnTe (Figure 3), the present fit comprises the set of $C_p^{\text{low}}(T)$ data (○) given by Kremer et al. [46] in combination with the upper section ($102\,\text{K} < T < 327\,\text{K}$) of the set of $C_p^{\text{low}}(T)$ data points (□) given by Gavrichev et al. (in Table 1 of [38]), including a set of approximate $C_p^{\text{high}}(T)$ data points (■) due to Gadzhiev et al. (which we have redigitalized from Figure 1 shown in [80]), and a series of equidistant (smoothed) $C_p^{\text{high}}(T)$ data points (∗) available from the SGTE data review [81]. The resulting total set of adjusted parameters is listed in Table 1. For comparisons of the presently analyzed data with some former ones given by various authors we have still shown in Figure 3 the $C_p^{\text{low}}(T)$ data points due to Adachi et al. [10, 33] (◊), Demidenko and Maltsev [31] (+), and the compatible sections (up to 600 K) of a series of smoothed $C_p^{\text{high}}(T)$ values (□) quoted by Gavrichev et al. [82].

The fitted (continuous) $C_p(T)$ and $C_{Vh}(T)$ dependencies, which are implied by the material-specific sets of fitted parameter values (listed for ZnSe and ZnTe in the upper part of Table 1), are represented by solid and dashed curves, respectively, in Figures 2 and 3. In the insets of these figures we have represented the respective magnitudes of $\rho_{p/h}(T) \equiv C_{p/Vh}(T)/T^3$ ratios. The series of empty circles [46] as well as the corresponding solid curves, $\rho_p(T)$, show pronounced non-Debye behaviours (maxima) in the vicinities of $T_{\max} = 16.9\,\text{K}$ (for ZnSe) and $T_{\max} = 13.6\,\text{K}$ (for ZnTe). The c_3 levels indicate the estimated $T \to 0$ limiting values, $c_3 = \rho_h(T \to 0)$, of the $\rho_h(T) \equiv C_{Vh}(T)/T^3$ curves, which are due exclusively to the *lattice* contributions to the measured heat capacities; that is, $C_{Vh}(T) \cong C_p(T) - c_1 T$ (within the liquid-helium-hydrogen region).

In the lower part of Table 1 we have still quoted a series of characteristic, material-specific quantities, the actual values of which are resulting in unambiguous way from the fitted parameter sets. This concerns in particular the associated magnitudes (due to (19)) of the coefficients c_3 and c_5 (referring to the conventional cryogenic odd-order series expansion (6)), the first- and second-order moments of the PDOS model function, $\mu_P^{(1)}$ and $\mu_P^{(2)}$ (due to (13)), the respective average phonon temperatures [14, 21, 50], $\Theta_P \equiv \mu_P^{(1)}/k_B$, the dispersion coefficients [14, 21, 50], $\Delta_P \equiv \sqrt[2]{\mu_P^{(2)}/(\mu_P^{(1)})^2 - 1}$, and the high-temperature ($T \to \infty$) limiting values of the true

Debye temperatures [14, 21, 50], $\Theta_{Dh}(\infty) \equiv \sqrt[2]{(5/3)\mu_P^{(2)}}/k_B$ (cf. (42) below). Furthermore, with respect to the frequently considered (thermochemical) reference temperature, $T_r \equiv 298.15$ K, we have still quoted the fitted (smoothed) isobaric heat capacity values, $C_p(T_r)$ [14, 50], and the respective entropy values and enthalpy differences defined by [14, 15, 50]

$$S_p(T_r) = \int_0^{T_r} dT \frac{C_p(T)}{T}, \qquad H_p(T_r) - H_p(0) = \int_0^{T_r} dT\, C_p(T). \tag{24}$$

We have given in Table 2, for a series of selected T-values, a list of smoothed isobaric lattice heat capacity values, $C_p^{(L)}(T) = C_p(T) - c_1 T$, including the corresponding (effective) lattice Debye temperatures, $\Theta_D^{(L)}(T)$, which have been calculated on the basis of precision formulas to be presented below (in Section 5).

5. Transformation of Heat Capacities into Debye Temperatures

Expressive visualizations of typical *non*-Debye features of the temperature dependencies of heat capacities of solids are based above all on the results of point-by-point transformations of isobaric heat capacity data, $C_p(T)$, into the respective (effective) Debye temperature values, $\Theta_D(T)$. The latter are well known from Debye's classical paper [1] to be defined in implicit way [1–4, 13–15, 56] by an integral of the form (1), where the magnitudes of the characteristic (dimensionless) variable, $x_D(T) \equiv \Theta_D(T)/T$, are representing the upper limits of integration. The original task of calculating temperature dependencies of heat capacities within the frame of Debye's theory [1] involved thus, primarily, the necessity of preparing good approximations for the dependence of Debye's model-specific heat capacity function, $C_D(T) \to C_{Vh}(\infty)\kappa_D(T)$, on $x_D(T)$. A detailed analytical study of the corresponding $\kappa_D(x_D)$ function (see (A.1)) is performed in Appendix A. Crucial results of this study are given, above all, in terms of approximate algebraic expressions for the respective low- and high-capacity dependencies, $\kappa_{Dl}(x_D > 14)$ and $\kappa_{Dh}(x_D < 12)$, which can be, fortunately, transformed in *exact* way into corresponding algebraic expressions for the respective *inverse* dependencies, $x_D^{l_0}$ $(0 < \kappa < 0.03)$ (see (B.1)) and $x_D^{h_0}$ $(0.05 < \kappa \leq 1)$ (see (B.3)). Starting from these approximate expressions for low- and high-capacity curve sections, we succeeded to construct in Appendix B an unprecedented, integral-free analytical apparatus for approximate (more or less rough) estimations up to high-precision calculations of the complete x_D $(0 < \kappa \leq 1)$ dependence. Let us display in explicit form the integral-free formulas which are actually to be used here for the transformations of the heat capacity data under study (for ZnSe and ZnTe) into the respective Debye temperatures.

5.1. Asymptotic Low- and High-Temperature $\Theta_D(T)$ Formulas. The $x_D \to \infty$ asymptote of Debye's low-temperature expansion [1, 3], $\kappa(x_D) \to (4\pi^4/5x_D^3)$ (cf. (A.4)), was already found in [13] to be of primary importance for a proper interpretation of the $\Theta_D(T)$ behaviour in the liquid-helium-hydrogen region. From the latter followed readily for the *inverse* function, $x_D(\kappa)$, an asymptotic dependence of the form $x_D(\kappa \to 0) \to x_D^{l_0}(\kappa) = \sqrt[3]{4\pi^4/5\kappa}$ (see (B.1)). The respective limiting $(T \to 0)$ dependence of the Debye temperature, $\Theta_D(T) \equiv T \cdot x_D(T)$, was thus explicitly given by the asymptotic expression [13, 15]

$$\Theta_D^{l_0}(T) = T \cdot \left(\frac{4\pi^4}{5\kappa(T)}\right)^{1/3} = T \cdot \left(\frac{4\pi^4 C_{Vh}(\infty)}{5C_p(T)}\right)^{1/3} = T \cdot \Theta_D(0)\left(\frac{c_3}{C_p(T)}\right)^{1/3} \tag{25}$$

(cf. Eq. (9) in [13]), where the last representation of the $\Theta_D^{l_0}(T)$ dependence in (25) is in accordance with the general relation $c_3 = (4/5)\pi^4 C_{Vh}(\infty)/(\Theta_D(0))^3$ (cf. Eq. (6) in [13]). However, the inherent deviations (overestimations) of approximate $\Theta_D^{l_0}(T)$ values (due to (25)) from exact $\Theta_D(T)$ values are found to be lower than 1% only for very low magnitudes of heat capacities, $0 \leq \kappa(T) = C_p(T)/C_{Vh}(\infty) < 0.03$, which are corresponding to $x_D(T) = \Theta_D(T)/T$ values higher than 14 (cf. Table 9 and Figure 9).

A good basis for establishing an appropriate algebraic formula for the qualitatively different behaviour at considerably smaller x_D values is provided by the unprecedented Taylor series expansion $x_D^{h_0}(\kappa_D)$ (see (B.3)), which we have shown here to follow in unambiguous way for the reciprocal version, $\kappa_{Dh}^{m_R=2}(x_D) \equiv (\kappa_{DRh}^{m_R=2}(x_D))^{-1}$ (see (A.11) and (B.2)), of the $\kappa_{DRh}^{m_R=2}(x_D)$ function (see (A.10)). Accordingly, the temperature dependence of the Debye temperature, $\Theta_D(T) \equiv T \cdot x_D(T)$, for moderate-to-high heat capacities, is clearly confirmed to be given by the approximate algebraic expression [15]

$$\Theta_D(T) \cong \Theta_D^{h_0}(T) = T \cdot \sqrt[2]{-35 + 5 \cdot \sqrt[2]{49 + 56\left(\frac{C_{Vh}(\infty)}{C_p(T)} - 1\right)}}. \tag{26}$$

Note that the latter formula corresponds just to Eq. (26) in [15], where it had already been taken into consideration for $\Theta_D(T)$ data discussions of various III–V materials. This $\Theta_D^{h_0}(T)$ formula is seen to provide a reasonable approximation (within possible deviations up to ±0.7%) to the exact $\Theta_D(T)$ curve throughout the range $0.05 < \kappa(T) = C_p(T)/C_{Vh}(\infty) \leq 1$ (as shown in the upper inset of Figure 9 by the relative deviations, $\delta x_D/x_D$, of the approximate $x_D^{h_0}(\kappa)$ curve from the exact $x_D(\kappa)$ dependence).

5.2. Couple of High-Precision Formulas for Effective Debye Temperatures. In view of the sample character of the numerical analyses performed within the present study for the unusually fine and comprehensive $C_p^{\text{low}}(T)$ data provided

Table 2: Smoothed values of the isobaric *lattice* heat capacities, $C_p^{(L)}(T) = C_p(T) - c_1 T$, and of the corresponding (effective) Debye temperatures, $\Theta_D^{(L)}(T)$ (due to (27) and (28)).

T (K)	ZnSe $C_p^{(L)}$ (J·K^{-1} mol^{-1})	ZnSe $\Theta_D^{(L)}$ (K)	ZnTe $C_p^{(L)}$ (J·K^{-1} mol^{-1})	ZnTe $\Theta_D^{(L)}$ (K)
2	0.0015949	269.16	0.0029434	219.43
4	0.013255	265.76	0.025043	214.98
6	0.048985	257.84	0.10158	202.19
8	0.13589	244.67	0.31540	184.80
10	0.32260	229.26	0.75959	172.32
12	0.65875	216.85	1.44623	166.75
14	1.15330	209.86	2.3390	165.39
16	1.78730	207.06	3.3764	166.53
18	2.5396	206.75	4.4832	169.36
20	3.3793	208.10	5.6015	173.32
25	5.6132	216.48	8.3098	185.43
30	7.7928	228.48	10.903	197.66
35	9.9059	240.78	13.439	208.50
40	12.002	251.94	15.931	217.65
45	14.096	261.58	18.361	225.22
50	16.176	269.74	20.704	231.40
55	18.221	276.58	22.938	236.40
60	20.210	282.27	25.045	240.41
65	22.125	287.00	27.014	243.60
70	23.952	290.91	28.841	246.12
75	25.682	294.13	30.525	248.07
80	27.309	296.78	32.071	249.57
85	28.832	298.95	33.485	250.68
90	30.251	300.72	34.775	251.47
95	31.569	302.14	35.950	251.99
100	32.790	303.28	37.021	252.28
110	34.964	304.85	38.822	252.29
120	36.819	305.68	40.429	251.69
130	38.401	305.94	41.721	250.58
140	39.755	305.73	42.807	249.01
150	40.916	305.14	43.727	247.01
160	41.917	304.19	44.512	244.57
170	42.784	302.91	45.188	241.67
180	43.540	301.31	45.774	238.29
190	44.203	299.38	46.288	234.38
200	44.787	297.12	46.740	229.88
220	45.766	291.54	47.503	218.84
240	46.522	284.38	48.122	204.41
260	47.197	275.36	48.640	185.43
280	47.735	264.13	49.083	159.77
300	48.194	250.18	49.471	122.68
350	49.100	197.64	50.283	—
400	49.792	77.81	50.961	—
450	50.363	—	51.573	—
500	50.863	—	52.153	—
550	51.321	—	52.722	—
600	51.757	—	53.293	—

by Kremer et al. [46], we have constructed in Appendix B (Section B.3) a high-precision framework for the $x_D(\kappa)$ dependence. This consists in a combination of a highly elaborate (12-parameter) low-capacity formula, $x_D^{l_{12}}$ ($\kappa \leq \kappa_c$) (see (B.8)), with a duly modified (11-parameter) high-capacity formula, $x_D^{h_{11}}$ ($\kappa \geq \kappa_c$) (see (B.9)) (where $\kappa_c = 0.26$ represents the position of the point of crossing between the two complementary $x_D(\kappa)$ branches). The respective low- and high-temperature formulas for the Debye temperature, $\Theta_D^{l/h}(T) = T \cdot x_D^{l_{12}/h_{11}}(\kappa(T))$, are thus given explicitly by the couple of analytical expressions of the form

$$\Theta_{D(h)}^{l}(T) = T \cdot \left[\left(\frac{4\pi^4}{5\kappa_{(h)}(T)} \right)^{1/3} - \left(\sum_{n=2}^{10} \frac{a_n}{\left(\kappa_{(h)}(T)\right)^{n/2}} \right)^{-1} - \exp\left(-\frac{p}{\kappa_{(h)}(T)} \right) \sum_{n=2}^{3} g_n \left(\kappa_{(h)}(T)\right)^n \right], \quad (27)$$

for the interval $0 \leq C_p(T)/C_{Vh}(\infty) = \kappa_{(h)}(T) \leq \kappa_c = 0.26$ (with expansion coefficients a_n, p, and g_n listed in the second column of Table 8), and

$$\Theta_{D(h)}^{h}(T) = T \cdot \frac{\sqrt[2]{-35 + 5 \cdot \sqrt[2]{49 + 56\left(1/\kappa_{(h)}(T) - 1\right)}}}{\left[1 + \sum_{n=2}^{12} b_n \left(1 - \kappa_{(h)}(T)\right)^n\right]}, \quad (28)$$

for the interval $0.26 = \kappa_c \leq \kappa_{(h)}(T) = C_{p(Vh)}(T)/C_{Vh}(\infty) \leq 1$ (with expansion coefficients b_n listed in the third column of Table 8). Note that we have still endowed the $\kappa(T)$ functions in (27) and (28) by an additional subscript "h" (included in brackets) in order to indicate that precisely the same formulas can also be used, alternatively, for high-accuracy calculations of the "true" Debye temperatures, $\Theta_{Dh}(T)$, which are visualizing the characteristic non-Debye features of the *harmonic* parts of lattice heat capacity curves, $C_{Vh}(T)$ (see (4); Figures 4 and 5).

By means of this couple of the high-accuracy formulas (27) and (28) we have performed, first of all, the point-by-point transformations, $C_p(T) \Rightarrow \Theta_D(T)$, for all those isobaric heat capacity data points, $C_p(T)$, with respect to which the concept of *effective* Debye temperatures is actually applicable, that is, for $C_p(T) < C_{Vh}(\infty) = 6R$ (cf. Figures 2 and 3). The respective Debye temperature values, $\Theta_D(T) > 0$, are represented for ZnSe (in Figure 4) and for ZnTe (in Figure 5) by the same symbols as the original $C_p(T)$ data points (shown in Figures 2 and 3, resp.).

Furthermore, we have transformed in the same way (i.e., via (27) or (28)) the continuous $C_p(T)$ curves (solid curves shown in Figures 2 and 3), which resulted from the preceding numerical fittings in Section 4. These continuous $\Theta_D(T) > 0$ curves (solid curves, in Figures 4 and 5) are naturally ending

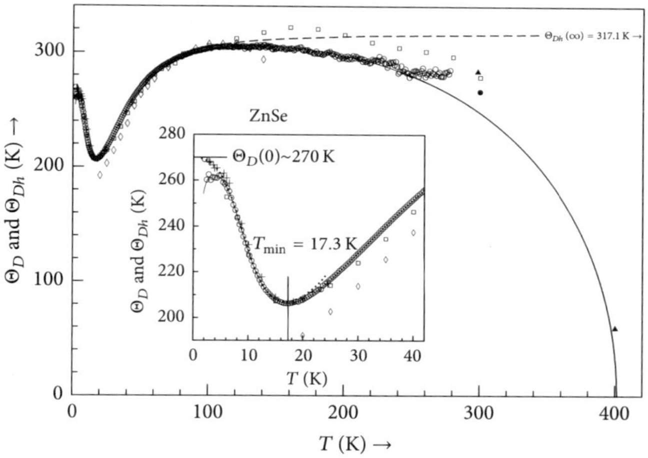

FIGURE 4: Effective Debye temperature values, $\Theta_D(T)$, as resulting for ZnSe from measured (isobaric) heat capacities, $C_p(T)$ ([46] (○), [34] (+), [10] (◊), [79] (□), [80] (●), and [78] (▲)). The continuous curve for the true Debye temperature, $\Theta_{Dh}(T)$ (- - - - -), is due to (28) in combination with (9) for the isochoric (harmonic) heat capacity shape function, $\kappa_h(T) \equiv C_{Vh}(T)/6R$ (according to the parameters listed in Table 1). The continuous curve for the effective Debye temperature, $\Theta_D(T)$ (solid curve), is due to (27) and (28) in combination with (5) and (9) for the associated isobaric (anharmonic) heat capacity shape function, $\kappa(T) \equiv C_p(T)/6R$. The dotted curve (in the inset) shows the low-temperature dependence of the true Debye temperature, $\Theta_{Dh}(T)$, according to (30).

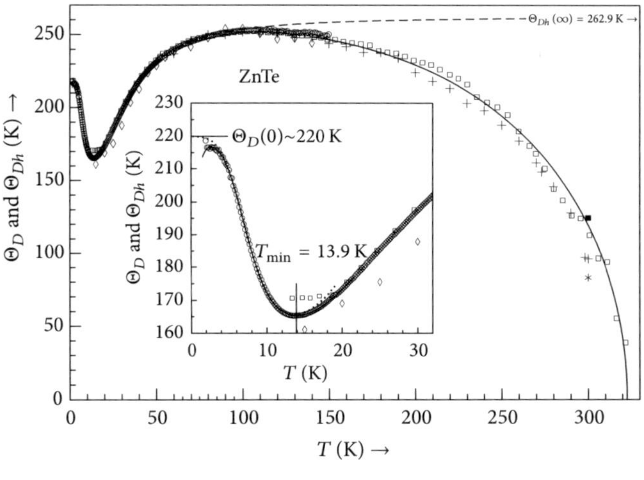

FIGURE 5: Effective Debye temperature values, $\Theta_D(T)$, as resulting for ZnTe from measured (isobaric) heat capacities, $C_p(T)$ ([46] (○), [38] (□), [31] (+), [10] (◊), [80] (■), and [81] (∗)). The continuous curve for the true Debye temperature, $\Theta_{Dh}(T)$ (- - - - -), is due to (28) in combination with (9) for the isochoric (harmonic) heat capacity shape function, $\kappa_h(T) \equiv C_{Vh}(T)/6R$ (according to the parameters listed in Table 1). The continuous curve for effective Debye temperature, $\Theta_D(T)$ (solid curve), is due to (27) and (28) in combination with (5) and (9) for the associated isobaric (anharmonic) heat capacity shape function, $\kappa(T) \equiv C_p(T)/6R$. The dotted curve (in the inset) shows the low-temperature dependence of the true Debye temperature, $\Theta_{Dh}(T)$, according to (30).

at those temperature points, T_f, where the fitted $C_p(T)$ curves (in Figure 2 or 3) are crossing just the classical Delong-Petit heat capacity value, $C_p(T_f) = C_{Vh}(\infty) = 6R$, that is, at about 401 K for ZnSe (cf. Figures 2 and 4) and about 323 K for ZnTe (cf. Figures 3 and 5).

5.3. Calculation of "True" (Harmonic Lattice) Debye Temperatures. Of considerable interest, particularly from theoretical points of view [3, 14, 21, 28, 29, 32, 36, 56, 72], is the determination of the "true" ($C_{Vh}(T)$-related) Debye temperature curves, $\Theta_{Dh}(T)$. The latter are defined with respect to the harmonic (isochoric) lattice heat capacities, $C_{Vh}(T)$ (see (4)), by [14, 21]

$$\frac{C_{Vh}(T)}{C_{Vh}(\infty)} \equiv \kappa_h(T) = \frac{3}{(x_{Dh}(T))^3}\int_0^{x_{Dh}(T)} dx \frac{x^4 e^x}{(e^x-1)^2} \le 1, \qquad (29)$$

where the dimensionless variables $x_{Dh}(T) \equiv \Theta_{Dh}(T)/T$ are representing the respective upper boundaries of integration. Comparing (29) with (A.1) we see that the analytical connection of the upper boundaries of integration, x_{Dh}, with the respective normalized (harmonic) heat capacity ratios, κ_h (in (29)), is just the same as the preceding connection between x_D and κ (in (A.1)). Consequently, all the approximation formulas derived in Appendix B, for low- and high-capacity branches of the $x_D(\kappa)$ function (cf. Figure 9), can be readily used for calculations of the $x_{Dh}(\kappa_h)$ dependence implied by (29). This means, in particular, that (27) and (28) can be simultaneously used (as already indicated by the inclusion of the subscript "h") for high-precision calculations of the dependencies of the "true" Debye temperatures, $\Theta_{Dh}(T) = T \cdot x_{Dh}(T)$, on the respective harmonic (isochoric) heat capacities, $C_{Vh}(T) = \kappa_h(T) \cdot C_{Vh}(\infty)$. The corresponding material-specific $\Theta_{Dh}(T)$ curves are represented by dashed curves (for ZnSe and ZnTe in Figures 4 and 5, resp.).

With respect to this alternative use of (27) and (28), for calculations of $\Theta_{Dh}(T)$ curves, we would still like to point out that, apart from the liquid-helium-hydrogen region (see below), appreciable differences between $\Theta_{Dh}(T)$ and $\Theta_D(T)$ values, similarly to the differences between the underlying $C_{Vh}(T)$ and $C_p(T)$ values (cf. Section 4), are encountered as a rule only in ranges of moderately low to higher temperatures, $T > T_h$, where the heat capacities in question are higher than about 50% of the classical Dulong-Petit limiting value, $\kappa_h(T) = C_{Vh}(T)/C_{Vh}(\infty) > 0.5$. This region is automatically comprised by the range of validity of (28). Thus it is mostly not necessary to involve (27) into separate calculations of $\Theta_{Dh}(T)$ curve sections.

On the other hand, significant differences between $\Theta_{Dh}(T)$ and $\Theta_D(T)$ values may also occur at very low (liquid-helium-hydrogen) temperatures, when we are (accidentally) concerned with nonnegligible contributions, $C^{(E)}(T) = c_1 T$, of a degenerate electronic system to the measured heat capacities, $C_p(T) \to C_{Vh}(T) + c_1 T$ (cf. (5)). Consequently, in order to determine the respective temperature dependence of the $\Theta_{Dh}(T)$ curve within the liquid-helium-hydrogen region, we have to insert $\kappa_h(T) = (C_p(T) - c_1(T))/C_{Vh}(\infty)$, instead of $\kappa(T) = C_p(T)/C_{Vh}(\infty)$, into the respective asymptotic (low-T) expression (see (25)). The monotonically decreasing sections, $d\Theta_{Dh}(T)/dT < 0$, of the "true" Debye temperature curves at sufficiently low temperatures, $T < T_{\min}$ (see the insets of Figures 4 and 5), are thus adequately described by a corresponding asymptotic expression of the form [13]

$$\begin{aligned}\Theta_{Dh}^{l_0}(T) &= T \cdot \left(\frac{4\pi^4}{5\kappa_h(T)}\right)^{1/3} \\ &= T \cdot \left(\frac{4\pi^4 C_{Vh}(\infty)}{5C_{Vh}(T)}\right)^{1/3} \\ &= \Theta_D(0)\left(\frac{\rho_h(0)}{\rho_h(T)}\right)^{1/3},\end{aligned} \qquad (30)$$

where $\rho_h(T) \equiv C_{Vh}(T)/T^3$ and $\rho_h(0) = c_3 = (4/5)\pi^4 C_{Vh}(\infty)/(\Theta_D(0))^3$ (see also [13] and cf. the insets of Figures 2 and 3). These asymptotic $\Theta_{Dh}^{l_0}(T)$ dependencies are represented by dotted curve sections in the insets of Figures 4 and 5.

6. Discussion

The present investigations were devoted above all to the finding of practicable solutions for longstanding computational problems which are frequently emerging within interpretations of the results of experimental heat capacity studies utilizing the somewhat troublesome concept of Debye temperatures [1, 3, 4, 6, 10–15]. The notorious computational complications are due to circumstance that the respective $\Theta_D(T)$ values are involved into the definition of the *upper limits* of integration, $x_D(T) \equiv \Theta_D(T)/T$, due to the integral expression for isobaric heat capacities, $C_p(T)$ (see (1)). Consequently, in view of the previous lack of integral-free formulas for $C_p(T)$ dependencies on $x_D(T)$ values, it was thus necessary either to calculate the individual $C_p(T)$ values via numerical integration procedures (using (1)) for all the individual $x_D(T)$ points in consideration or to determine the $C_p(T) = C_{Vh}(\infty)\kappa(T)$ values approximately, via point-by-point interpolations, using some of the available $\kappa_D(x_D)$ tables [6, 8, 47–49].

Within our detailed study of the analytical properties of the conventional $\kappa_D(x_D)$ dependence (in Appendix A) we succeeded to derive, first of all, an unprecedented, rapidly converging high-temperature Taylor series expansion, $\kappa_{Dh}^{(m_R=7)}(x_D)$ (see (A.11)). Combining the latter with Debye's conventional low-temperature expansion [1], $\kappa_{Dl}^{(n_D=15)}(x_D)$ (see (A.4)), it became possible to perform high-accuracy calculations of $\kappa_D(x_D)$ values (up to at least 10 significant figures; cf. Table 7) without involving any numerical integration procedure. The subsequent adoption of respective $\kappa_{Dh}(x_D)$ and $\kappa_{Dl}(x_D)$ formulas to the Debye-like component, $\kappa_{C1}(x_1(T))$ (see (10)), of the long-wave TA phonon contribution to the cryogenic heat capacity, including the derivation of analogous formulas for the associated non-Debye component, $\kappa_{C2}(x_1(T))$ (see (10)), enabled us to display (in Section 3.2) an integral-free version of the multioscillator hybrid model [14, 21, 50]. This has been used here (in Section 4) for numerical fittings of the $C_p(T)$ data sets under study.

The high-accuracy results obtained in Appendix A for the $\kappa(x_D)$ dependence provided us, among other things, with the numerical basis for the solution of another and practically even more important mathematical problem, namely, the construction of appropriate integral-free formulas for the *inverse* Debye function dependence, that is, the $x_D(\kappa)$ function (see Figure 9 and Table 9), which describes the functional dependence of the characteristic Debye versus lattice temperature ratios, $x_D(T) = \Theta_D(T)/T$, on the ratios, $\kappa(T) = C_p(T)/C_{Vh}(\infty)$, of given heat capacities versus Dulong-Petit classical heat capacity limit.

We have derived (in Appendix B) a series of more or less elaborate formulas for low- and high-capacity curve sections of the $x_D(\kappa)$ curve (solid curve in Figure 9). The simplest one of these approximate $x_D(\kappa)$ expressions is given in form of a junction formula, $\widetilde{x}_D(\kappa)$ (see (B.4a)/(B.4b)), the actual use of which, however, is still connected with deviations from the exact $x_D(\kappa)$ dependence up to 0.6% (cf. the upper inset of Figure 9). Reductions of residual deviations to ±0.13%, or even to ±0.0153%, could be achieved (in Sections B.1 and B.2) via least-mean-square fittings of the exact $x_D(\kappa)$ dependence, adopting either the still relatively simple (2-parameter) formula, $x_D^{l_2}$ $(0 < \kappa \le \kappa_2)$ (see (B.5a) and (B.5b)), or the refined (9-parameter) formula, $x_D^{l_9}$ $(0 < \kappa \le \kappa_9)$ (see (B.6)) (cf. the upper and middle insets of Figure 9, resp.). Furthermore, for the sake of high-precision calculations of Debye temperatures, we have still constructed (in Section B.3) an elaborate couple of formulas consisting of a significantly refined low-capacity (12-parameter) formula, $x_D^{l_{12}}$ $(0 < \kappa \le 0.26)$ (see (B.8)), in combination with a similarly accurate high-capacity (11-parameter) formula, $x_D^{h_{11}}$ $(0.26 \le \kappa \le 1)$ (see (B.9)). From the lower inset of Figure 9 we see that this couple of high-precision formulas effectuates an enormous reduction of residual deviations from the exact $x_D(\kappa)$ dependence by several orders of magnitude (down to ±0.000016%). A set of corresponding high-precision $x_D(\kappa)$ values (consisting of 7 significant figures) is given, for a series of equidistant κ-points, in Table 9.

Another important aim of the present study was an exhibition of the significant progress (in comparison with [14]) in the analytical and numerical description of the heat capacity properties of ZnSe and ZnTe, which we could achieve here owing to the availability of the unusually comprehensive and informative $C_p(T)$ data sets given by Kremer et al. [46]. A remarkable improvement for the case of ZnSe is due, above all, to the dense set of novel $C_p(T)$ data points (○) for the region $2\,\mathrm{K} \le T \le 277\,\mathrm{K}$, which can be seen from Figure 2 to differ significantly, for $T > 100\,\mathrm{K}$, from the relatively scarce (and obviously less accurate) set of former $C_p(T)$ data points (□) due to [79], which had been considered previously in [14]. The qualitative difference between the novel and the former $C_p(T)$ data sets for ZnSe is especially apparent from the visualization of the respective Debye temperature points, $\Theta_D(T)$, in Figure 4. As a consequence of this qualitative difference we note, above all, that the present numerical analyses (in Section 4) led to a $T \to \infty$ limiting value for the "true" (harmonic lattice) Debye temperature of $\Theta_{Dh}(\infty) = 317.1\,\mathrm{K}$ (cf. Table 1 and Figure 4), which is by 1.5% lower than the former estimate (of about 322 K; cf. [14])

Another significant progress is brought about by the unusually dense, informative sets of $C_p^{\mathrm{low}}(T)$ data points available from [46] for the liquid-helium-hydrogen region $(2\,\mathrm{K} \le T \le 20\,\mathrm{K})$. These novel cryogenic data were of great importance particularly for the case of ZnTe, because in experimental papers published up to 2011 no $C_p(T)$ data at all had been given for temperatures lower than 13 K (and those quoted in Table 1 of [38] for the interval $13\,\mathrm{K} < T < 18\,\mathrm{K}$ were obviously not sufficiently accurate; cf. the insets of Figures 3 and 5). This was the reason why it was practically impossible, prior the year 2012, to perform a direct determination of the *caloric* value for the $T \to 0$ limiting Debye temperature value, $\Theta_D(0)$ (cf. [13]).

Just with respect to the latter, however, we are now concerned, for ZnTe, with a significant difference (of about 4.5%) between the present estimate of $\Theta_D(0) \cong 220\,\mathrm{K}$ (see the inset of Figure 5) and a former estimate of $\Theta_D(0) \approx 230\,\mathrm{K}$ due to Kremer et al. [46]. Even a larger difference (of about 7%) is an analogous discrepancy, for ZnSe, between the present value of $\Theta_D(0) \cong 270\,\mathrm{K}$ (see the inset of Figure 4) and a former estimate of $\Theta_D(0) \approx 289\,\mathrm{K}$ due to Kremer et al. [46]. Let us examine in due detail the obvious causes of the discrepancies between $\Theta_D(0)$ values determined for ZnSe and ZnTe within the present study versus the former estimates quoted in [46].

6.1. Limiting Debye Temperatures and Electronic Heat Capacity Contributions. A provisional estimation of $\Theta_D(0)$ values can be made in simple way by viewing the sets of cryogenic Debye temperature points, $\Theta_D(T)$ (empty circles visualized in the insets of Figures 4, 5, and 6), which are exactly corresponding to the experimental $C_p(T)/T^3$ data points provided for ZnSe and ZnTe by Kremer et al. [46] (as shown above by empty circles in the insets of Figures 2 and 3). Concerning the lower part of the liquid-helium-hydrogen region $(2\,\mathrm{K} < T < 10\,\mathrm{K})$, we see readily that the respective $\Theta_D(T)$ values are throughout ranging *below* the following upper boundaries:

$$\begin{aligned} \Theta_D(T) &< 270\,\mathrm{K} \quad \text{for ZnSe}, \\ \Theta_D(T) &< 220\,\mathrm{K} \quad \text{for ZnTe}. \end{aligned} \tag{31}$$

It is remarkable that these strict limitations refer even to the lowest experimental points (which are located in the vicinity of 2 K, in both cases).

In accordance with these mere visual estimations, it followed also from our detailed analyses (in Section 4) that the limiting $(T \to 0)$ values of Debye temperatures, $\Theta_D(0)$ (cf. Table 1), are nearly coinciding (within possible uncertainties of order ±2 K) with these upper boundaries (31); that is,

$$\begin{aligned} \Theta_D(0) &\cong 270\,K\ (\pm 2\,\mathrm{K}), \quad \text{for ZnSe}, \\ \Theta_D(0) &\cong 220\,K\ (\pm 2\,\mathrm{K}), \quad \text{for ZnTe}. \end{aligned} \tag{32}$$

These results are, of course, in clear contrast to the former $\Theta_D(0)$ estimates by Kremer et al. [46] (of about 289 K for ZnSe and 230 K for ZnTe). Let us reveal the basic cause of these discrepancies.

It is well known from various earlier studies of low-temperature heat capacities of nonmetals that, within relatively small temperature regions (from 0 up to about $\Theta_D(0)/30$), the isobaric heat capacities can as a rule be well approximated by odd-order power series expansions of the conventional form [13, 16, 18, 23, 35, 39, 65]

$$C_p(T) = \sum_{n=0}^{3} c_{2n+1} T^{2n+1}, \quad (33)$$

where the $c_1 T$ term [4, 6, 13, 23, 60–63] represents a possible electronic contribution and the subsequent three odd-order terms are due to harmonic lattice contributions. Of special interest in this connection is the lowest-order lattice contribution, $c_3 T^3$, which corresponds to Debye's classical ($T \to 0$ limiting) cubic law [1, 3, 6, 8, 21], $C_{Vh}(T \to 0) \propto T^3$. The corresponding proportionality factor, c_3, is known to be connected in a simple way with the limiting Debye temperature, $\Theta_D(0)$, by the relation [4, 6, 13, 15, 46, 59, 62]

$$\Theta_D(0) = \sqrt[3]{\frac{4\pi^4 C_{Vh}(\infty)}{5c_3}}. \quad (34)$$

Consequently, within the frame of the frequently adopted $C_p(T)/T$ versus T^2 representation [4, 6, 13, 15, 46, 62, 63, 83]

$$\frac{C_p(T)}{T} = c_1 + c_3 T^2 + c_5 T^4 + c_7 T^6, \quad (35)$$

the coefficient c_3 corresponds just to the slopes of the $C_p(T)/T$ versus T^2 curves in the $T \to 0$ limit (which are represented in Figure 6 by dash-dot lines). However, these asymptotic lines used to be nearly coincident with the actual $C_p(T)/T$ versus T^2 curves only within very narrow intervals, from 0 to about $\Theta_D(0)/100$ [16, 18, 35, 83, 84] (i.e., here up to only about 3 K for ZnSe and 2 K for ZnTe). Consequently, in view of the complete lack of $C_p(T)$ data for the interval from 0 to 2 K, there remains only the possibility of determining c_3 in combination with the other expansion coefficients (i.e., c_1, c_5, and c_7) by careful fittings via (35) of the somewhat larger $C_p(T)/T$ versus T^2 curve sections extending from 2 K up to about $\Theta_D(0)/30$ (see the solid curves in Figure 6). The parameter values obtained in this way are listed in Table 3. Viewing the respective $\Theta_D(0)$ values (listed in the last line of Table 3) we satisfy ourselves that, within the existing uncertainty of ±2 K (as indicated above, in (32)), they are in accordance with the results of the analyses performed in Section 4 (see Table 1 and cf. Figures 4 and 5).

On the background of the curves shown in Figure 6 one can now easily see why the rougher simulations of $C_p(T)/T$ data (in [46]) via a shortened series expansion of type

$$\frac{C_p(T)}{T} \cong c_3 T^2 + c_5 T^4, \quad (36)$$

for T up to 10 K, were a priori *in*capable of giving adequate results. We see that a shortened $C_p(T)/T$ expansion of type (36) (due to taking $c_7 = 0$, in (35)) corresponds to the dash-double-point curves (shown in Figure 6), which begin to deviate significantly from the exact (solid) curves already in the vicinity of 6 K, for ZnSe, or 4 K, for ZnTe (i.e., at temperatures of order $\Theta_D(0)/50$). Thus one comes unavoidably to inadequate parameter values when a shortened (two-parameter) series expansion like (36) is carelessly applied to a markedly larger interval (up to 10 K [46]). In accordance with this observation we have still verified that, if necessary, even by means of a shortened series expansion like (36) it is, in principle, possible to detect physically reasonable $\Theta_D(0)$ values (closely approaching to those quoted in Table 3), provided the fitting ranges had been strictly limited to $T <$ 6 K, for ZnSe, and $T <$ 4 K, for ZnTe.

TABLE 3: Conventional Taylor series expansion coefficients determined by least-mean square fittings of cryogenic $C_p(T)/T$ data [46] for ZnSe and ZnTe (see Figure 6) via (35). The respective $\Theta_D(0)$ values are resulting from fitted c_3 values via (34).

	ZnSe	ZnTe
Intervals (K)	2.0 to 8.7	1.9 to 6.4
c_1 (mJ·K^{-2} mol^{-1})	0.1339	0.0409
c_3 (mJ·K^{-4} mol^{-1})	0.1937	0.3686
c_5 (mJ·K^{-6} mol^{-1})	0.602×10^{-3}	0.537×10^{-3}
c_7 (mJ·K^{-8} mol^{-1})	0.815×10^{-5}	0.638×10^{-4}
$\Theta_D(0)$ (K)	271.7	219.3

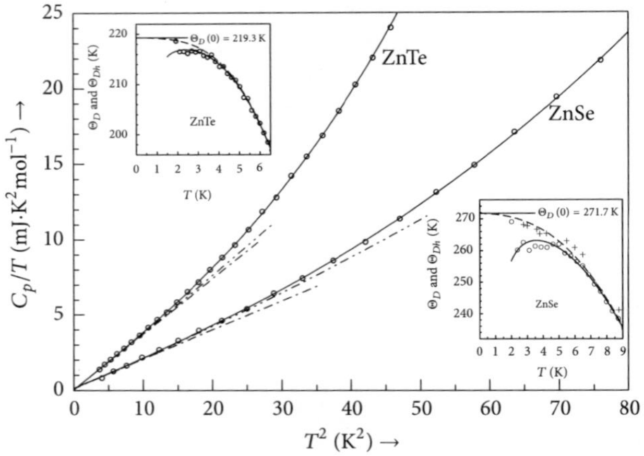

FIGURE 6: Fittings via (35) of the $C_p(T)/T$ data points (○) corresponding to the cryogenic $C_p(T)$ data provided by Kremer et al. [46] for ZnSe and ZnTe [—, least-mean square fittings from 2 K to 8.7 K or to 6.4 K, resp.; ··–··–··–··, tail behaviour, for $c_7 \to 0$; ·–·–·–·, $T \to 0$ asymptote, for $c_7 = 0$ and $c_5 = 0$]. The insets show the effective $\Theta_D(T)$ values (○) following directly (via (25)) from Kremer's cryogenic $C_p(T)$ data [46] for ZnSe and ZnTe. For comparison we have still included the $\Theta_D(T)$ values (+) following from former $C_p(T)$ data quoted for ZnSe by Birch [34], which show no indication for an appreciable electronic contribution. The continuous curves in the insets are resulting from (25), implying the cryogenic $C_p(T)$ series expansion (see (33)), with fitted Taylor series expansion coefficients listed in Table 3. [—, $\Theta_D(T)$ curve, for $c_1 > 0$; - - - - -, $\Theta_{Dh}(T)$ curve, for $c_1 \to 0$.]

In summary one can assess that the obviously erroneous $\Theta_D(0)$ values quoted in [46] are the inevitable consequence of a careless application of an unnecessarily shortened series expansion (of type (36)) to disproportionately extended temperature intervals (up to 10 K [46]). At the same time we have found that, even within the frame of careful applications of the full-blown low-temperature series expansion (of type (35)), one is still concerned with residual uncertainties (of about ±1%) for estimated $\Theta_D(0)$ values. These moderate uncertainties are obviously due to certain contributions of degenerate electronic systems to the measured heat capacities, $C_p(T)$, which found their numerical expression in nonvanishing magnitudes for the coefficients c_1 (cf. Tables 1 and 3). Concerning the actual magnitudes of the latter we see, among other things, that the respective c_1 value for ZnSe is by a factor of about 3 higher than its counterpart for ZnTe. Accordingly, the electronic contribution for ZnSe finds its graphical representations in form of a sufficiently well discernible nonmonotonic behaviour in the vicinity of 3.7 K, whereas the indications for a similar behaviour in the case of ZnTe, in the vicinity of 2.6 K, are markedly weaker (see the more or less pronounced local minimum behaviours of the $\rho_p(T) \equiv C_p(T)/T^3$ data points [46] in the insets of Figures 2 and 3 and the respective local maximum behaviours of the corresponding $\Theta_D(T)$ data points in the insets of Figures 4, 5, and 6). Our analyses have shown that the presently available cryogenic $C_p(T)$ data sets are not yet comprehensive enough for a more incisive quantification of the electronic contributions to the heat capacities of ZnSe and ZnTe. A significant refinement would require rather detailed heat capacity data to be available especially for the interval from 1 K to 2 K. The complete lack of such data is clearly the reason of the still existing uncertainties (of order ±2 K.)

We satisfy ourselves, at least, that the present estimation of $\Theta_D(0) \cong 270$ K (±2 K), including the particular value of $\Theta_D(0) = 271.4$ K (Table 3), for ZnSe, is in good accordance with earlier caloric values of $\Theta_D(0) \cong 271$ K (±1 K or ±2 K) due to Birch [34] and Collins et al. [35]. Furthermore we can make the satisfactory statement that the present estimation of $\Theta_D(0) \cong 220$ K (±2 K), including the particular value of $\Theta_D(0) = 219.3$ K (Table 3), for ZnTe, is in good agreement with several $\Theta_D^{el}(0)$ values derived from elastic constants. This concerns in particular the $\Theta_D^{el}(0)$ values of 223.2 K (±5 K) due to [35], 223.3 K due to [75], and 219.3 K due to [76]. On the other hand, one can also find in literature somewhat higher as well as markedly lower $\Theta_D^{el}(0)$ values for ZnTe, such as 225.3 K due to [74] and 228 K due to [77] or 190 K (±20 K) [38], 210 K due to [73], and 204.5 K due to [85]. Yet, in striking contrast to this multitude of earlier (mutually more or less compatible) estimates of $\Theta_D^{el}(0)$ values, for ZnTe, we have still found in contemporary literature a drastically deviating quotation of 758 K by Guo et al. [86], which appears to be classifiable as a physically obviously absurd one.

Concerning the $\Theta_D(T)$ and $\Theta_{Dh}(T)$ curves shown in the insets of Figure 6 we would still like to note that, by inserting the conventional odd-order power series expansions (see (33)) for the low-temperature $C_p(T)$ dependence into the low-capacity formula (see (25)) for effective Debye temperatures, one comes readily to a relatively simple (algebraic) $\Theta_D(T)$ expression of type [13]

$$\Theta_D^{l_0}(T) \longrightarrow \sqrt[3]{\frac{\left(4\pi^4/5\right)\cdot C_{Vh}(\infty)}{c_1/T^2 + c_3 + c_5T^2 + c_7T^4}} = \frac{\Theta_D(0)}{\sqrt[3]{1 + \left(c_1/T^2 + c_5T^2 + c_7T^4\right)/c_3}}, \quad (37)$$

in accordance with (34). This formula has been used here, with the parameter values quoted in Table 3, for calculations of the low-temperature sections of the $\Theta_D^{l_0}(T)$ dependencies (solid curves in the insets of Figure 6). The associated $\Theta_{Dh}^{l_0}(T)$ dependencies (due to (30)) for the "true" Debye temperatures, which are represented by dashed curves in the insets of Figure 6, have equally been calculated via a formula of type (37) where, merely, the electronic coefficient has been removed ($c_1 \to 0$). With respect to this characteristic $\Theta_{Dh}^{l_0}(T \to 0)$ behaviour it is necessary, of course, to point out that simple algebraic formulas [13] of type (37), for $\Theta_D^{l_0}(T)$ and $\Theta_{Dh}^{l_0}(T)$, apply only to very narrow cryogenic intervals, $0 < T < \Theta_D(0)/30$, where the respective Debye temperature curves show an exclusively *convex* shape (cf. the insets of Figure 6). Equivalently one can state that algebraic formula of the type (37) is *not* applicable to temperatures higher than $\Theta_D(0)/30$ (i.e., to $T > 9$ K for ZnSe or $T > 7$ K ZnTe), where the (nearly coinciding) $\Theta_D^{l_0}(T)$ and $\Theta_{Dh}^{l_0}(T)$ dependencies are showing *concave* shapes (cf. the insets of Figures 4 and 5).

6.2. Minima of Debye Temperature Curves and Their Interpretation. The monotonic fall of the true Debye temperatures, $\Theta_{Dh}(T)$, from their limiting ($T \to 0$) levels, $\Theta_D(0)$, to the respective minima, $\Theta_{D\min} \equiv \Theta_D(T_{\min})$, is described in good approximation by the relatively simple (algebraic) low-temperature formula, $\Theta_{Dh}^{l_0}(T)$ (see (30)). We see from the graphical representation of the latter by dotted curves (in the insets of Figures 4 and 5) that this $\Theta_{Dh}^{l_0}(T)$ expression provides, indeed, close approaches to the exact $\Theta_{Dh}(T)$ curves from 0 to $T_{\min} \approx \Theta_D(0)/15$. With respect to these limited cryogenic regions, $0 < T < \Theta_D(0)/15$, one can thus describe the monotonic decrease of the ratios $\Theta_D(0)/\Theta_{Dh}(T)$ in good approximation by the relationship [13, 14]

$$\frac{\Theta_D(0)}{\Theta_{Dh}(T)} \cong \frac{\Theta_D(0)}{\Theta_{Dh}^{l_0}(T)} = \sqrt[3]{\frac{\rho_h(T)}{\rho_h(0)}}, \quad (38)$$

in accordance with (30). The latter allows us to establish a useful approximate relationship between the minima, $\Theta_{D\min} = \Theta_{Dh}(T_{\min})$, of the $\Theta_{Dh}(T) \to \Theta_D(T)$ curves, on the one hand, and the maxima, $\rho_{\max} = \rho_h(T_{\max})$ of the $\rho_h(T) \to \rho_p(T)$ curves, on the other hand. Observing that the relative shifts of $T_{\min}^{(\Theta)}$ positions (indicated in the insets of Figures 4 and 5) versus the $T_{\max}^{(\rho)}$ positions (indicated in the insets of Figures 2 and 3) are smaller than 2.5%, in both cases (cf. also Table 4), we can hence approximate the minima of the $\Theta_{Dh}(T)$ curves by $\Theta_{Dmin} \cong \Theta_{Dh}(T_{max}^{(\rho)})$. Accordingly, by specializing (38) for $T \to T_{max}^{(\rho)} \cong T_{min}^{(\Theta)}$, we come to an approximate relationship between the extreme values of the $\Theta_{Dh}(T)$ versus $\rho_h(T)$ curves of the form [14]

$$\frac{\Theta_D(0)}{\Theta_{Dmin}} \cong \sqrt[3]{\frac{\rho_{max}}{\rho_h(0)}} = \sqrt[3]{\frac{\rho_{max}}{c_3}}. \qquad (39)$$

Comparing the actual $\Theta_D(0)/\Theta_{Dmin}$ and $\sqrt[3]{\rho_{max}/\rho_h(0)} = \sqrt[3]{\rho_{max}/c_3}$ values (quoted in Table 4) we see that approximate relation (39) is indeed well fulfilled (within deviations smaller than 0.25%) by the empirical Θ_{Dmin} and ρ_{max} values for ZnSe and ZnTe.

It is instructive to compare the presently obtained (empirical) Θ_{Dmin} values with their theoretical counterparts that had been estimated in various earlier papers. For the case of ZnSe we see that the previously estimated (theoretical) Θ_{Dmin} values of about 207 K (cf. Figure 2 in [34]) and 218 K (cf. Figure 14 in [36]) are in reasonable agreement with the presently determined (empirical) Θ_{Dmin} value of 206.3 K (cf. Figure 4 and Table 4). In contrast to the latter, however, the theoretical $\Theta_D(T)$ curve presented in Figure 7 of [87] shows an exclusively monotonic behaviour for any $T > 10$ K, without giving any indication of a possible local minimum behaviour in the vicinity of $T_{min} \approx 17$ K (cf. the inset of Figure 4). It is suggested by Figure 7 of [87] that the magnitude of the Debye temperature at 10 K should be of order 120 K, which is about 40% lower than the actual Θ_{Dmin} value. A detailed numerical comparison with Figure 4 in the present paper (including Table 2) reveals, consequently, that the $\Theta_D(T)$ curve shown for ZnSe in Figure 7 of [87] is quite *in*adequate within a rather large cryogenic interval, from 10 K up to about 80 K (note that an analogous criticism applies also to the obviously erroneous, monotonic Θ_D ($T > 16$ K) curve shown for ZnS in Figure 7 of the same paper).

For the case of ZnTe, we see that the previously estimated Θ_{Dmin} values of about 157 K (cf. Figure 5 in [32]), 162 K (cf. Figure 1 in [35]), 159 K (cf. Figure 15 in [36]), and 168 K (cf. Figure 3 in [38]) are throughout in reasonable agreement with the presently determined (empirical) Θ_{Dmin} value of 165.4 K (cf. Figure 5 and Table 4).

Concerning the physical interpretation of the occurrence of the closely correlated maxima and minima of the respective $\rho_{p/h}(T)$ and $\Theta_{D/Dh}(T)$ curves, we reconsider briefly the typical experimental manifestation of a heat capacity contribution, $C_E(T) = (\Theta_E/2T)^2/(\sinh(\Theta_E/2T))^2$, which is due to an individual Einstein oscillator located at $\varepsilon_E = k_B\Theta_E$. It can easily be shown that the maximum of the respective $\rho_E(T) \equiv C_E(T)/T^3$ curve is located at $T_{max}^{(E)} = 0.203\Theta_E$ (according to Figure A1 of the supporting information to [50]). Thus, when we connect the maximum of a given $\rho_{p/h}(T)$ curve in roughest approximation with the contribution of a single Einstein oscillator, we can estimate its characteristic phonon temperature to amount to $\Theta_E = 4.93 \cdot T_{max}^{(E)}$. Accordingly, one can readily derive from the known T_{max} positions (of 16.9 K for ZnSe and 13.6 K for ZnTe) values of $\Theta_E = 83.3$ K, for ZnSe, and $\Theta_E = 70.0$ K, for ZnTe, for the characteristic phonon temperatures of apparently dominating low-energy Einstein oscillators. Comparing these Θ_E values with the set of Θ_n values obtained by means of the detailed analyses via the four-oscillator hybrid model (in Section 4) we see that they are ranging, for both materials, within the lower parts of the intervals between the first and second Einstein oscillators,

$$\Theta_{1(TA)} < \Theta_E < \Theta_{2(TA)} \qquad (40)$$

(cf. Table 1). Moreover, from the TA phonon dispersion relations shown for ZnSe and ZnTe in Figures 1(a) and 2(a) of [46] we infer that the characteristic TA phonon temperatures associated with the L and X points of the Brillouin zone boundaries, $\Theta_{TA(L)}$ and $\Theta_{TA(X)}$, are about 75 K and 93 K, for

TABLE 4: Pieces of evidence of the close correlation between the characteristic cryogenic non-Debye features of the $\rho_{p/h}(T) \equiv C_{p/Vh}(T)/T^3$ curves (see the insets of Figures 2 and 3) and the respective Debye temperature curves, $\Theta_D(T)$ (see Figures 4 and 5). The prominent features of the $\rho_{p/h}(T)$ curves are characterized by the $T \to 0$ levels for the lattice heat capacity, $\rho_h(0) = c_3$, the coordinates of their *maxima*, T_{max} and $\rho_{max} \equiv \rho(T_{max})$, and the respective *elevation* ratios, $\rho_{max}/\rho_h(0) > 1$. The $\Theta_{D(h)}(T)$ curves are characterized above all by the limiting ($T \to 0$) levels, $\Theta_D(0)$, the coordinates of their *minima*, T_{min} and $\Theta_{Dmin} \equiv \Theta_D(T_{min})$, and the corresponding *drop* ratios, $\Theta_{D(min)}/\Theta_D(0) < 1$. The material-specific ratios $\Theta_{D(min)}/T_{min}$ are quoted for comparisons with the general condition $\Theta_{D(min)}/T_{min} > 14$ (cf. Section 5) of close approach to asymptotic behaviour due to (25). Furthermore, the maximum values of the effective Debye temperatures, $\Theta_{D(max)}$, the theoretical high-temperature limiting value of the true Debye temperature, $\Theta_{Dh}(\infty)$ (42), and several characteristic ratios, the approximate equality of which, for both materials, is indicating their high degree of similarity, are quoted. Particularly the large deviations from unity of the cryogenic drop ratios, $\Theta_{D(min)}/\Theta_D(0) \approx 0.76$, and of the ratios for the subsequent rise to maxima, $\Theta_{D(max)}/\Theta_{D(min)} \approx 1.5$, illustrate the considerable strength of T-related variations of Debye temperatures for both materials under study.

Material	ZnSe	ZnTe
$\rho_h(0) = c_3$ (μJ·K^{-4} mol^{-1})	197.5	365.0
T_{max} (K)	**16.9**	**13.6**
ρ_{max} (μJ·K^{-4} mol^{-1})	440.0	855.7
$\rho_{max}/\rho_h(0)$	2.228	2.344
$\sqrt[3]{\rho_{max}/\rho_h(0)}$	**1.306**	**1.328**
$\Theta_D(0)$ (K)	270	220
T_{min} (K)	**17.3**	**13.9**
$\Theta_D(0)/T_{min}$	15.6	15.8
$\Theta_{D(min)}$ (K)	206.3	165.4
$\Theta_D(0)/\Theta_{Dmin}$	**1.309**	**1.330**
$\Theta_{D(min)}/T_{min}$	11.9	11.9
$\Theta_{D(max)}$ (K)	305.7	252.3
$\Theta_{D(max)}/\Theta_{D(min)}$	1.482	1.525
$\Theta_{Dh}(\infty)$ (K)	317	262
$\Theta_D(0)/\Theta_{Dh}(\infty)$	0.852	0.840
$\Theta_{D(min)}/\Theta_{Dh}(\infty)$	0.651	0.631

ZnSe, and about 59 K and 77 K, for ZnSe. Accordingly, we find that the effective Einstein oscillator temperatures, Θ_E, are located close to the centre of gravity of the intervals between the $\Theta_{TA(L)}$ and $\Theta_{TA(X)}$ values; that is, $\Theta_E \approx (\Theta_{TA(L)}+\Theta_{TA(X)})/2$ (for both materials). These findings can be considered as a clear confirmation of various earlier assessments [14, 20, 84], according to which the occurrences of the maximum behaviours of $\rho_{p/h}(T)$ curves (like those shown in the insets of Figures 2 and 3) as well as the closely related minimum behaviours of the associated $\Theta_{D(h)}(T)$ curves (shown in the insets of Figures 4 and 5) are characteristic manifestations of strongly increasing contributions of short-wave TA phonons to cryogenic heat capacities.

6.3. Debye Temperature Behaviour at Higher Temperatures. In view of the analogy of the analytical connections (1) and (29), between the $x_D(T) = \Theta_D(T)/T$ ratios with $C_p(T)$ values, on the one hand, and the $x_{Dh}(T) = \Theta_{Dh}(T)/T$ ratios with $C_{Vh}(T)$ values, on the other hand, it is obvious that the physical causes of the general differences, $\Theta_{Dh}(T) > \Theta_D(T)$ (see Figures 4 and 5), between "true" and effective Debye temperatures, must be sought in the mechanisms which are responsible for the inherent differences, $C_p(T) - C_{Vh}(T) > 0$ (Section 2), between measured (isobaric) and theoretical (isochoric/harmonic) lattice heat capacities. For nonmagnetic materials these differences are well known to be due to

(i) possible electronic contributions, $C^{(E)}(T) = c_1 T$ (in (5)), which are responsible for qualitatively different asymptotic behaviours, that is, $C_p(T) \to c_1 T$ versus $C_{Vh}(T) \to c_3 T^3$, in the $T \to 0$ limit,

(ii) the cumulative effects of lattice expansion and lattice anharmonicities, which are responsible for markedly increasing differences at $T > T_h$ (cf. Figures 2 and 3).

Concerning the latter, it has been found within a series of earlier studies [14–18, 21, 71, 88] that a relatively simple way for detection of temperature dependencies of "true" Debye temperatures, $\Theta_{Dh}(T)$, is offered via fittings of properly confined (i.e., exclusively concave) sections of $\Theta_D(T)$ versus T^{-2} curves on the basis of (C.1) (see Figure 10, for the presently considered cases of ZnSe and ZnTe). This conventional method is as a rule capable of providing good results for materials showing a sufficiently high degree of phonon dispersion like the present ones, $\Delta_P > 1/3$ (see Table 1; note that this frequently encountered state of affairs is in analogy to that found, for example, for silicon and germanium [21] as well as many II–VI and III–V materials [14, 50]). This *large* dispersion regime is expressively visualized by $\Theta_D(T)$ curves showing pronounced *non*monotonic behaviours in the vicinities of their minima (like those shown in Figures 4 and 5). The respective large degree of variability of $\Theta_D(T)$ values, particularly with respect to the region $T > T_{\min}$, is characterized by rise ratios, $\Theta_{D\max}/\Theta_{D\min}$, that are significantly *higher* than unity (like $\Theta_{D\max}/\Theta_{D\min} \approx 1.5$, for ZnSe and ZnTe; cf. Table 4). In contrast to this, for materials with relatively *small* phonon dispersion, $1/3 > \Delta_P > 1/4$ (like diamond and BN [50], or MgO [17]), for which the characteristic (material-specific) $\Theta_{D\max}/\Theta_{D\min}$ ratios are known to be closely approaching to unity, this conventional method [14–18, 21, 71, 88] usually does not give good results for high-temperature behaviours, $\Theta_{Dh}(T \to \infty)$, of "true" Debye temperatures (for an illustration of significant residual uncertainties involved by a tentative application of this conventional $\Theta_{Dh}(T)$ estimation method to a low-dispersion material like MgO see [17]).

In view of the limited applicability of this conventional $\Theta_{Dh}(T) \to \Theta_D(T)$ fitting method (cf. Appendix C) we have displayed here in due detail (in Sections 2–4) a more general, alternative method, which is based on a properly adapted multioscillator hybrid model. This model involves, above all, a physically plausible decomposition of isobaric heat capacities, $C_p(T)$ (see (5)), into prevailing harmonic lattice heat capacity parts, $C_{Vh}(T) \propto \kappa_h(T)$ (see (4)), and a rapidly increasing rest term, $C_p(T) - C_{Vh}(T) \propto (\kappa_h(T))^2 T$, which represents the cumulative effects of lattice expansion and anharmonicities. Using this analytical framework for careful least-mean-square fittings of comprehensive sets of measured $C_p(T)$ data, we have determined the respective parameter sets (Table 1). On the basis of the latter it was then possible to calculate in unambiguous way the corresponding *harmonic* lattice heat capacities, $C_{Vh}(T) = C_{Vh}(\infty)\kappa_h(T)$ (using (9), in combination with (14) to (17)). Consequently, we could here readily determine the T-dependencies of the "true" Debye temperatures, $\Theta_{Dh}(T)$, using the high-precision formulas (27) and (28). The respective $\Theta_{Dh}(T)$ dependencies are shown (by dashed curves) in Figures 4 and 5.

Comparing the limiting $\Theta_{Dh}(T \to \infty)$ values listed in Table 4 (indicated by arrows in Figures 4 and 5) with their counterparts resulting from the conventional estimation procedure listed in Table 10 (indicated in Figure 10 by the intercepts of the dashed curves with the ordinate), we can discern only minor differences (of less than 0.3%) between the limiting $\Theta_{Dh}(\infty)$ values estimated via the two alternative analysis procedures. The slightly different results due to the alternative analyses procedures can thus be summarized in the form

$$\Theta_{Dh}(\infty) \cong 317\,\text{K}\ (\pm 1\,\text{K}) \quad \text{for ZnSe},$$
$$\Theta_{Dh}(\infty) \cong 263\,\text{K}\ (\pm 1\,\text{K}) \quad \text{for ZnTe}. \tag{41}$$

The rather small size of residual uncertainties can be considered as clear indications of an obviously high degree of internal consistency and compatibility of the two qualitatively different $\Theta_{Dh}(\infty)$ estimation procedures.

An additional confirmation of the apparent adequacy of these caloric $\Theta_{Dh}(\infty)$ values (see (41)) can be obtained by comparison with independent theoretical estimations based on first-principle calculations of PDOS spectra. It is well known that the magnitudes of $\Theta_{Dh}(\infty)$ are connected in unambiguous way with the second moments, $\mu_P^{(2)}$, by [14, 16, 17, 21, 54, 56, 71, 72]

$$\Theta_{Dh}(\infty) = \frac{\sqrt[2]{(5/3)\,\mu_P^{(2)}}}{k_B} \equiv \frac{\sqrt[2]{5/3}\cdot \varepsilon_P(2)}{k_B}, \tag{42}$$

where $\varepsilon_P(2) \equiv \sqrt[2]{\mu_P^{(2)}}$ represents the associated (second order) phonon energy. We have performed within a former study [56] evaluations of moments, $\mu_P^{(m)}$ (see (12)), on the basis of theoretically calculated PDOS spectral functions, $g_P(\varepsilon)$, which had been published for a variety of group-IV, III–V, and II–VI materials in a variety of earlier papers. Among the numerical results listed in Table I of [56] one can find, among other things, corresponding $\Theta_{Dh}(\infty)$ values (due to (42)) for both materials here under study, namely, 324 K or 319 K for ZnSe (detected from the spectra presented in Figure 8 of [36] and Figure 2 of [66], resp.), and 264 K or 259 K for ZnTe (detected from the spectra presented in Figure 9 of [36] and Figure 3 of [32], resp.). Instructive are also comparisons with graphical representations of precalculated $\Theta_{Dh}(T)$ curves, from which one can detect limiting $\Theta_{Dh}(T \rightarrow \infty)$ values of about 324 K for ZnSe and 259 K for ZnTe (as indicated by Figures 14 and 15 in [36]) and 259 K for ZnTe (as indicated by Figure 5 in [32]). In this way we satisfy ourselves that the presently obtained (caloric) $\Theta_{Dh}(\infty)$ values (41) show only moderate deviations (<2.5%) from former theoretical estimates [32, 36].

Concerning the high-temperature dependencies of the *effective* (anharmonicity-related) Debye temperatures, $\Theta_D(T)$, we would still like to point out that, owing to the strict limitation of their definition (see (1)) to ratios $C_p(T)/C_{Vh}(\infty) \leq 1$, their magnitudes are automatically dropping to zero at those positions, T_f, where the individual $C_p(T)$ curves are crossing the classical Delong-Petit level; that is,

$$\Theta_D\left(T_f\right) = 0 \quad \text{for } C_p\left(T_f\right) = C_{Vh}(\infty) = 6R. \tag{43}$$

This basic property is manifested by the rapid $\Theta_D \rightarrow 0$ drops of the solid $\Theta_D(T)$ curves (in Figures 4 and 5), which are located just at those positions, $T_f = 401$ K (for ZnSe) or 321 K (for ZnTe), where the respective $C_p(T)$ curves are crossing the classical Dulong-Petit level, $C_{Vh}(\infty) = 6R$ (as shown in Figures 2 and 3).

6.4. Characteristic PDOS-Related Phonon Energy Values. Within the frame of the presently considered four-oscillator hybrid model, the totality of quantitative information on the material-specific PDOS spectra is given (in Table 1) in terms of the adjusted sets of effective Einstein oscillator temperatures, Θ_n ($n = 1$ to 4), and the respective weights, w_n ($n = 1$ to 4), including the weights w_{C1} and w_{C2} for the associated continuous low-energy curve section components (of Debye and non-Debye type, resp.). The corresponding Einstein peak *energy* positions, $\varepsilon_n \equiv k_B\Theta_n$ ($n = 1$ to 4), are quoted in Table 5. On the basis of these sets of fitted parameter values one can readily calculate a variety of characteristic dispersion-related quantities, some of which are directly comparable with their theoretical counterparts being detectible from theoretically calculated PDOS spectra presented in various former papers [32, 36, 45, 46, 66, 89].

Of importance for a physical understanding of the prominent role of short-wave TA phonons, which are mainly responsible for the occurrence of the characteristic non-Debye (maximum) behaviours of $C_p(T)/T^3$ curves in the liquid hydrogen region (cf. the insets of Figures 2 and 3), is also the knowledge of the centres of gravity of TA phonon sections, $\overline{\varepsilon}_{\mathrm{TA}}$. The magnitudes of the latter are given, in accordance with the respective TA phonon normalization condition (see (23)), by [14]

$$\overline{\varepsilon}_{\mathrm{TA}} = 3\left[\left(\frac{3w_{C1}}{4} + \frac{5w_{C2}}{6} + w_1\right)\varepsilon_1 + w_2\varepsilon_2\right] \tag{44}$$

(cf. Eq. (18) in [14]). Concerning the presently estimated $\overline{\varepsilon}_{TA}$ values of 8.48 meV for ZnSe and 6.80 meV for ZnTe (as quoted in Table 5), we can make the satisfying statement that they are in good agreement with their counterparts (of about 8.4 meV or 8.5 meV, for ZnSe, and 6.8 meV or 6.9 meV, for ZnTe), which we have detected from the theoretically calculated PDOS spectra presented for ZnSe and ZnTe in Figure 3 of [45] or Figure 4 of [46], respectively. Accidentally, these $\overline{\varepsilon}_{\mathrm{TA}}$ positions are even nearly coinciding with the positions of the highest TA peaks shown within the same figures (which are readily seen to be located at about 8.5 meV or 8.6 meV, for ZnSe, and between 6.8 meV and 6.9 meV, for ZnTe).

Similarly informative are also comparisons of the average values of LA phonon energies, $\overline{\varepsilon}_{\mathrm{LA}}$, which are represented within the present formalism (due to (22)) by the third discrete oscillator, $\overline{\varepsilon}_{\mathrm{LA}} = \varepsilon_3$. The presently estimated $\overline{\varepsilon}_{\mathrm{LA}}$ values of 18.36 meV for ZnSe and 14.40 meV for ZnTe (cf. Table 5) are in reasonable agreement with their counterparts (of about 18.5 meV or 19.2 meV, for ZnSe, and 14.5 meV or 15.0 meV, for ZnTe), which we have detected, again, from the theoretically calculated PDOS spectra presented in Figure 3 of [45] or Figure 4 of [46], respectively.

The average values for all acoustical (i.e., TA and LA) phonon energies, $\overline{\varepsilon}_{\mathrm{TA\&LA}}$, are another characteristic quantity for comparisons between the results of numerical analyses of heat capacity data with theoretically calculated PDOS spectra. The analytical expression for the latter, which thus comprises the whole *lower* halves of PDOS spectral sections, is given within the present formalism by [14]

$$\begin{aligned}\overline{\varepsilon}_L &= \overline{\varepsilon}_{\mathrm{TA\&LA}} \\ &= 2\left[\left(\frac{3w_{C1}}{4} + \frac{5w_{C2}}{6} + w_1\right)\varepsilon_1 + w_2\varepsilon_2 + \frac{\varepsilon_3}{6}\right] \\ &= \frac{\left[2\overline{\varepsilon}_{\mathrm{TA}} + \overline{\varepsilon}_{\mathrm{LA}}\right]}{3}.\end{aligned} \tag{45}$$

The presently estimated $\overline{\varepsilon}_L = \overline{\varepsilon}_{\mathrm{TA\&LA}}$ values of 11.77 meV for ZnSe and 9.33 meV for ZnTe (cf. Table 5) are found again to be in reasonable agreement with their counterparts of about 11.8 meV or 12.1 meV, for ZnSe, and 9.4 meV or 9.6 meV, for ZnTe, which followed immediately via (45) from the couples of $\overline{\varepsilon}_{\mathrm{TA}}$ and $\overline{\varepsilon}_{\mathrm{LA}}$ values for the average TA and LA phonon energies (which had been detected from the theoretically calculated PDOS spectra presented in Figure 3 of [45] or Figure 4 of [46], resp.). Furthermore we observe that the presently estimated (caloric) $\overline{\varepsilon}_L = \overline{\varepsilon}_{\mathrm{TA\&LA}}$ values are in good agreement with our former estimations [56], which were based on theoretically calculated PDOS spectral functions published in various earlier papers, namely, 11.7 meV due to

TABLE 5: Phonon energy values for the 4 discrete (Einstein) oscillators, $\varepsilon_n = k_B\Theta_n$ ($n = 1$ to 4, corresponding to the Θ_n values given in Table 1), the average phonon energies for the TA phonon sections, $\bar{\varepsilon}_{TA}$ (44), and the whole acoustic phonon energy sections, $\bar{\varepsilon}_{TA\&LA}$ (45), including a series of equivalent (moment-related) phonon energies, $\varepsilon_P(m)$ (due to (46) and (47)).

Material	ZnSe	ZnTe
$\varepsilon_1 = \varepsilon_{TA1}$ (meV)	6.75	5.34
$\varepsilon_2 = \varepsilon_{TA2}$ (meV)	9.43	7.58
$\varepsilon_3 = \bar{\varepsilon}_{LA}$ (meV)	**18.36**	**14.40**
$\varepsilon_4 = \bar{\varepsilon}_{TO\&LO} \equiv \bar{\varepsilon}_U$ (meV)	**27.10**	**22.69**
$\bar{\varepsilon}_{TA}$ (meV)	**8.48**	**6.80**
$\bar{\varepsilon}_{TA\&LA} = \bar{\varepsilon}_L$ (meV)	**11.77**	**9.33**
$\varepsilon_P(-5/2)$ (meV)	10.13	8.18
$\varepsilon_P(-2)$ (meV)	11.78	9.49
$\varepsilon_P(-3/2)$ (meV)	13.12	10.58
$\varepsilon_P(-1)$ (meV)	14.44	11.68
$\varepsilon_P(-1/2)$ (meV)	15.79	12.83
$\varepsilon_P(0)$ (meV)	17.11	13.97
$\varepsilon_P(1) = \bar{\varepsilon}_{L\&U}$ (meV)	**19.43**	**16.01**
$\varepsilon_P(2)$ (meV)	21.17	17.55
$\varepsilon_P(4)$ (meV)	23.22	19.36
$\varepsilon_P(6)$ (meV)	24.27	20.29
$\varepsilon_P(8)$ (meV)	24.89	20.83
$\varepsilon_P(10)$ (meV)	25.30	21.18

[66, 89], or 12.5 meV due to [36], for ZnSe, and 9.5 meV or 9.4 meV due to [32, 36], for ZnTe.

Let us still consider the *upper* part of the PDOS spectra in consideration, which are consisting of one LO and two TO branches. The corresponding average optical phonon energy is given within the present formalism (due to (21)) just by the fourth discrete oscillator; that is, $\bar{\varepsilon}_U = \bar{\varepsilon}_{TO\&LO} = \varepsilon_4$. The presently estimated $\bar{\varepsilon}_U$ values (cf. Table 5) of 27.10 meV, for ZnSe, and 22.69 meV, for ZnTe, are found to be in more or less good agreement with their counterparts (of about 27.3 meV or 25.6 meV, for ZnSe, and 23.1 meV or 21.7 meV, for ZnTe), which we have detected again from the theoretically calculated PDOS spectra presented in Figure 4 of [46] and Figure 3 of [45], respectively. Informative is also a comparison of present (caloric) results with our former estimations [56] of $\bar{\varepsilon}_U$ values on the basis of PDOS spectral functions published in various earlier papers, namely, 27.3 meV, 27.1 meV, or 28.4 meV due to [36, 66, 89], for ZnSe, and 22.0 meV or 22.7 meV due to [32, 36], for ZnTe. Thus we are concerned in the case of the upper (optical) PDOS spectral sections with differences between approximate $\bar{\varepsilon}_U$ values of different origin up to a maximum order of 5% (for both ZnSe and ZnTe). At the same time we can see that the presently estimated (caloric) $\bar{\varepsilon}_U$ values (Table 5) are ranging near the middle between the highest and lowest $\bar{\varepsilon}_U$ estimates referring to the variety of PDOS spectra in consideration for the two materials under study. This shows that one has good chances to get reliable information on various prominent features of material-specific PDOS spectra even via dispersion-related analyses of heat capacity data.

Very informative in this respect are, above all, comparisons between moments $\mu_P^{(m)}$, of material-specific PDOS spectral functions [3, 4, 16, 18, 19, 21, 50, 52, 56, 69–72], which are generally defined by (12) (for $-3 < m \neq 0$). Within the present formalism of the four-oscillator hybrid PDOS model function, $g_P(\varepsilon)$ (i.e., (7), for $n_E = 4$), one obtained for the corresponding moments, $\mu_P^{(m)}$ (see (12)), the model-specific expression (13) [14, 21]. Yet, for visualizations of the m-dependencies of such PDOS spectra moments it is convenient to represent the individual mth-order moments, $\mu_P^{(m)}$, in terms of their equivalent mth-order phonon energies [14, 16, 21, 50, 56], $\varepsilon_P(m) \equiv \sqrt[m]{\mu_P^{(m)}}$. The latter are thus given within the frame of the present four-oscillator hybrid model (in accordance with (13)) by an expression of the form [14, 21, 50]

$$\varepsilon_P(m) = \varepsilon_1 \left[\frac{3w_{C1}}{3+m} + \frac{5w_{C2}}{5+m} + w_1 + \sum_{n=2}^{4} w_n \left(\frac{\varepsilon_n}{\varepsilon_1} \right)^m \right]^{1/m}, \quad \text{for } -3 < m < 0,\ 0 < m. \tag{46}$$

By means of a $m \longrightarrow 0$ limiting transition one obtains from (46) in particular

$$\varepsilon_P(m \longrightarrow 0) = \varepsilon_1 \exp\left[-\frac{w_{C1}}{3} - \frac{w_{C2}}{5} + \sum_{n=2}^{4} w_n \ln\left(\frac{\varepsilon_n}{\varepsilon_1} \right) \right]. \tag{47}$$

Collections of corresponding material-specific $\varepsilon_P(m)$ values for a series of integers and half-integers (within the interval $-3 < m \leq 10$) are listed in Table 5 and are represented by open circles in Figure 7. In view of the fact that the whole m-dependence of the $\varepsilon_P(m)$ curve (due to (46)) divides into a couple of *continuous* curve sections (for $-3 < m < 0$ and $0 < m$), the sequence of discrete $\varepsilon_P(m)$ values (○) are perfectly connected by the respective solid curve sections. Furthermore we have shown in Figure 7 several sets of discrete $\varepsilon_P(m)$ values which we have obtained via detailed numerical analyses of the theoretically calculated PDOS spectra presented for ZnSe and/or ZnTe in [32] (×), [36] (∗), [45] (◊), [46] (△), [66] (+), and [89] (☆). The comparisons between the different $\varepsilon_P(m)$ values, in Figure 7, show a multitude of good agreements between the discrete $\varepsilon_P(m)$ values due to the PDOS spectra published by different authors, including present (caloric) results (○). This concerns especially the region of negative orders ($m < 0$), where the actual $\varepsilon_P(m)$ values are predominantly due to the contributions of TA phonons. At the same time we are concerned with some markedly larger deviations in the region of positive orders ($m > 0$), where the monotonically increasing $\varepsilon_P(m)$ values tend to be the stronger ones governed by optical phonons the higher their orders are. The maximum difference between estimated $\varepsilon_P(10)$ values for ZnTe (due to [46] (△) versus [45]

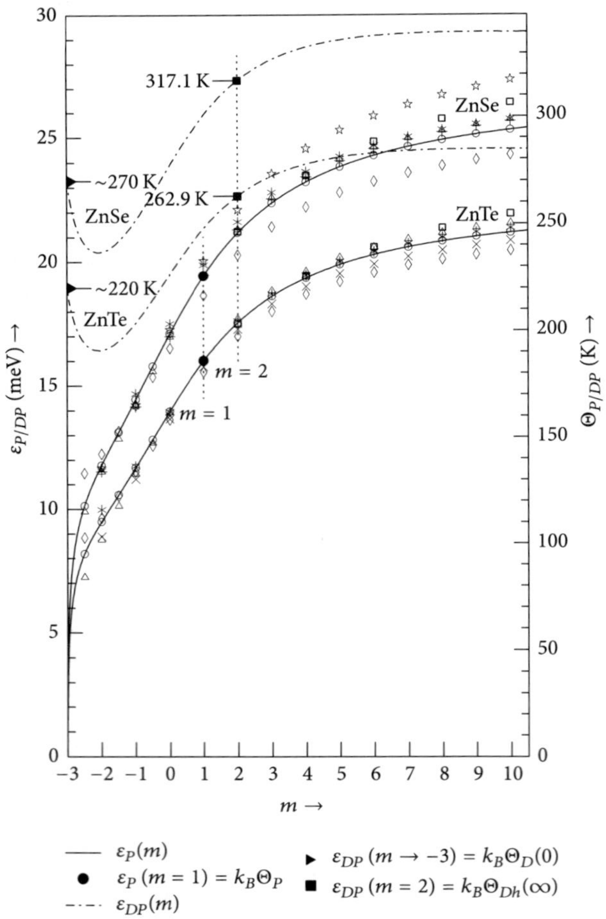

FIGURE 7: Visualization of the effective phonon energy curves, $\varepsilon_P(m > -3)$ (see (46); solid curves), for ZnSe and ZnTe. Included are the corresponding discrete $\varepsilon_P(m)$ points ((○); cf. Table 5), for integer and several negative half integer orders. For comparisons with counterparts following from former theoretically calculated PDOS spectra we have shown here also the respective series of discrete $\varepsilon_P(m)$ values obtained via detailed numerical analyses of the theoretically calculated PDOS spectra presented for ZnSe and/or ZnTe in [32] (×), [36] (∗), [45] (◊), [46] (△), [66] (+), and [89] (☆). The series of even-order $\varepsilon_P(m)$ values (for $m = 2$ to 10, listed in Table 10), which have been determined in Appendix C via the conventional analyses procedure for Debye temperature values representing the harmonic lattice, $\Theta_{Dh}(T) \cong \Theta_D(T)$ (cf. Figure 10), are represented by squares (□).

(◊)) amounts to about 6%, whereas, for ZnSe, the maximum difference (due to [89] (☆) versus [45] (◊)) amounts even to about 13%. Relatively large differences like the latter ones are obviously mainly due to significant qualitative differences between the upper (optical) sections of the theoretically calculated PDOS spectra presented by the corresponding authors (of [45] (◊) versus [89] (☆)). At the same time we can make the satisfactory statement that the presently obtained (caloric) $\varepsilon_P(10)$ values (Table 5) differ from those we have derived from several other PDOS spectra in consideration for ZnSe ([36] (∗), [46] (△), and [66] (+)) and for ZnTe ([32] (×), [36] (∗), and [46] (△)) by less than 2%.

The latter findings underscore the above assessment, according to which one has as a rule good chances to get reliable information of various characteristic features of material-specific PDOS spectra by means of detailed dispersion-related analyses of heat capacity data. The rather fine quantification abilities of our elaborate analytical framework (displayed in Section 4) concern obviously also such delicate quantities like *higher*-order phonon energy values, $\varepsilon_P(m \gg 1)$, and, consequently, even the corresponding higher-order moments, $\mu_P^{(m)} = (\varepsilon_P(m))^m$, of PDOS spectra.

Just with respect to the latter it might be instructive to consider still the alternative series of five even-order $\varepsilon_P(m)$ values (for $m = 2$ to 10), which have been determined in Appendix C by means of the conventional analyses procedure [14–18, 21, 71, 88] for Debye temperatures representing the harmonic lattice properties, $\Theta_{Dh}(T) \cong \Theta_D(T)$ (the range of which is limited to $T < 96$ K, for ZnSe, and to $T < 80$ K, for ZnTe; cf. Figure 10 and Table 10). The corresponding alternative $\varepsilon_P(m)$ values (listed in Table 10) are represented in Figure 7 by squares (□). Comparing the latter with their hybrid-model-related counterparts (○) (listed in the last five rows of Table 5) we find excellent agreements, at least, for the lowest two orders ($m = 2$ and 4). Towards higher orders ($m = 6$ to 10), however, we are concerned with growing deviations between the two alternative series of $\varepsilon_P(m)$ values, which are seen to increase for ZnSe from 2.3% to 4.4% and for ZnTe from 1.5% to 3.6% (cf. Figure 7).

Finally, let us still consider the behaviour (m-dependence) of the conventional *Debye*-model-related energy curves, which are defined (in analogy to [16–19, 64, 71]) by an expression of the form

$$\varepsilon_{DP}(m) \equiv \left(\frac{(3+m)\,\mu_P^{(m)}}{3}\right)^{1/m} = \left(\frac{(3+m)}{3}\right)^{1/m} \varepsilon_P(m). \tag{48}$$

in accordance with the general definition, $\varepsilon_P(m) \equiv (\mu_P^{(m)})^{1/m}$, of the mth-order phonon energies [14, 21, 50, 56]. Observing that the m-dependencies of the two complementary $\varepsilon_P(m)$ curve sections (for $-3 < m < 0$ and $0 < m$) are given by (46), we come thus readily (via (48)) to the material-specific $\varepsilon_{DP}(m)$ dependencies, for ZnSe and ZnTe, we have visualized by corresponding dash-dot curves in Figure 7 (note that, for the special point $m \to 0$, the particular $\varepsilon_{DP}(0)$ value follows from a corresponding limiting transition to adopt just the value $\varepsilon_{DP}(m \to 0) = e^{1/3}\varepsilon_P(0)$ (cf. [16, 17]), where $\varepsilon_P(0)$ is given by (47)).

A prominent feature of these (conventional) Debye-model-related energy curves, $\varepsilon_{DP}(m)$ (see (48)), is due to the well-known property [16, 17, 19], according to which the $m \to -3$ limit of this function is automatically (exactly) equal to the $T \to 0$ limit of the cut-off energy for Debye's fictive phonon energy spectrum; that is,

$$\varepsilon_{DP}(m \longrightarrow -3) = (w_{C1})^{-1/3}\,\varepsilon_1 = k_B\Theta_D(0) = \varepsilon_D(0), \tag{49}$$

in accordance with (20). The corresponding magnitudes of $\varepsilon_D(0) = 23.3$ meV, for ZnSe, and $\varepsilon_D(0) = 19.0$ meV, for ZnTe, are indicated by solid triangles (▶) in Figure 7.

Another characteristic $\varepsilon_{DP}(m)$ value is that for $m = 2$, which follows readily from (48), in combination with (42), to be just equal to the $T \to \infty$ limit of the cut-off energy for Debye's fictive (harmonic) phonon energy spectrum; that is,

$$\varepsilon_{DP}(m = 2) = \sqrt[2]{\frac{5}{3}}\varepsilon_P(2) = k_B\Theta_{Dh}(\infty) \equiv \varepsilon_{Dh}(\infty). \quad (50)$$

The corresponding magnitudes of $\varepsilon_{Dh}(\infty) = 27.3$ meV, for ZnSe, and $\varepsilon_{Dh}(\infty) = 22.6$ meV, for ZnTe, are indicated by solid squares (■) in Figure 7. Accordingly, the ratios between these two limiting Debye energy (or temperature) values, $\varepsilon_D(0)/\varepsilon_{Dh}(\infty) = \Theta_D(0)/\Theta_{Dh}(\infty)$, amount to about 0.85 or 0.84 for ZnSe and ZnTe, respectively (cf. Table 4). The significant differences (of about 15%) between these two limiting Debye temperature values are already, per se, clear indications for a pronounced *non*-Debye character of both PDOS spectra in consideration.

Actually, for the hypothetical case of a properly normalized phonon energy spectrum of Debye's type, that is, for the fictive (exclusively quadratic) PDOS function $g_D(\varepsilon) = 3\varepsilon^2/\varepsilon_{Dc}^3$ [14, 21] (within the interval $0 \le \varepsilon \le \varepsilon_{Dc}$, and $g_D(\varepsilon) = 0$, for $\varepsilon_{Dc} < \varepsilon < \infty$), the corresponding moments are readily seen to be given just by $\mu_D^{(m)} = 3\varepsilon_{Dc}^m/(3+m)$ [14]. Substituting hence the actual (material-specific) moments $\mu_P^{(m)}$ in (48) by the moments $\mu_D^{(m)}$ due to Debye's fictive model, we see readily that expression (48) for $\varepsilon_{DP}(m)$ reduces just to a constant

$$\varepsilon_{DP}(m) \longrightarrow \varepsilon_{DD}(m) \equiv \left(\frac{(3+m)\,\mu_D^{(m)}}{3}\right)^{1/m} = \varepsilon_{Dc}, \quad (51)$$

$$\text{for any order } (-3 < m < \infty).$$

Such a constancy of an $\varepsilon_{DP}(m)$ function (of type (48)), however, has never been observed in physical reality (not even in form of a relatively close approach of one or the other $\varepsilon_{DP}(m)$ curve to a horizontal line). The rather strong variations of $\varepsilon_{DP}(m)$ values shown in Figure 7, particularly the pronounced *non*monotonic (minimum) behaviours of both $\varepsilon_{DP}(m)$ curves in the vicinity of $m \approx -2$, can be looked upon as clear visual manifestations (and a global confirmation) of the obviously very pronounced *non*-Debye character of the PDOS spectra [32, 36, 45, 46, 66, 89] for both materials under study.

7. Conclusions

The most important progress of the present study in comparison with all earlier papers, which involved in one or the other way the somewhat delicate concept of material-specific Debye temperatures, $\Theta_D(T)$, consists obviously in our elaboration (in Appendix B) of appropriate integral-*free* formulas for direct calculations of $\Theta_D(T)$ values associated with given heat capacities, $C_p(T)$. In view of the sample character of the present applications of this novel analytical apparatus to given isobaric and/or isochoric (harmonic) heat capacity data sets, we have performed the corresponding $C_{p/Vh}(T) \Rightarrow \Theta_{D/Dh}(T)$ transformations here (in Section 5) on the basis of a couple of high-accuracy formulas (27) and (28), the residual deviations from exact Debye temperature values can be seen (from the lower inset of Figure 9) to be extremely small (limited to an order of ±0.000016%). Such a high degree of accuracy, however, did not use to be actually required in practice, owing to the well-known fact that typical experimental uncertainties of measured heat capacities used to be of order ±1%. Couples of somewhat simpler low- and high-capacity formulas might thus also be largely adequate for future practical applications. In this respect we would like to draw attention above all to the somewhat simpler low-capacity formulas of type (B.5a) and (B.5b) or (B.6) which, in combination with the complementary high-capacity formula (B.3), are providing already good results (within maximum deviations of orders ±0.13% or ±0.015%, resp.; cf. Figure 9). In particular the latter choice appears to be quite sufficient for forthcoming numerical applications.

A crucial prerequisite for our analytical constructions and tests of appropriate analytical formulas suitable for $C_{p/Vh}(T) \Rightarrow \Theta_{D/Dh}(T)$ transformations (within Appendix B) consisted in a preliminary, high-accuracy determination of the general dependence of heat capacities, $C_{p/Vh}(T)$, on given Debye temperatures, $\Theta_{D/Dh}(T)$. An excellent solution of this conventional problem has been found in Appendix A in form of a couple of an unprecedented high-capacity series expansion (see (A.11)) and Debye's conventional low-capacity (low-temperature) representation [1–3] (see (A.4)). We have shown (in Appendix A) that it is rather easy to perform high-accuracy $\Theta_{D/Dh}(T) \Rightarrow C_{p/Vh}(T)$ transformations by means of this couple of integral-free expressions. A collection of corresponding high-accuracy sample values (consisting of 10 significant figures) for Debye's model-specific $\kappa_D(x_D)$ function (see (A.5)) is given in Table 7. In view of this extraordinarily comfortable way offered here (in Section A.1 of Appendix A) for high-accuracy calculations of heat capacities, $C_p(T) \equiv \kappa_D(T)C_{Vh}(\infty)$, at given magnitudes of Debye temperature versus lattice temperature ratios, $x_D(T) = \Theta_D(T)/T$, it will hence no longer be necessary either to perform point-by-point calculations of $\kappa_D(x_D)$ values (via numerical integrations according to (A.5)) or to resort to cumbersome point-by-point interpolations between neighbouring $\kappa_D(x_D)$ values utilizing an available table with discrete $\kappa_D(x_D)$ values (like Table 7 in the present paper or one of those published in [6, 8, 47–49]).

Another significant progress with respect to the intended elaboration of a more easy-to-use analytical apparatus for heat capacity analyses within the frame of the hybrid model (Section 3) has been achieved here (in Section A.2 of Appendix A) by the derivation of couples of appropriate high- and low-capacity formulas for the normalized low-energy TA phonon contributions of Debye and non-Debye type, $\kappa_{C1}(\Theta_1/T)$ and $\kappa_{C2}(\Theta_1/T)$ (see (10)), to the measured cryogenic heat capacities. Incorporating the corresponding couples of high- and low-temperature expressions (i.e., (14) to (17)) into the general formalism of the multioscillator hybrid model we have established a completely integral-*free* version of our analytical framework. Its actual use effectuated a significant acceleration of the multiparameter least-mean-square fitting processes performed (in Section 4) for fine simulations of the temperature dependencies of the comprehensive heat capacity data sets under study.

An additional important progress in quantitative respect (in comparison with our previous study [14]) could be made here owing to the availability of the unusually comprehensive and informative low-temperature heat capacity data sets given by Kremer et al. [46] for natural ZnSe and ZnTe. Of great importance just for ZnTe have been in this connection above all the unprecedented cryogenic $C_p(T)$ data available for the liquid-helium-hydrogen region ($2\,\mathrm{K} < T < 20\,\mathrm{K}$), which gave us here for the first time the chance to perform a realistic estimation of the *caloric* $T \rightarrow 0$ limiting value of the Debye temperature, $\Theta_D(0) \cong 220\,\mathrm{K}$ ($\pm 2\,\mathrm{K}$). Moreover, for the case of ZnSe, the novel $C_p(T)$ data implied a noticeable refinement (in comparison with [14]) of the $\Theta_D(T)$ behaviour in the region between 100 K and 300 K (cf. Figure 4 in the present paper with Figure 12 in [14]).

Finally we would still like to point out that, in comparison with a possible use of the global (apparently universal) non-Debye formula displayed in [15], a decisive advantage of the multioscillator hybrid model used within the present study consists in the delivery of physically important byproducts in form of a multitude of adequate quantitative information (like those presented in Section 6.4) on the PDOS functions of the materials under study. At the same time one must make the reservation that the hybrid model in the specialized *four*-oscillator version (as displayed in Section 4) used to function quite well just for systems, the PDOS spectra of which are consisting of 3 acoustical in combination with only 3 optical phonon branches. This favourable state of affairs is encountered above all within a large variety of III–V and II–VI materials with cubic structure, including silicon and germanium. However, for systems with PDOS spectra comprising *more* than 3 optical branches (e.g., for binary systems with hexagonal structure, or ternary systems), it will as a rule be necessary to envisage more or less different constellations of discrete oscillators in order to come again to fine numerical simulations of corresponding heat capacity data sets.

Appendix

A. Integral-Free Formulas for Debye and Non-Debye Functions

Within Debye's original model [1], the temperature dependencies of lattice heat capacities, $C_D^{(L)}(T) = C_{Vh}(\infty)\kappa_D(x_D(T))$, had been assumed to be unambiguously controlled by the ratios, $x_D(T) = \Theta_D/T$, of a *fixed* (material-specific) Debye temperature versus lattice temperature. These dimensionless ratios are playing just the role of the upper limits of integration for the corresponding normalized heat capacity shape function of Debye's type [1]

$$\kappa_D(x_D) = \frac{3}{x_D^3}\int_0^{x_D} dx \frac{x^4 e^x}{(e^x - 1)^2} \le 1, \tag{A.1}$$

in analogy to (1). Though this conventional model is as a rule not directly applicable to physically realistic $C_{Vh}(T)$ dependencies (like those shown by the dashed curves in Figures 2 and 3), it is worth to perform here a detailed analytical study of Debye's dimensionless heat capacity shape function, $\kappa_D(x_D)$ (see (A.1)), for the following reasons:

(i) Firstly, high-accuracy results available for the $\kappa_D(x_D)$ dependence are the prerequisite for successful constructions of hitherto lacking analytical formulas for its *inverse* functional dependence, $x_D(\kappa_D)$ (in Appendix B).

(ii) Secondly, elaborate analytical expressions for the dependence of the $\kappa_D(\kappa_D)$ function (see (A.1)) on their upper limits of integration (ratios $x_D(T) = \Theta_D/T$) can be directly used for representing, alternatively, the functional dependence of the characteristic low-temperature heat capacity component of Debye's type, $\kappa_{Ck=1}(x_1)$ (see (10)), on the respective ratios $x_1(T) = \Theta_1/T$ (see (11)).

A.1. Conventional and Alternative High-Precision Debye Function Formulas. First of all one obtains, via an integration by parts of the integral in (A.1), for the $\kappa_D(x_D)$ function the equivalent expression

$$\kappa_D(x_D) = 4D_3(x_D) - \frac{3x_D}{(\exp(x_D) - 1)}, \tag{A.2}$$

where $D_3(x_D)$ represents the familiar Debye integral [90] for the 3-dimensional case

$$\begin{aligned} D_3(x_D) &\equiv \frac{3}{x_D^3}\int_0^{x_D} dx \frac{x^3}{e^x - 1} \\ &= \frac{3}{x_D^3}\int_0^{\infty} dx \frac{x^3}{e^x - 1} - \frac{3}{x_D^3}\int_{x_D}^{\infty} dx \frac{x^3}{e^x - 1} \\ &= \frac{\pi^4}{5x_D^3} - \frac{3}{x_D^3}\int_{x_D}^{\infty} dx \frac{x^3}{e^x - 1}. \end{aligned} \tag{A.3}$$

Expanding then the denominator in the integrand of (A.3) into a Taylor series, $(e^x - 1)^{-1} = \sum_{n=1}^{\infty} e^{-nx}$, and performing the integrations for the individual $x^3 e^{-nx}$ terms from x_D to infinity (via three subsequent integrations by parts), one obtains finally for the $\kappa_D(x_D)$ function (A.2) Debye's conventional low-capacity (low-temperature) series expansion [1–3]

$$\kappa_{Dl}^{n_D}(x_D) = \frac{4\pi^4}{5x_D^3} - \frac{3x_D}{e^{x_D}-1} - 12x_D \sum_{n=1}^{n_D(\to\infty)} e^{-nx_D}\left(\frac{1}{nx_D} + \frac{3}{(nx_D)^2} + \frac{6}{(nx_D)^3} + \frac{6}{(nx_D)^4}\right). \quad \text{(A.4)}$$

We see from Figure 8, among other things, that even strongly truncated versions of this low-temperature (low-capacity) expansion (A.4), for example, $\kappa_{Dl}^{n_D\le 3}(x_D)$, are already capable of providing rather close approaches to the exact $\kappa_D(x_D)$ curve through relatively large x_D-regions. In particular we see from the upper inset of Figure 8 that, for example, for $n_D = 2$, the relative errors, $\delta\kappa_D/\kappa_D$, are smaller than 0.2 % for any $x_D > 3$. At the same time one can state that the accuracy of numerical calculations by means of this conventional $\kappa_{Dl}^{n_D}(x_D)$ expansion (see (A.4)) increases enormously when we take into account a markedly longer series of summation terms ($n_D > 10$). Actually, we see in particular from the lower inset of Figure 8 that, for example, for $n_D = 20$, the relative deviations, $\delta\kappa_D/\kappa_D$, of (A.4) from the exact $\kappa_D(x_D)$ values (A.1) are smaller than 10^{-11} through the interval $1.4 < x_D < \infty$. Furthermore we see that, by extending the summation in (A.4), for example, up to $n_D = 100$, the high-accuracy range (where $\delta\kappa_D/\kappa_D < 10^{-11}$) comprises even a much more extended interval ($0.26 < x_D < \infty$). At the same time it is obvious that the expansion (A.4) is divergent in the $x_D \to 0$ limit. A reasonable calculation of $\kappa_D(x_D)$ values for relatively small x_D values requires thus a qualitatively different $\kappa_D(x_D \to 0)$ expansion to be available.

Rewriting for convenience the original expression (A.1) for Debye's heat capacity shape function, $\kappa_D(x_D)$, in the equivalent form [13, 14, 21]

$$\kappa_D(x_D) = \frac{3}{4x_D^3}\int_0^{x_D} dz \frac{x^4}{(\sinh(x/2))^2} \le 1, \quad \text{(A.5)}$$

and using for the $(\sinh(x/2))^{-2}$ function (in analogy to [16, 21, 54]) the Taylor series expansion

$$\left(\sinh\left(\frac{x}{2}\right)\right)^{-2} = 4x^{-2} + 4\sum_{m=1}^{\infty}\frac{(-1)^m(2m-1)|B_{2m}|x^{2m-2}}{(2m)!}, \quad \text{(A.6)}$$

which is convergent within the region $|x| < 2\pi$, we satisfy ourselves that the respective high-capacity (high-temperature) behaviour of the $\kappa_D(x_D)$ function (see (A.1)/(A.5)) can be represented (primarily) by a Taylor series expression of the conventional form [1, 2, 16, 21, 54]

$$\kappa_{Dh}^{m_D}(x_D) = 1 + \sum_{m=1}^{m_D(\to\infty)}(-1)^m\left|c_{2m}^D\right|x_D^{2m} = 1 + \sum_{m=1}^{m_D(\to\infty)} c_{2m}^D x_D^{2m}, \quad \text{(A.7)}$$

where the expansion coefficients for the individual even-order x_D^{2m} power terms (with alternating signs), $c_{2m}^D = (-1)^m|c_{2m}^D|$, are given by

$$c_{2m}^D = (-1)^m\frac{3(2m-1)|B_{2m}|}{(2m+3)\cdot(2m)!}, \quad m = 1,2,3,\ldots. \quad \text{(A.8)}$$

Here we have denoted by $|B_{2m}|$ (in analogy to [16, 21, 54, 90]) the absolute values of the respective Bernoulli numbers [90] (i.e., $|B_2| = 1/6$, $|B_4| = 1/30$, $|B_6| = 1/42$, $|B_8| = 1/30$, $|B_{10}| = 5/66$, $|B_{12}| = 691/2730$, $|B_{14}| = 7/6$, $|B_{16}| = 3617/510$, etc.). Accordingly, one finds the numerical values of the expansion coefficients, c_{2m}^D, for the first eight x_D^{2m} power terms (in (A.7)) to be explicitly given by the following fractions:

$$c_2^D = \frac{-1}{20},$$

$$c_4^D = \frac{1}{560},$$

$$c_6^D = \frac{-1}{18144},$$

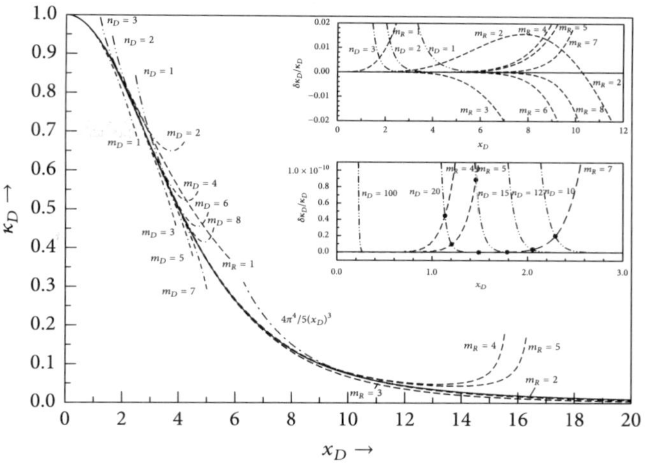

FIGURE 8: Visualization of the unusually rapid approaches of truncated versions of the presently developed (unprecedented) high-temperature x_D^{2m}-series expansion, $\kappa_{Dh}^{m_R}(x_D)$ (- - - - -; see (A.11), for $m_R = 1$ to 8), to the exact $\kappa_D(x_D)$ function (—), in comparison with the markedly slower approaches of truncated versions of Debye's conventional high-temperature x_D^{2m}-series expansion [1, 2, 16, 21, 54], $\kappa_{Dh}^{m_D}(x_D)$ (- - · - - · -; see (A.7), for $m_D = 1$ to 8).

$$c_8^D = \frac{1}{633600},$$
$$c_{10}^D = \frac{-1}{23063040},$$
$$c_{12}^D = \frac{691}{594397440000},$$
$$c_{14}^D = \frac{-1}{32572108800},$$
$$c_{16}^D = \frac{3617}{4505374089216000}, \ldots . \tag{A.9}$$

Note that the first 5 expansion coefficients, that is, c_2^D up to c_{10}^D, had already been quoted in [1, 2].

However, due to the alternating signs of the c_{2m}^D coefficients (A.9), this conventional high-capacity (high-temperature) expansion (see (A.7)) turned out to be only a slowly converging one (as shown, in Figure 8, by the rapidly seceding $\kappa_{Dh}^{1\le m_D\le 8}(x_D)$ curves following from truncated series expansions of type (A.7)). As already mentioned by Debye [1], one finds, for example, that expansions of type (A.7), which have been truncated after the x_D^6 or x_D^{10} term, show deviations of less than 1% from the exact $\kappa_D(x_D)$ curve (A.1) only within rather small x_D intervals (from 0 up to about 2 or 3, resp.). In particular one finds that the conventional high-capacity (high-temperature) expansion (A.7) is obviously not capable of providing high-precision results for the vicinity of $x_D \approx 2$.

We had already found in [54] that a considerable improvement of the convergence properties of the high-capacity (high-temperature) Taylor series expansion of type $\kappa_{Dh}^{m_D}(x_D)$ (see (A.7)) can be achieved by a straightforward transformation of the latter, for example, into an associated (equivalent) exponential series representation. Actually, by considering (in the appendix of [54]) just the logarithm, $\kappa_{DLh}(x_D) \equiv \log(\kappa_{Dh}(x_D))$, of the original $\kappa_{Dh}^{m_D}(x_D)$ function and performing a straightforward transformation of the expansion coefficients, $\{c_{2m}^D\} \to \{c_{2m}^{DL}\}$ (in analogy to the general infinite series transformation procedure sketched in Section 4.5.4 of [91]), we succeeded to construct a corresponding exponential Taylor series representation (see Eqs. (A.2) and (A.7) in [54]), which showed markedly better convergence properties than the original series expansion, $\kappa_{Dh}^{1\le m_D\le 5}(x_D)$, of type (A.6) (cf. Figure 6 in [54]).

Furthermore, within a more comprehensive (hitherto unpublished) analytical study performed in connection with the derivation of Eq. (3) in [15], we have found that an even faster convergence of a conveniently transformed version of the conventional high-capacity Taylor series expansion [1, 2, 16, 21, 54], $\kappa_{Dh}^{m_D}(x_D)$ (see (A.7)), can be achieved when we change from the latter to a corresponding Taylor series for *reciprocal* κ_{Dh}^{-1} values, $\kappa_{DRh}(x_D) \equiv 1/\kappa_{Dh}(x_D)$. Performing the straightforward transformation of the expansion coefficients, $\{c_{2m}^D\} \to \{c_{2m}^{DR}\}$ [91] (which can be done in comfortable way by using, for example, a mathematical standard program like Maple), we come to an unprecedented Taylor series expansion for the reciprocal values, $\kappa_{DRh}(x_D)$, of the form

$$\kappa_{DRh}^{m_R}(x_D) = 1 + \sum_{m=1}^{m_R(\to\infty)} c_{2m}^{DR} x_D^{2m} = 1 + \frac{x_D^2}{20} + \frac{x_D^4}{1400} + \frac{x_D^6}{648000} - \frac{73x_D^8}{3492720000} + \cdots . \tag{A.10}$$

Note that a more extended list of the 8 lowest-order expansion coefficients, $c_{2\le 2m\le 16}^{DR}$, is given in Table 6.

The relevant $\kappa_{Dh}(x_D)$ function in question can be thus represented, henceforth, in terms of the reciprocal magnitudes, $[\kappa_{DRh}^{m_R}(x_D)]^{-1}$, of $\kappa_{DRh}^{m_R}(x_D)$ values (see (A.10)); that is, explicitly

$$\kappa_{Dh}^{m_R}(x_D) = \left[1 + \sum_{m=1}^{m_R(\to\infty)} c_{2m}^{DR} x_D^{2m}\right]^{-1} = \left[1 + \frac{x_D^2}{20} + \frac{x_D^4}{1400} + \frac{x_D^6}{648000} - \frac{73x_D^8}{3492720000} + \cdots\right]^{-1}. \tag{A.11}$$

We see from Figure 8 that the truncated (low-order) series expansions of the latter type, $\kappa_{Dh}^{1\le m_R\le 5}(x_D) \equiv [\kappa_{DRh}^{1\le m_R\le 5}(x_D)]^{-1}$ (see (A.11)), are already much more rapidly converging to the exact $\kappa_D(x_D)$ dependence (solid curve) than their conventional predecessors, $\kappa_{Dh}^{1\le m_D\le 5}(x_D)$ (see (A.7)).

One can easily recognize the cause of the significantly improved convergence properties to be due to a qualitatively quite different constellation, of both signs and magnitudes, of expansions coefficients. Comparing in particular the magnitudes of the original expansion coefficients with the transformed ones, that is, the $|c_{2m}^D|$ values (in (A.9)) with the $|c_{2m}^{DR}|$ values (listed in Table 6), we see that the latter are decreasing much more rapidly with increasing order ($2m = 2, 4, 6, \ldots$) of the individual x_D^{2m} power terms. In particular we see that, for example, for $m = 8$, the corresponding coefficient, $c_{16}^{DR} = 4.952 \times 10^{-17}$ (in (A.11); cf. Table 6), is by more than 4 orders of magnitude smaller than its counterpart, $c_{16}^D = 3617/4505374089216000 = 8.028 \times 10^{-13}$, due to Debye's original high-temperature Taylor series expansion (cf. (A.7)). The unusually rapid decrease of magnitudes of the transformed coefficients, $|c_{2m}^{DR}|$, with increasing order of the x_D^{2m} power terms, is the main reason of the significantly improved convergence properties of the unprecedented high-capacity (high-temperature) Taylor series expansion (A.11) (as shown in Figure 8).

The truncated $\kappa_{Dh}^{m_R=7}(x_D)$ version of the novel expansion (see (A.11)), which is seen (from the lower inset of Figure 8) to intersect the conventional (low-capacity) $\kappa_{Dl}^{10\le n_D\le 20}(x_D)$ curves somewhere within the interval $1.4 < x_D < 2.3$ (as indicated by solid circles), turned out to be especially useful for calculations of high-precision $\kappa_D(x_D)$ values. Consider in particular the point of intersection between the $\kappa_{Dh}^{m_R=7}(x_D)$

TABLE 6: The 8 lowest-order expansion coefficients, $c^{DR}_{2\leq 2m\leq 16}$, implied by the unprecedented high-capacity (high-temperature) x_D^{2m}-series expansion, $\kappa^{m_R}_{Dh}(x_D)$ (A.11), for the conventional Debye function, $\kappa_D(x_D)$ (A.1). The same expansion coefficients, $c^{DR}_{2\leq 2m\leq 16}$, apply also the analogous x_1^{2m}-series expansion, $\kappa^{m_R}_{C1h}(x_1)$ (A.15), for the Debye-like component, $\kappa_{C1}(x_1)$ (A.13), of the low-energy TA phonon section of the lattice heat capacity. With respect to the latter we give here also the set of the 8 lowest-order expansion coefficients, $c^{NR}_{2\leq 2m\leq 16}$, which are implied by the analogous x_1^{2m}-series expansion, $\kappa^{m_R}_{C2h}(x_1)$ (A.23), for the associated non-Debye heat capacity component, $\kappa_{C2}(x_1)$ (A.17).

m	c^{DR}_{2m}	c^{NR}_{2m}
1	$\frac{1}{20} = 0.05$	$\frac{5}{84} = 0.05952380952$
2	$\frac{1}{1400} = 7.14285714286 \times 10^{-4}$	$\frac{13}{10584} = 1.228269085 \times 10^{-3}$
3	$\frac{1}{648000} = 1.54320987654 \times 10^{-6}$	$\frac{205}{19559232} = 1.048098412 \times 10^{-5}$
4	$\frac{-73}{3492720000} = -2.09006161387 \times 10^{-8}$	$\frac{4559}{160190110080} = 2.845993424 \times 10^{-8}$
5	$\frac{1}{82555200000} = 1.21131073512 \times 10^{-11}$	$\frac{-503}{4485323082240} = -1.121435381 \times 10^{-10}$
6	$\frac{5591}{5148967824000000} = 1.08584869650 \times 10^{-12}$	$\frac{-478871}{1585247736956083200} = -3.020795986 \times 10^{-13}$
7	$\frac{-1879}{212199886080000000} = -8.85485866515 \times 10^{-15}$	$\frac{-48299}{20240431054552702 9760} = -2.386261987 \times 10^{-17}$
8	$\frac{563743}{113831092222848000000000} = 4.9524518213 \times 10^{-17}$	$\frac{887853331}{1657692290336786637 3734400} = 5.355959827 \times 10^{-17}$

and $\kappa^{n_D=12}_{Dl}(x_D)$ curves (being located at $x_c = 2.052$), where the relative deviation, $\delta\kappa_D/\kappa_D$, of both curves from the exact $\kappa_D(x_c)$ value is $\kappa^{m_R=7}_{Dh}(x_c) - \kappa_D(x_c) = \kappa^{n_D=12}_{Dl}(x_c) - \kappa_D(x_c) = 0.37 \times 10^{-11}$. Thus, we can readily determine high-precision values for the whole $\kappa_D(x_D)$ function by using, alternatively, the functional values $\kappa^{m_R=7}_{Dh}(x_D) \rightarrow \kappa_D(x_D)$, for $0 \leq x_D \leq x_c$, and $\kappa^{n_D=12}_{Dl}(x_D) \rightarrow \kappa_D(x_D)$, for $x_c \leq x_D < \infty$. A series of corresponding high-precision $\kappa_D(x_D)$ sample values (limited to 10 significant figures) is presented in Table 7. The first 4 to 6 significant figures of our $\kappa_D(x_D)$ values (as well as the associated $D_3(x_D)$ values) are, of course, in complete accordance with corresponding earlier results presented in [6, 47, 90].

By the way, the precise knowledge of the actual crossing point positions is not really necessary in view of the generally monotonic increase or decrease of the deviations of the $\kappa^{m_R=7}_{Dh}$ ($x_D < 10$) function and the $\kappa^{n_D}_{Dl}$ ($x_D > 0$) functions, respectively, from the true $\kappa_D(x_D)$ curve. It is thus quite adequate to take (with respect, e.g., to the interval $0 \leq x_D \leq 3$) simply the lower ones of the numerical values following from the two complementary $\kappa_D(x_D)$ branches; that is,

$$\kappa_D(x_D) \longrightarrow \min\left\{\kappa^{m_R=7}_{Dh}(x_D), \kappa^{n_D\geq 12}_{Dl}(x_D)\right\}, \quad \text{for } 0 \leq x_D \leq 3, \tag{A.12a}$$

$$\kappa_D(\kappa) \longrightarrow \kappa^{n_D\geq 12}_{Dl}(x_D), \quad \text{for } 3 \leq x_D < \infty. \tag{A.12b}$$

Thus, at variance with the partly rather laborious procedures (see, e.g., [47]) involved by earlier $\kappa_D(x_D)$ calculations, we have devised here a rather comfortable way for calculations of $\kappa_D(x_D)$ values with an even considerably higher degree of accuracy (up to at least 10 significant figures), via the couple of the integral-free formulas (A.4) and (A.11). In view of this unforeseen and rather handy calculation scheme for the $\kappa_D(x_D)$ function, a use of extensive $\kappa_D(x_D)$ tables (like those presented in [6, 47–49]) appears to be completely unnecessary within the frame of possible future Debye function calculations (for 3-dimensional systems, at least).

The $m_R = 2$ version of the $\kappa^{m_R}_{Dh}(x_D)$ expansion (see (A.11)), that is, the function $\kappa^{m_R=2}_{Dh}(x_D) = [1+x_D^2/20+x_D^4/1400]^{-1}$, just turns out to be of considerable importance within the present context for the following reasons:

(i) Firstly, as we can see from the upper inset of Figure 8, this simple function deviates by less than 2% from the exact $\kappa_D(x_D)$ dependence within an unusually large x_D-interval ($0 \leq x_D < 11$).

(ii) Secondly, it provides already a relatively good approximation for the *inverse* dependence of $x_D(T) \equiv \Theta_D(T)/T$ values on heat capacity ratios, $\kappa(T) \equiv C_p(T)/C_{Vh}(\infty) > 0.05$ (see (B.3) below). The inverse form of the $\kappa^{m_R=2}_{Dh}(x_D)$ function plays thus an important role within the analytical apparatus (displayed in Appendix B) for constructions of reasonable $x_D(\kappa)$ formulas.

A.2. Integral-Free Formulas for Low-Phonon-Energy Heat Capacity Contributions. The normalized contributions, $\kappa_{Ck=1}(x_1(T))$ and $\kappa_{Ck=2}(x_1(T))$, of the low-phonon energy

TABLE 7: High-accuracy sample values for Debye's normalized heat capacity shape function, $\kappa_D(x_D)$ (A.1), which are directly comparable with their somewhat more restricted precedents listed in [6, 47] (up to 6 or 8 significant figures, resp.). The individual high-accuracy $\kappa_D(x_D)$ values have been calculated here for the interval $0 \le x_D \le 1.8$, via the unprecedented high-capacity (high-temperature) Taylor series expansion, $\kappa_{Dh}^{m_R=7}(x_D)$ (A.11) and, for the interval $1.8 \le x_D \le 40$, by means of Debye's conventional low-capacity (low-temperature) expansion, $\kappa_{Dl}^{n_D=15}(x_D)$ (A.4) (note that a simple asymptotic expression, $\kappa_D(x_D) \to 77.92727286/x_D^3$, applies to $x_D > 40$). For comparisons with numerical results quoted in mathematical standard literature (see, e.g., Table 27.1 in [90]) we have listed here also the respective magnitudes of the conventional Debye function, $D_3(x_D)$ (for 3-dimensional systems), $D_3(x_D) = [\kappa_D(x_D) + 3x_D/(\exp(x_D) - 1)]/4$ (according to (A.2) and (A.3)).

x_D	$\kappa_D(x_D)$	$D_3(x_D)$	x_D	$\kappa_D(x_D)$	$D_3(x_D)$
0.0	1.0000000000	1.0000000000	5.0	0.3686348236	0.1175974118
0.1	0.9995001785	0.9629999405	5.5	0.3132547732	0.09524080272
0.2	0.9980028536	0.9269990483	6.0	0.2655968891	0.07758132473
0.3	0.9955144242	0.8919951865	6.5	0.2250561623	0.06360434235
0.4	0.9920454895	0.8579848069	7.0	0.1908564046	0.05250585097
0.5	0.9876107521	0.8249629690	7.5	0.1621687997	0.04365502115
0.6	0.9822288834	0.7929233677	8.0	0.1381873779	0.03656029567
0.7	0.9759223556	0.7618583673	8.5	0.1181717647	0.03084031600
0.8	0.9687172408	0.7317590428	9.0	0.1014670944	0.02619989259
0.9	0.9606429816	0.7026152269	9.5	0.08750918665	0.02241065587
1.0	0.9517321357	0.6744155641	10.0	0.07582100303	0.01929576569
1.1	0.9420200996	0.6471475686	11.0	0.05773082402	0.01457049734
1.2	0.9315448136	0.6207976880	12.0	0.04478011684	0.01125032746
1.3	0.9203464535	0.5953513709	13.0	0.03534728217	0.008858858804
1.4	0.9084671129	0.5707931370	14.0	0.02835178383	0.007096677015
1.5	0.8959504801	0.5471066514	15.0	0.02307128730	0.005771263228
1.6	0.8828415127	0.5242747994	16.0	0.01901817415	0.004755893959
1.7	0.8691861161	0.5022797638	17.0	0.01585873694	0.003965212076
1.8	0.8550308264	0.4811031024	18.0	0.01336097602	0.003340449610
1.9	0.8404225040	0.4607258256	19.0	0.01136091855	0.002840309476
2.0	0.8254080384	0.4411284737	20.0	0.009740756601	0.002435220067
2.2	0.7943467192	0.4041938106	21.0	0.008414505229	0.002103638250
2.4	0.7622123728	0.3701368830	22.0	0.007318466926	0.001829621334
2.6	0.7293549760	0.3387926100	23.0	0.006404797367	0.001601201112
2.8	0.6961028658	0.3099951582	24.0	0.005637096943	0.001409274915
3.0	0.6627580442	0.2835798281	25.0	0.004987344231	0.001246836318
3.2	0.6295928776	0.2593847028	26.0	0.004433731487	0.001108432971
3.4	0.5968481116	0.2372520497	28.0	0.003549893920	0.0008874734947
3.6	0.5647320637	0.2170294741	30.0	0.002886195280	0.0007215488222
3.8	0.5334208309	0.1985708314	32.0	0.002378151635	0.0005945379091
4.0	0.5030593309	0.1817369138	34.0	0.001982680461	0.0004956701152
4.2	0.4737629932	0.1663959309	36.0	0.001670251904	0.0004175629760
4.4	0.4456199253	0.1524238050	38.0	0.001420164616	0.0003550411541
4.6	0.4186933946	0.1397043074	40.0	0.001217613638	0.0003044034095
4.8	0.3930244869	0.1281290579			

terms, $\propto \varepsilon^2$ and $\propto \varepsilon^4$ (in the PDOS model function (7)), to the harmonic lattice heat capacities are generally given by the integrals (10). Yet, in order to avoid, henceforth, the corresponding integration procedures also within the frame of least-mean-square fitting processes (like those ones performed in Section 4), it is desirable to substitute the corresponding integrals by appropriate integral-free formulas. This substitution can be immediately performed, for the $\kappa_{Ck=1}(x_1(T))$ function (of Debye type) at least, on the basis of the results obtained in Section A.1. Actually, by comparing the integral representations (10) for the $\kappa_{Ck=1}(x_1(T))$ function

$$\kappa_{C1}(x_1) = \frac{3}{x_1^3}\int_0^{x_1} dx \frac{x^4 e^x}{(e^x - 1)^2} = \frac{3}{4x_1^3}\int_0^{x_1} dx \frac{x^4}{(\sinh(x/2))^2} \tag{A.13}$$

with the integral representations (A.1) and (A.5) for Debye's normalized heat capacity function, $\kappa_D(x_D)$, we see that the only difference consists in different limits of integration (and the respective denominators in prefactors), namely, $x_1(T) \equiv \Theta_1/T$ (in (A.13)), instead of $x_D(T) \equiv \Theta_D/T$ (in Section A.1). Thus one can directly use the final formulas derived in Section A.1 for the $\kappa_D(x_D(T))$ dependence (like (A.4) and (A.11)) by making but the corresponding substitutions of the characteristic (dimensionless) phonon versus lattice temperature ratios, $x_D(T) \rightarrow x_1(T)$. Accordingly, the low- and high-temperature branches of the $\kappa_{C1}(x_1(T))$ function are explicitly given by

$$\kappa_{C1l}^{n_D}(x_1) = \frac{4\pi^4}{5x_1^3} - \frac{3x_1}{e^{x_1} - 1} - 12x_1 \sum_{n=1}^{n_D(\to\infty)} e^{-nx_1}\left(\frac{1}{nx_1} + \frac{3}{(nx_1)^2} + \frac{6}{(nx_1)^3} + \frac{6}{(nx_1)^4}\right), \tag{A.14}$$

$$\kappa_{C1h}^{m_R}(x_1) = \left[1 + \sum_{m=1}^{m_R(\to\infty)} c_{2m}^{DR} x_1^{2m}\right]^{-1} = \left[1 + \frac{x_1^2}{20} + \frac{x_1^4}{1400} + \frac{x_1^6}{648000} - \frac{73x_1^8}{3492720000} + \cdots\right]^{-1}, \tag{A.15}$$

in analogy to (A.4) and (A.11), respectively. An especially useful approximate version of the latter expansion (see (A.15)) is due again to a cut just after the x_1^{14} term (i.e., for $m_R = 7$). Observing further that the contributions of components $\kappa_{C1}(x_1)$ of Debye type (as well as the contributions of the components $\kappa_{C2}(x_1)$ of non-Debye type; see below) to the complete $\kappa_h(T)$ function (see (9)) are usually very small (owing to their relatively small weights, $w_{Ck} \ll 1$; cf. Table 1), it used to be sufficient for practical applications to determine only the first 4 significant figures of these functions. We have found that this reduced degree of precision is achieved when we cut the low-temperature series expansion (A.14) just after the $n_D = 1$ term. Consequently, we have used for the $\kappa_{C1}(x_1)$ component (in Section 4) an approximation of the form

$$\kappa_{C1}^{\text{appr.}}(x_1) \longrightarrow \min\left\{\kappa_{C1h}^{m_R=7}(x_1), \kappa_{C1l}^{n_D=1}(x_1)\right\}, \quad \text{for } 0 \le x_1 \le 10, \tag{A.16}$$

in combination with $\kappa_{C1}^{\text{appr.}}(x_1) \rightarrow \kappa_{C1l}^{n_D=1}(x_1)$, for $10 \le x_1 < \infty$. We see from the upper inset of Figure 1 (in Section 3) that the respective crossing point between the $\kappa_{C1h}^{m_R=7}(x_1)$ and $\kappa_{C1l}^{n_R=1}(x_1)$ branches is located at $x_{c1(D)} = 6.5135$, where the deviation of the approximate values, $\kappa_{C1}^{\text{appr.}}(x_{c1(D)})$ (due to (A.16)) from the exact $\kappa_{C1}(x_{c1(D)})$ value (A.13), reaches a (maximum) value of 0.748×10^{-4}.

Consider finally the associated non-Debye function, $\kappa_{Ck=2}(x_1(T))$ (see (10)), whose dependence on $x_1(T)$ was described hitherto by the respective integral representations

$$\kappa_{C2}(x_1) = \frac{5}{x_1^5}\int_0^{x_1} dx \frac{x^6 e^x}{(e^x - 1)^2} = \frac{5}{4x_1^5}\int_0^{x_1} dx \frac{x^6}{(\sinh(x/2))^2}. \tag{A.17}$$

Comparing the latter with (A.13) we see that the difference consists in different prefactors (i.e., $5/x_1^5$, instead of $3/x_1^3$) in combination with a higher-order power of the integration variable (i.e., x^6, instead of x^4), in the integrand of (A.17) versus (A.13). These are the causes of a pronounced qualitative difference of the x_1-dependence of the $\kappa_{C2}(x_1)$ versus $\kappa_{C1}(x_1)$ curve (cf. Figure 1, in Section 3).

A corresponding high-accuracy series expansion for the $\kappa_{C2l}(x_1)$ dependence can be derived, again, via several subsequent integrations by parts of the first integral version in (A.17) (in analogy to the derivation of the $\kappa_{Dl}^{n_D}(x_1)$ series expansion (see (A.4)) from the original integral form (see (A.1)). In this way we have obtained for the low-temperature branch of the $\kappa_{C2}(x_1)$ function (see (A.17)) an expansion of the form

$$\kappa_{C2l}^{n_N}(x_1) = \frac{80\pi^6}{21x_1^5} - \frac{5x_1}{e^{x_1} - 1} - 30x_1 \sum_{n=1}^{n_N(\to\infty)} e^{-nx_1}\left(\frac{1}{nx_1} + \frac{5}{(nx_1)^2} + \frac{20}{(nx_1)^3} + \frac{60}{(nx_1)^4} + \frac{120}{(nx_1)^5} + \frac{120}{(nx_1)^6}\right). \tag{A.18}$$

For the complementary high-temperature behaviour one can use, again, the Taylor series expansion (A.6) for the $(\sinh(x/2))^{-2}$ function in the second version of (A.17). In this way one comes readily to an infinite Taylor series expansion of the form

$$\kappa_{C2h}^{m_N}(x_1) = 1 + \sum_{m=1}^{m_N(\to\infty)} (-1)^m \left|c_{2m}^N\right| x_1^{2m} = 1 + \sum_{m=1}^{m_N(\to\infty)} c_{2m}^N x_1^{2m}, \tag{A.19}$$

where the respective expansion coefficients for the individual even-order x_1^{2m} power terms (with alternating signs), $c_{2m}^N = (-1)^m |c_{2m}^N|$, are given by

$$c_{2m}^N = (-1)^m \frac{5(2m-1)\left|B_{2m}\right|}{(2m+5)\cdot(2m)!}, \quad m = 1, 2, 3, \ldots. \tag{A.20}$$

Accordingly, one finds the numerical values of the expansion coefficients, c_{2m}^N, for the first eight x_1^{2m} power terms to be explicitly given by the following fractions:

$$c_2^N = \frac{-5}{84},\quad c_4^N = \frac{1}{432},\quad c_6^N = \frac{-5}{66528},\quad c_8^N = \frac{1}{449280},\quad c_{10}^N = \frac{-1}{15966720},\quad c_{12}^N = \frac{691}{404190259200},\quad c_{14}^N = \frac{-1}{21842472960},\quad c_{16}^N = \frac{3617}{298777439600400}. \tag{A.21}$$

However, due to the alternating signs of the c_{2m}^N coefficients (A.21), this high-temperature expansion (see (A.19)) turned out to be, again, a rather slowly converging one (in analogy to the conventional high-temperature expansion (see (A.7)) for the Debye function).

We have found that a considerably faster convergence can be achieved also for the latter case when we change from the original $\kappa_{C2h}^{m_N}(x_1)$ expansion (see (A.19)) to an associated Taylor series for *reciprocal* κ_{C2h}^{-1} values, $\kappa_{C2Rh}(x_D) \equiv 1/\kappa_{C2h}(x_D)$. Performing the corresponding transformation of the expansion coefficients, $\{c_{2m}^N\} \rightarrow \{c_{2m}^{NR}\}$ (in analogy, again, to the transformation procedure sketched in Section 4.5.4 of [91]), we came to a Taylor series expansion for the reciprocal values, $\kappa_{C2Rh}(x_D)$, of the form

$$\kappa_{C2Rh}^{m_R}(x_1) = 1 + \sum_{m=1}^{m_R(\rightarrow\infty)} c_{2m}^{NR} x_1^{2m} = 1 + \frac{5x_1^2}{84} + \frac{13x_1^4}{10584} + \frac{205x_1^6}{19559232} + \cdots. \tag{A.22}$$

Note that a list of the 8 lowest-order expansion coefficients, $c_{2\le 2m\le 16}^{NR}$, is given in Table 6. Accordingly, the $\kappa_{C2h}^{m_N}(x_1)$ function in question can be represented, alternatively, in the form of the reciprocal of (A.22); that is, explicitly

$$\kappa_{C2h}^{m_R}(x_1) = \left[1 + \sum_{m=1}^{m_R(\rightarrow\infty)} c_{2m}^{NR} x_1^{2m}\right]^{-1} = \left[1 + \frac{5x_1^2}{84} + \frac{13x_1^4}{10584} + \frac{205x_1^6}{19559232} + \cdots\right]^{-1}. \tag{A.23}$$

One can recognize the cause of significantly improved convergence properties of the latter series expansion (see (A.23)), in comparison with (A.19), to be due, again, to the qualitatively different constellations of both signs and magnitudes of respective expansions coefficients. Comparing in particular the magnitudes of the original with the transformed expansions coefficients, that is, the $|c_{2m}^N|$ values (in (A.21)) with the $|c_{2m}^{NR}|$ values (listed in Table 6), we see that the latter are decreasing much more rapidly with increasing order ($2m = 2, 4, 6, \ldots$) of the individual x^{2m} power terms. In particular we see that, for example, for $m = 8$, the corresponding coefficient, $c_{16}^{NR} = 5.356 \times 10^{-17}$ (in (A.23)); cf. Table 6), is by more than 5 orders of magnitude smaller than its counterpart, $c_{16}^N = 3617/298777439600400 = 1.211 \times 10^{-11}$, due to the original high-temperature Taylor series expansion (A.19). The unusually rapid decrease of magnitudes of the transformed coefficients, $|c_{2m}^{NR}|$, with increasing order of the x_1^{2m} power terms is the main reason of significantly improved convergence properties of the high-temperature Taylor series expansion, $\kappa_{C2h}^{m_R}(x_1)$ (see (A.23)), for the $\kappa_{C2}(x_1)$ function (see (A.19)) of non-Debye type.

An especially convenient version of the latter expansion within the present context has been found to be given by a cut just after the x_1^{12} term (i.e., for $m_R = 6$). Furthermore it turns out to be sufficient to limit the associated low-temperature expansion (see (A.18)) to the lowest order ($n_N = 1$). Thus we can combine the couple of the corresponding $\kappa_{C2h}^{m_R=6}(x_1)$ and $\kappa_{C2l}^{n_N=1}(x_1)$ branches to provide a reasonable approximation for the $\kappa_{C2}(x_1)$ function (see (A.17)) of the form

$$\kappa_{C2}^{\text{appr.}}(x_1) \longrightarrow \min\left\{\kappa_{C2h}^{m_R=6}(x_1), \kappa_{C2l}^{n_N=1}(x_1)\right\}, \quad \text{for } 0 \le x_1 \le 10, \tag{A.24}$$

in combination with $\kappa_{C2}^{\text{appr.}}(x_1) \rightarrow \kappa_{C2l}^{n_N=1}(x_1)$, for $10 \le x_1 < \infty$.

We see from the lower inset of Figure 1 (in Section 3) that the respective crossing point between the $\kappa_{C2h}^{m_R=6}(x_1)$ and $\kappa_{C2l}^{n_N=1}(x_1)$ branches is located at $x_{c2(N)} = 7.0962$, where the deviation of the approximate values $\kappa_{C2}^{\text{appr.}}(x_{c2(N)})$ (due to (A.24)) from the exact $\kappa_{C2}(x_{c2(N)})$ value (see (A.13)) reaches a (maximum) value of 1.306×10^{-4}.

B. Integral-Free Formulas for the Inverse Debye Function

Let us now study in due detail the hitherto almost completely ignored (though practically highly important) *inverse* problem of devising appropriate, integral-free formulas for the dependence of the upper limits of integration, $x_D(T) = \Theta_D(T)/T$ (in (A.1)), on the given magnitudes of capacity ratios, $\kappa(T) = C_p(T)/C_{Vh}(\infty)$.

B.1. Basic Models. For very low (liquid-helium-hydrogen) temperatures, the first term, $\kappa_{Dl}^{\lim}(x_D) = 4\pi^4/5x_D^3$, in the low-capacity expansion (A.4) is the strongly dominating one. Inverting this asymptotic relationship between κ and x_D one obtains readily for the corresponding asymptotic dependence of x_D on κ the simple (cubic root) expression formula

$$x_D(\kappa \longrightarrow 0) \longrightarrow x_D^{l_0}(\kappa) = \sqrt[3]{\frac{4\pi^4}{5\kappa}}. \quad \text{(B.1)}$$

Yet, a comparison (in Figure 9) of the asymptotic (dotted) $x_D^{l_0}(\kappa)$ curve with the exact $x_D(\kappa)$ dependence (solid curve) shows that the deviations, $x_D^{l_0}(\kappa) - x_D(\kappa) > 0$, are smaller than 1% only for very low magnitudes of heat capacities, namely, $0 < \kappa < 0.03$ (cf. the upper inset of Figure 9). This corresponds to a confinement of the applicability of (B.1) to a limiting region of relatively high x_D values ($x_D > 14$; see also [3, 13] and cf. Table 7). At the same time it is clear that, for higher κ values, it is impossible to derive a more general $x_D(\kappa)$ formula directly from Debye's rather complex low-capacity (low-temperature) expression [1–3], $\kappa_{Dl}^{n_D}(x_D)$ (see (A.4)).

In order to come to a practicable algebraic expression for the qualitatively different $x_D(\kappa)$ behaviour adopted at sufficiently high κ values (including the $\kappa \rightarrow 1$ limit) we start from the unprecedented Taylor series expansion for reciprocal values $(\kappa(x_D))^{-1} = \kappa_{DRh}^{m_R}(x_D)$ (see (A.10)); that is,

$$\kappa^{-1} = 1 + \frac{x_D^2}{20} + \frac{x_D^4}{1400} + \frac{x_D^6}{648000} - \cdots. \quad \text{(B.2)}$$

Limiting ourselves, for convenience, to the two lowest-order x_D^{2n} terms (which corresponds to the limitation of the summation in (A.10) to $m_R = 2$), we readily obtain via a solution of the respective quadratic equation, $z_D^2/1400 + z_D/20 = \kappa^{-1} - 1$ (where $z_D \equiv x_D^2$), for the dependence of $x_D \equiv \sqrt[2]{z_D}$ on κ^{-1} in exact way a simple algebraic expression of the form [15]

$$x_D^{h_0}(\kappa) = \sqrt[2]{-35 + 5 \cdot \sqrt[2]{49 + 56 \cdot (\kappa^{-1} - 1)}}. \quad \text{(B.3)}$$

This highly useful approximation for the high-capacity (high-temperature) dependence, $x_D^h(\kappa)$, had already preliminarily been used for some numerical comparisons with the exact $x_D(\kappa)$ dependence in our preceding paper [15] (though without indicating, there, its actual origin; cf. Eq. (26) in [15]). Now we have displayed, within the frame of the closely interrelated Appendices A and B, the underlying analytical basis. Furthermore we would like to note that the behaviour of the $x_D^{h_0}(\kappa)$ function close to the upper limit, $\kappa \rightarrow 1$, follows readily from (B.3) to tend to the asymptote $x_D^{h_0}(\kappa \rightarrow 1) \rightarrow \sqrt[2]{20 \cdot (\kappa^{-1} - 1)}$ (in accordance with Eq. (27) in [15]).

The whole $x_D^{h_0}(\kappa)$ dependence (B.3) is represented by the dashed curve (labelled by h_0 ($m_R = 2$)) in Figure 9. Within the large scale representation, at least, this $x_D^{h_0}(\kappa)$ curve appears to be almost indistinguishable from the exact (solid) $x_D(\kappa)$ curve, throughout a very large interval of κ-values ($0.05 < \kappa \leq 1$). On the other hand we see from the upper inset (to Figure 9) that the approximate $x_D^{h_0}(\kappa)$ curve shows within this interval deviations, $\delta x_D/x_D$, from the exact $x_D(\kappa)$ curve up to about 0.6%.

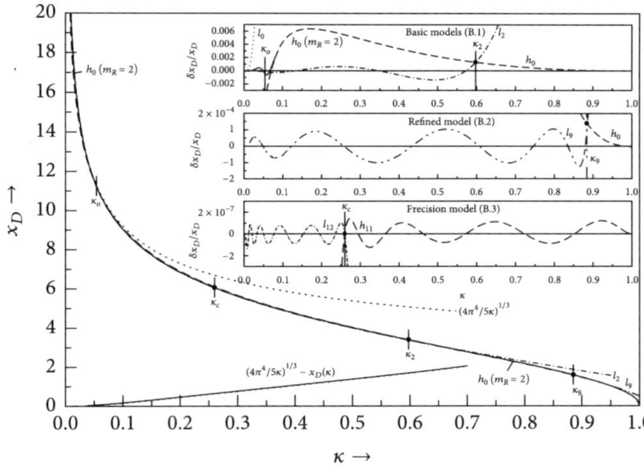

Figure 9: Exact dependence (—) of the characteristic Debye variable x_D ($=\Theta_D(T)/T$) on the relative (normalized) magnitudes of heat capacities, κ ($= C_p(T)/C_{Vh}(\infty)$), in comparison with the low-κ asymptote, $x_D^{l_0}(\kappa) = (4\pi^4/5\kappa)^{1/3}$ ($\cdots$; see (B.1)), and the characteristic high-κ approximation, $x_D^{h_0}(\kappa)$ (- - - - -; see (B.3)). Considerable reductions of maximum deviations to ±0.13% (see the upper inset) or ±0.0153% (see the middle inset) are achieved by means of combinations of two more elaborate, qualitatively different versions of low-to-intermediate-κ approximations, $x_D^{l_2}$ ($0 \leq \kappa \leq \kappa_2$) (–·– · –·–; see (B.5a) and (B.5b)) and $x_D^{l_9}$ ($0 \leq \kappa \leq \kappa_9$) (– – · – –·–; see (B.6)), in combination with the complementary high-κ curve sections, that is, $x_D^{h_0}$ ($\kappa_2 \leq \kappa \leq 1$) or $x_D^{h_0}$ ($\kappa_9 \leq \kappa \leq 1$), respectively. From the lower inset we see that the residual deviations of the complementary high-precision formulas $x_D^{l_{12}}$ ($0 \leq \kappa \leq \kappa_c$) (see (B.8)) and $x_D^{h_{11}}$ ($\kappa_c \leq \kappa \leq 1$) (see (B.9)) are throughout smaller than ±0.000016%.

Considering in more detail the characteristic deviations of both the $x_D^{l_0}(\kappa)$ curve (see (B.1)) and of the $x_D^{h_0}(\kappa)$ curve (see (B.3)) from the exact $x_D(\kappa)$ dependence in low-capacity region ($0 < \kappa < 0.07$) we see that they have different signs, and their magnitudes are reaching the same order (of about ±1.5%) in the vicinity of $\kappa \approx 0.035$. For such a constellation of noncrossing curve branches it is possible to achieve a significant reduction (partial mutual compensation) of deviations via a connection of the two complementary curve branches, $x_D^{l_0}(\kappa)$ and $x_D^{h_0}(\kappa)$, by an appropriate junction formula of type

$$\widetilde{x}_D(\kappa) = \frac{x_D^{h_0}(\kappa) + \exp\left(\left(\kappa_o - \kappa\right)/\Delta\kappa\right) x_D^{l_0}(\kappa)}{1 + \exp\left(\left(\kappa_o - \kappa\right)/\Delta\kappa\right)}. \quad \text{(B.4a)}$$

An optimal approach of $\widetilde{x}_D(\kappa)$ to the exact $x_D(\kappa)$ function, within the interval $0 < \kappa < 0.1$, has been found via a corresponding least-mean square fitting process. This led to the following couple of junction parameter values:

$$\kappa_o = 0.054, \qquad \Delta\kappa = 0.006. \quad \text{(B.4b)}$$

The respective $\widetilde{x}_D(\kappa)$ curve is represented in the upper inset of Figure 9 by the slightly oscillating solid curve section, for $\kappa < 0.07$ (corresponding to an x_D-region of $x_D > 10$). In view of the resulting limitation of residual deviations to on order of ±0.06% (within the interval $0 < \kappa < 0.07$) it is obvious that the adoption of the junction formula, $\tilde{x}_D(\kappa)$ (see (B.4a) and (B.4b)), effectuates here a significantly closer approach to the exact $x_D(\kappa)$ curve than their individual components per se, $x_D^{l_0}(\kappa)$ (see (B.1)) and $x_D^{h_0}(\kappa)$ (see (B.3)). On the other hand we see also from the upper inset of Figure 9 that, outside this relatively narrow low-capacity interval (i.e., for $\kappa > 0.07$), the junction formula, $\tilde{x}_D(\kappa)$ (see (B.4a) and (B.4b)), is still implying deviations from the exact $x_D(\kappa)$ curve up to about 0.6% (owing to the asymptotic approach $\tilde{x}_D(\kappa) \rightarrow x_D^{h_0}(\kappa)$, which is implied by the actual junction parameter constellation, κ_o and $\Delta\kappa$ (B.4b), in (B.4a)).

An important chance for significant improvements of the analytical description of the $x_D(\kappa)$ dependence within the middle region ($0.1 < \kappa < 0.7$) is provided by the unfamiliar observation that the difference, $x_D^{l_0}(\kappa) - x_D(\kappa) \geq 0$, between the $\kappa \rightarrow 0$ asymptote, $x_D^{l_0}(\kappa) = \sqrt[3]{4\pi^4/5\kappa}$ (dotted curve), and the exact $x_D(\kappa)$ dependence (solid curve) corresponds, roughly, to a *linear* dependence on κ. This peculiarity is apparent from the respective $\sqrt[3]{4\pi^4/5\kappa} - x_D(\kappa) \geq 0$ curve (as shown in the lower part of Figure 9). Estimating the slope of this curve (e.g., within an interval of $0.2 < \kappa < 0.5$) we have obtained an average value of $d(x_D^{l_0}(\kappa) - x_D(\kappa))/d\kappa \cong 3.036$. At the same time we see (from Figure 9) that the slope of this curve decreases monotonically, with decreasing κ, and finally tends to 0 in the $\kappa \rightarrow 0$ limit. This behaviour of the slope can be simulated, for example, by some "slope reduction factor" of exponential type, $\exp(-p/\kappa)$, so that the difference in question has a chance to be reasonably approximated by a function of type $g\kappa\exp(-p/\kappa)$ (where the factor g should be approximately equal to the effective slope (quoted above) for the middle region). Accordingly, we choose for an approximate simulation of the $x_D(\kappa)$ dependence in the region of low-to-intermediate κ values an effective 2-parameter formula of type

$$x_D^{l_2}(\kappa) = \sqrt[3]{\frac{4\pi^4}{5\kappa}} - g\kappa \exp\left(-\frac{p}{\kappa}\right). \quad \text{(B.5a)}$$

Performing a least-means square process of the exact $x_D(\kappa)$ curve by means of this approximate 2-parameter formula (B.5a), within the range $0 < \kappa < 0.63$, we have obtained the parameter values

$$g = 3.145972, \quad p = 0.07037359. \quad \text{(B.5b)}$$

This extended low-κ curve branch, $x_D^{l_2}(\kappa)$ (see (B.5a) and (B.5b)), crosses the prefabricated high-κ curve branch, $x_D^{h_0}(\kappa)$ (see (B.3)), at the point $\kappa_2 = 0.5975$ (see the upper inset of Figure 9). At this point the deviations of both approximate curves from the exact one are found to amount to about 0.13%. This maximum overestimation at the crossing point, κ_2, is seen to be of the same order as the maximum underestimation of the exact $x_D(\kappa)$ function by the $x_D^{l_2}(\kappa)$ function (see (B.5a) and (B.5b)), in the vicinity of $\kappa \approx 0.5$. Henceforth, by switching from the extended low-κ formula $x_D^{l_2}(\kappa)$ (see (B.5a) and (B.5b)) to the high-κ formula $x_D^{h_0}(\kappa)$ (see (B.3)) just at the crossing point, $\kappa_2 = 0.5975$, we are obtaining a complete sequence of approximate x_D ($0 \leq \kappa \leq 1$) values; the maximum deviations from the exact ones are not exceeding an order of ±0.13%. Such approximate results may already appear to be largely acceptable with respect to various practical applications.

In this connection we would still like to note that, even within the low-temperature region ($0 < T < 300$ K), the uncertainties of experimentally measured heat capacities, $C_p(T)$, used to be of order 0.5 to 1%. Thus the inherent experimental uncertainties used to have a significantly stronger influence on residual uncertainties of final results than the computational inaccuracies that are involved by the use of a couple of complementary approximation formulas, like $x_D^{l_2}$ ($0 < \kappa \leq \kappa_2$) (see (B.5a) and (B.5b)) in combination with $x_D^{h_0}$ ($\kappa_2 \leq \kappa \leq 1$) (see (B.3)).

B.2. Refined Model. A significant refinement of the analytical model for low-to-intermediate heat capacities can be achieved when we take properly into account that the slope of the difference, $\sqrt[3]{4\pi^4/5\kappa} - x_D(\kappa) \geq 0$, tends to be slightly increasing in the region $\kappa > 0.5$ (cf. the corresponding curve in the lower part of Figure 9). One can readily simulate these successively increasing deviations from linearity by including a series of higher-order power terms, κ^n ($n = 2, 3, \ldots$), into (B.5a). Limiting ourselves here to a series of 7 additional power terms, and adopting a unique slope reduction factor for the whole power series, we can well represent the $x_D^l(\kappa)$ dependence by a significantly refined 9-parameter expression of the form

$$x_D^{l_9}(\kappa) = \sqrt[3]{\frac{4\pi^4}{5\kappa}} - \exp\left(-\frac{p}{\kappa}\right)\sum_{n=1}^{8} g_n \kappa^n. \quad \text{(B.6)}$$

Performing a least-means square fitting process of the exact $x_D(\kappa)$ curve by means of this 9-parameter formula (B.6) within the range $0 < \kappa < 0.89$ we have obtained the parameter values listed in the first column of Table 8. The respective $x_D^{l_9}(\kappa)$ curve is seen to cross the prefabricated high-κ curve branch, $x_D^{h_0}(\kappa)$ (see (B.3)), at the point $\kappa_9 = 0.8844$ (see the middle inset of Figure 9). At this point the deviations of the two complementary (approximate) curve branches, $x_D^{l_9}(\kappa)$ and $x_D^{h_0}(\kappa)$, from the exact $x_D(\kappa)$ curve are seen to be of order 0.015%. This slight overestimation at the crossing point is seen to be of the same order of magnitude as the maximum deviations, $(\delta x_D/x_D)_{\max} \approx \pm 0.015\%$, of the significantly refined 9-parameter formula, $x_D^{l_9}(\kappa)$ (see (B.6)) from the exact $x_D(\kappa)$ dependence throughout the whole interval of its possible applicability, $0 \leq \kappa \leq \kappa_9$. Consequently, by switching from the latter to the prefabricated high-κ formula, $x_D^{h_0}(\kappa)$ (see (B.3)) just at the respective crossing point, $\kappa_9 = 0.8844$, one is coming to a significantly refined sequence of approximate x_D ($0 \leq \kappa \leq 1$) values; the maximum deviations

TABLE 8: Fitted values of the model-specific coefficients involved by the refined low-to-intermediate capacity $x_D(\kappa)$ formula, $x_D^{l_9}$ $(0 < \kappa \le \kappa_9)$ (B.6), and for the couple of complementary high-accuracy low- and high-capacity Debye temperature formulas, $x_D^{l_{12}}$ $(0 < \kappa \le \kappa_c)$ (B.8) and $x_D^{h_{11}}$ $(\kappa_c \le \kappa \le 1)$ (B.9), which have been determined by least-mean-square fittings of a sufficiently comprehensive set of couples of high-accuracy κ versus x_D curve points (in accordance with Table 7).

$x_D^{l_9}(\kappa)$ (see (B.6)) $0 < \kappa \le \kappa_9 = 0.8844$	$x_D^{l_{12}}(\kappa)$ (see (B.8)) $0 < \kappa \le \kappa_c = 0.260$	$x_D^{h_{11}}(\kappa)$ (see (B.9)) $0.260 = \kappa_c \le \kappa \le 1$
$p = 0.071797185$	$a_2 = 2.8743354 \times 10^{-1}$	$b_2 = 6.0777673 \times 10^{-3}$
$g_1 = 3.042348$	$a_3 = 1.1377662 \times 10^{-1}$	$b_3 = 5.1466371 \times 10^{-3}$
$g_2 = 1.816742$	$a_4 = -9.5999179 \times 10^{-2}$	$b_4 = -6.8442924 \times 10^{-3}$
$g_3 = -6.892814$	$a_5 = 4.3062699 \times 10^{-2}$	$b_5 = 2.6051894 \times 10^{-2}$
$g_4 = 6.954053$	$a_6 = -4.1060421 \times 10^{-3}$	$b_6 = -3.6011454 \times 10^{-2}$
$g_5 = 5.456090$	$a_7 = -1.1955013 \times 10^{-3}$	$b_7 = 3.6169365 \times 10^{-2}$
$g_6 = -11.20018$	$a_8 = 4.9953204 \times 10^{-4}$	$b_8 = -3.7204523 \times 10^{-2}$
$g_7 = 2.022044$	$a_9 = -6.2681506 \times 10^{-5}$	$b_9 = 1.6804524 \times 10^{-2}$
$g_8 = 2.824376$	$a_{10} = 3.3449835 \times 10^{-6}$	$b_{10} = -3.0967328 \times 10^{-3}$
—	$p = 1.2969$	$b_{11} = 5.2219439 \times 10^{-2}$
—	$g_2 = -3.1541530$	$b_{12} = -6.2935315 \times 10^{-2}$
—	$g_3 = 8.7184026$	—

from the exact ones are not surpassing the rather narrow inaccuracy range of ±0.015%.

B.3. Precision Model. Notwithstanding the circumstance that the refined model just displayed (within the preceding Section B.2) may already be looked upon as a quite adequate one for most practical applications, it appears useful from theoretical points of view to envisage still the construction of a highly elaborate analytical model, the numerical accuracy of which should be by several orders of magnitude higher than the preceding one. To this end we observe, first of all, that a detailed study of the $\kappa \to 0$ limiting behaviour of the exact $x_D(\kappa)$ dependence does not really show a perfect plateau behaviour (as suggested above by the global exponential slope reduction factors, $\exp(-p/\kappa)$, in (B.5a), (B.5b), and (B.6)). Actually, we have found within a careful numerical analysis that the difference between the $\kappa \to 0$ asymptote, $x_D^{l_0}(\kappa)$ (see (B.1)), and the exact $x_D(\kappa)$ dependence at $\kappa \ll 0.01$ is tending just to a 5th-order power dependence, $[x_D^{l_0}(\kappa) - x_D(\kappa)] \propto \kappa^5$. Towards higher κ values, the effective power function exponent decreases successively until the difference reaches a nearly linear dependence, in the vicinity of $\kappa \approx 0.2$. We have found that a very fine simulation of this gradual change from a κ^5 to a κ^1 dependence can be achieved via the reciprocal values of a truncated power series consisting of κ^{-5} to κ^{-1} power terms (including, for convenience, even terms with half integer exponents); that is,

$$\left[x_D^{l_0}(\kappa) - x_D(\kappa)\right] \longrightarrow \left[\sum_{n=2}^{10} \frac{a_n}{\kappa^{n/2}}\right]^{-1}, \tag{B.7}$$

within the range $0 < \kappa < 0.2$.

For the sake of an extension of the applicability of the envisaged high-accuracy formula up to $\kappa \approx 0.25$ (at least) we shall still include (in analogy to (B.6)) a quadratic and a cubic term in combination, again, with an exponential slope reduction factor for these two additional terms. In summary, we shall represent the $x_D(\kappa)$ dependence within the region $0 < \kappa < 0.26$ by the elaborate 12-parameter formula

$$x_D^{l_{12}}(\kappa) = \sqrt[3]{\frac{4\pi^4}{5\kappa}} - \left[\sum_{n=2}^{10} \frac{a_n}{\kappa^{n/2}}\right]^{-1} - \exp\left(-\frac{p}{\kappa}\right)\sum_{n=2}^{3} g_n \kappa^n. \tag{B.8}$$

Performing a least-mean-square fitting process of the exact $x_D(\kappa)$ curve by means of this 12-parameter formula (B.8) within the range $0 < \kappa < 0.26$ we have obtained the adjusted parameter values (a_n, $n = 2$ to 10, g_2, g_3, and p) listed in Table 8. From the lower inset of Figure 9 we see that the still remaining deviations of the respective $x_D^{l_{12}}(\kappa)$ curve (see (B.8)) from the exact $x_D(\kappa)$ dependence are limited to an order of ±0.000013%. This means an increase of accuracy of numerical results due to (B.8) by 3 orders of magnitude in comparison with (B.6). A set of corresponding high-accuracy $x_D^{l_{12}}(\kappa)$ values (for $0 \le \kappa \le 0.26$) is presented in Table 9.

In order to achieve a similarly strong increase of accuracy also for the rest interval of intermediate-to-high κ values ($0.26 \le \kappa \le 1$) we endow the prefabricated high-capacity formula $x_D^{h_0}(\kappa)$ (B.3) by an appropriate power series of $(1-\kappa)^n$ terms ($n = 2$ to 12) occurring in the denominator. This model corresponds thus to an 11-parameter expression of the form

$$x_D^{h_{11}}(\kappa) = \frac{\sqrt[2]{-35 + 5 \cdot \sqrt[2]{49 + 56 \cdot (\kappa^{-1} - 1)}}}{\left[1 + \sum_{n=2}^{12} b_n (1-\kappa)^n\right]}. \tag{B.9}$$

Table 9: High-precision sample values of the characteristic Debye variable, x_D ($\equiv \Theta_D(T)/T$), which have been calculated for a series of discrete (equidistant) κ-values via the couple of complementary low- and high-capacity formulas, $x_D^{l_{12}}$ ($0 < \kappa \le \kappa_c$) (B.8) and $x_D^{h_{11}}$ ($\kappa_c \le \kappa \le 1$) (B.9), using the model-specific coefficients listed in Table 8 (note that we have obtained precisely the same $x_D(\kappa)$ values also in alternative way via numerical integration procedures based directly on (A.1), for integration steps chosen to be of order 10^{-7}).

κ	$x_D(\kappa)$	κ	$x_D(\kappa)$	κ	$x_D(\kappa)$	κ	$x_D(\kappa)$	κ	$x_D(\kappa)$
0.01	19.825638	0.21	6.709531	0.41	4.666635	0.61	3.319275	0.81	2.100219
0.02	15.733335	0.22	6.568697	0.42	4.590077	0.62	3.258266	0.82	2.035428
0.03	13.737134	0.23	6.434348	0.43	4.514891	0.63	3.197532	0.83	1.969683
0.04	12.468384	0.24	6.305852	0.44	4.440994	0.64	3.137032	0.84	1.902849
0.05	11.557161	0.25	6.182662	0.45	4.368306	0.65	3.076725	0.85	1.834773
0.06	10.854261	**0.26**	**6.064298**	0.46	4.296752	0.66	3.016570	0.86	1.765274
0.07	10.285969	0.27	5.950343	0.47	4.226263	0.67	2.956525	0.87	1.694144
0.08	9.810966	0.28	5.840425	0.48	4.156772	0.68	2.896548	0.88	1.621130
0.09	9.403982	0.29	5.734216	0.49	4.088217	0.69	2.836595	0.89	1.545932
0.10	9.048539	0.30	5.631424	0.50	4.020538	0.70	2.776622	0.90	1.468177
0.11	8.733353	0.31	5.531788	0.51	3.953679	0.71	2.716581	0.91	1.387399
0.12	8.450382	0.32	5.435072	0.52	3.887587	0.72	2.656425	0.92	1.302998
0.13	8.193715	0.33	5.341064	0.53	3.822209	0.73	2.596103	0.93	1.214179
0.14	7.958888	0.34	5.249572	0.54	3.757496	0.74	2.535560	0.94	1.119852
0.15	7.742451	0.35	5.160421	0.55	3.693400	0.75	2.474740	0.95	1.018444
0.16	7.541692	0.36	5.073452	0.56	3.629874	0.76	2.413582	0.96	0.907538
0.17	7.354438	0.37	4.988520	0.57	3.566875	0.77	2.352021	0.97	0.783057
0.18	7.178928	0.38	4.905490	0.58	3.504357	0.78	2.289985	0.98	0.637031
0.19	7.013711	0.39	4.824241	0.59	3.442279	0.79	2.227398	0.99	0.448821
0.20	6.857583	0.40	4.744657	0.60	3.380599	0.80	2.164173	1.00	0.000000

Performing a least-means square fitting process of the exact $x_D(\kappa)$ curve by means of this 11-parameter formula (B.9) within the range $0.25 < \kappa < 1$ we have obtained the adjusted parameter values (b_n, $n = 2$ to 12) listed in Table 8. From the lower inset of Figure 9 we see that the residual deviations of the respective $x_D^{h_{11}}(\kappa)$ curve (see (B.9)) from the exact $x_D(\kappa)$ dependence are limited to an order of $\pm 0.000016\%$ (which are only slightly higher than the residual deviations due to the complementary low-to-intermediate κ high-accuracy formula $x_D^{l_{12}}(\kappa)$ (see (B.8))). A set of corresponding high-accuracy $x_D^{h_{11}}(\kappa)$ values (for $0.26 \le \kappa \le 1$) is presented in Table 9.

C. Conventional Estimation Method for True Debye Temperatures

An effective procedure for determining high-temperature limiting values of the "true" (harmonic-lattice-related) Debye temperatures, $\Theta_{Dh}(T \to \infty)$, on the basis of measured (isobaric) low-temperature heat capacity data, $C_p(T) \cong C_{Vh}(T)$, had been invented years ago by Barron et al. [16, 17]. A crucial feature of this conventional method is due to the circumstance that, in analogy to the exponential series analysis scheme [54] for heat capacities, its use for analyses of appropriate $\Theta_D(T)$ data sets does *not* assume a preliminary analytical modelling of PDOS spectra functions. The corresponding analysis procedure is based on the observation that certain sections of *low*-temperature dependencies of the squares, $(\Theta_{Dh}(T))^2$, of "true" Debye temperatures can be generally represented in form of Taylor series expansions involving exclusively *even*-order power terms of the reciprocal temperature with alternating signs [14–18, 21, 71, 88]; that is, explicitly

$$(\Theta_{Dh}(T))^2 = (\Theta_{Dh}(\infty))^2 \left(1 + \sum_{s=1}^{\infty} (-1)^s a_{2s} \left(\frac{\Theta_{Dh}(\infty)}{T}\right)^{2s}\right). \tag{C.1}$$

The symbol $\Theta_{Dh}(\infty)$ represents here again the $T \to \infty$ limit of the "true" (harmonic) Debye temperature, and the expansion coefficients, a_{2s}, used to be positive (and strongly decreasing in magnitude, with increasing order; $1 \gg a_2 \gg a_4 \gg a_6, \ldots$). The applicability of (C.1) is naturally limited to those regions of (usually relatively low) temperatures where the isochoric/harmonic and isobaric heat capacities are nearly coincident, $C_{Vh}(T) \cong C_p(T)$, which implies, as assumed, an approximate equality of "true" Debye temperatures with their effective ($C_p(T)$-related) counterparts, $\Theta_{Dh}(T) \cong \Theta_D(T)$. This requirement finds its visual realization (within a Θ_D versus T^{-2} representation like Figure 10) in limitations of fitting ranges to those regions where the sequences of the $\Theta_D(T)$ points show exclusively *concave* shapes.

The corresponding material-specific limitations of fitting ranges are indicated in the insets of Figure 10 by vertical (broken) lines and are quoted in Table 10. Accordingly, the upper limits of harmonic lattice behaviour are corresponding

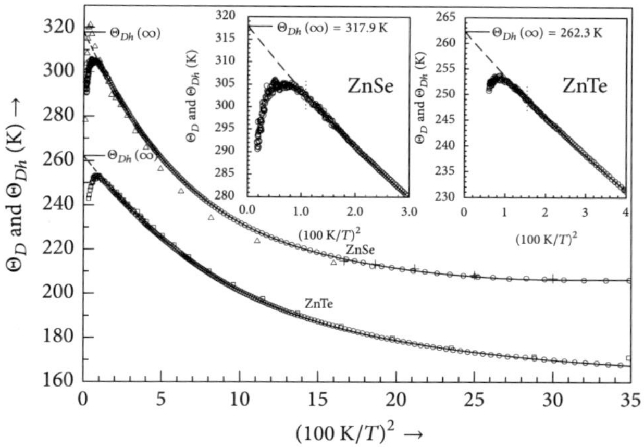

FIGURE 10: Fitting by means of (C.1) of the temperature dependencies of the harmonic sections (solid curves) of Debye temperature data, $\Theta_{Dh}(T) \cong \Theta_D(T)$, that were directly resulting from the high-accuracy transformations (via (27) or (28)) of the experimental $C_p(T)$ data points provided by Kremer et al. (○ [46]), for natural ZnSe and ZnTe. The extrapolations of the harmonic (concave) $\Theta_{Dh}(T)$ versus $\propto T^{-2}$ curve sections to the $T \to \infty$ limit are represented by dashed curve sections. For comparison we have also shown the more or less deviating $\Theta_D(T)$ data points due to $C_p(T)$ data that have been available from earlier studies for ZnSe by Birch (+ [34]) and Sirota et al. (△ [60]) and for ZnTe by Gavrichev et al. (□ [02G]).

to cuts of the relevant T-intervals at $T \le 96$ K, for ZnSe, and at $T \le 80$ K, for ZnTe (i.e., at temperatures of the order $(4/3)T_h$, in both cases; cf. also Figures 2 and 3). Performing hence the least-mean-square fittings of the duly limited series of $\Theta_D(T)$ points (○) for ZnSe and ZnTe (due to the measured $C_p(T)$ data provided by Kremer et al. [46]) via (C.1), we have obtained the respective magnitudes (listed in Table 10) of limiting Debye temperature, $\Theta_{Dh}(\infty)$, in combination with material-specific series of even-order expansion coefficients, a_{2s}.

On the basis of these sets of fitted parameter values it is possible to calculate a series of even-order PDOS spectra moments, $\mu_P^{(2s)}$, via the following equations [14]:

$$\mu_P^{(2)} = \frac{3}{5} (\varepsilon_{Dh}(\infty))^2, \tag{C.2a}$$

$$\mu_P^{(4)} = \left[\frac{3}{7} + 12 a_2\right] (\varepsilon_{Dh}(\infty))^4, \tag{C.2b}$$

$$\mu_P^{(6)} = \left[\frac{3}{9} + \frac{108}{5} a_2 + \frac{1512}{5} a_4\right] (\varepsilon_{Dh}(\infty))^6, \tag{C.2c}$$

$$\mu_P^{(8)} = \left[\frac{3}{11} + \frac{200}{7} a_2 + \frac{2160}{7} \left(2a_4 + (a_2)^2\right) + 8640 a_6\right] (\varepsilon_{Dh}(\infty))^8, \tag{C.2d}$$

$$\mu_P^{(10)} = \left[\frac{3}{13} + \frac{168}{5} a_2 + 880 \left(a_4 + (a_2)^2\right) + 19008 \left(a_6 + a_2 a_4\right) + 266112 a_8\right] (\varepsilon_{Dh}(\infty))^{10} \tag{C.2e}$$

TABLE 10: Limiting values of the "true" (harmonic lattice) Debye temperatures, $\Theta_{Dh}(\infty)$, including adjusted magnitudes of several even-order expansion coefficients, a_{2s}, that are representing the weights of the individual $(\Theta_{Dh}(\infty)/T)^{2s}$ power terms in (C.1). These sets of parameter values have been determined via fittings (shown in Figure 10) of an appropriate selection of usable Debye temperature points, $\Theta_{Dh}(T) \cong \Theta_D(T)$, implied by the experimental $C_p(T)$ data points given by Kremer et al. [46] for natural ZnSe and ZnTe. Listed are also the magnitudes of the respective (Debye-model-related) cut-off phonon energies, $\varepsilon_{Dh}(\infty) \equiv k_B\Theta_{Dh}(\infty)$, and several *even*-order phonon energy values, $\varepsilon_P(2s) \equiv (\mu_P^{(2s)})^{1/2s}$ (for $2s = 2$ to 10), which are resulting from the corresponding PDOS spectra moments, $\mu_P^{(2s)}$, calculated on the basis of the analytical expressions (C.2a) to (C.2e) (note that the latter equations are exactly corresponding to the series of $\mu_P^{(2s)}$ formulas originally derived within the frame of the supporting information to [21]; see also appendix B of [14] and/or the appendix B of [15]).

	ZnSe	ZnTe
Intervals (K)	17 to 96	15 to 80
$\Theta_{Dh}(\infty)$ (K)	317.9	262.3
a_2	0.9259×10^{-2}	0.9995×10^{-2}
a_4	0.6895×10^{-4}	0.7851×10^{-4}
a_6	0.2869×10^{-6}	0.3410×10^{-6}
a_8	0.6533×10^{-9}	0.7714×10^{-9}
a_{10}	(0.7135×10^{-12})	(0.7038×10^{-12})
a_{12}	(0.2510×10^{-15})	—
$\varepsilon_{Dh}(\infty)$ (meV)	27.40	22.60
$\varepsilon_P(2)$ (meV)	21.22	17.51
$\varepsilon_P(4)$ (meV)	23.48	19.45
$\varepsilon_P(6)$ (meV)	24.83	20.60
$\varepsilon_P(8)$ (meV)	25.75	21.38
$\varepsilon_P(10)$ (meV)	26.42	21.95

in complete accordance with the set of $\mu_P^{(2s)}$ formulas originally derived within the supporting information to [21]. Inserting into (C.2a) to (C.2e) the estimated magnitudes (listed in Table 10) for the cut-off phonon energies, $\varepsilon_{Dh}(\infty) \equiv k_B\Theta_{Dh}(\infty)$, and the associated expansion coefficients, a_2 to a_8, we have calculated the respective even-order moments, $\mu_P^{(2)}$ to $\mu_P^{(10)}$. From these moments follow readily the material-specific magnitudes of the corresponding even-order phonon energies, $\varepsilon_P(2s) = (\mu_P^{(2s)})^{1/2s}$. The latter are listed in Table 10 and represented by open squares in Figure 7.

Conflicts of Interest

The author declares that there are no conflicts of interest regarding the publication of this paper.

Acknowledgments

The author would like to thank Dr. R. K. Kremer, at the Max-Planck-Institut für Festkörperphysik, Stuttgart, Germany, for having sent him, in digital form, the experimental heat capacity data published for natural ZnSe and ZnTe in [46].

References

[1] P. Debye, "Zur Theorie der spezifischen Wärmen," *Annalen der Physik*, vol. 344, no. 14, pp. 789–839, 1912.

[2] E. Schrödinger, "'Spezifische Wärme', in Handbuch der Physik," in *Thermische Eigenschaften der Stoffe*, C. Drucker, E. Grüneisen, Ph. Kohnstamm et al., Eds., pp. 275–320, Springer-Verlag, Berlin, Heidelberg, 1926.

[3] M. Blackman, "The Specific Heat of Solids," in *Encyclopedia of Physics*, vol. 3 of *part 1: Crystal Physics 1*, pp. 325–382, Springer Berlin Heidelberg, Berlin, Heidelberg, 1955.

[4] P. H. Keesom and N. Pearlman, "Low Temperature Heat Capacity of Solids," in *Encyclopedia of Physics*, S. Flügge, Ed., vol. 14 of *Low temperature physics I*, pp. 282–337, Springer-Verlag, Berlin-Göttingen-Heidelberg, Berlin, Heidelberg, 1956.

[5] A. J. Dekker, *Solid State Physics*, Macmillan Education UK, London, 1981.

[6] E. S. Gopal, *Specific Heats at Low Temperatures*, Springer US, Boston, MA, 1966.

[7] S. Kobe, *Crystal Research and Technology*, Group III, vols. 17a/b, 22a/b, and 41a/b, Springer-Verlag, Berlin, Germany, 1982.

[8] T. H. Barron and G. K. White, *Heat Capacity and Thermal Expansion at Low Temperatures*, Springer US, Boston, MA, 1999.

[9] S. Adachi, *Handbook on Physical Properties of Semiconductors*, vol. vols. 1, 2, and 3, Kluwer Academic Publishers, Boston, Dordrecht, New York, London, 2004.

[10] S. Adachi, *Properties of Group-IV, III-V, and II-VI Semiconductors*, Wiley Sons Ltd, The Atrium, Southern Gate, Chichester, West Sussex, England, UK, 2005.

[11] U. Piesbergen, "Heat Capacity and Debye Temperatures," in *Semiconductors and Semimetals*, R. K. Willardson and A. C. Beer, Eds., vol. 2 of *Physics of III-V Compounds, Chapter 3*, pp. 49–60, Academic Press New York and London, 1966.

[12] J. S. Blakemore, "Semiconducting and other major properties of gallium arsenide," *Journal of Applied Physics*, vol. 53, no. 10, pp. R123–R181, 1982.

[13] R. Pässler, "Limiting Debye temperature behavior following from cryogenic heat capacity data for group-IV, III-V, and II-VI materials," *Physica Status Solidi (B) Basic Research*, vol. 247, no. 1, pp. 77–92, 2010.

[14] R. Pässler, "Non-Debye behaviours of heat capacities of cubic IIVI materials," *Journal of Physics and Chemistry of Solids*, vol. 72, no. 11, pp. 1296–1311, 2011.

[15] R. Pässler, "Non-Debye heat capacity formula refined and applied to GaP, GaAs, GaSb, InP, InAs, and InSb," *AIP Advances*, vol. 3, no. 8, Article ID 082108, 2013.

[16] T. H. Barron, W. T. Berg, and J. A. Morrison, "The thermal properties of alkali halide crystals. II. Analysis of experimental results," in *Proceedings of the Royal Society*, vol. 242, pp. 478–492, London, UK, 1957.

[17] T. H. Barron, W. T. Berg, and J. A. Morrison, "On the heat capacity of crystalline magnesium oxide," in *Proceedings of the Royal Society*, vol. 250, pp. 70–83, London, UK, 1959.

[18] P. Flubacher, A. J. Leadbetter, and J. A. Morrison, "The Heat Capacity of Pure Silicon and Germanium and Properties of their Vibrational Frequency Spectra," *The Philosophical Magazine*, vol. 4, no. 39, pp. 273–294, 1959.

[19] G. Dolling and R. A. Cowley, "The thermodynamic and optical properties of germanium, silicon, diamond and gallium arsenide," *Proceedings of the Physical Society*, vol. 88, no. 2, article no. 318, pp. 463–494, 1966.

[20] W. Schnelle and E. Gmelin, "Heat capacity of germanium crystals with various isotopic composition," *Journal of Physics Condensed Matter*, vol. 13, no. 27, pp. 6087–6094, 2001.

[21] R. Pässler, "Dispersion-related theory for heat capacities of semiconductors," *Physica Status Solidi (B) Basic Research*, vol. 244, no. 12, pp. 4605–4623, 2007.

[22] U. Piesbergen, "Die durchschnittlichen Atomwärmen der A^{III} B^{V}-Halbleiter ALSb, GaAs, GaSb, InP, InAs, InSb und die Atomwärme des Elements Germanium zwischen 12 und 273 ∘K," *Zeitschrift fur Naturforschung - Section A Journal of Physical Sciences*, vol. 18, no. 2, pp. 141–147, 1963.

[23] T. C. Cetas, C. R. Tilford, and C. A. Swenson, "Specific heats of Cu, GaAs, GaSb, InAs, and InSb from 1 to 30∘K," *Physical Review*, vol. 174, no. 3, pp. 835–844, 1968.

[24] A. F. Demidenko, V. I. Koshchenko, A. S. Pashinkin, and V. E. Yachmenev, "Temperature dependences of the specific heat and thermodynamic properties of gallium phosphide," *Inorganic Materials*, vol. 17, pp. 677–680, 1981.

[25] N. N. Sirota, V. V. Novikov, and A. M. Antyukhov, "Heat Capacity and Thermodynamic Functions of (GaAs)x(InAs)1-x Solid Solutions in the 5-300 K Temperature Range," *Doklady Akademii Nauk SSSR*, vol. 263, no. 1, pp. 96–100, 1982.

[26] N. N. Sirota, A. M. Antyukhov, V. V. Novikov, and A. A. Sidorov, "Temperature Dependence of the Heat Capacity and Characteristic Thermodynamic Functions of the Solid Solutions of $(GaAs)_x(InP)_{1-x}$ in the Temperature Range of 5 to 300 K," *Doklady Akademii Nauk SSSR*, vol. 266, no. 3, pp. 105–108, 1982.

[27] N. N. Sirota, A. M. Antyukhov, V. V. Novikov, and V. A. Fyodorov, "Thermodynamic functions of (InP)x(InAs)1-x from 5 to 300 K," *Crystal Research and Technology*, vol. 17, no. 3, pp. 279–287, 1982.

[28] J. C. Nipko and C.-K. Loong, "Phonon excitations and related thermal properties of aluminum nitride," *Physical Review B—Condensed Matter and Materials Physics*, vol. 57, no. 17, pp. 10550–10554, 1998.

[29] J. C. Nipko, C.-K. Loong, C. M. Balkas, and R. F. Davis, "Phonon density of states of bulk gallium nitride," *Applied Physics Letters*, vol. 73, no. 1, pp. 34–36, 1998.

[30] V. V. Novikov, "The lattice heat capacity of binary compounds in the approximation of three-dimensional debye sublattices," *Russian Journal of Physical Chemistry A*, vol. 80, no. 9, pp. 1456–1460, 2006.

[31] A. F. Demidenko and A. K. Maltsev, "Heat capacity of Zinc Telluride in the interval 56-300_0K," *Neorganicheskie Materialy*, vol. 5, pp. 158–160, 1969.

[32] N. Vagelatos, D. Wehe, and J. S. King, "Phonon dispersion and phonon densities of states for ZnS and ZnTe," *The Journal of Chemical Physics*, vol. 60, no. 9, pp. 3613–3618, 1974.

[33] J. C. Irwin and J. Lacombe, "Specific heats of ZnTe, ZnSe, and GaP," *Journal of Applied Physics*, vol. 45, no. 2, pp. 567–573, 1974.

[34] J. A. Birch, "Heat capacities of ZnS, ZnSe and CdTe below 25K," *Journal of Physics C: Solid State Physics*, vol. 8, no. 13, article no. 013, pp. 2043–2047, 1975.

[35] J. G. Collins, G. K. White, J. A. Birch, and T. F. Smith, "Thermal expansion of ZnTe and HgTe and heat capacity of HgTe at low temperatures," *Journal of Physics C: Solid State Physics*, vol. 13, no. 9, article no. 011, pp. 1649–1656, 1980.

[36] D. N. Talwar, M. Vandevyver, K. Kunc, and M. Zigone, "Lattice dynamics of zinc chalcogenides under compression: Phonon dispersion, mode Grneisen, and thermal expansion," *Physical Review B*, vol. 24, no. 2, pp. 741–753, 1981.

[37] H. -M. Kagaya and T. Soma, "Temperature dependence of specific heat and Debye temperature of tetrahedral compounds," *Physica Status Solidi B*, vol. 134, no. 2, pp. K101–K104, 1986.

[38] K. S. Gavrichev, V. N. Guskov, J. H. Greenberg, T. Feltgen, M. Fiederle, and K. W. Benz, "Low-temperature heat capacity of ZnTe," *Journal of Chemical Thermodynamics*, vol. 34, no. 12, pp. 2041–2047, 2002.

[39] W. T. Berg and J. B. Morrison, "The thermal properties of alkali halide crystals. I. The heat capacities of potassium chloride, potassium bromide, potassium iodide and sodium iodide between 2.8 and 270°K," in *Proceedings of the Royal Society A: Mathematical, Physical and Engineering Sciences*, vol. 242, 1231, pp. 467–477, 1957.

[40] M. P. Tosi and F. G. Fumi, "Temperature dependence of the Debye temperatures for the thermodynamic functions of alkali halide crystals," *Physical Review*, vol. 131, no. 4, pp. 1458–1465, 1963.

[41] I. I. Guseinov and B. A. Mamedov, "Calculation of integer and noninteger n-dimensional debye functions using binomial coefficients and incomplete gamma functions," *International Journal of Thermophysics*, vol. 28, no. 4, pp. 1420–1426, 2007.

[42] A. E. Dubinov and A. A. Dubinova, "Exact integral-free expressions for the integral Debye functions," *Technical Physics Letters*, vol. 34, no. 12, pp. 999–1001, 2008.

[43] S. Sönmezoğlu, "Investigation of the entropy and specific heat capacity of GaN using incomplete gamma functions," *International Journal of Modern Physics B*, vol. 22, no. 30, pp. 5349–5355, 2008.

[44] B. A. Mamedov, E. Eser, H. Koç, and I. M. Askerov, "Accurate evaluation of the specific heat capacity of solids and its application to mgo and zno crystals," *International Journal of Thermophysics*, vol. 30, no. 3, pp. 1048–1054, 2009.

[45] Y. Yu, H. L. Han, M. J. Wan, T. Cai, and T. Gao, "Structural, vibrational and thermodynamics propertiesof Zn-based semiconductors," *Solid State Sciences*, vol. 11, no. 8, pp. 1343–1349, 2009.

[46] R. K. Kremer, M. Cardona, R. Lauck, G. Siegle, and A. H. Romero, "Vibrational and thermal properties of ZnX (X=Se, Te): Density functional theory (LDA and GGA) versus experiment," *Physical Review B - Condensed Matter and Materials Physics*, vol. 85, no. 3, Article ID 035208, 2012.

[47] J. A. Beattie, "Six Place Tables of the Debye Energy and Specific Heat Functions," *Journal of Mathematics and Physics*, vol. 6, no. 1—4, pp. 1–38, 1926.

[48] "Landoldt-börnstein: physikalisch-chemische tabellen," in *Die Einsteinschen und Debyeschen Funktionen*, W. Roth and K. Scheel, Eds., vol. 1, 1, pp. 702–707, Springer-Verlag, Berlin, Germany, 5th edition, 1927.

[49] P. Harteck, "Landolt-Börnstein: Zahlenwerte und Funktionen aus Physik, Chemie, Astronomie, Geophysik und Technik," in *Debye-Funktionen*, vol. 67, pp. 767-768, 23 edition, 1955.

[50] R. Pässler, "Representative hybrid model used for analyses of heat capacities of group-IV, III-V, and II-VI materials," *Physica Status Solidi (B) Basic Research*, vol. 248, no. 4, pp. 904–920, 2011.

[51] A. A. Maradudin and R. F. Wallis, "Lattice-dynamical calculation of the surface specific heat of a crystal at low temperatures," *Physical Review*, vol. 148, no. 2, pp. 945–961, 1966.

[52] A. A. Maradudin, E. W. Montroll, G. H. Weiss, and I. P. Ipatova, "Theory of Lattice Dynamics in the Harmonic Approximation," in *Solid State Physics*, H. E. Ehrenreich et al., Ed., 3, p. viii+319,

Academic Press, New York-London, New York, NY, USA, 2nd edition, 1971.

[53] C. Lee and X. Gonze, "Ab initio calculation of the thermodynamic properties and atomic temperature factors of SiO2 α-quartz and stishovite," *Physical Review B*, vol. 51, no. 13, pp. 8610–8613, 1995.

[54] R. Pässler, "Exponential series representation for heat capacities of semiconductors and wide-bandgap materials," *Physica Status Solidi (B) Basic Research*, vol. 245, no. 6, pp. 1133–1146, 2008.

[55] W. Xue, Y. Yu, Y. Zhao, H. Han, and T. Gao, "First principles calculations of the electronic, dynamical, and thermodynamic properties of the rocksalt ScX (X = N, P, As, Sb)," *Computational Materials Science*, vol. 45, no. 4, pp. 1025–1030, 2009.

[56] R. Pässler, "Basic moments of phonon density of states spectra and characteristic phonon temperatures of group IV, III-V, and II-VI materials," *Journal of Applied Physics*, vol. 101, no. 9, Article ID 093513, 2007.

[57] H. Neumann, G. Kühn, and W. Möller, "Heat Capacity and Lattice Anharmonicity in Cu-III-VI2 Chalcopyrite Compounds," *physica status solidi (b)*, vol. 144, no. 2, pp. 565–573, 1987.

[58] W. Nernst and F. A. Lindemann, "Spezifische Wärme und Quantentheorie," *Zeitschrift für Elektrochemie*, vol. 17, pp. 817–827, 1911.

[59] R. Pässler, "Characteristic non-Debye heat capacity formula applied to GaN and ZnO," *Journal of Applied Physics*, vol. 110, no. 4, Article ID 043530, 2011.

[60] A. Sommerfeld, "Zur Elektronentheorie der Metalle auf Grund der Fermischen Statistik. I. Teil: Allgemeines, Strömungs- und Austrittsvorgänge," *Zeitschrift für Physik*, vol. 47, no. 1-2, pp. 1–32, 1928.

[61] N. Pearlman and P. H. Keesom, "The atomic heat of silicon below 100◦K," *Physical Review*, vol. 88, no. 2, pp. 398–405, 1952.

[62] C. A. Bryant and P. H. Keesom, "Low-temperature specific heat of germanium," *Physical Review*, vol. 124, no. 3, pp. 698–700, 1961.

[63] M. Wagner-Reetz, D. Kasinathan, W. Schnelle et al., "Phonon-drag effect in FeGa3," *Physical Review B - Condensed Matter and Materials Physics*, vol. 90, no. 19, Article ID 195206, 2014.

[64] T. H. K. Barron and J. A. Morrison, "The Thermal Properties of Alkali Halide Crystals. III. The Inversion of the Heat Capacity," *Proceedings of the Royal Society A: Mathematical, Physical and Engineering Sciences*, vol. 256, no. 1286, pp. 427–439, 1960.

[65] J. C. Holste, "Specific heats of GaSb, GaAs, InSb, InAs, Bi, Cd, Sn, and Zn below 30 K," *Physical Review B*, vol. 6, no. 6, pp. 2495–2497, 1972.

[66] B. Hennion, F. Moussa, G. Pepy, and K. Kunc, "Normal modes of vibrations in ZnSe," *Physics Letters A*, vol. 36, no. 5, pp. 376–378, 1971.

[67] H. Bilz and W. Kress, *Phonon Dispersion Relations in Insulators*, vol. 10 of *Springer Series in Solid-State Sciences*, Springer, Heidelberg, Berlin, Germany, 1979.

[68] A. Einstein, "Die Plancksche Theorie der Strahlung und die Theorie der spezifischen Wärme," *Annalen der Physik*, vol. 327, no. 1, pp. 180–190, 1906.

[69] H. Thirring, "Zur Theorie der Raumgitterschwingungen und der spezifischen WΣrme fester K÷rper," *Physikalische Zeitschrift*, vol. 14, pp. 876-873, 1913.

[70] E. Gmelin, "Thermal Properties of Alkaline Earth Oxides: II. Analysis of Experimental Results for MgO, CaO, SrO, and BaO," *Zeitschrift fur Naturforschung - Section A Journal of Physical Sciences*, vol. 25, no. 6, pp. 887–893, 1970.

[71] A. P. Rusakov, Y. K. Vekilov, and A. E. Kadyshevich, "Specific Heats of CdTe and HgTe and their Vibrational Frequency Spectra," *Fizika Tverdogo Tela*, vol. 12, no. 11, pp. 3238–3243, 1970.

[72] R. Pässler, "Moments of phonon density of states spectra and characteristic phonon temperatures of wide band gap materials," *Physica Status Solidi (B) Basic Research*, vol. 243, no. 12, pp. 2719–2727, 2006.

[73] H. Wagini, "Systematik charakteristischer Temperaturen von Halbleitern mit Zinkblende-Gitter," *Zeitschrift fur Naturforschung - Section A Journal of Physical Sciences*, vol. 22, no. 7, pp. 1135-1136, 1967.

[74] B. H. Lee, "Elastic constants of ZnTe and ZnSe between 77◦-300◦K," *Journal of Applied Physics*, vol. 41, no. 7, pp. 2984–2987, 1970.

[75] J. K. D. Verma and M. D. Aggarwal, "An averaging method for the shear modulus and Debye temperatures of cubic solids," *Journal of Applied Physics*, vol. 46, no. 7, pp. 2841–2844, 1975.

[76] D. Singh and Y. P. Varshni, "Debye temperatures for hexagonal crystals," *Physical Review B*, vol. 24, no. 8, pp. 4340–4347, 1981.

[77] H. Siethoff and K. Ahlborn, "The Dependence of the Debye Temperature on the Elastic Constants," *Physica Status Solidi (B)*, vol. 190, no. 1, pp. 179–191, 1995.

[78] A. S. Pashinkin and A. S. Malkova, "Heat capacity of Zinc Selenide," *Russian Journal of Physical Chemistry A*, vol. 77, no. 12, pp. 2068-2069, 2003.

[79] N. N. Sirota, Z. K. Petrova, and T. D. Sokolovsky, "Heat Capacity of Zinc Selenide over the Temperature Range 4.2 to 300 K," *Doklady Akademii Nauk Belorusi*, vol. 24, no. 3, pp. 214–217, 1980.

[80] G. G. Gadzhiev, Sh. M. Ismailow, and A. I. Dadashev, "Thermal Properties of AIIBVI-Based Ceramics," *High Temperature*, vol. 31, no. 3, pp. 350–354, 1993.

[81] I. Hurtado and D. Neuschütz, *Landolt-Börnstein: Numerical Data and Functional Relationships in Science and Technology, Group IV, (Thermodynamic Properties of Inorganic Materials)*, vol. 19 of *Subvolume A, Parts 1-4*, Springer-Verlag, Berlin, Heidelberg, 1999.

[82] K. S. Gavrichev, G. A. Sharpataya, V. N. Guskov et al., "Thermodynamic properties of ZnTe in the temperature range 15-925 K," *Physica Status Solidi (B) Basic Research*, vol. 229, no. 1, pp. 133–135, 2002.

[83] T. H. K. Barron and J. A. Morrison, "On the specific heat of solids at low temperatures," *Canadian Journal of Physics*, vol. 35, no. 7, pp. 799–810, 1957.

[84] M. Cardona, R. K. Kremer, R. Lauck, G. Siegle, J. Serrano, and A. H. Romero, "Heat capacity of PbS: Isotope effects," *Physical Review B - Condensed Matter and Materials Physics*, vol. 76, no. 7, Article ID 075211, 2007.

[85] F. Kelemen, E. Cruceanu, and D. Niculescu, "Untersuchung einiger thermischer Eigenschaften der Verbindungen HgSe, HgTe und ZnTe," *physica status solidi (b)*, vol. 11, no. 2, pp. 865–872, 1965.

[86] L. Guo, G. Hu, W.-J. Feng, and S.-T. Zhang, "Structural, elastic, electronic and optical properties of zinc-blende MTe (M=Zn/Mg)," *Wuli Huaxue Xuebao/ Acta Physico—Chimica Sinica*, vol. 29, no. 5, pp. 929–936, 2013.

[87] A. K. Kushwaha, "Phonon spectrum and thermal properties of semiconducting compounds ZnS and ZnSe," *Physica B: Condensed Matter*, vol. 405, no. 6, pp. 1638–1642, 2010.

[88] U. C. Boehnke, G. Kühn, F. I. Frolova, I. E. Paukov, and H. Neumann, "Heat capacity of LiInTe2," *Journal of Thermal Analysis*, vol. 33, no. 1, pp. 205–209, 1988.

[89] K. Kunc, M. Balkanski, and M. A. Nusimovici, "Lattice dynamics of several $a^N B^{8-N}$ compounds having the zincblende structure. II. Numerical Calculations," *Physica Status Solidi B*, vol. 72, no. 1, pp. 249–254, 1975.

[90] M. Abramowitz and I. A. Stegun, "Pocketbook of mathematical functions , Verlag Harri Deutsch," *Thun-Frankfurt/Main*, 1984.

[91] H. J. Schell, "Unendliche Reihen," in *Mathematik für Ingenieure, Naturwissenschaftler, Ökonomen und Landwirte*, K. Manteuffel, Ed., vol. 3, BSB B. G. Teubner Verlagsgesellschaft, Leipzig, Germany, 1st edition, 1974.

24

An Analytical Theory of Piezoresistive Effects in Hall Plates with Large Contacts

Udo Ausserlechner

Infineon Technologies, Villach, Austria

Correspondence should be addressed to Udo Ausserlechner; udo.ausserlechner@infineon.com

Academic Editor: Yuri Galperin

Four-terminal transducers can be used to measure the magnetic field via the Hall effect or the mechanical stress via the piezoresistance effect. Both effects are described by an anisotropic conductivity tensor with small offdiagonal elements. This has led other authors to the conclusion that there is some kind of analogy. In both cases the output voltage depends on the geometry of the device and the size of the contacts. For Hall plates this influence is accounted for by the Hall-geometry factor. The alleged analogy proposes that the Hall-geometry factor also applies to four-terminal stress transducers. This paper shows that the analogy holds only for a limited class of devices. Moreover, it is shown that devices of different geometries may have identical magnetic field sensitivity but different mechanical stress sensitivities. Thus, shape optimization makes sense for mechanical stress sensors. In extreme cases the output voltages of vertical Hall-effect devices may have notable magnetic field sensitivity but zero mechanical stress sensitivity. As byproduct, exact analytical formulae for the equivalent resistor circuit of rectangular and circular devices with two perpendicular mirror symmetries are given. They allow for an accurate description of how mechanical stress and deformation affect the output offset voltage and the magnetic sensitivity of Hall-effect devices.

1. Introduction

Resistive devices with four terminals have been proposed as mechanical stress sensors for a long time[1, 2]. They are called four-terminal piezotranducers [3] or van der Pauw stress sensors [4]. The very same phenomenon is observed in Hall plates at zero magnetic field, where the zero-point error (= initial offset error prior to spinning current scheme [5]) is very sensitive to mechanical stress. This was occasionally called Kanda effect [6] or pseudo-Hall effect [7]. An alternative view on the same topic is how to determine anisotropic resistivity from measurements on van der Pauw samples [8]. Over the past five decades a large number of papers have been published on various aspects of this bundle of problems. However, it seems that no analytical closed form expression was found, which relates the anisotropic conductivity to the output voltage of such a device with four contacts of arbitrary size. Conversely, the analogous problems for isotropic van der Pauw measurement at zero magnetic field and for Hall plates at small magnetic field have been solved recently [9–12]. In the following we apply similar methods to the piezoresistive effect in these devices. Thereby we use the affine transformation of van der Pauw [8] in order to replace the original device having anisotropic resistivity by an equivalent device having isotropic conductivity. For these devices with isotropic conductivity we compute the equivalent resistor circuit (ERC), which describes the behavior of the devices under the action of mechanical stress and magnetic field. For the ERC of rectangular devices we use our own results of former works [9–12], whereas we derive new formulae for the ERC of circular devices.

The outline of this work is as follows. Section 2 sums up basic phenomenological facts of the Hall effect and of piezoresistivity as they are known in the literature. The goal is to contrast similarities against differences of both effects. Section 3 briefly repeats the affine transformation of van der Pauw and specifies it for (100)-planes of cubic crystals, like silicon. With these formulae it becomes apparent that vertical Hall-effect devices are very slightly affected by mechanical stress. Section 4 elaborates on how rectangular and circular devices respond to biaxial inplane stress. There we make ample use of the ERC as it was developed in prior works [9, 10]. Section 4 closes with a comparison of several device shapes versus Wheatstone bridge circuits of resistor stripes in orthogonal directions: which one has the largest sensitivity to mechanical stress? Section 5 discusses the effect of biaxial inplane stress on the Hall-geometry factor of Hall plates. This leads to a drift of magnetic sensitivity versus mechanical stress, which is caused by piezoresistance and not by piezo-Hall effect. Finally, Section 6 compares the proposed analytical theory with results of other authors.

2. Similarities of Piezoresistance and Hall Effect

In a thin plate-like material with isotropic ohmic resistivity the Hall effect is described by the relation $\vec{E} = \underline{\underline{\rho}}\vec{J}$ between electric field $\vec{E}$ and current density $\vec{J}$ with the odd symmetric resistivity tensor in (x, y)-coordinates

$$\underline{\underline{\rho}} = \rho_0 \begin{pmatrix} 1 & \mu_H B_z \\ -\mu_H B_z & 1 \end{pmatrix} \tag{1a}$$

ρ_0 is the resistivity at zero magnetic field, $\mu_H > 0$ is the Hall mobility, and B_z is the magnetic flux density component perpendicular to the Hall plate, which lies in the (x, y)-plane [13]. The inverse relation $\vec{J} = \underline{\underline{\kappa}}\vec{E}$ uses the conductivity tensor

$$\underline{\underline{\kappa}} = \underline{\underline{\rho}}^{-1} = \frac{\kappa_0}{1 + \mu_H^2 B_z^2} \begin{pmatrix} 1 & -\mu_H B_z \\ \mu_H B_z & 1 \end{pmatrix}. \tag{1b}$$

$\kappa_0 = 1/\rho_0$ is the conductivity at zero magnetic field. Equations (1a) and (1b) assume negative charge carriers. Alternatively to μ_H the literature also uses the Hall coefficient $C_H = \mathrm{sgn}(q)\rho_0\mu_H$, where $\mathrm{sgn}(q)$ is the sign of the charge carriers. The static electric field is longitudinal, i.e., its curl vanishes. Therefore it can be expressed as the negative gradient of a scalar electric potential $\vec{E} = -\nabla\phi$. In the quasistatic case the divergence of the current density vanishes $\nabla\vec{J} = 0$. Combining these equations with (1b) gives

$$\frac{\partial^2\phi}{\partial x^2} + \mu_H B_z \left(\frac{\partial^2\phi}{\partial x \partial y} - \frac{\partial^2\phi}{\partial y \partial x} \right) + \frac{\partial^2\phi}{\partial y^2} = 0 \tag{2}$$

whereby we assumed that κ_0, μ_H, and B_z are constant versus spatial coordinates. The mixed-order terms $\partial^2\phi/\partial x\partial y = \partial^2\phi/\partial y\partial x$ cancel out and therefore (2) is the Laplace equation. Thus, the Hall effect does not enter (2); with and without magnetic field it is the same partial differential equation for the potential. The magnetic field affects only the boundary conditions: the potential at the contacts or the current flow via the contacts is forced by the external circuitry. At the isolating boundary the current density must flow parallel to the boundary and this means $J_n = 0$ or in other words $E_n/E_p = (\partial\phi/\partial n)/(\partial\phi/\partial t) = \pm\mu_H B_z$ (where n and t denote normal and tangential direction and the plus/minus sign depends on the direction of the magnetic field, the sign of the charge carriers, and if the boundary is left or right to the current flow). In other words, at the isolating boundary there is the Hall angle $\theta_H = \arctan(\mu_H B_z)$ between the electric field and the boundary. To sum up, the potential ϕ in a Hall plate and the output voltage V_{out} of a Hall plate can be computed by solving the Laplace equation and satisfying these boundary conditions. For Hall plates with two perpendicular mirror symmetries this was done in the seminal work [14] and the result is

$$V_{\mathrm{out}} = S_i I_{\mathrm{in}} B_z \quad \text{with } S_i = R_{\mathrm{sh}}\mu_H G_H, \tag{3}$$

where S_i is the current related magnetic sensitivity, R_{sh} is the sheet resistance (equal to one over the product of conductivity times thickness of the Hall plate), I_{in} is the current flowing through the Hall plate, and G_H is the Hall geometry factor. At small magnetic field we write $G_H \to G_{H0}$ and then the Hall geometry factor depends only on the lateral geometry (= the layout) of the Hall plate, not on its thickness and not on any material parameters. We can compute G_{H0} for any geometry of Hall plates from scratch (in rare cases analytically and in general numerically), but it is simpler to measure the input and output numbers of squares as shown in [9, 10] and to use the following formula (which is derived in [12]):

$$G_{H0} = \frac{1}{K\left(\sqrt{L(\lambda_p)}\right) K\left(\sqrt{L(\lambda_f)}\right)} \cdot \int_{\alpha=0}^{\pi/2} \frac{F\left(\sin\alpha, \sqrt{1 - L(\lambda_p)}\right)}{\sqrt{\sin^2\alpha + L(\lambda_f)\cos^2\alpha}} d\alpha \tag{4}$$

Thereby $F(w,k) = \int_0^w (1-\alpha^2)^{-1/2}(1-k^2\alpha^2)^{-1/2} d\alpha$ is the incomplete elliptic integral of the first kind, $K(k) = F(1,k)$ is the complete elliptic integral of the first kind, and $L(y)$ is the modular lambda function defined by $L(K'(k)/K(k)) = k^2$ for $0 \le k \le 1$ and $K'(k) = K(\sqrt{1-k^2})$. The input number of squares is $\lambda_f = R_{\mathrm{in}}/R_{\mathrm{sh}}$ and the output number of squares is $\lambda_p = R_{\mathrm{out}}/R_{\mathrm{sh}}$, whereby the input resistance R_{in} is measured between the two current supplying contacts of a Hall plate (while the sense contacts are floating) and the output resistance R_{out} is measured between the two sense contacts (while the other two contacts are floating). The role of input and output contacts is arbitrary – G_{H0} does not change if we swap it [12]. Equation (4) expresses the Hall geometry factor as a function of input and output numbers of squares without need to know any geometrical details of the device. This means that Hall plates with different geometry have identical G_{H0} if they have identical λ_f and λ_p. The reason why different geometries may have the same number of squares and the same Hall output voltage at identical supply current is that we can transform one into the other via conformal mapping [14]. Since the conformal mapping is done by analytic functions which satisfy Cauchy-Riemann differential equations, the mapping functions are also solutions of the Laplace equations. Therefore, if we have a solution of the potential in one Hall plate, we can apply a conformal transformation to this geometry and to the solution. Thereby the same current will flow through the transformed device and the same potentials will appear at its contacts. Consequently, we can study a specific device geometry (e.g., a circular disk) and we can be sure that there is no other ingeniously shaped device boundary (e.g., Greek cross or octagon. . .) that leads to better device performance. This was already noted by Wick in [14]. Of course this holds only within the limited scope of linear electrostatic theory; beyond that scope devices will be different: different shapes lead to more or less homogeneous electric fields and therefore to more or less pronounced velocity saturation effects. Although their total power may be identical, the spatial distribution of power density differs and this leads to different temperature distributions. The temperature coefficient of the resistivity acts back on the current distribution. Finally the contacts may have different temperatures, which leads to different thermoelectric voltages and different residual offsets after spinning current schemes are applied [15].

In the case there is piezoresistance at zero magnetic field, the conductivity tensor $\underline{\underline{\kappa}}$, and its inverse the resistivity tensor $\underline{\underline{\rho}}$ becomes anisotropic due to mechanical stress

$$\underline{\underline{\kappa}} = \begin{pmatrix} \kappa_{xx} & \kappa_{xy} \\ \kappa_{xy} & \kappa_{yy} \end{pmatrix} = \underline{\underline{\rho}}^{-1} = \begin{pmatrix} \rho_{xx} & \rho_{xy} \\ \rho_{xy} & \rho_{yy} \end{pmatrix}^{-1} \tag{5}$$

The significant difference between (1a), (1b), and (5) is that the conductivity tensor of the Hall effect is odd symmetric, whereas it is even symmetric for the piezoresistance. In the language of nonequilibrium thermodynamics both Hall effect and piezoresistance are irreversible dissipative processes, but the relation between current density and electric field is reciprocal for piezoresistance whereas it is antireciprocal for the Hall effect [16]. Therefore, a rotation of the coordinate system around the z-axis does not change the Hall-resistivity tensor (1a), yet it changes the piezoresistance resistivity tensor (5) in such a way that for certain orientations (the principal axes) the elements in the minor diagonal even vanish. In network theory it is well known that any reciprocal (i.e., symmetric) impedance matrix of a linear, lumped, and passive N-port can be realized by a network of resistors, inductances, capacitors, and ideal transformers [17], however a nonreciprocal impedance matrix needs at least one additional linear, lumped, passive element: the gyrator [18]. Instead of a gyrator one may also use the series connection of two current controlled voltage sources (CCVS) or the parallel connection of two voltage controlled current sources (VCCS). Therefore the equivalent lumped circuit of a Hall effect device comprises resistors and at least one gyrator [19, 20], whereas the equivalent lumped circuit of a piezoresistive device consists only of resistors. Hence, the gyrator makes an essential difference between piezoresistance devices and Hall effect devices, and this already denies any perfect analogy between piezoresistance and Hall effect.

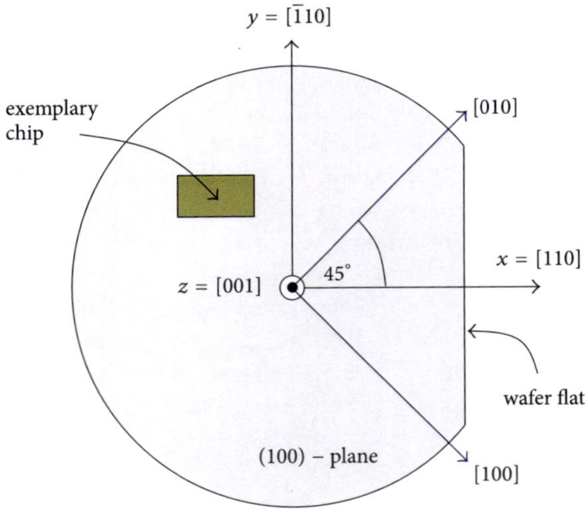

FIGURE 1: Reference frame ⟨100⟩ of the single crystal and chip reference frame $\Sigma = (x, y, z)$ for a (100)-silicon wafer.

In a conventional plastic encapsulated package the surface of a semiconductor chip is exposed to inplane mechanical stress [21, 22]. For inplane normal stress σ_{xx}, σ_{yy} and inplane shear stress components σ_{xy} on a (100) silicon chip it holds [23]

$$\rho_{xx} = \rho_0 \left(1 + \frac{\pi_{11} + \pi_{12} + \pi_{44}}{2}\sigma_{xx} + \frac{\pi_{11} + \pi_{12} - \pi_{44}}{2}\sigma_{yy}\right), \tag{6a}$$

$$\rho_{yy} = \rho_0 \left(1 + \frac{\pi_{11} + \pi_{12} - \pi_{44}}{2}\sigma_{xx} + \frac{\pi_{11} + \pi_{12} + \pi_{44}}{2}\sigma_{yy}\right), \tag{6b}$$

$$\rho_{xy} = \rho_0 \left(\pi_{11} - \pi_{12}\right)\sigma_{xy} \tag{6c}$$

with π_{11}, π_{12}, and π_{44} being the three piezoresistive coefficients of the cubic m3m single crystal. Thereby the x- and y-axes of the chip are aligned along ⟨110⟩-crystal directions as shown in Figure 1 (see also [24]).

Computing the conductivities from (6a) to (6c) gives

$$\kappa_{xx} = \frac{\rho_0}{\rho_{\text{iso}}^2}\left(1 + \frac{\pi_{11}+\pi_{12}-\pi_{44}}{2}\sigma_{xx} + \frac{\pi_{11}+\pi_{12}+\pi_{44}}{2}\sigma_{yy}\right), \tag{7a}$$

$$\kappa_{yy} = \frac{\rho_0}{\rho_{\text{iso}}^2}\left(1 + \frac{\pi_{11}+\pi_{12}+\pi_{44}}{2}\sigma_{xx} + \frac{\pi_{11}+\pi_{12}-\pi_{44}}{2}\sigma_{yy}\right), \tag{7b}$$

$$\kappa_{xy} = \frac{\rho_0}{\rho_{\text{iso}}^2}\left(\pi_{12}-\pi_{11}\right)\sigma_{xy}, \tag{7c}$$

$$\rho_{\text{iso}} = \sqrt{\rho_{xx}\rho_{yy}-\rho_{xy}^2} = \rho_0\sqrt{\left(1+\frac{\pi_{11}+\pi_{12}}{2}\left(\sigma_{xx}+\sigma_{yy}\right)\right)^2 - \left(\frac{\pi_{44}}{2}\right)^2\left(\sigma_{xx}-\sigma_{yy}\right)^2 - \left(\pi_{11}-\pi_{12}\right)^2\sigma_{xy}^2}. \tag{7d}$$

ρ_{iso} is the determinant of the resistivity matrix, and its name will become apparent later in (12a)–(12c). For $\sigma_{xx} = \sigma_{yy} = 0$ we get

$$\underline{\underline{\kappa}} = \rho_0^{-1}\left(1-\delta_{12}^2\right)^{-1}\begin{pmatrix} 1 & \delta_{12} \\ \delta_{12} & 1 \end{pmatrix} \tag{8}$$

with $\delta_{12} = (\pi_{12} - \pi_{11})\sigma_{xy}$ and $|\delta_{12}| \ll 1$. In the general case $\sigma_{xx}, \sigma_{yy}, \sigma_{xy} \neq 0$ we can apply the Mohr circle construction [25] to the second-rank conductivity tensor in (5): we rotate the (x, y)-coordinate system Σ by an angle $\widetilde{\varphi}$ in mathematically positive direction (counterclockwise)

$$\tan\left(2\widetilde{\varphi}\right) = \frac{\kappa_{yy}-\kappa_{xx}}{2\kappa_{xy}} = \frac{\pi_{44}}{\pi_{12}-\pi_{11}}\,\frac{\sigma_{xx}-\sigma_{yy}}{2\sigma_{xy}} \tag{9a}$$

to obtain the rotated coordinate system $\widetilde{\Sigma}(\widetilde{x}, \widetilde{y})$ [26]. In $\widetilde{\Sigma}$ the conductivity tensor has equal values on the main diagonal and maximum value on the minor diagonal

$$\underline{\underline{\widetilde{\kappa}}} = \frac{1}{2}\begin{pmatrix} \kappa_{xx}+\kappa_{yy} & \pm\left(\operatorname{sgn}\kappa_{xy}\right)\sqrt{\left(\kappa_{xx}-\kappa_{yy}\right)^2+\left(2\kappa_{xy}\right)^2} \\ \pm\left(\operatorname{sgn}\kappa_{xy}\right)\sqrt{\left(\kappa_{xx}-\kappa_{yy}\right)^2+\left(2\kappa_{xy}\right)^2} & \kappa_{xx}+\kappa_{yy} \end{pmatrix} \tag{9b}$$

In (9b) the plus sign is valid for $|\widetilde{\varphi}| < 45^\circ$ and the minus sign is valid for $45^\circ \le |\widetilde{\varphi}| < 90^\circ$ and $\operatorname{sgn}\kappa_{xy}$ denotes the sign of κ_{xy}. Inserting (7a)–(7d) into (9b) gives

$$\underline{\underline{\widetilde{\kappa}}} \cong \frac{\rho_0}{\rho_{\text{iso}}^2}\left(1+\left(\pi_{11}+\pi_{12}\right)\frac{\sigma_{xx}+\sigma_{yy}}{2}\right)\begin{pmatrix} 1 & \widetilde{\delta}_{12} \\ \widetilde{\delta}_{12} & 1 \end{pmatrix} \tag{10a}$$

$$\widetilde{\delta}_{12} = \pm\operatorname{sgn}\left(\pi_{12}-\pi_{11}\right)\operatorname{sgn}\left(\sigma_{xy}\right) \cdot \frac{\sqrt{4\left(\pi_{12}-\pi_{11}\right)^2\sigma_{xy}^2+\pi_{44}^2\left(\sigma_{xx}-\sigma_{yy}\right)^2}}{2+\left(\pi_{11}+\pi_{12}\right)\left(\sigma_{xx}+\sigma_{yy}\right)}. \tag{10b}$$

So in any case we get a symmetric conductivity tensor with identical values in the major diagonal and with small values in the minor diagonal. With (10a) and analogous to the Hall effect we get a second-order partial differential equation for the electric potential

$$\frac{\partial^2\phi}{\partial\widetilde{x}^2} + 2\widetilde{\delta}_{12}\frac{\partial^2\phi}{\partial\widetilde{x}\partial\widetilde{y}} + \frac{\partial^2\phi}{\partial\widetilde{y}^2} = 0. \tag{11}$$

(11) is not Laplace's equation, because the mixed derivative does not vanish and this is a consequence of the piezoresistive conductivity tensor being even symmetric. As for the boundary conditions they are similar to the Hall effect: the potential has constant values at the contacts, whereas it holds $E_n/E_p = (\partial\phi/\partial n)/(\partial\phi/\partial t) = \widetilde{\delta}_{12}$ at isolation boundaries parallel to $\widetilde{x}$-axis or $\widetilde{y}$-axis. So $\widetilde{\delta}_{12}$ seems to take over the role of $\tan\theta_H = \mu_H B_z$ – at least at these isolation boundaries. However, inside the sample there is a striking difference between Hall effect and piezoresistance [27, 28]: inside the Hall-effect region the angle between electric field vector $\vec{E}$ and current density vector $\vec{J}$ is homogeneous versus spatial position and equal to the Hall angle $\angle\{\vec{E},\vec{J}\} = \theta_H = \arctan(\mu_H B_z)$, whereas in the piezoresistive region in general the angle between $\vec{E}$ and $\vec{J}$ is inhomogeneous versus spatial position $\angle\{\vec{E},\vec{J}\} = \arctan(\widetilde{\delta}_{12}(E_{\widetilde{x}}^2 - E_{\widetilde{y}}^2)/(E_{\widetilde{x}}^2 + E_{\widetilde{y}}^2 + 2\widetilde{\delta}_{12}E_{\widetilde{x}}E_{\widetilde{y}}))$.

The similarities between Hall effect and piezoresistance led some authors to presume an analogy [3, 29–31], since $|\widetilde{\delta}_{12}| \ll 1$ it is tempting to drop the mixed derivatives in (11). Then the piezoresistive potential would become a solution of Laplace's equation just like the Hall potential and since both fulfill the same boundary conditions, we would only have to replace $\mu_H B_z$ by $\widetilde{\delta}_{12}$ to fulfill the skew-derivative boundary conditions. Then the output voltage of a van der Pauw stress

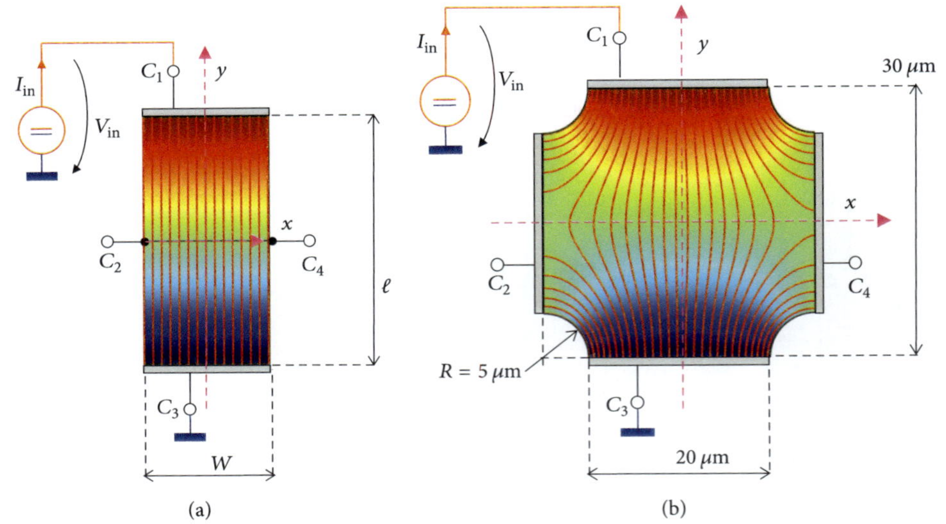

FIGURE 2: (a) Rectangular device with large supply contacts C_1–C_3 and point-sized sense contacts C_2–C_4 with potential distribution and current streamlines at $\sigma_{xy} = 1$ MPa. At zero stress it has two squares input resistance $\ell/W = 2$ and according to (4) this gives $G_{H0} = 0.93006$. (b) Cross-shaped device with large supply contacts C_1–C_3 and large sense contacts C_2–C_4 with potential distribution and current streamlines at $\sigma_{xy} = 1$ MPa. At zero stress it has 1.0359 squares input resistance and according to (4) this gives $G_{H0} = 0.45364$. The plots were obtained by a finite element simulation (FEM) with commercial software code COMSOL MULTIPHYSICS. This FEM simulation gives a stress output signal of device (a) which differs only 0.4 ppm from $V_{out} = R_{sh}I_{in}(\pi_{12} - \pi_{11})\sigma_{xy}G_{H0}$ from [2]. An FEM calculation of device (b) gives an output signal that is 57% larger than $V_{out} = R_{sh}I_{in}(\pi_{12} - \pi_{11})\sigma_{xy}G_{H0}$.

sensor would be $V_{out} \cong R_{sh}I_{in}\tilde{\delta}_{12}G_{H0}$ analogous to (3), whereby the $\cong$ sign denotes small stress and there we may neglect the stress dependence of R_{sh}. If both normal and shear stress are present in (10b) they would mix quadratically. If only inplane shear stress is present on (100)-silicon we would get $V_{out} \cong R_{sh}I_{in}(\pi_{12} - \pi_{11})\sigma_{xy}G_{H0}$. These relations may hold for certain device geometries, like the one in [2] with small sense contacts and homogeneous current density (which is shown in Figure 2(a)), or for devices with $G_{H0} \cong 1$ (as in [32]), but they do not hold in the general case. Figure 2(b) shows a device geometry with 90° symmetry and large contacts, where these simple formulae lead to an error of 36% in the output voltage when compared to an accurate finite element simulation.

Where does this error come from? In (11) we had to neglect the mixed derivative in order to obtain the analogy with the Hall effect, and this was not allowed. Of course we may neglect the $\tilde{\delta}_{12}$-term in (11) if we want to compute the potential ϕ up to zero-th order in $\tilde{\delta}_{12}$. However, if we compute the output voltage of the pseudo-Hall device, we subtract the potential at both output contacts in order to get the small imbalance of potential due to $\tilde{\delta}_{12}$. Hence, this output voltage is proportional to $\tilde{\delta}_{12}$; it is of first-order in $\tilde{\delta}_{12}$ and so we must not neglect terms in (11) which are linear in $\tilde{\delta}_{12}$. Therefore in general we cannot use the Hall geometry factor G_{H0} for van der Pauw stress sensors. However, in the special case of Figure 2(a) at small stress the potential is constant versus x. Therefore the mixed derivative in (11) vanishes and (11) becomes the Laplace equation. For this particular symmetry (11) and (2) are identical and the piezoresistive output is analogous to the Hall output when we replace $\tilde{\delta}_{12}$ by $\mu_H B_z$.

Figures 3(a) and 3(b) show the angle $\angle\{\vec{E}, \vec{J}\}$ between electric field and current density for the device of Figure 2(b) operated in the same way as in Figure 2(b). However in Figure 3(a) there is small shear stress $(\pi_{12} - \pi_{11})\sigma_{xy} = \tan 0.446^\circ$ and no magnetic field whereas in Figure 3(b) there is large magnetic field $\mu_H B_z = \tan 51.078^\circ$ and no mechanical stress. Obviously, in Figure 3(a) the angle $\angle\{\vec{E}, \vec{J}\}$ is small but inhomogeneous: it varies ±100% from −0.446° at the supply contacts to +0.446° at the sense contacts. Conversely, in Figure 3(b) the angle $\angle\{\vec{E}, \vec{J}\}$ is large and perfectly homogeneous (51.078°). Figures 4(a) and 4(b) show the same angle $\angle\{\vec{E}, \vec{J}\}$ for the device of Figure 2(a). In Figure 4(a) there is again small shear stress $(\pi_{12} - \pi_{11})\sigma_{xy} = \tan 0.446^\circ$ at zero magnetic field and in Figure 4(b) there is a large Hall angle $\theta_H = 51.078^\circ$ at zero stress. Figure 4(b) shows again a perfectly homogeneous large Hall angle. In Figure 4(a) the angle $\angle\{\vec{E}, \vec{J}\}$ is within a very narrow range of −0.445766° and −0.445752°; nevertheless the color coding reveals a systematic pattern of this angle. This is a striking contrast to Figure 3(a). The specific geometry of the device of Figure 4(a) gives a homogeneous current density and thus a homogeneous angle $\angle\{\vec{E}, \vec{J}\}$ at small mechanical stress just like for the Hall effect. This seems to be the deeper reason why one may use the Hall geometry factor in Figure 4(a) to compute the output voltage at shear stress, although this is not allowed for other device geometries like in Figure 3(a). At larger stress on the device in Figure 4(a) the spatial dependence of $\angle\{\vec{E}, \vec{J}\}$ gets more pronounced, yet this inhomogeneity is still smaller than for Figure 3(a) and the piezoresistive angle $\angle\{\vec{E}, \vec{J}\}$ will never change sign as in Figure 3(a).

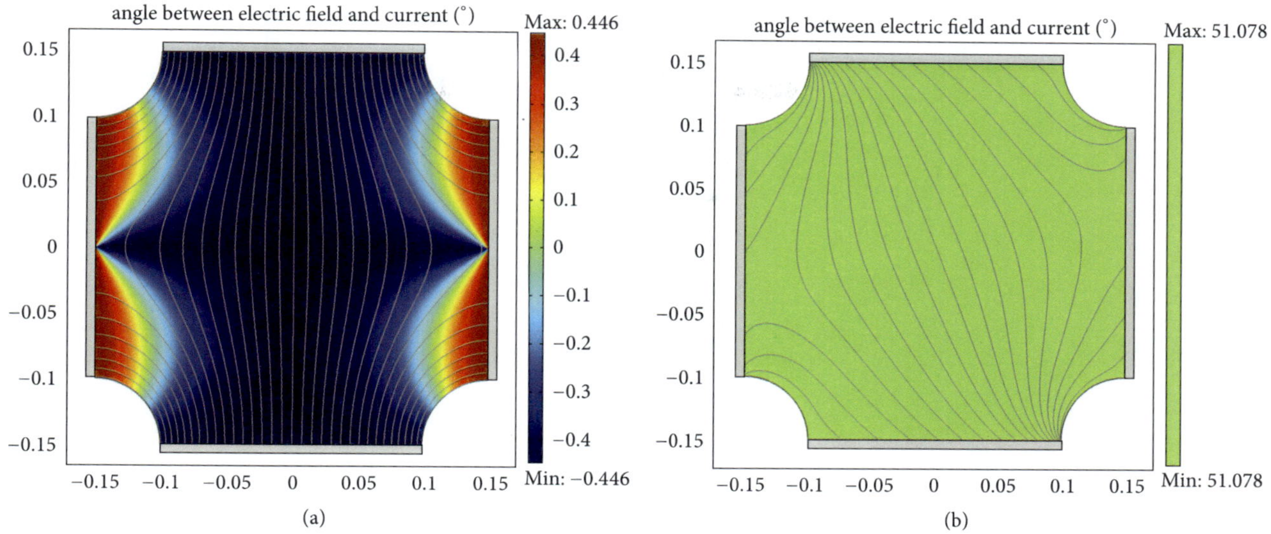

FIGURE 3: (a, b) Angle between electric field and current density in a device of Figure 2(b) operated like in Figure 2(b). The plots were obtained by FEM simulations with COMSOL MULTIPHYSICS. In (a) we have zero magnetic field and small mechanical stress $(\pi_{12} - \pi_{11})\sigma_{xy} = \tan 0.446^{\circ}$. In (b) we have zero stress and even large magnetic field with 51.078° Hall angle. As predicted by [28] the Hall angle is perfectly homogeneous throughout the device (b), whereas in the piezoresistive case (a) the angle $\angle\{\vec{E}, \vec{J}\}$ varies between −0.446° at the supply contacts and in the major portion of the device and +0.446° at the sense contacts. Therefore the analogy [2] between Hall effect and piezoresistive effect does not apply in this case.

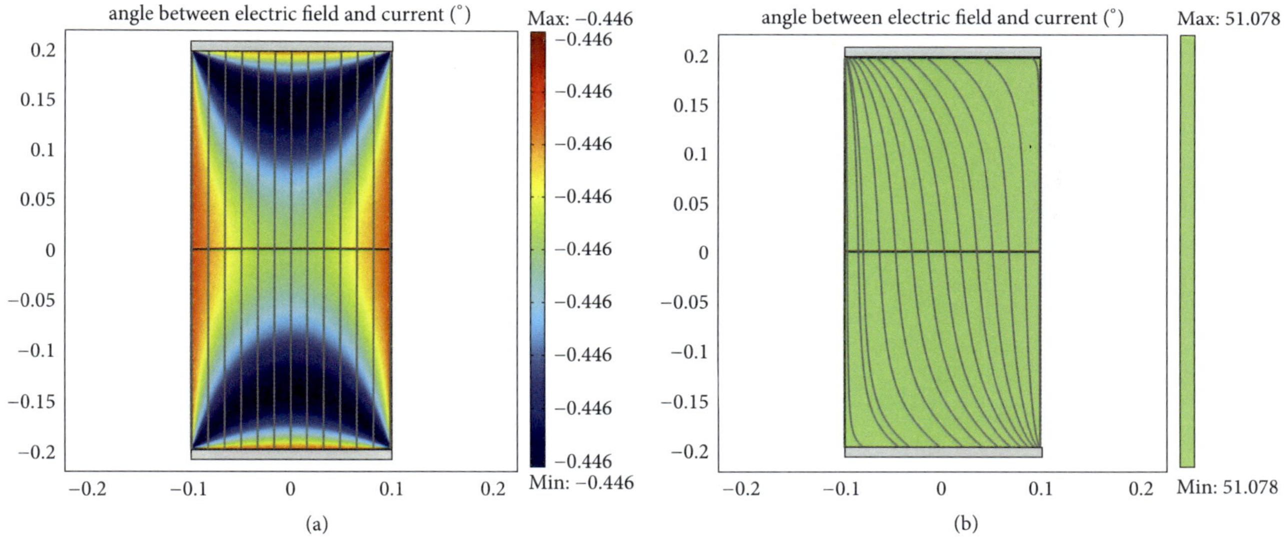

FIGURE 4: (a, b) Angle between electric field and current density in a device of Figure 2(a) operated like in Figure 2(a). The plots were obtained by FEM simulations with COMSOL MULTIPHYSICS. In (a) we have zero magnetic field and small mechanical stress $(\pi_{12} - \pi_{11})\sigma_{xy} = \tan 0.446^{\circ}$. In (b) we have zero stress and even large magnetic field with 51.078° Hall angle. As predicted by [28] the Hall angle is perfectly homogeneous throughout the device (b), but surprisingly in the piezoresistive case (a) the angle $\angle\{\vec{E}, \vec{J}\}$ is also highly homogeneous (it varies only between −0.445766° and −0.445752°). This seems to be the reason why the analogy [2] between Hall effect and piezoresistive effect works for this device.

3. The Affine Transformation

Since (11) is not Laplace's equation, we cannot directly apply a conformal transformation to solve it. However, (11) can be transformed into Laplace's equation by an affine transformation $\Sigma(x, y) \rightarrow \Sigma'(x', y')$. This procedure has been worked out in [8, 24, 33] (the affine transformation is also used in [34]; however, it seems that conductivities in x- and y-directions are mixed up. [7] uses again a different transformation which does not seem to comply with

(12a)–(12c).). We summarize it here briefly in a notation that will we used in the rest of this work.

$$\begin{pmatrix} x' \\ y' \end{pmatrix} = \underline{\underline{M}}^{\text{piezo}} \begin{pmatrix} x \\ y \end{pmatrix}$$

$$\text{with } \underline{\underline{M}}^{\text{piezo}} = \begin{pmatrix} \frac{1}{\sqrt{\rho_{\text{iso}}\kappa_{xx}}} & 0 \\ -\kappa_{xy}\sqrt{\frac{\rho_{\text{iso}}}{\kappa_{xx}}} & \sqrt{\rho_{\text{iso}}\kappa_{xx}} \end{pmatrix} \quad (12a)$$

In the transformed geometry the equivalent device has an isotropic conductivity

$$\underline{\underline{\kappa}}' = \frac{1}{\rho_{\text{iso}}} \begin{pmatrix} 1 & 0 \\ 0 & 1 \end{pmatrix}, \quad (12b)$$

$$\rho_{\text{iso}} = \sqrt{\rho_{xx}\rho_{yy} - \rho_{xy}^2} = \frac{1}{\kappa_{\text{iso}}} = \frac{1}{\sqrt{\kappa_{xx}\kappa_{yy} - \kappa_{xy}^2}} \quad (12c)$$

The potentials in corresponding points before and after the transformation are identical: $\phi(x, y) = \phi'(x', y')$. The matrix $\underline{\underline{M}}^{\text{piezo}}$ is chosen such that the currents through corresponding contacts remain constant under the transformation, because only then the transformation preserves the resistances. Also the total power dissipation in the device and the power density are not affected by the transformation. This implies that the thickness and the area of the device remain constant under the transformation [24].

Inserting (7a)–(7c) into (12a) and developing into a MacLaurin series for small stress terms gives the approximation

$$\underline{\underline{M}}^{\text{piezo}} \cong \begin{pmatrix} 1 + \frac{\pi_{44}(\sigma_{xx} - \sigma_{yy})}{4} & 0 \\ (\pi_{11} - \pi_{12})\sigma_{xy} & 1 - \frac{\pi_{44}(\sigma_{xx} - \sigma_{yy})}{4} \end{pmatrix}, \quad (13a)$$

$$\rho_{\text{iso}} \cong \rho_0 \left(1 + \frac{\pi_{11} + \pi_{12}}{2}(\sigma_{xx} + \sigma_{yy})\right), \quad (13b)$$

$$\rho_{\text{iso}}\kappa_{xx} \cong 1 - \frac{\pi_{44}}{2}(\sigma_{xx} - \sigma_{yy}). \quad (13c)$$

The affine transformation (12a), (12b), and (12c) applies to all crystal classes; however, (13a), (13b), and (13c) use (7a), (7b), (7c), and (7d) and therefore they apply only to (100)-planes of cubic crystal classes (e.g., silicon).

Note that an affine transformation is definitely not a conformal transformation: conformal transformations preserve the angles between two curves, whereas affine transformations do not, because they skew the geometry. Equation (13a) shows that the geometry is skewed for shear stress. The difference of normal inplane stresses scales the x- and y-axes differently. This also skews most devices except those ones with edges parallel to x-y-axes. The sum of normal inplane stresses scales the equivalent isotropic conductivity. Isotropic stress $\sigma_{xx} = \sigma_{yy}$ does not change the shape of the equivalent device; it only changes $\rho_0 \to \rho_{\text{iso}}$.

So far we neglected the pure deformation of the geometry (= strain) caused by the stress on the device. This is commonly justified because for many materials the piezoresistance effect dominates. However, the affine transformation (12a) already describes a deformation of the device, and so we can add the strain terms without unduly increasing the complexity of the calculation. A homogeneous two-dimensional strain with zero displacement of the origin is described by the transformation [35]

$$\begin{pmatrix} x' \\ y' \end{pmatrix} = \underline{\underline{M}}^{\text{strain}} \begin{pmatrix} x \\ y \end{pmatrix} \quad (14)$$

$$\text{with } \underline{\underline{M}}^{\text{strain}} = \begin{pmatrix} 1 + \varepsilon_{xx} & \varepsilon_{xy} \\ \varepsilon_{xy} & 1 + \varepsilon_{yy} \end{pmatrix}$$

whereby ε_{xx}, ε_{yy}, and ε_{xy} are the tensor strains (as opposed to the engineering strains [36]). The strain is related to the stress via the fourth rank compliance tensor, which in silicon has only three nonvanishing terms $s_{11} = 7.68 \times 10^{-12}$ Pa, $s_{12} = -2.14 \times 10^{-12}$ Pa, and $s_{44} = 12.6 \times 10^{-12}$ Pa [37]. Transforming the tensors into the chip coordinate system $\Sigma = (x, y, z)$ of Figure 1 gives

$$\varepsilon_{xx} = (s_{11} + s_{12})\frac{\sigma_{xx} + \sigma_{yy}}{2} + s_{44}\frac{\sigma_{xx} - \sigma_{yy}}{4}, \quad (15a)$$

$$\varepsilon_{yy} = (s_{11} + s_{12})\frac{\sigma_{xx} + \sigma_{yy}}{2} - s_{44}\frac{\sigma_{xx} - \sigma_{yy}}{4}, \quad (15b)$$

$$\varepsilon_{xy} = (s_{11} - s_{12})\sigma_{xy}, \quad (15c)$$

$$\varepsilon_{zz} = s_{12}(\sigma_{xx} + \sigma_{yy}). \quad (15d)$$

The last equation (15d) means that the thickness of the conducting region changes, too. We take account of this by keeping the unstrained thickness but changing the resistivity accordingly $\rho_{\text{iso}} \to \rho_{\text{iso}}/(1 + \varepsilon_{zz})$. The combined effect of piezoresistance and strain is obtained by multiplying both matrices $\underline{\underline{M}}^{\text{piezo}}$ and $\underline{\underline{M}}^{\text{strain}}$, whereby the order of multiplication is irrelevant for small stress.

$$\underline{\underline{M}}^{\text{strain}}\underline{\underline{M}}^{\text{piezo}} \cong \begin{pmatrix} 1+(\pi_{44}+s_{44})\dfrac{\sigma_{xx}-\sigma_{yy}}{4}+(s_{11}+s_{12})\dfrac{\sigma_{xx}+\sigma_{yy}}{2} & (s_{11}-s_{12})\sigma_{xy} \\ (\pi_{11}-\pi_{12})\sigma_{xy}+(s_{11}-s_{12})\sigma_{xy} & 1-(\pi_{44}+s_{44})\dfrac{\sigma_{xx}-\sigma_{yy}}{4}+(s_{11}+s_{12})\dfrac{\sigma_{xx}+\sigma_{yy}}{2} \end{pmatrix}, \quad (16a)$$

$$\rho_{\text{iso}} \cong \rho_0 \frac{1+(\pi_{11}+\pi_{12})(\sigma_{xx}+\sigma_{yy})/2}{1+s_{12}(\sigma_{xx}+\sigma_{yy})}. \quad (16b)$$

In the literature of the Hall geometry factor it is a common technique to map finite device geometries like disks, squares, rectangles, crosses, or octagons via conformal mapping onto the infinite upper half space of the (x, y)-plane [11, 14, 38]. If the original devices have contacts only at their perimeter, the conformal mapping transfers these contacts onto the x-axis boundary. Let us for a moment assume an *original* device with an arbitrary number of contacts on the x-axis, its active region being the entire upper half plane with anisotropic resistivity (6a), (6b), and (6c). We can transform it via affine transformation (12a), (12b), and (12c) into an equivalent problem with isotropic resistivity ρ_{iso}. However, since all contacts are on the x-axis and the geometry is infinite, matrix $\underline{\underline{M}}$ just scales the x-axis by the factor $\sqrt{\rho_{\text{iso}}\kappa_{xx}}$. The different scaling factor along the y-axis is irrelevant, because there are no features (neither contacts nor boundaries) except on the x-axis. So the affine transformation (12a), (12b), and (12c) effectively just scales the entire geometry by a single factor, and this does not change ratios of voltages on (or currents through) the contacts or ratios of resistances between the contacts. In the end the absolute values of resistances of the transformed device indeed change due to ρ_{iso} in (7d). This shows a clear difference between Hall effect and piezoresistance effect: for the Hall effect we can transform devices onto the upper half plane with conformal transformation and preserve the Hall output voltage, whereas for the piezoresistance effect any direct conformal transformation onto the upper half plane would make the piezoresistive output voltage vanish. The correct procedure for a piezoresistive potential problem is to apply the affine transformation (12a), (12b), and (12c) prior to conformal transformation(s): the affine transformation establishes the validity of Laplace's equation and then it is allowed to go on with conformal transformations. Figure 5 shows numerical simulations of such a device in the upper half plane with piezoresistive coefficients of low n-doped silicon. The input voltage is forced by an external voltage supply between the outer contacts. Then the change of output voltage at the center contact caused by mechanical stress is close to zero (except for numerical inaccuracies) while the input resistance changes with stress according to ρ_{iso} in (7d). This geometry has also some practical importance, because it is similar to vertical Hall devices [39, 40], where all contacts are on the top surface of a rectangular Hall region. In the early days this Hall region was the entire wafer and therefore it was close to infinite. The offset voltage of such a device can drift over lifetime due to drift of mechanical stress on the chip. With the above discussion we can say that for large Hall tubs this drift will get very small and for small Hall tubs it is caused by the isolating boundary at the sidewalls and bottom of the tub but not by the top surface. This was already stated in [41] based on the same theory using affine and conformal transformations. It was also experimentally observed that the offset of vertical Hall devices is more stable than the offset of conventional horizontal Hall plates [42].

4. Devices with Mirror Symmetries to x- and y-Axes and Contacts Not Lying on the x- and y-Axes, at Zero Shear Stress $\sigma_{xy} = 0$

In standard microelectronic packages the main part of the chip (except its perimeter) is under pure inplane biaxial stress in the order of 100 MPa [22]. Hence, this case is most relevant in practice. Then it holds for a (100)-silicon chip with (7a), (7d)

$$\rho_{\text{iso}} = \frac{\rho_0}{2}\sqrt{2+(\pi_{11}+\pi_{12}-\pi_{44})\sigma_{xx}+(\pi_{11}+\pi_{12}+\pi_{44})\sigma_{yy}}\sqrt{2+(\pi_{11}+\pi_{12}+\pi_{44})\sigma_{xx}+(\pi_{11}+\pi_{12}-\pi_{44})\sigma_{yy}}, \quad (17a)$$

$$\rho_{\text{iso}}\kappa_{xx} = \sqrt{\frac{2+(\pi_{11}+\pi_{12}-\pi_{44})\sigma_{xx}+(\pi_{11}+\pi_{12}+\pi_{44})\sigma_{yy}}{2+(\pi_{11}+\pi_{12}+\pi_{44})\sigma_{xx}+(\pi_{11}+\pi_{12}-\pi_{44})\sigma_{yy}}}. \quad (17b)$$

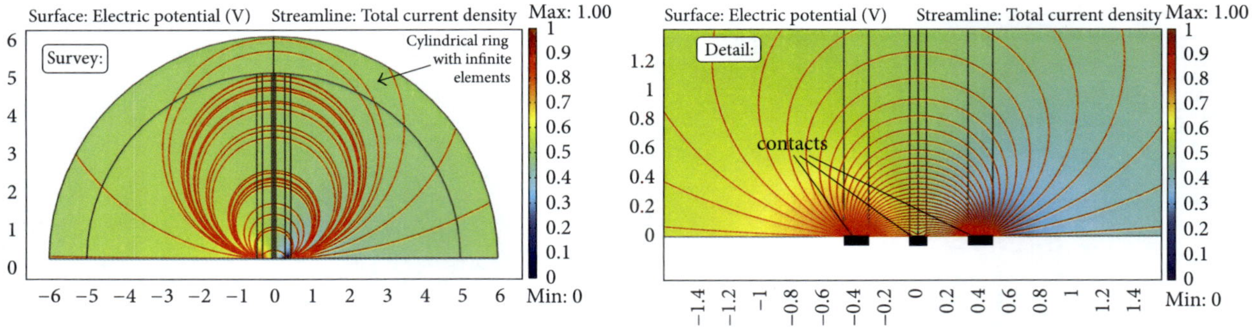

Figure 5: Finite element model of a resistive device with three contacts at the lower boundary (= x-axis). The plots were obtained by FEM simulations with COMSOL MULTIPHYSICS. They show the potential distribution and the current streamlines. The entire upper half plane is conductive with piezoresistance according to (5) and (6a)–(6c) with $\rho_0 = 1\,\Omega$m and the piezoresistive coefficients for low n-doped silicon ($\pi_{11} = -1.022$/GPa, $\pi_{12} = 0.534$/GPa, $\pi_{44} = -0.136$/GPa). Pure strain is neglected. The outer contacts are 0.15 m long, the center contact is 0.1 m long, and the spacing between contacts is 0.25 m. The thickness into the drawing plane is 1 m. The left contact is forced to 1 V, the right contact is forced to 0 V, and the center contact is floating (i.e., no current flows in or out). The infinite boundary is modeled with infinite elements. The mesh has 1.4 million elements. At zero mechanical stress the center contact is at 0.5 V and the input resistance is 1.88262 Ω. At $\sigma_{xx} = 100$ MPa the output voltage increases 57 nV and the input resistance decreases −2.443%. At $\sigma_{yy} = 100$ MPa the output voltage decreases −56 nV and the input resistance decreases −2.441%. At $\sigma_{xy} = 100$ MPa the output voltage decreases −2.5 μV and the input resistance decreases −1.218%. All these resistance changes deviate less than 10 ppm from theory (12a)–(12c). The changes in output voltage should be zero according to theory and the small deviations are probably due to insufficiently fine mesh.

For very large stress difference $\sigma_{xx} - \sigma_{yy}$ in the GPa-range the terms under the square-roots in (17a), (17b) become negative, and either ρ_{xx} or ρ_{yy} reverse sign (see (6a), (6b)), which is impossible for passive devices ($\vec{E} \cdot \vec{J} > 0$); there the first-order piezoresistive theory breaks down.

4.1. Rectangular Devices with Four Large Contacts. With (6a)–(6c), (12a), and (12b) and with the results of [10, 11] we can immediately give the equivalent resistor circuit (ERC) and its dependence on stress for rectangular devices shown in Figure 6. We summarize these findings in (18a)–(18m):

$$\overline{R}_{Hf} = \frac{4R_{\text{sh}}}{4\lambda_x + \lambda_f - \lambda_p}, \tag{18a}$$

$$\overline{R}_{Hp} = \frac{4R_{\text{sh}}}{4\lambda_x - \lambda_f + \lambda_p}, \tag{18b}$$

$$\overline{R}_{D} = \frac{2R_{\text{sh}}}{\lambda_f + \lambda_p - 4\lambda_x}, \tag{18c}$$

$$\lambda_f = \frac{K'\left((1-f)/(1+f)\right)}{K\left((1-f)/(1+f)\right)}, \tag{18d}$$

$$\lambda_p = \frac{K'\left(2\sqrt{p}/(1+p)\right)}{K\left(2\sqrt{p}/(1+p)\right)}, \tag{18e}$$

$$4\lambda_x = \frac{K'\left(((1-f)/(1+f))\left(2\sqrt{p}/(1+p)\right)\right)}{K\left(((1-f)/(1+f))\left(2\sqrt{p}/(1+p)\right)\right)}, \tag{18f}$$

$$\frac{1-f}{1+f} = \sqrt{\frac{1-k}{1+k}\frac{1+kw_3}{1-kw_3}}, \tag{18g}$$

$$\frac{2\sqrt{p}}{1+p} = \sqrt{\frac{1+k}{2}\frac{1+w_3}{1+kw_3}}, \tag{18h}$$

$$w_3 = \operatorname{sn}\left(K(k)\left(\frac{2s_0}{\ell_0} - 1\right), k\right), \tag{18i}$$

$$k = \sqrt{L\left(\frac{2W}{\ell}\right)}, \tag{18j}$$

$$W = W_0\sqrt{\rho_{\text{iso}}\kappa_{xx}}, \tag{18k}$$

$$\ell = \frac{\ell_0}{\sqrt{\rho_{\text{iso}}\kappa_{xx}}}, \tag{18l}$$

$$R_{\text{sh}} = \frac{\rho_{\text{iso}}}{t_d}. \tag{18m}$$

The parameters ℓ_0, W_0, and s_0 are length, width, and spacing between contacts, respectively, of the device at zero stress (see Figure 6). They undergo the affine transformation (12a) $\ell_0 \to \ell$ and $W_0 \to W$ to account for the mechanical stress. The ratio $s_0/\ell_0 = s/\ell$ is independent of stress, because both lengths are along x-direction and the common scaling factor of the affine transformation cancels out. t_d is the constant thickness of the device. R_{sh} is the sheet resistance of the equivalent device with isotropic resistivity ρ_{iso} after the affine transformation. $k \in [0,1]$, $w_3 \in [-1,1]$, $f \in [0,1]$, and $p \in [0,1]$ are dimensionless parameters which appeared in the sequence of conformal transformations in [10, 11]. In (18i) the Jacobi sine-amplitude function is used. It is defined by $x = \operatorname{sn}(F(x,y), y)$. Equations (18a)–(18f) are given in (14a–c) and (A11–13) in [10]; (18g)–(18j) are equivalent to (8, 9, 14) in [11]. The parameters $\lambda_f, \lambda_p, \lambda_x$ depend on mechanical stress; if we refer to their values at zero stress, we add a zero in the indices: λ_{f0}, λ_{p0}, and λ_{x0}. The ERC in Figure 6 and (18a)–(18m) is not

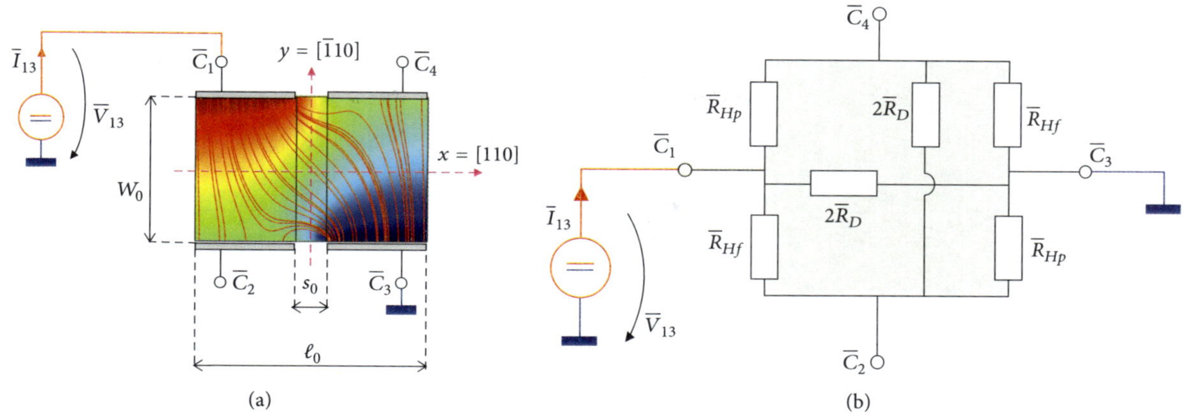

FIGURE 6: Rectangular $\sigma_{xx} - \sigma_{yy}$ stress sensor in (100)-silicon with two perpendicular mirror-symmetry axes $x = [110]$, $y = [\overline{1}10]$. No magnetic field is applied. An input voltage is forced between contacts $\overline{C}_1$–$\overline{C}_3$ and an output voltage is tapped between $\overline{C}_2$–$\overline{C}_4$. All contacts have the same size $(\ell_0 - s_0)/2$. (a) shows the top view of the potential distribution and the current streamlines in the plane device. (b) shows the equivalent resistor circuit (ERC).

an approximation; they are valid in a strict sense even for large mechanical stress and arbitrary size of contacts as long as we stay in the linear electrostatic and piezoresistive theory. With the ERC one can compute the input resistances, the ratios of output over input voltages, and the transimpedances of Hall- and van der Pauw-operating modes (all at zero magnetic field). For the Hall-operating mode it holds that

$$\left.\frac{\overline{V}_{13}}{\overline{I}_{13}}\right|_{\overline{I}_2=\overline{I}_4=0} = \overline{R}_{13} = \left.\frac{\overline{V}_{24}}{\overline{I}_{24}}\right|_{\overline{I}_1=\overline{I}_3=0} = \overline{R}_{24}$$
$$= 2\overline{R}_D \frac{\overline{R}_{Hf}\overline{R}_{Hp} + \overline{R}_D\overline{R}_{Hf} + \overline{R}_D\overline{R}_{Hp}}{\left(\overline{R}_{Hf} + 2\overline{R}_D\right)\left(\overline{R}_{Hp} + 2\overline{R}_D\right)} \tag{19a}$$
$$= R_{\mathrm{sh}}\left(\frac{1}{\lambda_p} + \frac{1}{\lambda_f}\right),$$

$$\left.\frac{\overline{V}_{24}}{\overline{V}_{13}}\right|_{\overline{I}_2=\overline{I}_4=0} = \left.\frac{\overline{V}_{13}}{\overline{V}_{24}}\right|_{\overline{I}_1=\overline{I}_3=0}$$
$$= \overline{R}_D \frac{\overline{R}_{Hp} - \overline{R}_{Hf}}{\overline{R}_{Hf}\overline{R}_{Hp} + \overline{R}_D\overline{R}_{Hf} + \overline{R}_D\overline{R}_{Hp}} \tag{19b}$$
$$= \frac{\lambda_f - \lambda_p}{\lambda_f + \lambda_p},$$

$$\left.\frac{\overline{V}_{24}}{\overline{I}_{13}}\right|_{\overline{I}_2=\overline{I}_4=0} = \left.\frac{\overline{V}_{13}}{\overline{I}_{24}}\right|_{\overline{I}_1=\overline{I}_3=0}$$
$$= 2\overline{R}_D^2 \frac{\overline{R}_{Hp} - \overline{R}_{Hf}}{\left(\overline{R}_{Hf} + 2\overline{R}_D\right)\left(\overline{R}_{Hp} + 2\overline{R}_D\right)} \tag{19c}$$
$$= R_{\mathrm{sh}}\left(\frac{1}{\lambda_p} - \frac{1}{\lambda_f}\right).$$

For the van der Pauw-operating mode it holds that

$$\left.\frac{\overline{V}_{43}}{\overline{I}_{43}}\right|_{\overline{I}_1=\overline{I}_2=0} = \frac{\overline{R}_{Hf}}{2} \frac{\overline{R}_{Hf}\overline{R}_{Hp} + 2\overline{R}_D\left(\overline{R}_{Hf} + 2\overline{R}_{Hp}\right)}{\left(\overline{R}_{Hf} + 2\overline{R}_D\right)\left(\overline{R}_{Hf} + \overline{R}_{Hp}\right)}$$
$$= R_{\mathrm{sh}}\left(\frac{1}{\lambda_f} + \frac{1}{4\lambda_x}\right), \tag{19d}$$

$$\left.\frac{\overline{V}_{23}}{\overline{I}_{23}}\right|_{\overline{I}_1=\overline{I}_4=0} = \frac{\overline{R}_{Hp}}{2} \frac{\overline{R}_{Hf}\overline{R}_{Hp} + 2\overline{R}_D\left(2\overline{R}_{Hf} + \overline{R}_{Hp}\right)}{\left(\overline{R}_{Hp} + 2\overline{R}_D\right)\left(\overline{R}_{Hf} + \overline{R}_{Hp}\right)}$$
$$= R_{\mathrm{sh}}\left(\frac{1}{\lambda_p} + \frac{1}{4\lambda_x}\right), \tag{19e}$$

$$\left.\frac{\overline{V}_{43}}{\overline{I}_{12}}\right|_{\overline{I}_4=\overline{I}_3=0} = \frac{\overline{R}_{Hf}^2}{2} \frac{2\overline{R}_D - \overline{R}_{Hp}}{\left(\overline{R}_{Hf} + 2\overline{R}_D\right)\left(\overline{R}_{Hf} + \overline{R}_{Hp}\right)}$$
$$= R_{\mathrm{sh}}\left(\frac{1}{\lambda_f} - \frac{1}{4\lambda_x}\right), \tag{19f}$$

$$\left.\frac{\overline{V}_{43}}{\overline{V}_{12}}\right|_{\overline{I}_4=\overline{I}_3=0} = \overline{R}_{Hf} \frac{2\overline{R}_D - \overline{R}_{Hp}}{\overline{R}_{Hf}\overline{R}_{Hp} + 2\overline{R}_D\left(\overline{R}_{Hf} + 2\overline{R}_{Hp}\right)}$$
$$= \frac{4\lambda_x - \lambda_f}{4\lambda_x + \lambda_f}, \tag{19g}$$

$$\left.\frac{\overline{V}_{23}}{\overline{I}_{14}}\right|_{\overline{I}_2=\overline{I}_3=0} = \frac{\overline{R}_{Hp}^2}{2} \frac{2\overline{R}_D - \overline{R}_{Hf}}{\left(\overline{R}_{Hp} + 2\overline{R}_D\right)\left(\overline{R}_{Hf} + \overline{R}_{Hp}\right)}$$
$$= R_{\mathrm{sh}}\left(\frac{1}{\lambda_p} - \frac{1}{4\lambda_x}\right), \tag{19h}$$

$$\left.\frac{\overline{V}_{23}}{\overline{V}_{14}}\right|_{\overline{I}_2=\overline{I}_3=0} = \overline{R}_{Hp} \frac{2\overline{R}_D - \overline{R}_{Hf}}{\overline{R}_{Hf}\overline{R}_{Hp} + 2\overline{R}_D\left(\overline{R}_{Hp} + 2\overline{R}_{Hf}\right)}$$
$$= \frac{4\lambda_x - \lambda_p}{4\lambda_x + \lambda_p}, \tag{19i}$$

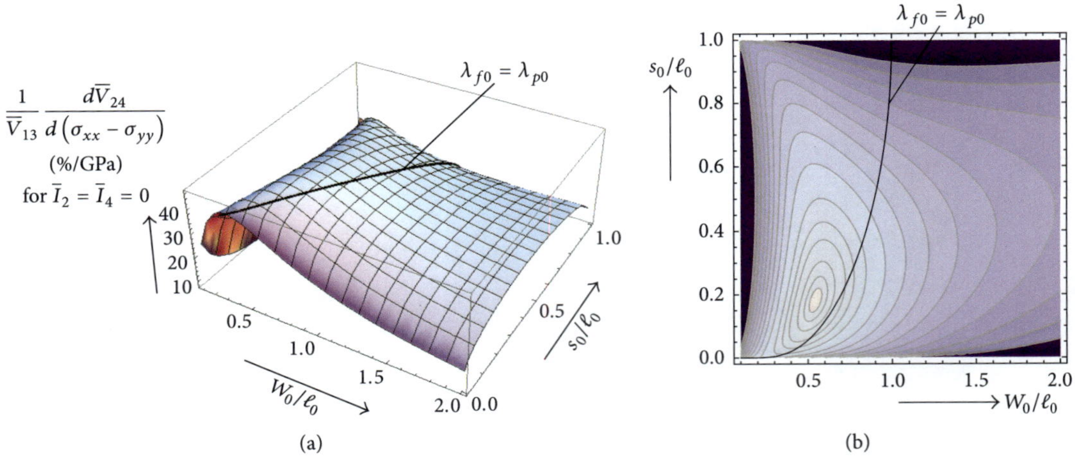

FIGURE 7: Sensitivity of output voltage $\overline{V}_{24}$ versus mechanical stress $\sigma_{xx} - \sigma_{yy}$ (normalized to supply voltage $\overline{V}_{13}$) for a device of Figure 6 operated in the same way as in Figure 6. All other components of the stress tensor vanish, as well as $\sigma_{xx} + \sigma_{yy} = 0$. (a) is a 3D-plot and (b) shows the contour lines thereof. In (a) the steep slopes of the surface near $W_0/\ell_0 < 0.1$ and $s_0/\ell_0 < 0.01$ and $s_0/\ell_0 > 0.99$ are only sketched due to numerical problems. The plots are graphical representations of (18a)–(18m) in the limit of small stress and with the piezoresistive coefficient $\pi_{44} = 1.381$/GPa for low p-doped silicon. Largest stress sensitivities are obtained for aspect ratios $W_0/\ell_0 \approx 0.55$. The dependence on contact size is small, but $s_0/\ell_0 \approx 0.18$ tends to be best. The black curve denotes offset-free devices with $\lambda_{f0} = \lambda_{p0}$, where $\overline{V}_{24} = 0$ at zero stress. The maximum of the surface does not lie on this curve. The contours in (b) are for 46.75, 46.5, 46, 45, 44, 42.5, 40, . . ., 22.5, 20%/GPa.

whereby $\overline{I}_{mn}$ is the current which enters the device through contact $\overline{C}_m$ and leaves through contact $\overline{C}_n$ and $\overline{V}_{mn}$ is the voltage across both contacts $\overline{V}_{mn} = \phi(\overline{C}_m) - \phi(\overline{C}_n)$.

Only for carefully chosen contact spacing s_0 does the output voltage $\overline{V}_{24}$ vanish at zero stress in the Hall-operating mode $\overline{I}_2 = \overline{I}_4 = 0$ and $\overline{I}_3 = -\overline{I}_1$: we get these offset-free devices with $\overline{R}_{Hf} = \overline{R}_{Hp}$ at zero stress (see (19b)) and this means $\lambda_{f0} = \lambda_{p0}$, from which follows the contact spacing (see [12])

$$\frac{s_0}{\ell_0} = \frac{1}{2}\left(1 + \frac{F\left(1/k_0 - 8/\left(1 + 4k_0 - k_0^2\right), k_0\right)}{K(k_0)}\right) \tag{20}$$

$$\text{with } k_0 = \sqrt{L\left(2\frac{W_0}{\ell_0}\right)} \text{ for } \frac{W_0}{\ell_0} \leq 1$$

Offset-free devices are only possible if the width W_0 is smaller or equal to the length ℓ_0. Figure 7 shows the sensitivity of the output voltage $\overline{V}_{24}$ in Hall-operating mode with respect to small changes in stress $\sigma_{xx} - \sigma_{yy}$ at fixed supply voltage $\overline{V}_{13}$ for devices of Figure 6 and operated according to Figure 6. There the stress sensitivity function has a flat maximum of 46.839%/GPa at $W_0/\ell_0 \cong 0.5444$, $s_0/\ell_0 \cong 0.1809$. At zero stress the number of squares between contacts $\overline{C}_1$–$\overline{C}_3$ or between contacts $\overline{C}_2$–$\overline{C}_4$ is 1.3386 (with $\lambda_{f0} \cong 1.7538$, $\lambda_{p0} \cong 1.3013$). The output voltage at zero stress is 14.81% of the supply voltage; this device is not offset-free ($\lambda_{f0} \neq \lambda_{p0}$). All these values were checked with a finite element simulation. If we restrict the search to offset-free devices the maximum stress sensitivity is only 2.5% smaller, namely, 45.674%/GPa at $W_0/\ell_0 \cong 0.6220$ and $s_0/\ell_0 \cong 0.1315$ (with $\lambda_{f0} = \lambda_{p0} \cong 1.5733$, see Figure 8); then the resistance between even or odd contacts has 1.2712 squares. The value of both maxima is computed for $\pi_{44} = 1.381$/GPa of low p-doped silicon; in case of other materials of a cubic crystal class the stress sensitivity scales linearly with π_{44} [43].

The theory was verified by finite element simulations at small and large stress (Figure 9). The device had the geometric parameters $W_0 = 0.62207$ m, $\ell_0 = t_d = 1$ m, and $s_0 = 0.13149$ m. For the conductance a linear material law was used according to (7a)–(7d) with $\rho_0 = 1\,\Omega$m and $\pi_{11} = 6.6 \times 10^{-11}$/Pa, $\pi_{12} = -1.1 \times 10^{-11}$/Pa, and $\pi_{44} = 138.1 \times 10^{-11}$/Pa. The two-dimensional simulation had a fine mesh of 1.76 million elements. It used Lagrange multiplier for improved accuracy of currents into the contacts. Table 1 compares the results of this numerical simulation with results of the analytical theory comprising (17a), (17b), (18a)–(18m), (19a), and (19b). Even up to very large stress levels FEM and the theory agrees better than 0.1%.

4.2. Rectangular Devices with Four Point-Sized Contacts. If the contacts of the device in Figure 6 get very small we have $s_0 \to \ell_0 - 2ds_0$, which we plug into (18i) and develop that into a series in small ds_0.

$$w_3 \longrightarrow 1 - 8\left(1 - k^2\right) K^2(k) \left(\frac{ds_0}{\ell_0}\right)^2 \tag{21a}$$

Inserting this into (18g), (18h), and (18d)–(18f) gives

$$\lambda_f \longrightarrow \frac{-\pi}{2} \frac{1}{\ln\left(\sqrt{k}K(k)\, ds_0/\ell_0\right)} \tag{21b}$$

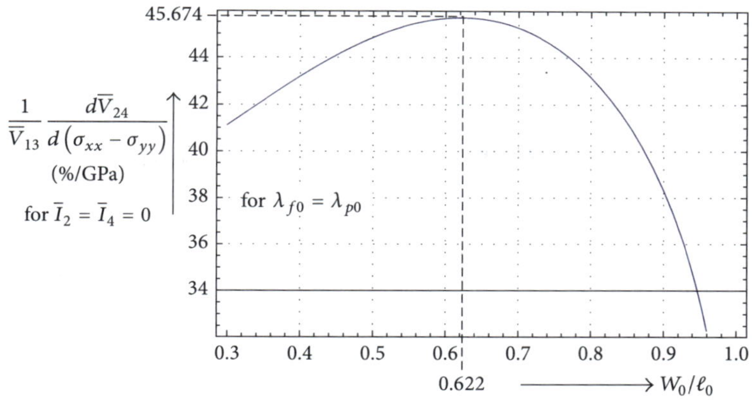

FIGURE 8: Sensitivity of output voltage $\overline{V}_{24}$ versus mechanical stress $\sigma_{xx} - \sigma_{yy}$ (normalized to supply voltage $\overline{V}_{13}$) for an offset-free device of Figure 6 operated in the same way as in Figure 6. All other components of the stress tensor vanish, as well as $\sigma_{xx} + \sigma_{yy} = 0$. The plot is a graphical representation of (18a)–(18m) in the limit of small stress and with the piezoresistive coefficient $\pi_{44} = 1.381/\text{GPa}$ for low p-doped silicon. Largest stress sensitivity of 45.674%/GPa is obtained for aspect ratios $W_0/\ell_0 \approx 0.622$. There $s_0/\ell_0 \cong 0.1315$ satisfies (20).

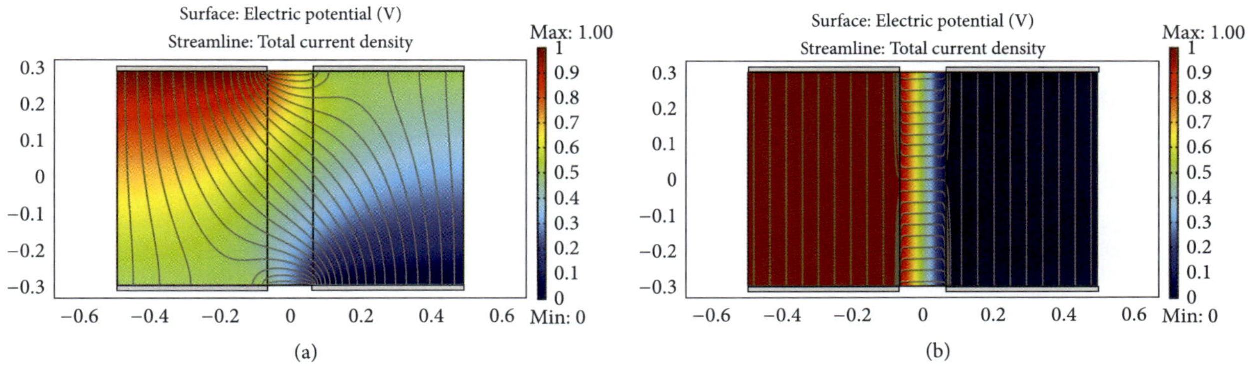

FIGURE 9: Rectangular device from Figure 6 with $W_0 = 0.62207$ m, $s_0 = 0.13149$ m for comparison of the analytical theory with a finite element simulation with COMSOL MULTIPHYSICS. The plots show the potential and current streamlines. (a) The left plot is for small mechanical stress difference $\sigma_{xx} - \sigma_{yy}$ (third line of Table 1); (b) the right plot is for large mechanical stress difference (last line of Table 1).

$$\lambda_p \longrightarrow 4\lambda_x \longrightarrow \frac{\pi}{2} \frac{1}{\ln 2 - \ln\left((1-k) K(k)\, ds_0/\ell_0\right)} \tag{21c}$$

Inserting this into (19a), (19b) gives

$$\left.\frac{\overline{V}_{13}}{\overline{I}_{13}}\right|_{\overline{I}_2=\overline{I}_4=0} \longrightarrow \frac{2}{\pi} \cdot R_{\text{sh}} \left(\ln 2 - \ln\left(\sqrt{k}(1-k) K^2(k) \left(\frac{ds_0}{\ell_0}\right)^2\right)\right) \tag{22a}$$

$$\frac{\overline{V}_{24}}{\overline{V}_{13}} \longrightarrow \frac{\ln\left(2\sqrt{k}\right) - \ln(1-k)}{\ln 2 - \ln\left(\sqrt{k}(1-k) K^2(k) \left(ds_0/\ell_0\right)^2\right)} \tag{22b}$$

$$\frac{\overline{V}_{24}}{\overline{I}_{13}} \longrightarrow \frac{2}{\pi} R_{\text{sh}} \ln\left(\frac{2\sqrt{k}}{1-k}\right) \tag{22c}$$

(22a)–(22c) hold for small contacts and arbitrary stress. k changes with stress according to (18j). For small stress we write $k = k_0 + dk$, where k_0 is the value at zero stress and dk is the small stress variation of k. For point-sized contacts $s_0/\ell_0 \to 1$ and then the only offset-free geometry with $\overline{V}_{24} = 0$ at zero stress is $k_0 = 3 - 2\sqrt{2}$, which gives $W_0/\ell_0 = 1$ (cf. (20)). From (18k), (18l), and (17b) we get for small stress

$$\frac{W}{\ell} = \frac{\rho_{\text{iso}} \kappa_{xx} W_0}{\ell_0} \longrightarrow 1 - \frac{\pi_{44}\left(\sigma_{xx} - \sigma_{yy}\right)}{2} \tag{23a}$$

Table 1: Mechanical stress sensitivity of rectangular device from Figure 9: comparison of numerical simulation (FEM) versus analytical theory. $\pi_{11} = 6.6 \times 10^{-11}$/Pa, $\pi_{12} = -1.1 \times 10^{-11}$/Pa, $\pi_{44} = 138.1 \times 10^{-11}$/Pa. The small residual output voltage of the FEM in the first line is probably due to insufficiently fine mesh. In the last line the stress difference $\sigma_{xx} - \sigma_{yy}$ is so large that ρ_{iso}/ρ_0 gets nearly zero. Then λ_f gets large (~48.9), λ_p gets small (~0.13), ρ_{yy}/ρ_0 gets small (~0.0017), and the output voltage approaches the supply voltage. For even larger stress difference the first-order piezoresistive theories (6a) and (6b) break down and also the numerical simulation gives meaningless results.

σ_{xx} [MPa]	σ_{yy} [MPa]	ρ_{iso}/ρ_0	$\overline{V}_{24}/\overline{V}_{13}$ FEM	$\overline{V}_{24}/\overline{V}_{13}$ theory	$\overline{V}_{13}/\overline{I}_{13}[\Omega]$ FEM	$\overline{V}_{13}/\overline{I}_{13}[\Omega]$ theory
0	0	1.00000	−0.000229	-4×10^{-8}	1.27104	1.27140
10	0	1.00025	0.004341	0.004570	1.26859	1.26895
0	10	1.00025	−0.004792	−0.004563	1.27416	1.27453
10	10	1.00055	−0.000229	-4×10^{-8}	1.27174	1.27210
1000	900	1.04998	0.043525	0.043750	1.30872	1.30910
−1000	−900	0.94523	−0.048074	−0.047841	1.23156	1.23190
856	−600	0.05801	0.994747	0.994751	0.45175	0.45216

Plugging this into (18j) and developing for small stress gives

$$k \longrightarrow k_0 + dk = k_0 - \left(\frac{2}{\pi}\right) k_0 \left(k_0^2 - 1\right) K^2\left(k_0\right) \pi_{44}\left(\sigma_{xx} - \sigma_{yy}\right) \tag{23b}$$

In (23b) we used the differentiation rule for the modular lambda function

$$\frac{dL(y)}{dy} = \left(\frac{4}{\pi}\right) L(y)\left(L(y) - 1\right) K^2\left(\sqrt{L(y)}\right) \tag{24a}$$

which can be proven by

$$\frac{dL(y)}{dy} = \frac{1}{dy/dL} = \left(\frac{d\left(K'(k)/K(k)\right)}{dk^2}\right)^{-1} = \frac{-2\left(1-k^2\right)k^2K^2(k)}{E(k)K'(k) - K(k)K'(k) + E'(k)K(k)} \tag{24b}$$

with the substitution $y = K'(k)/K(k)$ and $L(y) = k^2$. The denominator at the right hand side of (24b) is equal to $\pi/2$ according to Legendre's relation [44]. In (24b) $E(k)$ is the complete elliptic integral of the second kind, defined by $E(k) = \int_0^{\pi/2}\sqrt{1-k^2\sin^2\alpha}\,d\alpha$ and $E'(k) = E(\sqrt{1-k^2})$. Inserting (23b) into (22c) and developing for small stress gives

$$\frac{1}{\overline{I}_{13}}\frac{d\overline{V}_{24}}{d\left(\sigma_{xx} - \sigma_{yy}\right)} \longrightarrow \frac{128\pi}{\Gamma^4(-1/4)}\pi_{44}R_{\text{sh}} \cong 0.696602 \times \pi_{44}R_{\text{sh}} \cong 3.157251 \times \frac{\ln 2}{\pi}\pi_{44}\frac{\rho_0}{t_d} \tag{25a}$$

whereby $\Gamma(y)$ is the gamma function and we used $K(3 - 2\sqrt{2}) = (2\pi)^{3/2}(1+\sqrt{2})/\Gamma^2(-1/4)$ [45]. (25a) is identical with (16) in [46]. It was also checked by an FEM simulation. If the devices are not square but general rectangles with $W_0/\ell_0 \neq 1$ they are not offset-free and the transconductance is

$$\frac{\overline{V}_{24}}{\overline{I}_{13}} \longrightarrow \frac{2}{\pi}R_{\text{sh}}\left\{\ln\frac{2L_0^{1/4}}{1-\sqrt{L_0}} + \frac{1}{\pi}\frac{W_0}{\ell_0}\left(1+\sqrt{L_0}\right)^2 \cdot \left(K\left(\sqrt{L_0}\right)\right)^2 \pi_{44}\left(\sigma_{xx} - \sigma_{yy}\right)\right\} \tag{25b}$$

$$\text{with } L_0 = L\left(\frac{2W_0}{\ell_0}\right)$$

A plot of (25b) reveals that square devices have smaller stress sensitivity $d\overline{V}_{24}/d(\sigma_{xx} - \sigma_{yy})$ at given input current $\overline{I}_{13}$ than devices with large aspect ratio.

4.3. Circular Disk Shaped Devices with Four Large Contacts. Figure 10(a) shows a circular device with two pairs of contacts of different size: each contact covers an aperture angle of $2\theta_f$ and $2\theta_p$, respectively. Thus, its layout has two DoFs just like the device of Figure 6. The device has two perpendicular mirror symmetries, namely, the x- and y-axes. We assume that it is implemented in (100)-silicon with an alignment between the (x, y)-coordinate system and the crystal system like in Figure 1. Under a biaxial mechanical load along the symmetry axes the resistivity tensor becomes anisotropic ($\rho_{xx} \neq \rho_{yy}$) but the offdiagonal elements vanish ($\rho_{xy} = 0$). As explained above we may apply the affine transformation (12a), which makes an ellipse in the $\underline{z}'$-plane out of the circle in the original $\underline{z}$-plane (see Figure 10(b)). This device has the same potentials and currents at all contacts as the device in Figure 10(a), even though in Figure 10(b) the resistivity is isotropic and the potential is a solution of the Laplace equation, whereas in Figure 10(a) the resistivity is anisotropic and the potential is no solution of the Laplace equation. Thereby the stress σ_{xx} acts along a line, where the contacts with angles $2\theta_f$ are arranged. The stress σ_{yy} acts along a line, where the contacts with angles $2\theta_p$ are arranged. In a second step we use a conformal transformation $\underline{z}'' = -j\sqrt{\rho_{\text{iso}}\kappa_{xx}}\underline{z}'/\sqrt{(\rho_{\text{iso}}\kappa_{xx})^2 - 1}$ with j being the imaginary unit, which rotates the device by 90° in

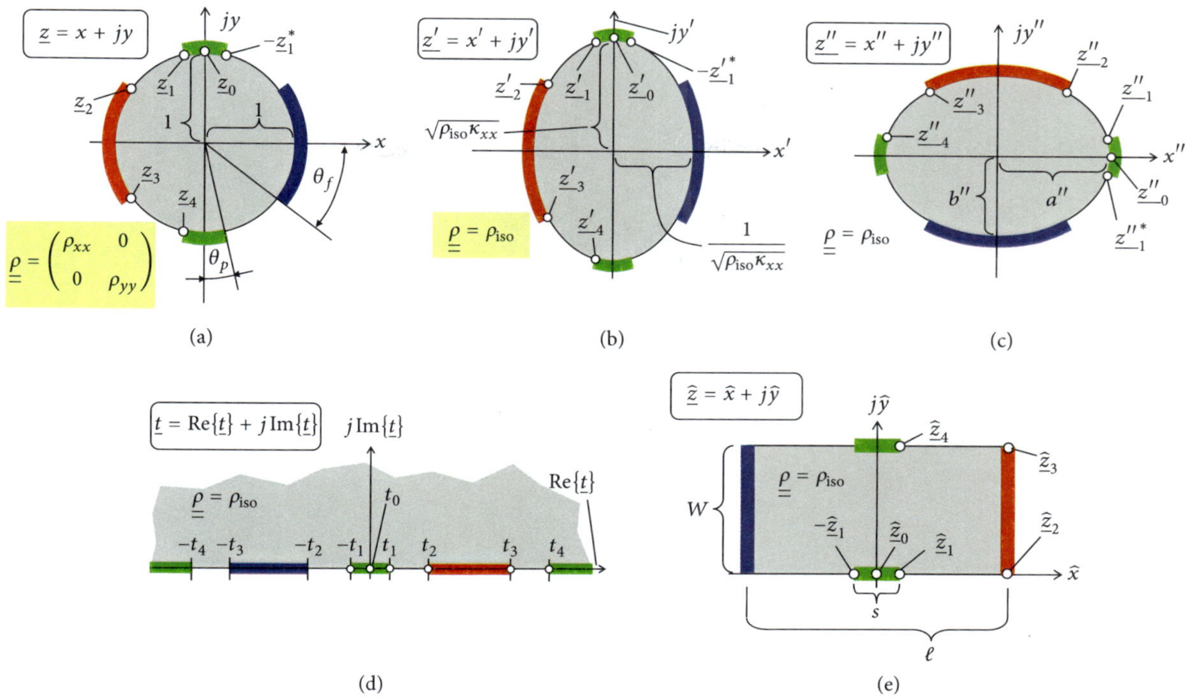

FIGURE 10: (a–d) Mapping a circular device with anisotropic conductivity by one affine and three conformal transformations onto an equivalent rectangular device with isotropic conductivity. (a) shows a circular disk shaped device with two pairs of contacts and two perpendicular mirror-symmetry axes x and y. Its resistivity is anisotropic with the principal axes x and y. This describes the action of biaxial load along x- and y-axes on chips in (100)-silicon aligned like in Figure 1. The elliptical device in (b) with isotropic resistivity is equivalent to the device in (a). The shape in (b) is obtained by the affine transformation $\underline{z} \to \underline{z}'$ (12a). The conformal transformation $\underline{z}' \to \underline{z}''$ rotates the device clockwise by 90° and scales it isotropically: $a'' = \rho_{\text{iso}}\kappa_{xx}/\sqrt{(\rho_{\text{iso}}\kappa_{xx})^2 - 1}$ and $b'' = 1/\sqrt{(\rho_{\text{iso}}\kappa_{xx})^2 - 1}$ with $a''^2 - b''^2 = 1$. A second-conformal transformation $\underline{z}'' \to \underline{t}$ (26a) and (26b) maps the interior of the ellipse in (c) onto the upper half of the t-plane in (d). A third-conformal transformation $\underline{t} \to \hat{\underline{z}}$ maps the upper half of the t-plane into the interior of the rectangle in (e). The contacts along the perimeter of the shapes relate to the contacts on the real t-axis. Corresponding points have equal indices. The sequence of points with indices 0, 1, 2, 3, and 4 is always counterclockwise while the conductive region is at the left hand side of this path. The asterisk means the conjugate complex.

clockwise direction and scales it isotropically with the factor $\sqrt{\rho_{\text{iso}}\kappa_{xx}/((\rho_{\text{iso}}\kappa_{xx})^2 - 1)}$ such that the foci of the ellipse are at $\underline{z}'' = \pm 1$. A second-conformal transformation $\underline{z}'' \to \underline{t}$ maps the interior of the ellipse onto the upper half of the $\underline{t}$-plane (Figure 10(d)), as explained in [48]. At the boundary of the elliptical region the mapping is

$$\underline{z}'' = \frac{\rho_{\text{iso}}\kappa_{xx}}{\sqrt{(\rho_{\text{iso}}\kappa_{xx})^2 - 1}} \cos\left(\frac{\pi}{K(\kappa)} F\left(\frac{t}{\sqrt{1+t^2}}, \kappa\right)\right) + \frac{j}{\sqrt{(\rho_{\text{iso}}\kappa_{xx})^2 - 1}} \sin\left(\frac{\pi}{K(\kappa)} F\left(\frac{t}{\sqrt{1+t^2}}, \kappa\right)\right), \tag{26a}$$

$$\kappa = \sqrt{L\left(\frac{1}{2\pi} \ln\left(\frac{\rho_{\text{iso}}\kappa_{xx} + 1}{\rho_{\text{iso}}\kappa_{xx} - 1}\right)\right)}; \tag{26b}$$

$\underline{z}''$ and $\underline{t}$ are complex numbers and $t = \text{Re}\{\underline{t}\}$. (26a) and (26b) are only valid for $\rho_{\text{iso}}\kappa_{xx} > 1$, which means $\pi_{44}(\sigma_{xx} - \sigma_{yy}) < 0$. If we move on the t-axis from $-\infty$ over $t_0 = 0$ to ∞ this means a walk in the $\underline{z}''$-plane in counterclockwise direction starting at $x'' = -\rho_{\text{iso}}\kappa_{xx}/\sqrt{(\rho_{\text{iso}}\kappa_{xx})^2 - 1}$ and moving through the lower $\underline{z}''$-plane over $x'' = \underline{z}''_0 = \rho_{\text{iso}}\kappa_{xx}/\sqrt{(\rho_{\text{iso}}\kappa_{xx})^2 - 1}$ back to $x'' = -\rho_{\text{iso}}\kappa_{xx}/\sqrt{(\rho_{\text{iso}}\kappa_{xx})^2 - 1}$ through the upper $\underline{z}''$-plane. Thus, positive real t corresponds to the upper half of the edge of the ellipse. Then the end points of the contacts in the $\underline{t}$-plane are

$$\left(1 - \kappa^2\right)^{-1/4} \geq t_1 = \text{sc}\left(\frac{\theta_p}{\pi} K(\kappa), \kappa\right) \geq 0, \tag{27a}$$

$$\left(1 - \kappa^2\right)^{-1/4} \geq t_2 = \text{sc}\left(\left(\frac{1}{2} - \frac{\theta_f}{\pi}\right) K(\kappa), \kappa\right) \geq t_1, \tag{27b}$$

$$t_3 = \text{sc}\left(\left(\frac{1}{2} + \frac{\theta_f}{\pi}\right) K(\kappa), \kappa\right) \geq \left(1 - \kappa^2\right)^{-1/4}, \tag{27c}$$

$$t_4 = \text{sc}\left(\left(1 - \frac{\theta_p}{\pi}\right) K(\kappa), \kappa\right) \geq t_3 \tag{27d}$$

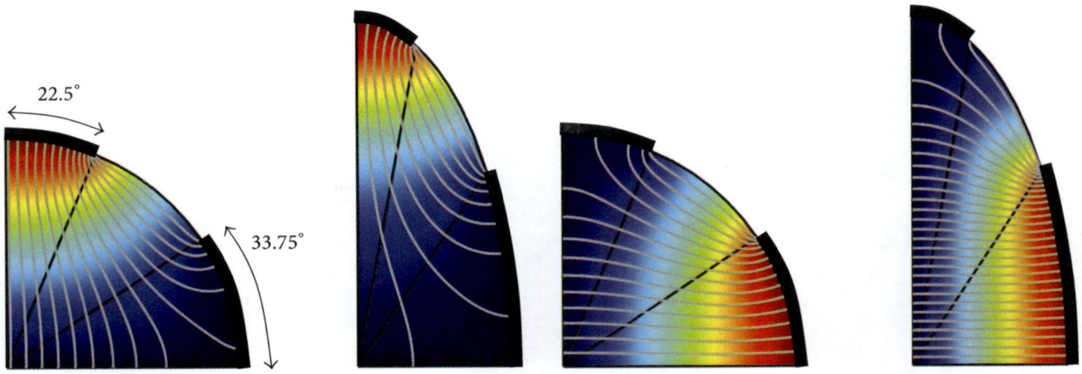

FIGURE 11: FEM-computation of input and output resistances of circular and elliptical devices with isotropic conductivity. Due to symmetry only a quarter of the real devices had to be modeled. The elliptical device is obtained from the circular one by stretching it times 1.5 in vertical direction and times 2/3 in horizontal direction. Contacts are thick black lines on the perimeter. The color coding denotes the electric potential (red = 1 V, blue = 0 V) and the gray lines are current streamlines.

for $0 < \theta_f + \theta_p < 90^\circ$ and $\mathrm{sc}(y,\kappa) = \mathrm{sn}(y,\kappa)/\mathrm{cn}(y,\kappa)$, $\mathrm{sn}(y,\kappa)$, and $\mathrm{cn}(y,\kappa)$ being sine and cosine of the Jacobi amplitude. Next we map the interior of the rectangular device of Figure 10(e) from its $\underline{\hat{z}}$-plane onto the upper half of the $\underline{t}$-plane in Figure 10(d) via the Schwartz-Christoffel transformation $\underline{\hat{z}} = WF(\underline{t}/t_2,k)/K'(k)$ with $\ell/W = 2K(k)/K'(k)$. Thereby the center $\underline{\hat{z}}_0$ of the lower partial contact is mapped onto the origin in the $\underline{t}$-plane. The device in Figure 10(d) is identical to the even symmetry device in [10]. Therefore the flush contacts at positions $\hat{x} = \pm\ell/2$ have the resistance $\lambda_f R_{\mathrm{sh}}$ with $R_{\mathrm{sh}} = \rho_{\mathrm{iso}}/t_d$ between them, and the resistance between the partial contacts is $\lambda_p R_{\mathrm{sh}}$ (see [10]). Applying the Schwartz-Christoffel transformation to $\underline{\hat{z}}_3$ gives $F(t_3/t_2,k) = K(k) + jK'(k)$. However, this is identical to $F(1/k,k)$. Therefore, it holds $k = t_2/t_3$, and with (27b) and (27c) we get

$$k = \frac{\mathrm{sc}\left(\left(1/2 - \theta_f/\pi\right)K(\kappa),\kappa\right)}{\mathrm{sc}\left(\left(1/2 + \theta_f/\pi\right)K(\kappa),\kappa\right)} \tag{28a}$$

With the half-period addition theorem $\mathrm{sc}(z-K(\kappa),\kappa) = -(1-\kappa^2)^{-1/2}/\mathrm{sc}(z,\kappa)$ it follows that [49]

$$k = \sqrt{1-\kappa^2}\,\mathrm{sc}^2\left(\left(\frac{1}{2} - \frac{\theta_f}{\pi}\right)K(\kappa),\kappa\right) \tag{28b}$$

Finally, we apply the Schwartz-Christoffel transformation to $\underline{\hat{z}}_1$: $s/2 = WF(t_1/t_2,k)/K'(k)$. With (27a) and (27b) this gives

$$\frac{s}{\ell} = \frac{1}{K(k)}F\left(\frac{\mathrm{sc}\left(\pi^{-1}\theta_p K(\kappa),\kappa\right)}{\mathrm{sc}\left(\left(1/2-\theta_f/\pi\right)K(\kappa),\kappa\right)},k\right) \tag{28c}$$

With (28b) and (28c) and with $s/\ell = s_0/\ell_0$ we immediately know w_3 from (18i), and with (18d)–(18h) we can compute λ_f, λ_p, and λ_x. With these parameters we know all resistors in the ERC of the even symmetry device in Figure 10(a) (see [10]) and also of the odd symmetry device in Figure 12(a). With (19a)–(19i) we get all impedance, voltages, and transresistances in Hall- and van der Pauw-operating modes.

We checked (28b), (28c), and (18d)–(18h) with finite element calculations of the circular device in Figure 11 with $\theta_p = 22.5^\circ$ and $\theta_f = 33.75^\circ$ with isotropic conductivity. For zero stress we set $\rho_{\mathrm{iso}}\kappa_{xx} \to 1$ and obtained $\lambda_p = 1.33951$ and $\lambda_f = 1.13789$ as input/output numbers of squares. Then we stretched the device in x- and y-directions according to Figure 10(b) with $\rho_{\mathrm{iso}}\kappa_{xx} \to 2.25$ and obtained $k = 0.0207218$, $s/\ell = 0.0852258$, $\ell/W = 0.596948$, $\lambda_p = 2.14874$, and $\lambda_f = 0.594916$. Both times the results of FEM and analytical theory matched up to 0.04% for λ_p and λ_f. Thereby, the FEM model had a mesh with 0.6 million elements per device.

For $\rho_{\mathrm{iso}}\kappa_{xx} < 1$ we must change the transformation between Figures 10(b) and 10(c) to $\underline{z}'' = \underline{z}'\sqrt{\rho_{\mathrm{iso}}\kappa_{xx}}/\sqrt{1-(\rho_{\mathrm{iso}}\kappa_{xx})^2}$. This means to swap θ_f with θ_p in (28a)–(28c) and to replace $\rho_{\mathrm{iso}}\kappa_{xx} \to (\rho_{\mathrm{iso}}\kappa_{xx})^{-1}$ in (26b). Since the new transformation $\underline{z}' \to \underline{z}''$ does not rotate the device, θ_f-contacts are now mapped onto the partial contacts of the rectangular device in Figure 10(e), so that we also have to swap λ_f with λ_p.

We checked the correctness of (26b), (28b), (28c), and (18d)–(18i) with a series of finite element simulations given in Table 2.

With these foundations we can study the mechanical stress sensitivity of symmetric circular devices with arbitrary ratio of input/output resistances. Output voltages of devices from Figure 10 do not respond to $\sigma_{xx} - \sigma_{yy}$, because their contacts are on the axes of mirror-symmetry and even after the affine transformation (12a) they will stay on these symmetry axes so that their output signals remain zero. One can prove this by inserting λ_f, λ_p, and λ_x from (18d) to (18f) into (2a–c) of [10] to compute the equivalent resistor circuit (ERC) that applies to a device from Figure 10. This ERC is shown in Figure 2b of [10] – it is different from the ERC of Figure 6 in this current work. For the measurement of $\sigma_{xx} - \sigma_{yy}$ we have to consider the complementary devices where contacts and isolating boundaries are swapped (see Figure 12(a)). In Figure 12(b) we checked the analytical theory against numerical simulations for various contact

Table 2: Mechanical stress sensitivity of a circular device from Figure 10(a): comparison of numerical finite element simulation (FEM) versus analytical theory. $\lambda_f R_{sh}$ is the resistance between the θ_f-contacts in Figure 10(a) and $\lambda_p R_{sh}$ is the resistance between the θ_p-contacts in Figure 10(a). The mesh of the FEM-simulation had 2.66 million elements and the model used Lagrange multiplier and the piezoresistivity matrix from (6a) to (6c) with $\pi_{11} = 6.6 \times 10^{-11}$/Pa, $\pi_{12} = -1.1 \times 10^{-11}$/Pa, and $\pi_{44} = 138.1 \times 10^{-11}$/Pa. The theory used (18d), (18e), (18g), (18h), (18i), (28b), (28c), and (26b) for $\rho_{iso}\kappa_{xx} > 1$. For $\rho_{iso}\kappa_{xx} < 1$ we swapped θ_f with θ_p in (28b) and (28c) and λ_f with λ_p, and we replaced $\rho_{iso}\kappa_{xx} \to (\rho_{iso}\kappa_{xx})^{-1}$ in (26b).

θ_f [°]	θ_p [°]	σ_{xx} [MPa]	σ_{yy} [MPa]	ρ_{iso}/ρ_0	$\rho_{iso}\kappa_{xx}$	$\lambda_f R_{sh}$ FEM	$\lambda_f R_{sh}$ Difference FEM vs. theory	$\lambda_p R_{sh}$ FEM	$\lambda_p R_{sh}$ Difference FEM vs. theory
17	55	0	0	1.000000	1.00000	1.264527	−0.04%	0.796063	−0.02%
17	55	−5	5	0.999976	1.006929	1.262001	−0.04%	0.800431	−0.02%
17	55	−50	50	0.997613	1.071608	1.236262	−0.04%	0.838836	−0.02%
17	55	−500	500	0.723332	2.337100	0.641383	−0.03%	1.068508	−0.03%
17	55	500	−500	0.723332	0.427881	1.180510	−0.06%	0.280305	−0.02%
17	55	−500	−400	0.972803	1.073496	1.204879	−0.04%	0.819107	−0.02%
17	55	400	500	1.022421	1.069814	1.267640	−0.04%	0.858564	−0.02%
17	55	−400	−500	0.972803	0.931536	1.254778	−0.04%	0.731615	−0.02%
17	55	500	400	1.022421	0.934742	1.317534	−0.04%	0.771065	−0.02%
55	17	−500	500	0.723332	2.337100	0.280305	−0.02%	1.180511	−0.06%

geometries in the wide range of mechanical stress up to ±640 MPa. In an experiment the silicon wafer would very likely break at smaller stress levels, and the linear piezoresistive theory would describe the measured results insufficiently. Nevertheless, we chose the large stress to show the excellent agreement between analytical theory and numerical simulation.

The mechanical stress sensitivity at small stress is plotted in Figure 13, where we note a symmetry when θ_f and θ_p are swapped. Offset-free devices have $\theta_f = \theta_p$. They correspond to the black line and this line goes right through the global maximum of the function at the point $\theta_f = \theta_p = 22.5°$. This is one distinct difference to rectangular devices from Figure 5. This maximum sensitivity is 48.3501%/GPa, which is more than 3% larger than the maximum sensitivity with systematic offset in Figure 5 and nearly 6% larger than for offset-free rectangular devices in Figure 5. The mechanical stress sensitivity of offset-free circular devices is shown in Figure 14. The examples of rectangular versus circular devices show that *different shapes of devices lead to different maximum achievable stress sensitivities*. This is another distinct difference to Hall plates, where all shapes of devices, which can be derived from each other by conformal mapping, lead to the same magnetic sensitivity. However, there is another similarity between Hall plates and circular pseudo-Hall devices: at constant supply voltage their sensitivity to magnetic field and mechanical stress, respectively, is maximized, if the input and output resistances are identical to $\sqrt{2}$ squares: $\lambda_{f0} = \lambda_{p0} = \sqrt{2}$ (compare Figure 14 with [12]). Yet this similarity does not hold for rectangular devices or inverted Greek crosses (see Figure 18(a)).

Figure 15 shows the relative change of resistances of the equivalent resistor circuit (ERC) for offset-free circular devices from Figure 12 with $\lambda_{f0} = \lambda_{p0}$, when $\sigma_{xx} - \sigma_{yy} = -1$ MPa is applied and $\sigma_{xx} + \sigma_{yy} = \sigma_{xy} = 0$. Low p-doped silicon is assumed with $\pi_{44} = 1.381$/GPa. Obviously $\overline{R}_D$ does not change while $\overline{R}_{Hf}$ and $\overline{R}_{Hp}$ change with opposite signs and equal magnitude. Their maximum stress sensitivity is ±69.6384%/GPa and it occurs at $\theta_f = \theta_p = 27.3364°$. For large contacts with $\theta_f = \theta_p = 5°$ the resistances $\overline{R}_{Hf}$ and $\overline{R}_{Hp}$ change less, only ±43.5873%/GPa. Both cases were checked and verified by finite element simulations. The stress sensitivities of long resistor stripes of the same doping are $\partial R_{[110]}/\partial(\sigma_{xx} - \sigma_{yy}) = -\partial R_{[\bar{1}10]}/\partial(\sigma_{xx} - \sigma_{yy}) = R_0\pi_{44}/2 = 69.05\%/\text{GPa} \times R_0$ (for stripes with current flowing in x, y-directions) and $\partial R_{[100]}/\partial(\sigma_{xx}-\sigma_{yy}) = \partial R_{[010]}/\partial(\sigma_{xx}-\sigma_{yy}) = 0$ (for stripes with current flowing in ⟨100⟩-directions). Hence, the diagonal resistors $\overline{R}_D$ in the ERC behave like resistor stripes in diagonal direction ⟨100⟩: they do not change versus stress. However, the "horizontal" resistors $\overline{R}_{Hp}$ in the ERC have a stress drift that depends on the size of the contacts. For medium sized contacts their stress drift can even slightly exceed the stress drift of resistor stripes in [110] direction, but for small or large contacts their stress drift is significantly lower. Therefore, *in general the stress dependence of the resistors in the ERC between any two contacts is different from the stress dependence of simple resistor stripes aligned in the same direction between said contacts*. Yet, for circular devices with medium sized contacts the difference is moderate.

4.4. *Offset-Free Circular Devices with Point-Sized Contacts at Small Stress.*

From Figure 12(a) we see that this limit means $\pi/2 - \theta_f - \theta_p \to 2d\theta$ with small $d\theta$. However, we discuss only the offset-free case $\theta_f = \theta_p \to \pi/4 - d\theta$. For small stress $\rho_{iso}\kappa_{xx} \to 1 - (\pi_{44}/2)(\sigma_{xx} - \sigma_{yy})$. Inserting this into (26b) gives the differential

$$\kappa \longrightarrow d\kappa = 2\left(-4\pi_{44}\left(\sigma_{xx} - \sigma_{yy}\right)\right)^{1/4}, \tag{29a}$$

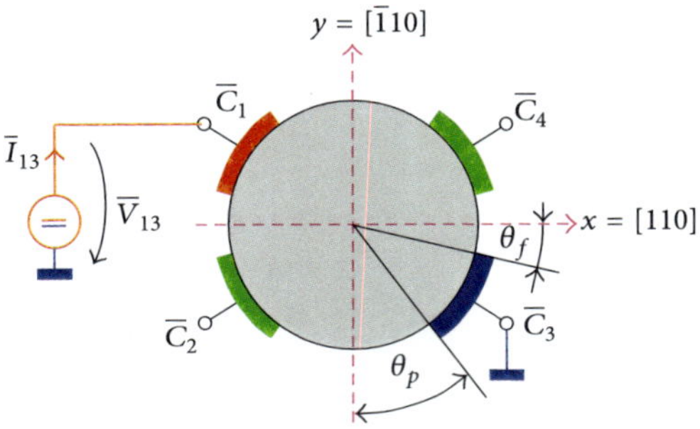

(a) Circular pseudo-Hall device with two perpendicular mirror-symmetry axes x and y. It can be used for the measurement of $\sigma_{xx} - \sigma_{yy}$ on a chip made of (100)-silicon. The device is supplied with electrical power at the contacts $\bar{C}_1$–$\bar{C}_3$. The output signal is tapped between $\bar{C}_2$ and $\bar{C}_4$. The equivalent resistor circuit (ERC) is identical to Figure 6

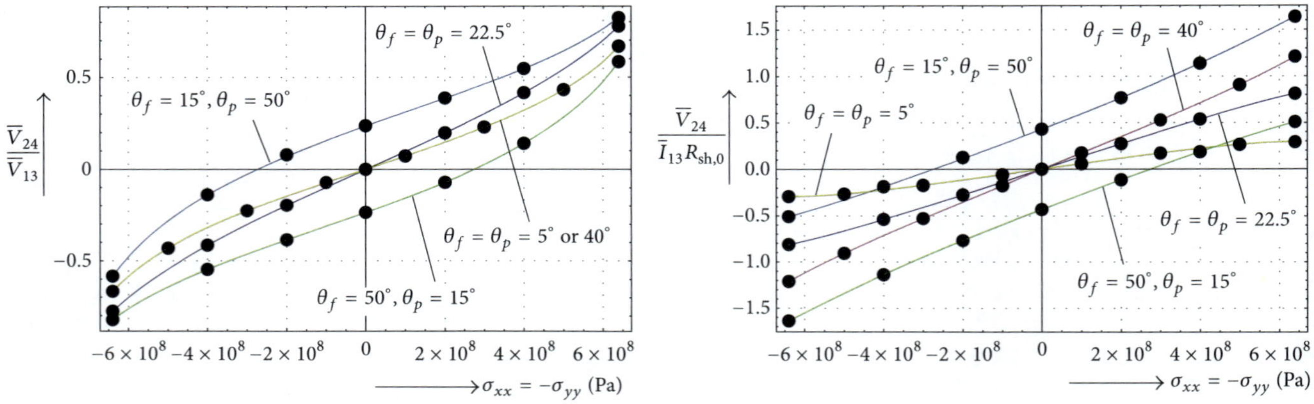

(b) Output voltage of the circular device from (a) versus biaxial inplane mechanical stress $\sigma_{xx} = -\sigma_{yy}$ with $\sigma_{xy} = 0$. The left plot shows the ratio of output over supply voltage. The right plot shows the ratio of transimpedance over sheet resistance at zero stress $R_{sh,0}$. The solid curves are computed with the analytical theory. The black dots denote results form 29 finite element simulations with $\pi_{11} = 6.6 \times 10^{-11}$/Pa, $\pi_{12} = -1.1 \times 10^{-11}$/Pa, and $\pi_{44} = 138.1 \times 10^{-11}$/Pa (as in low p-doped silicon)

Figure 12

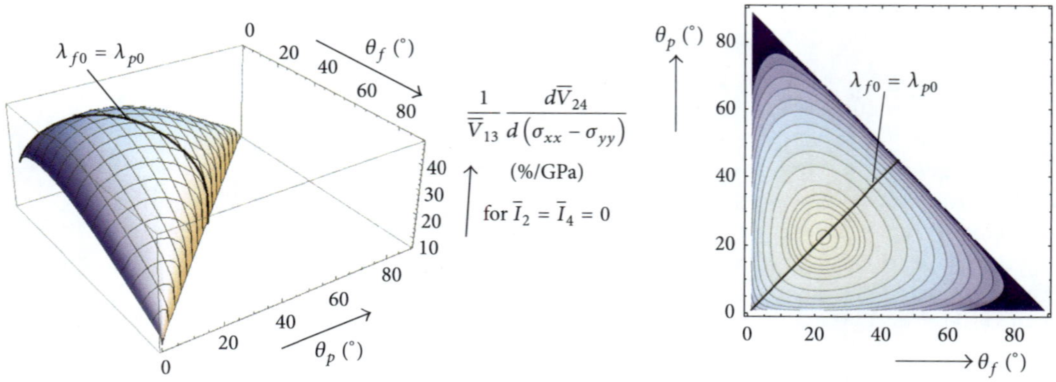

(a) (b)

Figure 13: Sensitivity of output voltage $\bar{V}_{24}$ versus mechanical stress $\sigma_{xx} - \sigma_{yy}$ (normalized to supply voltage $\bar{V}_{13}$) for a circular device of Figure 12. All other components of the stress tensor vanish, as well as $\sigma_{xx} + \sigma_{yy} = 0$. The plots are graphical representations of (26b), (28b), (28c), (18a)–(18i), and (19b) in the limit of small stress and with the piezoresistive coefficient $\pi_{44} = 1.381$/GPa for low p-doped silicon. Largest stress sensitivities are obtained for contact size $\theta_f = \theta_p = 22.5^\circ$ with a flat maximum. The black curve denotes offset-free devices with $\lambda_{f0} = \lambda_{p0}$ at zero stress; they are the only ones, where $\bar{V}_{24} = 0$ at zero stress. The black curve goes through the maximum of the surface. The contours in (b) are for 48.25, 48, 47.5, 47, 46.5, 46, 45, 42.5, 40, ..., 22.5, 20%/GPa. In (a) the steep slopes of the surface near $\theta_f < 1^\circ$ and $\theta_p < 1^\circ$ are only sketched due to numerical problems.

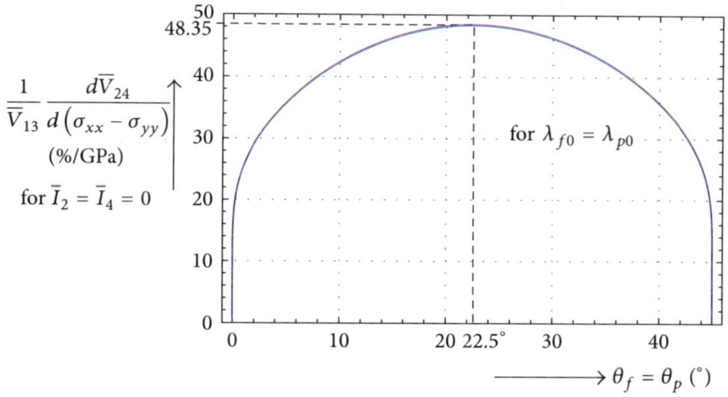

FIGURE 14: Sensitivity of output voltage $\overline{V}_{24}$ versus mechanical stress $\sigma_{xx} - \sigma_{yy}$ (normalized to supply voltage $\overline{V}_{13}$) for offset-free circular devices of Figure 13. All other components of the stress tensor vanish, as well as $\sigma_{xx} + \sigma_{yy} = 0$. The plot is a graphical representation of (26b), (28b), (28c), (18a)–(18i), and (19b) in the limit of small stress and with the piezoresistive coefficient $\pi_{44} = 1.381/\text{GPa}$ for low p-doped silicon. Largest stress sensitivity of 48.3501%/GPa is obtained for aspect ratios $\theta_f = \theta_p = 22.5°$ which means $\lambda_{f0} = \lambda_{p0} = \sqrt{2}$ and this is also the number of squares between supplies or sense contacts. The curve is mirror symmetric to $\theta_f = \theta_p = 22.5°$.

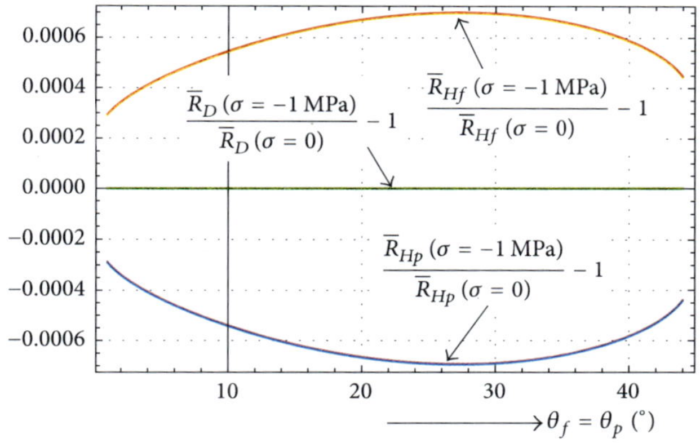

FIGURE 15: Relative change of resistances of the equivalent resistor circuit (ERC) for offset-free ($\lambda_{f0} = \lambda_{p0}$) symmetric circular devices from Figure 12. $\sigma = \sigma_{xx} - \sigma_{yy} = -1$ MPa is applied and $\sigma_{xx} + \sigma_{yy} = \sigma_{xy} = 0$. For $\pi_{44} = 1.381/\text{GPa}$.

whereby we used $L(x) \to 1/\cosh^2(\pi x/2 - \ln 2)$ for large x (see [10]) and we made a series in the small stress terms. With (28b) and (28c) we get

$$k \longrightarrow 3 - 2\sqrt{2} + \left(3\sqrt{2} - 4\right)\left(2d\theta - \frac{d\kappa^4}{64}\right) + \frac{\left(-16 + 11\sqrt{2}\right) d\theta d\kappa^4}{64} \quad (29b)$$

$$\frac{s}{\ell} = \frac{s_0}{\ell_0} \longrightarrow \frac{F\left(1 - 2\sqrt{2}d\theta - \sqrt{2}d\theta\left(\kappa^4/64\right), k\right)}{K(k)} \quad (29c)$$

We plug this into (18i) and get

$$w_3 \longrightarrow 1 - 8\sqrt{2}d\theta - \frac{\sqrt{2}d\theta\kappa^4}{16} \quad (30a)$$

Plugging (29b) and (30a) into (18d)–(18h) gives

$$\left.\begin{matrix}\lambda_f \\ \lambda_p\end{matrix}\right\} \longrightarrow \frac{\pi}{\ln(4/d\theta)}\left(1 + \left\{\begin{matrix}+ \\ -\end{matrix}\right\} \frac{\pi_{44}\left(\sigma_{xx} - \sigma_{yy}\right)}{\ln(4/d\theta)}\right) \quad (30b)$$

In the limit $d\theta \to 0$ we get the transimpedance with (19c)

$$\frac{1}{I_{13}} \frac{dV_{24}}{d\left(\sigma_{xx} - \sigma_{yy}\right)} \longrightarrow \frac{2}{\pi}\pi_{44}R_{\text{sh}} \cong 0.63662 \times \pi_{44}R_{\text{sh}} \quad (31)$$

Equation (31) was checked by an FEM simulation. Comparing (31) with (25a) we note that in the limit of point-sized contacts the current related mechanical stress sensitivity of the disk shaped device is 8.6% smaller than for the square shaped device. Conversely, the current related *magnetic* sensitivity of both devices is identical, namely, $\mu_H R_{\text{sh}}$, according to (3) ($G_H = 1$ for point-sized contacts).

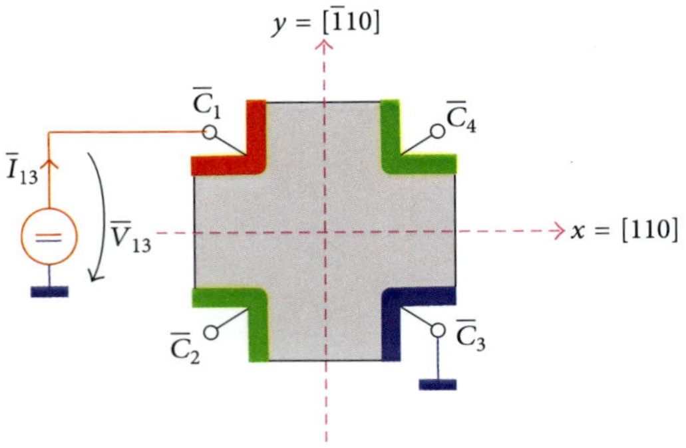

FIGURE 16: Pseudo-Hall device with two perpendicular mirror-symmetry axes x and y: inverted Greek cross geometry. It can be used for the measurement of $\sigma_{xx} - \sigma_{yy}$ on a chip made of (100)-silicon. The device is supplied with electrical power at the contacts $\overline{C}_1$–$\overline{C}_3$. The output signal is tapped between $\overline{C}_2$ and $\overline{C}_4$. The equivalent resistor circuit (ERC) is identical to Figure 6.

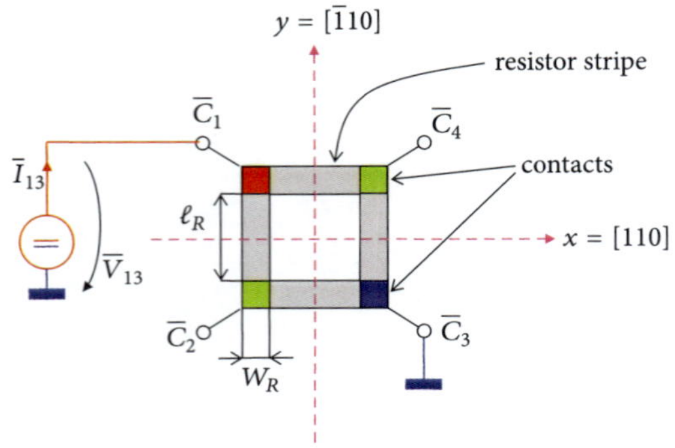

FIGURE 17: Wheatstone bridge with two perpendicular mirror-symmetry axes x and y comprises four resistor stripes. It can be used for the measurement of $\sigma_{xx} - \sigma_{yy}$ on a chip made of (100)-silicon. The device is supplied with electrical power at the contacts $\overline{C}_1$–$\overline{C}_3$. The output signal is tapped between $\overline{C}_2$ and $\overline{C}_4$. Note the similarity of the Wheatstone bridge with the inverted Greek cross of Figure 16 when the inner square of the bridge arrangement is filled with resistive material.

4.5. Inverted Greek Cross with Four Large Contacts. The device geometry is shown in Figure 16. In the piezoresistive case the device is effectively stretched in x-direction, which gives two perpendicular mirror symmetries. A numerical investigation with finite element simulation was done. Its results are shown in comparison with the other devices discussed so far in Figure 18.

4.6. Optimization of Stress Sensor Devices. Let us compare the stress sensitivity (25a) with a conventional Wheatstone bridge circuit with four simple stripes of resistors oriented in x- and y-directions (see Figure 17). The stress dependence of these resistors is given by (6a) and (6b) if we assume that the stripes are much longer than wide. Thus, at constant input current the bridge output voltage at small stress is

$$V_{\text{out}} = I_{\text{in}} \left(\sigma_{xx} - \sigma_{yy}\right) \pi_{44} R_{\text{sh}} \frac{1}{2} \frac{\ell_R}{W_R}, \qquad (32)$$

where ℓ_R/W_R is the number of squares of each resistor (i.e., the aspect ratio of each resistance stripe). A comparison of (25a) and (32) shows that the output voltage of the Wheatstone bridge is larger if each resistor has at least 1.4 squares. However, the ratio $V_{\text{out}}/I_{\text{in}}$ is not an appropriate measure to select optimum sensors, because the pseudo-Hall device of (25a) has infinite input and output resistance whereas the Wheatstone bridge has finite resistance. The resistance of the pseudo-Hall device diverges logarithmically when the contact size goes to zero (see (22a), (30b)). Hence, if we want to force a current through this sensor we would need infinite supply voltage and the thermal noise at the output would also increase unboundedly. Conversely, if the supply voltage is limited, the sensor output voltage goes logarithmically towards zero (see (22b)). In practice a circuit designer chooses the impedance level of the sensor circuit first to trade off current consumption and noise, and under this constraint he looks for the sensor geometry that maximizes the signal-to-noise ratio (this is explained in [12]). For the Wheatstone bridge sensor this means that $R_{\text{in}} = R_{\text{sh}}\ell_R/W_R$ is fixed and then V_{out} is maximized. If the sensor is operated from a battery or a regulated on-chip supply voltage, then the input voltage is also limited, and there is no degree of freedom left for optimization: $V_{\text{out}} = V_{\text{in}}(\sigma_{xx} - \sigma_{yy})\pi_{44}/2$. Conversely, if the sensor is powered by an off-chip supply with "unlimited" power, the relevant limit is maximum power dissipation P_d and thermal destruction inside the sensor while R_{in} can be chosen arbitrarily. Then we get $V_{\text{out}} = \sqrt{P_d R_{\text{in}}}(\sigma_{xx}-\sigma_{yy})\pi_{44}/2$ for the Wheatstone bridge and the output thermal noise voltage is proportional to $\sqrt{R_{\text{in}}}$.

For a device with finite contacts from Figure 12(a) we can set the impedance by choosing an appropriate thickness t_d (see (19a)). Alternatively we could connect two (or several) devices in parallel to effectively double (or multiply) their thickness. The resistivity should be low, because this means low doping and this maximizes the piezoresistive coefficients [50]. Yet, in silicon the doping should also not be lower than $10^{15}/\text{cm}^3$, because then mobile ions in the interfaces between various layers on the chip may give rise to lifetime drift [51]. For battery operation the signal should be maximized at fixed supply voltage and impedance. This leads to the maximum of Figure 14, with $V_{\text{out}} = V_{\text{in}}(\sigma_{xx} - \sigma_{yy})\pi_{44} \times 0.350109$. Thus, at equal supply voltage and supply current the output voltage of an offset-free pseudo-Hall device from Figure 12(a) is 30% smaller than the output voltage of a Wheatstone bridge sensor $(0.35/0.5 = 0.70 = 1 - 0.30)$. If the offset-free pseudo-Hall device does not have optimum contact size as in the maximum in Figure 14, its output voltage is even lower. Thus, stress sensitivity is not a strong argument for pseudo-Hall devices. Their advantage over Wheatstone bridges is their smaller size.

Figures 18(a) and 18(b) showthe output voltage over input voltage and input current, respectively, for several device geometries with finite contacts: rectangular devices from Figure 6, circular devices fromFigure 12, and the inverted Greek cross from Figure 16. Only offset-free devices are considered. The piezoresistive coefficient $\pi_{44} = 1.381/\text{GPa}$ for low p-doped silicon was used. We assume low stress $\sigma_{xx} = -\sigma_{yy}$ and vanishing stress σ_{xy}. Then the output signal depends only on $\sigma_{xx} - \sigma_{yy}$. In Figure 18(a) we see

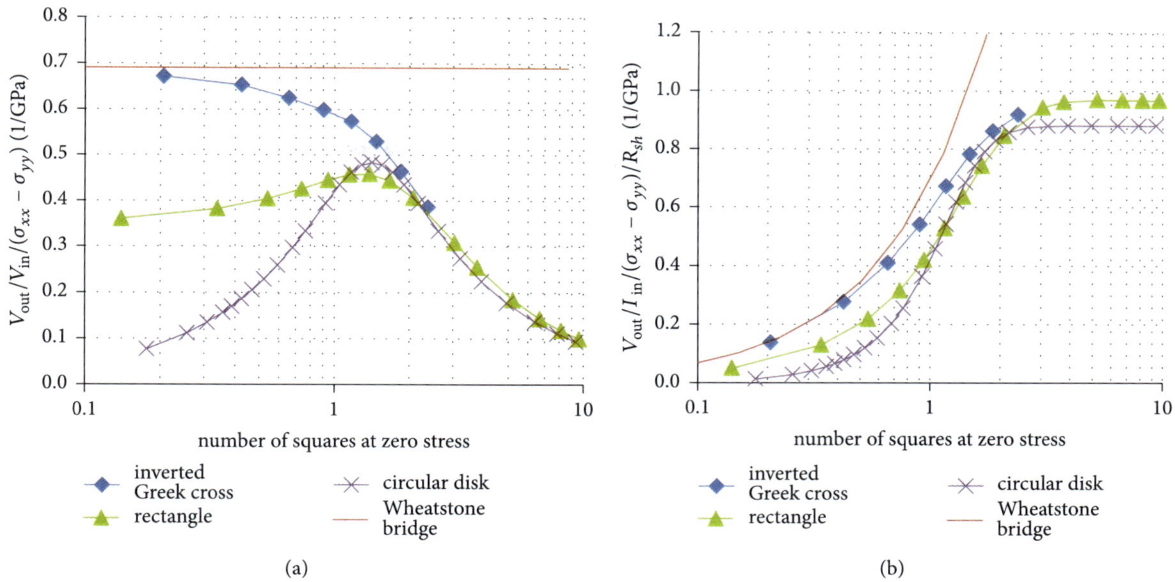

FIGURE 18: Comparison of (a) supply voltage related mechanical stress sensitivity, and (b) supply current related stress sensitivity of pseudo-Hall stress sensors with different shapes. Piezoresistive coefficient π_{44} = 1.381/GPa for low p-doped silicon was assumed. Only offset-free devices with $\lambda_{f0} = \lambda_{p0}$ were considered. For the pseudo-Hall devices the number of squares is $R_{in}/R_{sh} = 1/\lambda_f + 1/\lambda_p$ and for the Wheatstone bridge it is $R_{in}/R_{sh} = \ell_R/W_R$.

that the Wheatstone bridge has the largest voltage related stress sensitivity, independent of the number of squares of its resistors. In the limit of low number of squares (i.e., low input resistance) the inverted Greek cross has the same stress sensitivity as the Wheatstone bridge. This agrees with common sense, because in this limit of the inverted Greek cross the supply and sense contacts face each other with small spacing, so that the current flows in pure x- and y-directions like in a Wheatstone bridge circuit. The rectangular and the circular devices both show similar maxima at similar internal resistance. It is interesting to see that rectangular and circular piezoresistive devices behave similar, yet the inverted Greek cross is quite different, whereas for Hall plates their magnetic sensitivities are all identical when the abscissa axis gives the number of squares. In Figure 18(b) we see the current related stress sensitivity, which rises monotonously with the number of squares for all devices. However, the maximum of circular disks is smaller than the maxima of rectangle and inverted Greek cross (which are of course identical because in this limit their shapes are also identical). The smaller stress sensitivity of pseudo-Hall devices seems to have two reasons: (i) the diagonal resistor $2\overline{R}_D$ in the equivalent resistor circuit loads the output and (ii) depending on contact size the stress sensitivity of the other resistors $\overline{R}_{Hf}$ and $\overline{R}_{Hp}$ in the equivalent resistor circuit is smaller than for resistor stripes in pure x- or y-directions (see Figure 15).

At the end of this section, we mention how the devices of Figures 6, 12, and 16 can be used to measure σ_{xy}, if it holds $\sigma_{xx} - \sigma_{yy} = 0$. In a reference frame $\Sigma'(x', y')$, which is rotated against $\Sigma(x, y)$ by -45° the stress tensor reads $\sigma_{x'x'} = (\sigma_{xx} + \sigma_{yy})/2 - \sigma_{xy}$, $\sigma_{y'y'} = (\sigma_{xx} + \sigma_{yy})/2 + \sigma_{xy}$, and $\sigma_{x'y'} = (\sigma_{xx} - \sigma_{yy})/2$. Inserting this into the resistivity tensor of cubic crystals shows that we only need to rotate the devices by -45° and replace $(\sigma_{xx} - \sigma_{yy})\pi_{44}/2$ by $\sigma_{xy}(\pi_{12} - \pi_{11})$ in (17a) and (17b).

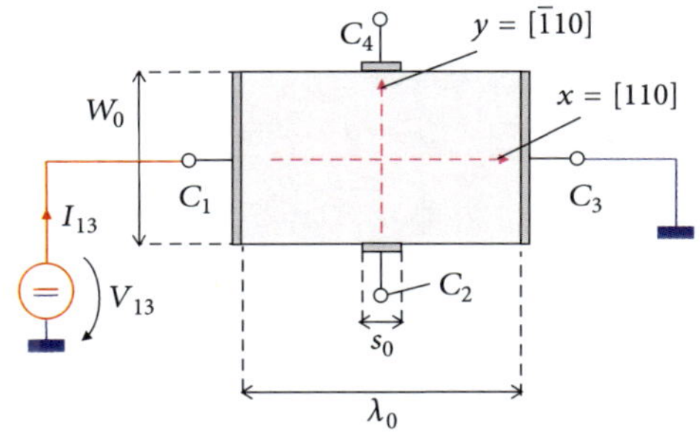

FIGURE 19: Rectangular Hall plate in (100)-silicon with two perpendicular mirror-symmetry axes x = [110] and y = [$\overline{1}$10]. An input voltage is forced between contacts C_1 and C_3 and an output voltage is tapped between C_2–C_4.

If both $\sigma_{xy} \neq 0$ and $\sigma_{xx} - \sigma_{yy} \neq 0$, then the proposed theory cannot describe the ERC and the output signals.

5. Devices with Mirror Symmetries to x- and y-Axes and Contacts Lying on the x- and y-Axes, at Zero Shear Stress $\sigma_{xy} = 0$

An example of such a device with rectangular geometry is shown in Figure 19. It is complementary to the device in Figure 6, because contacts and isolating boundary are swapped. Also a different equivalent resistor circuit applies

TABLE 3: Comparison of G_{H0} between this work and [47]. The reference data of [47] were obtained by the FEM simulation from Figure 20: it is the slopes of the curves in the origin inserted into (33) and solved for G_{H0}. In the FEM simulation the parameters of the Hall plate from Figure 19 were $\ell_0/W_0 = 2$ and $s_0/\ell_0 = 10^{-7}$. The difference between both is small and it seems to come from the nonlinearity of the curves (compare first and last lines).

$\sigma_{xx} = -\sigma_{yy}$ [MPa]	ρ_{xx}/ρ_{yy}	$\rho_{iso}\kappa_{xx}$	$C_H B_z/\rho_{yy}$	$V_{42}/(V_{13} C_H B_z/\rho_{yy})$	G_{H0} ([47])	G_{H0} (this work)	difference
4411.76	0.25	2	0.01	1.3504349	0.67521745	0.67531489	−144 ppm
0	1	1	0.01	0.4650187	0.93003740	0.93006014	−24.5 ppm
−4411.76	4	0.5	0.01	0.1246212	0.99696960	0.99697235	−2.76 ppm
4411.76	0.25	2	0.0001	1.3506300	0.67531500	0.67531489	0.159 ppm

to the new device (see [10]). Nevertheless, the quantities λ_f, λ_p are identical to (18d) and(18e) (see [10]). At zero magnetic field the voltage between contacts C_2–C_4 does not depend on σ_{xx} and σ_{yy} due to symmetry. However, λ_f and λ_p depend on σ_{xx} and σ_{yy}. And according to (4) the Hall-geometry factor G_{H0} for small magnetic field also depends on σ_{xx} and σ_{yy}. Hence, the magnetic sensitivity of the Hall plate changes with mechanical stress due to the piezoresistance effect! Note that this is a different effect than the well-known piezo-Hall effect [52], which describes the change of a material property—the Hall coefficient C_H—caused by mechanical stress. In our case the piezoresistivity leads to a small net change in Hall-plate geometry described by the affine transformation (12a)–(12c) and this slightly changes the Hall-geometry factor. Therefore Hall plates of identical material but different geometry also have slightly different drift of magnetic sensitivity versus mechanical stress.

We verify this model by comparison with Figure 3 of [47]. There the authors plotted the Hall output voltage versus magnetic field for several cases of highly anisotropic resistivity for a Hall plate with $\ell_0/W_0 = 2$ and $s_0/\ell_0 \to 0$, i.e., point-sized output contacts. For a meaningful comparison we need accurate numerical data, which were not given in [47]. Therefore we repeated three calculations of [47] with a finite element simulation (FEM) with COMSOL MULTIPHYSICS having 1.4 million elements and Lagrange multipliers. The results are shown in Figure 20: with the naked eye these curves match perfectly with the respective curves in Figure 3 of [47]. In the FEM we used the resistivity tensor of (5) with (6a)–(6c) and with the piezoresistive coefficients for low n-doped silicon ($\pi_{11} = -1.022$/GPa, $\pi_{12} = 0.534$/GPa, and $\pi_{44} = -0.136$/GPa). For a stress $\sigma_{xx} = -\sigma_{yy} = \pm 4.41176$ GPa this gives very strong anisotropy $4\rho_{xx} = \rho_{yy}$ and $\rho_{xx} = 4\rho_{yy}$, respectively. In the minor diagonal of the resistivity tensor we added the antireciprocal Hall-terms $\pm C_H B_z$. Next we used the FEM to get exact numerical data of the slopes of the curves at small magnetic field. According to theory these slopes are

$$\frac{V_{42}/V_{13}}{C_H B_z/\rho_{yy}} = \frac{W_0}{\ell_0}\frac{\rho_{yy}}{\rho_{xx}} G_{H0} \quad (33)$$

from which one can obtain G_{H0} for the three cases. Table 3 compares these values from the FEM with G_{H0} according to the analytical theory (18d), (18e), and (18g)–(18l) inserted into

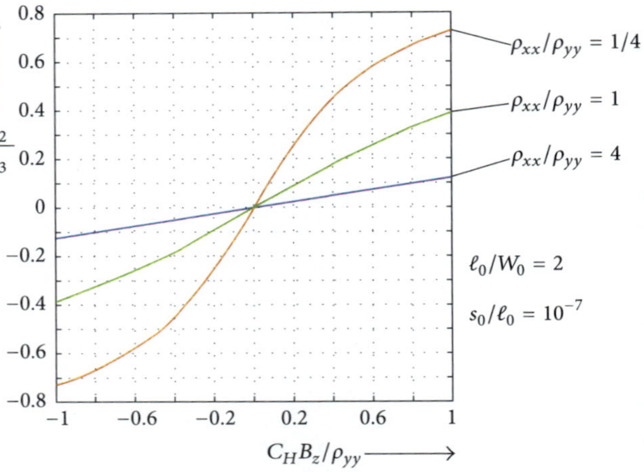

FIGURE 20: Output voltage over supply voltage of a Hall plate from Figure 19 with two squares input resistance and point-sized output contacts as obtained by a finite element simulation with COMSOL MULTIPHYSICS. The strong anisotropy of the conductivity tensor is obtained by large stress $\sigma_{xx} = -\sigma_{yy} = \pm 4.41176$ GPa in low n-doped silicon. The curves match with respective curves 1 (for $\gamma = 0$) in Figure 3 of [47]. The slopes of the curves in the weak field limit agree with the analytical theory of the present work (see Table 3).

(4). Despite the large stress the agreement is up to 10^{-6} for small magnetic field.

Next we compute the piezoresistive drift of current related magnetic sensitivity $\partial G_{H0}/\partial(\sigma_{xx} - \sigma_{yy})/G_{H0}$ for circular Hall plates of Figure 10(a) versus all possible contact geometries θ_f and θ_p (see Figure 21). This effect vanishes for symmetric devices with $\theta_f = \theta_p$ (except for very large stress). This was also checked by a finite element simulation for the case $\theta_f = \theta_p = 22.5°$.

Therefore, they should be a perfect choice if one wants to measure the piezo-Hall coefficient. Note that $\partial G_{H0}/\partial(\sigma_{xx} - \sigma_{yy}) = 0$ for any device with 90°-symmetry—not only for circular ones. The proof is short: if G_{H0} increases by $\sigma_{xx} - \sigma_{yy} > 0$, then it decreases by $\sigma_{xx} - \sigma_{yy} < 0$. However, the latter case is obtained by swapping σ_{xx} with σ_{yy}, and obviously this cannot change G_{H0} if the device is 90° symmetric. Therefore G_{H0} does not change at all under the action of $\sigma_{xx} - \sigma_{yy}$.

Conversely, if one pair of contacts is point-sized and the other one is close to 62.5°, Figure 21 shows a maximum

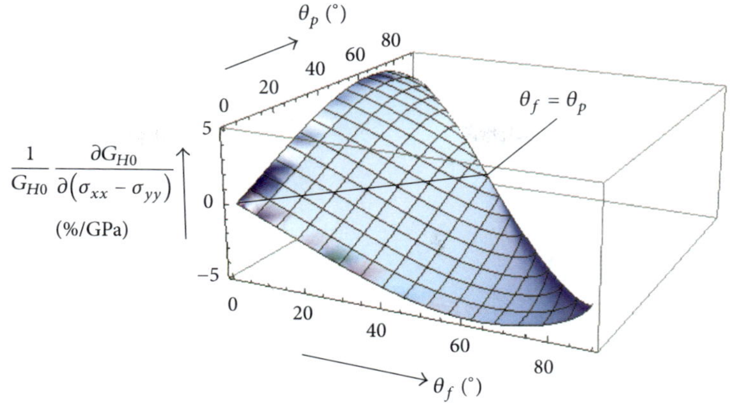

FIGURE 21: $\partial G_{H0}/\partial(\sigma_{xx}-\sigma_{yy})/G_{H0}$ for circular Hall plates of Figure 10(a) versus contact sizes θ_f and θ_p. This is the piezoresistive influence on the current related magnetic sensitivity of the Hall plate. It vanishes for symmetric devices with $\theta_f = \theta_p$ and it is strongest for devices, where one contact pair is point-sized and the other contact pair has a half-aperture angle of 62.5°. The plot assumes low n-doped (100)-silicon with $\pi_{44} = -0.136$/GPa. The plot region 89.5° $< \theta_f + \theta_p <$ 90° is not shown due to numerical problems in its evaluation – for $\theta_f + \theta_p = 90°$; it holds $G_{H0} = 0$ (see Figure 13) and $\partial G_{H0}/\partial(\sigma_{xx} - \sigma_{yy})/G_{H0} = 0$.

piezoresistive magnetic sensitivity drift of $\partial G_{H0}/\partial(\sigma_{xx} - \sigma_{yy})/G_{H0} = \pm 4.961\%$/GPa. A numerical FEM simulation with stresses 0.1 MPa, 1 MPa, 10 MPa, 100 MPa, and 1 GPa was conducted for two cases: first current was sent through the large contacts while voltage was tapped at the point-sized contacts, and then the roles of the contacts were swapped: the agreement with the analytical theory was around 0.01%. If such a device was used in a uniaxial stress state to characterize the piezo-Hall coefficient in low n-doped (100)-silicon—which is $P_{12} \approx 40\%$/GPa—the 5%/GPa from the piezoresistive effect would lead to a significant error in P_{12}, if one does not account explicitly for this effect. On the other hand, a comparison of magnetic sensitivities of two such Hall plates, one with large contacts along the x-axis and the other one with large contacts along the y-axis, could be used to determine the piezoresistive coefficient π_{44} without measuring resistances. If one uses an isotropic biaxial stress state $\sigma_{xx} = \sigma_{yy}$ (e.g., in a wafer bow experiment [53]) to characterize the piezo-Hall effect, the current related magnetic sensitivity does not depend on the piezoresistance effect. In smart silicon Hall sensors with mechanical stress compensation the piezoresistance effect leads to a $\sigma_{xx} - \sigma_{yy}$ dependence of the magnetic sensitivity, whereas the piezo-Hall effect has a $\sigma_{xx} + \sigma_{yy}$ dependence: this leads to increased complexity of the compensation circuit as described in chapter 16.6.3 in [22].

In the original paper by Hälg [52] the macroscopic Hall plates had the geometry of Figure 19 with $\ell_0 = 10$ mm, $W_0 = 1.6$ mm, $s_0 = 0.1$ mm, $t_d = 0.38$ mm with low n-doping. At zero stress the Hall-geometry factor G_{H0} is 0.9601759. He measured a piezo-Hall coefficient $P_{12} = 40 \cdots 45\%$/GPa depending on the doping concentration. The change of G_{H0} due to piezoresistance in the (100)-plane is 0.275%/GPa, which gives a relative error of 0.7% in P_{12}. If we include pure strain effects after (16a) the relative error reduces by one tenth. This error is so small because a device with small geometry effects was chosen by Hälg ($G_{H0} \approx 1$).

6. Comparison with Other Authors

With (18a)–(18m) and the ERC we can describe the behavior of the device versus arbitrarily large inplane normal stress in any operating mode: in the Hall-operating mode of Figure 6 (with the supply contacts $\overline{C}_1$–$\overline{C}_3$ and the sense contacts $\overline{C}_2$–$\overline{C}_4$), or in the van der Pauw operating mode (with the supply contacts $\overline{C}_1$-$\overline{C}_2$ and the sense contacts $\overline{C}_3$-$\overline{C}_4$ as shown in Figure 5 in [10]), or in the grounded diagonal operating mode (with the positive supply contact $\overline{C}_1$ while $\overline{C}_2$ and $\overline{C}_4$ are grounded and the sense contact is $\overline{C}_3$, cf. [10]).

However, the body of (18a)–(18m) seems much more complicated than piezoresistive theories of other authors [2, 3, 6, 54]. According to the principle of Occam's razor we should prefer the simplest way to explain phenomena in nature. Therefore, our theory only makes sense, if it has some advantage over simpler theories:

In Figure 2 it was already explained that the theory of [2] fails for devices with four extended contacts and low or moderate number of squares, whereas the theory proposed in this paper works for all rectangular and circular devices with two perpendicular mirror symmetries.

The theory of [54] uses an equivalent resistor circuit (ERC) that comprises only four resistors in a Wheatstone bridge configuration. Therefore it is not apt to describe van der Pauw measurements, where the device is supplied with energy at two neighboring contacts and the signal is tapped between the two other neighboring contacts. Instead, in [54] the authors focused on typical Hall-operating modes, where current flows between two nonneighboring contacts and signal is tapped between the other contacts. The stress dependence of the four resistors in the ERC is modeled identically to the stress dependence of individual resistor stripes, which are oriented at 45° to the global current flow direction. No justification is given for this assumption in [54]. Nevertheless, it seems to have worked reasonably well, because the authors claim fair agreement with experiments (deviations up to 35% were reported in their Table 1). However, empirical investigations were reported only for square Hall plates with point-sized output contacts (although various crystal planes and current flow directions were examined). An extension to rectangular Hall devices is mentioned in two sentences, but not verified. In case of more general device geometries with extended contacts it is not self-evident, which directions we should assign to each resistor in the ERC: e.g., in a symmetric device like in Figure 2(b) one might guess that according to [54] the equivalent resistors of the ERC are aligned at $\pm 45^{\circ}$ to the vertical line; however, in a less symmetric device like in Figure 6 it is unclear if an equivalent resistor should be aligned at $\pm 45^{\circ}$ to the vertical line or differently (especially if $\ell_0 \gg W_0 \gg s_0$).

Up to the author's knowledge, no one has so far described the effect of piezoresistance on the magnetic sensitivity. The plots in Figure 3 of [47] implicitly contain this information, but the scale is too coarse to note these subtle effects there, and the authors have not mentioned this phenomenon in their text.

7. Conclusion

This paper displayed similarities and differences between the reciprocal piezoresistance effect and the antireciprocal Hall effect. Alleged analogies of former authors were found to be valid only for special shapes of devices and contacts. An analytical theory was developed for devices with two perpendicular mirror symmetries exposed to biaxial stress along the symmetry axes. The theory was applied to rectangular and circular devices. In both cases it was found that the output signal is strongest for specific sizes of contacts, and the maximum output signals at given stress differ for various shapes of devices. Recurring to a statement of Wick in [14] we may say that for Hall plates there is no ingenious shape to maximize their magnetic sensitivity but for van der Pauw stress sensors and pseudo-Hall devices there definitely exist shapes which have larger mechanical stress sensitivity than others. For example, circular devices offer ~6% higher stress sensitivity than rectangular devices. An outstanding case is deep vertical Hall effect devices powered by voltage sources, where the potentials at floating contacts at zero magnetic field are not at all affected by mechanical stress in the cross-section plane.

It was also shown that the magnetic sensitivity of Hall plates with finite contacts has a small dependence on the piezoresistive effect, which adds to the piezo-Hall effect. The piezoresistance can be accounted for by an affine transformation, which slightly distorts the shape of the Hall plate. This leads to small changes in the number of squares of input and output resistances, which changes the Hall geometry factor and consequently also the magnetic sensitivity. This effect depends on the specific shape of the Hall plate. It vanishes for devices with 90° symmetry in cubic crystals, and therefore these devices are well suited to characterize the piezo-Hall effect with high accuracy.

Conflicts of Interest

The author declares that there are no conflicts of interest.

References

[1] W. G. Pfann and R. N. Thurston, "Semiconducting stress transducers utilizing the transverse and shear piezoresistance effects," *Journal of Applied Physics*, vol. 32, no. 10, pp. 2008–2019, 1961.

[2] V. A. Gridchin, "Influence of the specimen geometry on the magnitude of the transverse piezo emf," *Soviet Physics Journal*, vol. 19, no. 4, pp. 457–460, 1976.

[3] A. V. Gridchin and V. A. Gridchin, "The four-terminal piezotransducer: theory and comparison with piezoresistive bridge," *Sensors and Actuators A: Physical*, vol. 58, no. 3, pp. 219–223, 1997.

[4] A. K. M. Mian, J. C. Suhling, R. C. Jaeger, and B. M. Wilamowski, "Evaluation of die stress using van der Pauw sensors," *Applitcations of Experimental Mechanics to Electronic Packaging*, vol. 226, pp. 59–67, 1997.

[5] P. J. A. Munter, "A low-offset spinning-current Hall plate," *Sensors and Actuators A: Physical*, vol. 22, no. 1–3, pp. 743–746, 1989.

[6] Y. Kanda and A. Yasukawa, "Hall-effect devices as strain and pressure sensors," *Sensors and Actuators*, vol. 2, no. C, pp. 283–296, 1981.

[7] M. Doelle, D. Mager, P. Ruther, and O. Paul, "Geometry dependent sensitivity of planar plezoresistive stress sensors based on the pseudo-Hall effect," in *Proceedings of the 18th IEEE International Conference on Micro Electro Mechanical Systems (MEMS '05)*, pp. 267–270, February 2005.

[8] L. J. van der Pauw, "Determination of resistivity tensor and Hall tensor of anisotropic conductors," *Philips Research Reports*, vol. 16, pp. 187–195, 1961.

[9] U. Ausserlechner, "Closed form expressions for sheet resistance and mobility from Van-der-Pauw measurement on 90° symmetric devices with four arbitrary contacts," *Solid-State Electronics*, vol. 116, pp. 46–55, 2016.

[10] U. Ausserlechner, "Van-der-Pauw measurement on devices with four contacts and two orthogonal mirror symmetries," *Solid-State Electronics*, vol. 133, pp. 53–63, 2017.

[11] U. Ausserlechner, "A method to compute the Hall-geometry factor at weak magnetic field in closed analytical form," *Electrical Engineering*, pp. 1–8, 2015.

[12] U. Ausserlechner, "The signal-to-noise ratio and a hidden symmetry of Hall plates," *Solid-State Electronics*, vol. 135, pp. 14–23, 2017.

[13] H. T. Grahn, *Introduction to Semiconductor Physics*, chapter 6.3, World Scientific Publishing, 1999.

[14] R. F. Wick, "Solution of the field problem of the germanium gyrator," *Journal of Applied Physics*, vol. 25, no. 6, pp. 741–756, 1954.

[15] U. Ausserlechner, "Sensor arrangement having thermo-emf compensation," US patent US9829543, 2017.

[16] S. R. de Groot and P. Mazur, *Nonequilibrium thermodynamics*, chapter XIII, section 5, Dover Publications, New York, NY, USA, 1984.

[17] H. J. Carlin, "On the physical realizability of linear nonreciprocal networks," *Proceedings of the IRE*, vol. 43, no. 5, pp. 608–616, 1955.

[18] B. D. H. Tellegen, "The Gyrator, a new electric network element," *Philips Research Reports*, vol. 3, pp. 81–101, 1948.

[19] H. J. Carlin and A. B. Giordano, *Network Theory: An Introduction to Reciprocal and Nonreciprocal Circuits*, chapter 5, Prentice-Hall, New Jersey, NJ, USA, 1964.

[20] J. M. Garg and H. J. Carlin, "Network theory of semiconductor hall-plate circuits," *IEEE Transactions on Circuit Theory*, vol. 12, no. 1, pp. 59–73, 1965.

[21] Y. Zou, J. Suhling, R. Jaeger, and H. Ali, "Three dimensional die surface stress measurements in delaminated and non-delaminated plastic packages," in *Proceedings of the 48th Electronic Components and Technology Conference*, pp. 1223–1234, Seattle, WA, USA, 1998.

[22] M. Motz and U. Ausserlechner, "Electrical compensation of mechanical stress drift in precision analog circuits," in *Wideband Continuous-time ΣΔ ADCs, Automotive Electronics, and Power Management: Advances in Analog Circuit Design 2016*, chapter 16, pp. 297–326, Springer International Publishing, 2016.

[23] D. A. Bittle, J. C. Suhling, R. E. Beaty, R. C. Jaeger, and R. W. Johnson, "Piezoresistive stress sensors for structural analysis of electronic packages," *Journal of Electronic Packaging, Transactions of the ASME*, vol. 113, no. 3, pp. 203–215, 1991.

[24] A. Mian, J. C. Suhling, and R. C. Jaeger, "The van der Pauw stress sensor," *IEEE Sensors Journal*, vol. 6, no. 2, pp. 340–356, 2006.

[25] J. F. Nye, *Physical Properties of Crystals—Their Representation by Tensors and Matrices*, vol. 36, chapter II, section 4, p. 44, Oxford Science Publications, Oxford, UK, 1985.

[26] W. Schnell, D. Gross, and W. Hauger, *Technische Mechanik. Band 2: Elastostatik*, chapter 2.2.2, Springer Berlin Heidelberg, 4th edition, 1992.

[27] G. Heppke and F. Schneider, "Elektrische Eigenschaften von rechteckigen Proben anisotroper Leiter," *Annalen der Physik*, vol. 29, no. 3, pp. 271–288, 1973.

[28] R. Terakado and T. Sakakibara, "Analysis of two-dimensional anisotropic regions by affine and conformal mappings," *Proceedings of the IEEE*, vol. 75, no. 6, pp. 844–846, 1987.

[29] H. Reiche, "Der Halleffect als Anisotropie des elektrischen Leiters," *Wissenschaftliche Zeitschrift der Technischen Universität Dresden*, vol. 12, no. 1, pp. 179–182, 1963.

[30] A. Sutor and R. Lerch, "New Model for a CMOS-Compatible Stress Sensor Based on the Anisotropy of the Piezoresistive Effect," in *Proceedings of the 10th Intl Fair and Congress for Sensors, Transducers and Systems*, pp. 29–34, May 2001.

[31] G. O. Coraucci, R. V. D. de Oliveira, V. Garcia, and F. Fruett, "A study of the geometrical correction factor and the membrane thickness on the sensitivity of the transversal piezoresistive pressure sensor," *Journal of Integrated Circuits and Systems*, vol. 5, no. 2, pp. 140–147, 2010.

[32] V. Gridchin, A. Gridchin, V. Lubimsky, and V. Besedin, "Experimental investigation of the Greek cross-shaped piezoresistive element," in *Proceedings of the 1998 4th International Conference on Actual Problems of Electronic Instrument Engineering Proceedings (APEIE '98)*, vol. 1, pp. 113–118, Novosibirsk, Russia, September 1998.

[33] W. L. V. Price, "Electric potential and current distribution in a rectangular sample of anisotropic material with application to the measurement of the principal resistivities by an extension of van der Pauw's method," *Solid-State Electronics*, vol. 16, no. 7, pp. 753–762, 1973.

[34] K. A. Borup, K. F. F. Fischer, D. R. Brown, G. J. Snyder, and B. B. Iversen, "Measuring anisotropic resistivity of single crystals using the van der Pauw technique," *Physical Review B: Condensed Matter and Materials Physics*, vol. 92, no. 4, Article ID 045210, 2015.

[35] J. F. Nye, *Physical Properties of Crystals. Their Representation by Tensors and Matrices*, vol. 36, chapter VI, section 2, Oxford Science Publications, Oxford, UK, 1985.

[36] J. F. Nye, *Physical Properties of Crystals. Their Representation by Tensors and Matrices*, vol. 36, chapter VI, section 3.1, Oxford Science Publications, Oxford, UK, 1985.

[37] M. A. Hopcroft, W. D. Nix, and T. W. Kenny, "What is the Youngs Modulus of Silicon?" *Journal of Microelectromechanical Systems*, vol. 19, no. 2, pp. 229–238, 2010.

[38] W. Versnel, "Analysis of symmetrical Hall plates with finite contacts," *Journal of Applied Physics*, vol. 52, no. 7, pp. 4659–4666, 1981.

[39] R. S. Popovic, "The vertical Hall-effect device," *IEEE Electron Device Letters*, vol. 5, no. 9, pp. 357-358, 1984.

[40] U. Ausserlechner, "Hall Effect Devices with Three Terminals: Their Magnetic Sensitivity and Offset Cancellation Scheme," *Journal of Sensors*, vol. 2016, Article ID 5625607, 16 pages, 2016.

[41] R. Sunier, S. Taschini, O. Brand, T. Vancura, and H. Baltes, "Quasi-analytical study of offset voltage due to piezoresistive effect in vertical Hall devices by mapping techniques," in *Proceedings of the IEEE International Solid-State Sensors and Actuators Conference*, pp. 1582–1585, Boston, MA, USA, June 2003.

[42] K. Maenaka, "Piezoresistive effect in vertical Hall cell," *IEEE Electronics Letters*, vol. 29, no. 4, pp. 381-382, 1993.

[43] R. E. Newnham, *Properties of Materials. Anisotropy, Symmetry, Structure*, chapter 13.4, Oxford UP, Oxford, UK, 2005.

[44] J. M. Borwein and P. B. Borwein, *Pi and the AGM–A Study in Analytical Number Theory and Computational Complexity*, Canadian Mathematical Society Series of Monographs and Advanced Texts, John Wiley & Sons, Inc., New York, NY, USA, 1987.

[45] http://functions.wolfram.com/EllipticIntegrals/EllipticK/03/01/0003/.

[46] R. C. Jaeger, M. Motalab, S. Hussain, and J. C. Suhling, "Four-wire bridge measurements of silicon van der Pauw stress sensors," *Journal of Electronic Packaging, Transactions of the ASME*, vol. 136, no. 4, Article ID 041014, 10 pages, 2014.

[47] H. Shibata and J.-I. Oide, "Analysis of the Hall effect device using an anisotropic material," *Journal of Applied Physics*, vol. 88, no. 8, pp. 4813–4817, 2000.

[48] W. von Koppenfels and F. Stallmann, *Praxis der Konformen Abbildung*, chapter B.7, p. 318, Springer-Verlag, Berlin, Germany, 1959.

[49] http://functions.wolfram.com/09.34.16.0020.001.

[50] Y. Kanda, "Piezoresistance effect of silicon," *Sensors and Actuators A: Physical*, vol. 28, no. 2, pp. 83–91, 1991.

[51] R. S. Popović, "Hall-effect devices," *Sensors and Actuators*, vol. 17, no. 1-2, pp. 39–53, 1989.

[52] B. Hälg, "Piezo-Hall coefficients of n-type silicon," *Journal of Applied Physics*, vol. 64, no. 1, pp. 276–282, 1988.

[53] U. Ausserlechner, M. Motz, and M. Holliber, "Compensation of the piezo-Hall effect in integrated Hall sensors on (100)-Si," *IEEE Sensors Journal*, vol. 7, no. 11, pp. 1475–1482, 2007.

[54] Y. Kanda and M. Migitaka, "Effect of mechanical stress on the offset voltages of Hall devices in Si IC," *Physica Status Solidi (a)—Applications and Materials Science*, vol. 35, no. 2, pp. K115–K118, 1976.

Permissions

All chapters in this book were first published in ACMP, by Hindawi Publishing Corporation; hereby published with permission under the Creative Commons Attribution License or equivalent. Every chapter published in this book has been scrutinized by our experts. Their significance has been extensively debated. The topics covered herein carry significant findings which will fuel the growth of the discipline. They may even be implemented as practical applications or may be referred to as a beginning point for another development.

The contributors of this book come from diverse backgrounds, making this book a truly international effort. This book will bring forth new frontiers with its revolutionizing research information and detailed analysis of the nascent developments around the world.

We would like to thank all the contributing authors for lending their expertise to make the book truly unique. They have played a crucial role in the development of this book. Without their invaluable contributions this book wouldn't have been possible. They have made vital efforts to compile up to date information on the varied aspects of this subject to make this book a valuable addition to the collection of many professionals and students.

This book was conceptualized with the vision of imparting up-to-date information and advanced data in this field. To ensure the same, a matchless editorial board was set up. Every individual on the board went through rigorous rounds of assessment to prove their worth. After which they invested a large part of their time researching and compiling the most relevant data for our readers.

The editorial board has been involved in producing this book since its inception. They have spent rigorous hours researching and exploring the diverse topics which have resulted in the successful publishing of this book. They have passed on their knowledge of decades through this book. To expedite this challenging task, the publisher supported the team at every step. A small team of assistant editors was also appointed to further simplify the editing procedure and attain best results for the readers.

Apart from the editorial board, the designing team has also invested a significant amount of their time in understanding the subject and creating the most relevant covers. They scrutinized every image to scout for the most suitable representation of the subject and create an appropriate cover for the book.

The publishing team has been an ardent support to the editorial, designing and production team. Their endless efforts to recruit the best for this project, has resulted in the accomplishment of this book. They are a veteran in the field of academics and their pool of knowledge is as vast as their experience in printing. Their expertise and guidance has proved useful at every step. Their uncompromising quality standards have made this book an exceptional effort. Their encouragement from time to time has been an inspiration for everyone.

The publisher and the editorial board hope that this book will prove to be a valuable piece of knowledge for researchers, students, practitioners and scholars across the globe.

List of Contributors

Ahmed Mahmood
Department of Electrical Engineering, Salahaddin University, Erbil, Iraq
Department of Electrical and Electronics Engineering, University of Gaziantep, 27310 Gaziantep, Turkey

Gölge Ögücü Yetkin
Department of Electrical and Electronics Engineering, University of Gaziantep, 27310 Gaziantep, Turkey

Cumali Sabah
Department of Electrical and Electronics Engineering, Middle East Technical University-Northern Cyprus Campus (METU-NCC), Kalkanli, 99738 Guzelyurt, TRNC, Mersin 10, TurkeyKalkanli Technology Valley, Middle East Technical University-Northern Cyprus Campus (METU-NCC), Kalkanli, 99738 Guzelyurt, TRNC, Mersin 10, Turkey

Pedro Pereyra
Departamento de Ciencias Básicas, UAM Azcapotzalco, 02200 Mexico City, Mexico

Alberto Tufaile and Adriana Pedrosa Biscaia Tufaile
Soft Matter Laboratory, Escola de Artes, Ciências e Humanidades, Universidade de São Paulo, 03828-000 São Paulo, SP, Brazil

Timm A. Vanderelli
Ferrocell USA, 739 Route 259, Ligonier, PA 15658, USA

Yu Chen, Zhihao Chen and Xiaoyu Dai
International Collaborative Laboratory of 2D Materials for Optoelectronics Science and Technology, Key Laboratory of Optoelectronic Devices and Systems of Ministry of Education and Guangdong Province, College of Optoelectronic Engineering, Shenzhen University, Shenzhen 518060, China

Binyi Qin
School of Electronics and Communication Engineering, Yulin Normal University, Yulin 537000, Guangxi, China

Gul Gulpinar
Department of Physics, Dokuz Eylül University, 35160 Izmir, Turkey

Hongwei Liu, Guosheng Song, Chuanhui Zhang, Jinwen Tan and Wengang Feng
Key Laboratory of HV and EMC Beijing, Beijing 102206, China
School of Electric and Electronic Engineering, North China Electric Power University, Beijing 102206, China

Xiao-Le Yan, Li-Mei Hao and Mei-Ling Men
Department of Applied Physics, Xi'an University of Science and Technology, Xi'an 710054, China

Zhi Chen
Department of Applied Physics, Northwestern Polytechnical University, Xi'an 710129, China

A. M. H. R. Hakimi, F. Schoofs and M. G. Blamire
Department of Materials Science and Metallurgy, University of Cambridge, 27 Charles Babbage Road, Cambridge CB3 0FS, UK

S. Langridge
Science and Technology Facilities Council, Rutherford Appleton Laboratory, Harwell Science and Innovation Campus, Didcot OX11 0QX, UK

S. S. Dhesi
Diamond Light Source, Chilton, Didcot,Oxfordshire OX11 0DE, UK

A. L. Danilyuk and A. G. Trafimenko
Belarusian State University of Informatics and Radioelectronics, P. Browka 6, 220013 Minsk, Belarus

A. K. Fedotov and I. A. Svito
Belarusian State University, Nezalezhnastsi Av. 4, 220030 Minsk, Belarus

S. L. Prischepa
Belarusian State University of Informatics and Radioelectronics, P. Browka 6, 220013 Minsk, Belarus
National Research Nuclear University (MEPHI), Kashirskoe Highway 31, Moscow 115409, Russia

Anil Aryal, Abdiel Quetz, Sudip Pandey, Igor Dubenko and Naushad Ali
Department of Physics, Southern Illinois University, Carbondale, IL 62901, USA

Shane Stadler
Department of Physics & Astronomy, Louisiana State University, Baton Rouge, LA 70803, USA

Wenxia Zhao, Shiwei Liu, Hongjian Zhao and Yani Yan
School of Chemistry and Chemical Engineering, Ningxia Normal University, Guyuan 756000, China

Zenglin Wang
Key Laboratory of Applied Surface and Colloid Chemistry, Ministry of Education, School of Chemistry and Chemical Engineering, Shaanxi Normal University, Xi'an 710119, China

Liang Qiao
Department of Chemical and Biological Engineering, University at Buffalo (SUNY), Buffalo, NY 14260, USA

Hsien-Kuo Chin
Division of Cardiovascular, Department of Surgery, Kaohsiung Armed Forces General Hospital, Kaohsiung 802, Taiwan

Hui-Ying Kuo
Department of Optometry, Chung Shan Medical University, Taichung 402, Taiwan
Department of Ophthalmology, Chung Shan Medical University Hospital, Taichung 402, Taiwan

Yang-Chen Zheng, Jia-De Lin and Chia-Rong Lee
Department of Photonics, National Cheng Kung University, Tainan 701, Taiwan

J. Sun
Center of Applied Superconductivity and Sustainable Energy Research (CASER), Chubu University, 1200 Matsumoto-cho, Kasugai, Aichi 487-8501, Japan

H. Ohara
Department of Electrical Engineering, Chubu University, 1200 Matsumoto-cho, Kasugai, Aichi 487-8501, Japan

S. Yamaguchi
Center of Applied Superconductivity and Sustainable Energy Research (CASER), Chubu University, 1200 Matsumoto-cho, Kasugai, Aichi 487-8501, Japan
Department of Electrical Engineering, Chubu University, 1200 Matsumoto-cho, Kasugai, Aichi 487-8501, Japan

Ahmed Mahmood and Golge Ogucu Yetkin
Department of Electrical and Electronics Engineering, University of Gaziantep, 27310 Gaziantep, Turkey

Cumali Sabah
Department of Electrical and Electronics Engineering, Middle East Technical University-Northern Cyprus Campus (METU-NCC), Kalkanli, Guzelyurt, 99738 TRNC/Mersin 10, Turkey
Kalkanli Technology Valley, Middle East Technical University-Northern Cyprus Campus (METU-NCC), Kalkanli, Guzelyurt, 99738 TRNC/Mersin 10, Turkey

Nina A. Zharova
Institute of Applied Physics, Russian Academy of Sciences, Nizhny Novgorod 603950, Russia

Alexander A. Zharov and Alexander A. Zharov Jr.
Institute for Physics of Microstructures, Russian Academy of Sciences, Nizhny Novgorod 603950, Russia

Yanli Zhang, Xuejie Zhang and Jianqiang Zhu
Shanghai Institute of Optics and Fine Mechanics, Key Laboratory of High Power Laser and Physics, Chinese Academy of Sciences, Shanghai 201800, China
National Laboratory on High Power Laser and Physics, Shanghai 201800, China

Kewei You
Shanghai Institute of Optics and Fine Mechanics, Key Laboratory of High Power Laser and Physics, Chinese Academy of Sciences, Shanghai 201800, China
National Laboratory on High Power Laser and Physics, Shanghai 201800, China
University of Chinese Academy of Sciences, Beijing 100049, China

Pham Dinh Tam
Le Quy Don University of Technology, 100 Hoang Quoc Viet, Cau Giay District, Hanoi, Vietnam

Bui Duc Tinh
Institute of Research andDevelopment, Duy Tan University, 03 Quang Trung, Da Nang, Vietnam
Hanoi National University of Education, 136 XuanThuy Street, Cau Giay District, Hanoi, Vietnam

Nguyen Quang Hoc
Hanoi National University of Education, 136 XuanThuy Street, Cau Giay District, Hanoi, Vietnam

Pham Duy Tan
College of Armor, TamDao, Vinh Phuc, Vietnam

Man Wu and Ying Li
SZU-NUS Collaborative Innovation Center for Optoelectronic Science & Technology, International Collaborative Laboratory of 2D Materials for Optoelectronics Science and Technology of Ministry of Education, College of Optoelectronic Engineering, Shenzhen University, Shenzhen 518060, China

Allen Kelley, Kush Patel and Eric R. Bittner
Department of Chemistry, University of Houston, Houston, TX 77204, USA

Guobao Jiang and Ying Chen
School of Electronic Information and Electrical Engineering, Changsha University, Changsha 410003, China
Key Laboratory for Micro-/Nano-Optoelectronic Devices of Ministry of Education, School of Physics and Electronics, Hunan University, Changsha 410082, China

Yuan Zhou and Lulu Wang
School of Electronic Information and Electrical Engineering, Changsha University, Changsha 410003, China

Leilei Xu
School of Chemistry and Materials Science, Key Laboratory of Magnetic Molecules and Magnetic Information Materials, Ministry of Education, Shanxi Normal University, Linfen 41004, China
Key Laboratory of Nanodevices and Applications, Suzhou Institute of Nano-Tech and Nano-Bionics, Chinese Academy of Sciences, Ruoshui Road 398, Suzhou 215123, China

Jiafeng Feng and Xiufeng Han
Beijing National Laboratory for Condensed Matter Physics, Institute of Physics, Chinese Academy of Sciences, Beijing 100190, China

Kangkang Zhao, He Huang and Zhongming Zeng
Key Laboratory of Nanodevices and Applications, Suzhou Institute of Nano-Tech and Nano-Bionics, Chinese Academy of Sciences, Ruoshui Road 398, Suzhou 215123, China

Weiming Lv and Zhongyuan Liu
State Key Laboratory of Metastable Materials Science and Technology, Yanshan University, Qinhuangdao 066004, China

Xiaohong Xu
School of Chemistry and Materials Science, Key Laboratory of Magnetic Molecules and Magnetic Information Materials, Ministry of Education, Shanxi Normal University, Linfen 41004, China

Zlatko Koinov
Department of Physics and Astronomy, University of Texas at San Antonio, San Antonio, TX 78249, USA

Israel Chávez Villalpando
Instituto de Investigaciones en Materiales, Universidad Nacional Autónoma de México, Apartado Postal 20-364, 01000 Mexico Distrito Federal, Mexico

R. Pässler
Institut für Physik, Technische Universität Chemnitz, 09107 Chemnitz, Germany

Udo Ausserlechner
Infineon Technologies, Villach, Austria

Index